3rd International Conference on Advances in Mechanical Engineering and its Interdisciplinary Areas (ICAMEI 2021)

IOP Conference Series: Materials Science and Engineering
Volume 1080

Kolaghat, India
5-7 January 2021

Editors:

Dipak K. Mandal
Ali C. Benim

ISBN: 978-1-7138-4584-3
ISSN: 1757-8981

Printed from e-media with permission by:

Curran Associates, Inc.
57 Morehouse Lane
Red Hook, NY 12571

proceedings.com

Some format issues inherent in the e-media version may also appear in this print version.

Institute of Physics
Dirac House, Temple Back
Bristol BS1 6BE UK

Phone: 44 1 17 929 7481
Fax: 44 1 17 920 0979

techtracking@iop.org

Additional copies of this publication are available from:

Curran Associates, Inc.
57 Morehouse Lane
Red Hook, NY 12571 USA
Phone: 845-758-0400
Fax: 845-758-2633
Email: curran@proceedings.com
Web: www.proceedings.com

TABLE OF CONTENTS

IOP Conf. Series: Materials Science and Engineering 1080 (2021) 011001 doi:10.1088/1757-899X/1080/1/011001

PREFACE

New Technologies are being introduced every day, which is progressively transforming the future of this field. Vision of this conference titled "3rd **International Conference on Advances in Mechanical Engineering and its Interdisciplinary Areas"** is to promote excellence in scientific knowledge and innovations in the field of Mechanical engineering and related disciplines to motivate young researchers. The traditional field of Mechanical Engineering has become multidisciplinary and constantly evolving towards new concepts to optimize solutions to existing problems. Now-a-days, the academia and researchers are not only pondering but also experiencing the overwhelming outcomes of interdisciplinary researches. Moreover, it has been ubiquitously encouraged by the Governments, research agencies and by the academic institutions.

The context of the conference is to foster as well as exaggerate the research culture among academia and industry facilitated by sprinkled out ideas by the exchange of intellect during the conference. Furthermore, the intent of the activity is to let the participants acquaint with transcendental growth, recent trends, innovations, and security issues involved in the domain of Mechanical Engineering and their impact on societal applications through various brainstorming sessions.

The conference is structured as plenary lectures followed by parallel sessions and held virtually from 5th to 7th January 2021 by College of Engineering and Management, Kolaghat. The plenary lectures are delivered by the eminent personalities invited to introduce the theme of the conference. Each parallel session starts with an invited talk on specific topic followed by contributed papers. Papers are invited from the prospective authors from industries, academic institutions, R&D organizations and from professional engineers. 51 papers are accepted after peer reviewed by eminent academicians.

The distinguished keynote speakers

Prof. Suman Chakraborty, IIT Kharagpur

Prof S.K.Das, Director, IIT Ropar

Prof. Ali Cemal Benim, Duesseldorf University of Applied Sciences, Germany

Prof. Manoj Gupta, NUS

IOP Conf. Series: Materials Science and Engineering 1080 (2021) 011001 doi:10.1088/1757-899X/1080/1/011001

Prof. Himadri Chattopadhyay, Jadavpur University

Prof. Debabrata Nag, Jadavpur University

Prof. T. Asokan, IIT Madras

Prof. Sandipan Bandyopadhyay, IIT Madras

Prof. S. K. Saha, IIT Delhi

Prof. J. C. Mandal, IIT Bombay

Dr. Debabrata Chatterjee, Ex. CMERI, Durgapore

Prof. Santanu Das, Kalyani Govt. Engineering College

Prof Goutam Kr.Bose, HIT Haldia

Prof. Ashish Dutta, IIT Kanpur

Prof. Dipankar Narayan Basu, IIT Guwahati

Prof. Arup Kumar Das, IIT Roorkee

Prof. Nirmalendu Biswas, Jadavpur University

Prof. Samrat Rao, IIT Jammu

Prof. Nirmal Kr Manna, Jadavpur University

Prof. P K Panigrahi , IIT Kanpur

Prof. Sudipta De, Jadavpur University

Prof. Soumen Das, SMST, IIT Kharagpur

Prof. Rajeevlochana G. Chittawadigi, Amrita School of Engineering

Patron: Prof. (Col) Probhas Bose, Director- College of Engineering and Management, Kolaghat

Convenors: Dr. Dipak Kumar Mandal, HOD, ME, College of Engineering and Management, Kolaghat

Dr Tarun Kanti Pal, ME, Assistant Prof., College of Engineering and Management, Kolaghat

Co-ordinator:

Ms. Tanushree Jana, ME, Assistant Prof., College of Engineering and Management, Kolaghat

Proceedings editor:

Dr. Dipak Kumar Mandal, College of Engineering and Management, Kolaghat

Prof. Ali Cemal Benim, Duesseldorf University of Applied Sciences, Germany

IOP Conf. Series: Materials Science and Engineering 1080 (2021) 011002 doi:10.1088/1757-899X/1080/1/011002

Peer review declaration

All conference organisers/editors are required to declare details about their peer review. Therefore, please provide the following information:

- **Type of peer review: Single-blind / Double-blind / Triple-blind / Open / Other (please describe):minimum- Double-blind, maximum- Triple-blind**

- **Conference submission management system: Easychair (https://easychair.org/conferences/?conf=icamei2021)**

- **Number of submissions received: 76**

- **Number of submissions sent for review: 72**

- **Number of submissions accepted:51**

- **Acceptance Rate (Number of Submissions Accepted / Number of Submissions Received X 100):67%**

- **Average number of reviews per paper: Two to Three**

- **Total number of reviewers involved:118**

- **Any additional info on review process: No**

- **Contact person for queries:** **Dr Dipak Kumar Mandal, dkm@cemk.ac.in**
  Associate Prof. & Head
  Department of Mechanical Engineering
  College of Engineering and Management, Kolaghat
  PO:KTPP Township, East Midnapore
  Pin:721171, M:9434988752/6297123160

Please submit this form along with the rest of your files on the submission date written in your publishing agreement.

The information you provide will be published as part of your proceedings.

Modelling and Simulation of Series Parallel Hybrid Electric Vehicle

Patil Shubham[1],Aritra Ganguly[2]

[1] Under Graduate Student of Mechanical Engineering Department, IIEST Shibpur, West Bengal-711103

[2] Associate Professor of Mechanical Engineering Department, IIEST Shibpur, West Bengal-711103, India

Ph-+91-33-26684561-63, Fax-+91-33-26682916/4564

*Corresponding Author: aritra78@gmail.com

Abstract. The present paper intends to discuss the architectural structure and the different modes of operation of a series parallel hybrid electric vehicle during a drive cycle. Here an attempt has been made to simulate a standard series parallel HEV using the MATLAB and SIMULINK software to estimate the consumption of fuel for the different drive cycles. The results obtained from the model in terms of the fuel consumption have also been compared with that of a conventional internal combustion engine based vehicle. The study thus reinforces the need and viability of the use of hybrid electric vehicles as an intermediate transition and a pre-requisite to the full electric vehicle for the developing countries like India.

Keywords: Series Parallel Hybrid Electric Vehicle (SP HEV), Power Split Device (PSD)

1. Introduction

Automobile industry has seen rigorous boom in developing countries like India in the last two decades and the prognosis of this sector states even higher rise in manufacturing in the coming time. This results in an increase in the amount of fossil fuel consumption, dramatically followed with subsequent rise in emission of pollutant gases. One of the primary purposes to introduce the concept of a hybrid vehicle is to improve the fuel economy of the vehicle. Development in this area began, with Toyota launching its first Hybrid Electric Vehicle (HEV) under mass-production, gaining much market interests in Japan, named Toyota Prius long back in 1997. [1]

Koichiro Muta from **Toyota Motor** Corp carried out performance study in 2004 of SP HEV, where an improvement of car mileage of 7-8% was obtained. [2] This augmentation in the fuel economy is much higher than that of the previous modifications, which mainly focused on improvement in fuel injection and at most led to an enhancement ranging from 0.5 to 1%. [2]. Regardless of such significant benefits, attention towards this modification by Indian car manufacturers has not been as expected. Moreover, the uncompromising emission norms lay by Govt. of India (Bharat Stage VI) [9] demands automobiles to reduce emission level furthermore. Thus, it is important to address the developments in HEV till date and the ways to inculcate the benefits of this model in our country.

In any hybrid vehicle (series or parallel or series-parallel), the main objective is an attempt to operate the IC engine only in its most efficient band range. In series HEV the mechanical output from IC

IOP Conf. Series: Materials Science and Engineering 1080 (2021) 012001 doi:10.1088/1757-899X/1080/1/012001

engine is converted to electricity by the generator and this electric power can be used to charge the battery or operate motor and drive the vehicle. But for short trips IC engine is downsized, which sizes the motor. This makes the system expensive and economically not feasible. In parallel HEV, IC engine and motor output independently supplies power to the wheels. This design implies a mechanical coupling between the engine and the transmission, inhibiting the engine to always operate in its most efficient band, thus also not feasible for short ride passenger vehicles.[13]
In a Series-Parallel HEV model, both the power sources (electric and fuel based) are combined at the transmission (Epicyclic planetary gear) and then supplied to the wheels. This paper explains the modelling of the **series parallel** HEV model, supposed to be the most efficient design for the passenger vehicle. [3] The model discussed here has been referred from Mathworks official website [10]. The analysis of the model has been done using the parameters like fuel type and the drive cycle corresponding to that of Indian scenario to obtain the improvised fuel economy of the HEV.

2. Architectural Overview

The main components of a series parallel (SP) HEV are:-

1. Internal Combustion engine
2. Motor Generator 1 (MG1)- acting as generator
3. Motor Generator 2(MG2)- acting as motor
4. Power Split Device (PSD)- an Epicyclic compound planetary gear set

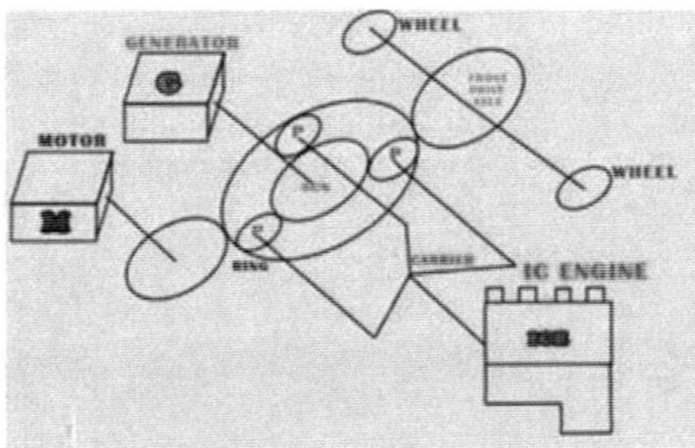

Fig 1:- Schematic of SP HEV Power train

As illustrated in Fig.1, the architectural structure of a series parallel HEV comprises of the carrier of ring gear connected to the output of an IC engine, followed by the sun gear connected to the shaft from the generator and lastly the ring gear is connected to the shaft from the motor. The motor and the generator are further connected, via an electric circuit, to a battery source. The ring gear is finally connected to the differential of the front drive axle, which has wheels mounted connected via a set of gears and shafts, inhibiting certain inertia and rotational damping. [4].

3. Modelling of SP HEV

A Series Parallel HEV model in MATLAB and SIMULINIK version 2013a is shown in Fig. 2. It comprises of several sub-systems which are connected to each other by means of mechanical and electrical linkage. [6] All of these subsystems operate in synergy for the vehicle to deliver the desired speed and acceleration from the driver, and minimize the fuel consumption while doing so.
The blocks in Fig.2 depict the following subsystems:

Combustion Engine	Electrical Systems	Power Split Device	Control System	Vehicle Dynamics	Input from the Dri Cycle

IOP Conf. Series: Materials Science and Engineering 1080 (2021) 012001 doi:10.1088/1757-899X/1080/1/012001

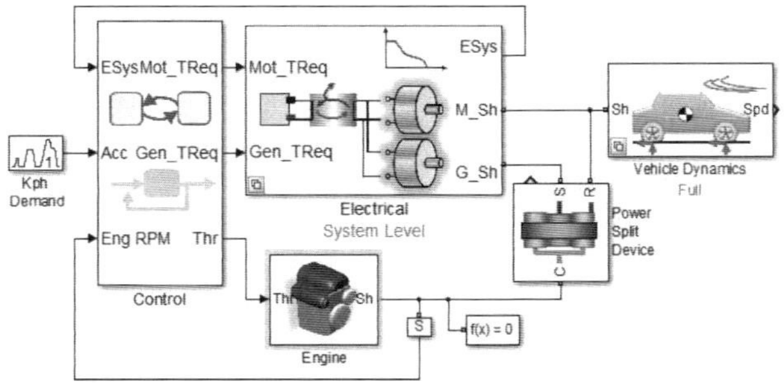

Figure 2: Schematic of the SP HEV model

Each of the sub-systems is discussed in the following section:

A. Engine:-

Based on input from control logic value, the engine produces torque and speed and the values are sent back to the control logic as feedback. The engine is connected to the carrier and the same drives the generator via Carrier-Sun gear connection. During the drive cycle, when the power demand is low, only the motor is operated to meet the driver requirements and engine is shut off. This is advantageous because the motor operates on DC current from the battery and is agile, unlike that of the engine, under unsteady load conditions.

However, during the high power demands (such as during sudden acceleration of the vehicle), the motor falls short to meet the driver demands and thus the engine needs to be operated. The amount of power to be delivered by engine is calculated by the control logic by subtracting power demand from the driver and the maximum power motor can provide. The torque demands of the vehicle are met by motor and hence, the engine only has to supply the balance torque such that the fuel economy is maximized. Also when the battery state of charge falls below the permissible level, the engine is operated to run the generator and charge the battery. [10]

In contrast to SI engines of conventional vehicles operating on Diesel cycle, the IC engine in HEV operates on Atkinson cycle. This is because of the fact that the engine in HEV does not operate on its full load capacity as compared to that in conventional passenger vehicles, but operates at part loads corresponding to maximum efficiency. Due to this phenomenon the engine cannot operate in Diesel cycle but on Atkinson cycle, which caters to the part load during engine operation in HEV, giving a 10% better fuel efficiency than SI engine. [12]

However the power density is reduced in operation of the Atkinson cycle due to reduction in input load. This is compensated by power input from the motor during cycle operation.

Table1:- Engine Specifications [7]

Cycle	Atkinson Cycle
Idling Speed	800 rpm
Hp	98 hp 1.8L 4cylinder

B. Electrical System

The electrical sub-system comprises of the following:

i. Generator:- Synchronous servo generator and drive
ii. Motor:- Synchronous servo motor and drive
iii. DC-DC converter and battery

IOP Conf. Series: Materials Science and Engineering 1080 (2021) 012001 doi:10.1088/1757-899X/1080/1/012001

Similar to that of the engine, the generator and the motor have input ports connected to the mode logic of the hybrid electric vehicle sending signals to it. The output shaft from the generator is connected to the Sun and that of the motor to the Ring gear of the Power Split Device (PSD).

Table 2:- Generator and the Motor Specifications [7]

Properties	Generator	Motor
Stator Resistance	0.0048	0.091
Shaft Inertia	0.2 kgm^{-2}	0.2 kgm^{-2}
Stator Mass (kg)	8.58	16
Rotor Mass (kg)	3.93	6.7
Magnet Mass(kg)	0.448	0.768
Efficiency		91%

For operation of motor in between 2000 RPM to 6500 RPM, high mechanical efficiency is available from 96- 99%. [7]

C. Power Split Device (PSD):-

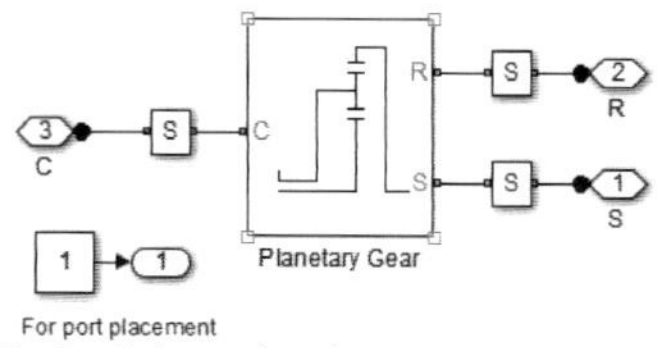

Fig 3:- Schematic of Power Split Device

Negligible meshing losses between the sun and the planet and between the ring and the planet gears have been considered in this model, for purpose of simplification. However values of (0.96 0.98) losses exhibit on-road characteristics. [7]. The equations governing the torque and speed (RPM) of the sun, planet and carrier in PSD are discussed here in brief. The transmission efficiency is given as $\eta t = \frac{1}{\eta m + \eta g} + (1 - \frac{1}{\eta m + \eta g}) \frac{1}{\rho + 1} \frac{1}{SR}$ where η_m and η_g are motor-generator eff. and SR is ratio of speed of engine and motor and ρ is gear ratio. [14]

The axle power P_a which is the sum of power from battery and the mechanical power from the IC Engine can be given by:

$$P_A = T_a\omega_a = P_B + T_e\,\omega_e \tag{1}$$

where, P_B denotes Battery power, T_e and ω_e denotes Engine torque & speed

And angular velocity as,

$$\omega_C = \frac{\rho\omega s}{\rho+1} + \frac{\omega r}{\rho+1} \tag{2}$$

where ω_C, ω_S and ω_r is velocity of carrier, sun and ring gear respectively. [4]

D. Control System

The control system is a continuous closed loop feedback system sub-divided into several components, as stated below

i. Controllers

a) Engine speed controller

b) Motor speed controller

c) Generator speed controller

d) Battery charge controller

4. Simulation of SP HEV

Consider the below drive cycle for simulation of the above modelled SP HEV [10]

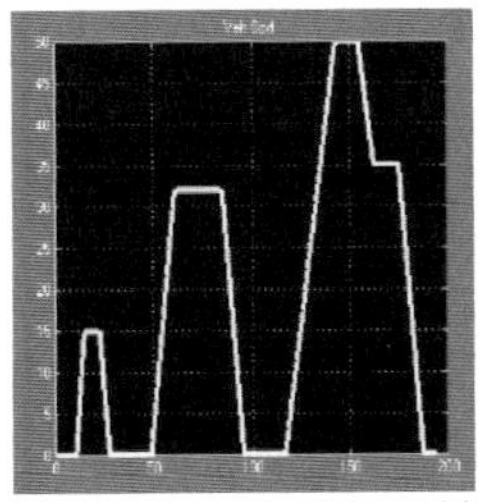

Fig 4:- Drive cycle of the vehicle

The scopes in the model provide detailed plots for speed and torque of different components in the vehicle like generator, engine and motor, as shown below.

A. Motor and Generator

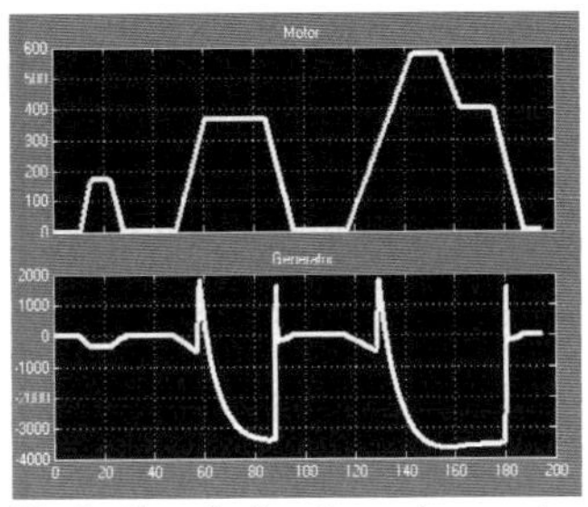

Fig 5:- Speed of motor and generator

It can be clearly seen from the above plot graph that the motor traces a similar path as that of the drive cycle (Fig. 4). When the motor falls short to supply the required power, the generator traces a peak at its power production. As the generator is driven by the IC engine, from the graph it can be seen that the motor encompasses majority of power supply and IC engine is driven only when necessary.

B. ICE

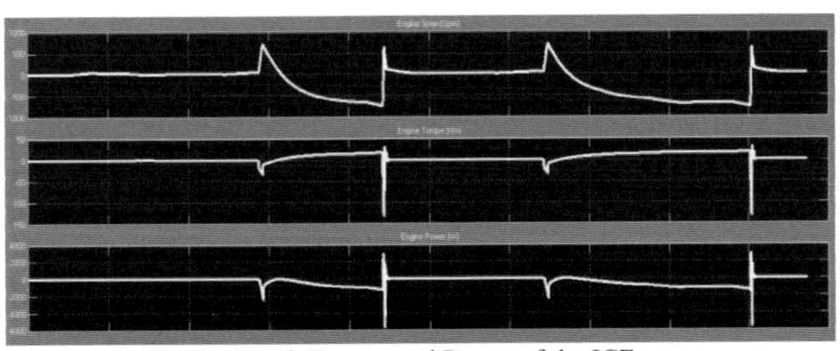

Fig 6:- Speed, Torque and Power of the ICE

Comparing Fig.5 (operation of motor and generator) with Fig.6 (operation of IC engine) for the given input drive cycle (Fig. 4), it is seen that the engine operation peaks only when the motor falls short of the required supply. Moreover, from Fig 6, engine can be seen to be operated only at the most efficient points.

As a result, the fuel economy of the engine improves significantly and results in less pollutant emissions.

C. Battery

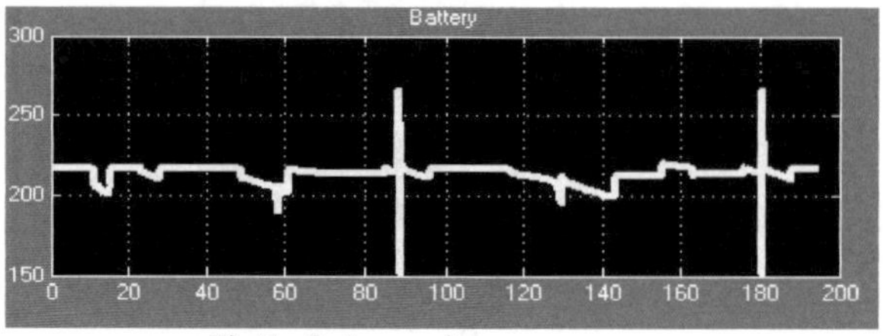

Fig 7:- Battery voltage

Comparing the battery output voltage with that of engine operation and the drive cycle, the peak of the drive cycle corresponds to peak of power demand. Here the battery supplies maximum0 power to motor for its operation. During vehicle retardation or constant speed movement, battery output is invariant and traces minor dips, as seen in Fig. 7.

For the considered drive cycle, we obtain the fuel economy as:-

Table 3:- Fuel economy of vehicle

litre /100 km	**3.124**
km / litre	**32.01**
Total fuel used (L)	0.03106

As it can be inferred from Table 3, the mileage obtained for the considered drive cycle is 32 km/litre for diesel. An average of 22 km/litre is observed in a conventional vehicle which corresponds to an increase of 50-65% This rise is due to the fact that the mode logic sends input to the engine such that it operated in its band of maximum efficiency as much as possible.

5. **Conclusion**

Thus in this paper, architectural structure and the modelling of a series parallel HEV has been discussed, in which the different major components and their individual roles were illustrated. The simulation performed was based on a model referred from Mathworks, while the input parameters for the analysis like fuel calorific value followed by engine and motor parameters and the drive cycle were corresponding to that of Indian scenario. It can be revealed from the study that: -

1. The fuel economy (32 km/litre) is significantly improved compared to that of a conventional IC engine based vehicle (diesel engine). The improvement is of the order of 50-65%.

It is being addressed that these developments have been established in the early 2000s in Japan and U.S however they still have not been explored in our country by automobile manufacturers to the needful extend.

6. References

1. Sasaki, S. (1998). Toyota's newly developed hybrid powertrain. *IEEE International Symposium on Power Semiconductor Devices & ICs (ISPSD)*, 17–22. https://doi.org/10.1109/ispsd.1998.702540

2. Muta, K., Yamazaki, M., & Tokieda, J. (2004). Development of new-generation hybrid system THS II - Drastic improvement of power performance and fuel economy. *SAE*

Technical Papers, *2004*(724). https://doi.org/10.4271/2004-01-0064

3. Kim, N., Kwon, J., & Rousseau, A. (2012). Comparison of powertrain configuration options for plug-in HEVs from a fuel economy perspective. *SAE Technical Papers*. https://doi.org/10.4271/2012-01-1027

4. Ehsani, M., Gao, Y., & Miller, J. M. (2007). Hybrid electric vehicles: Architecture and motor drives. *Proceedings of the IEEE*, *95*(4), 719–728. https://doi.org/10.1109/JPROC.2007.892492

5. Enang, W., & Bannister, C. (2017). Modelling and control of hybrid electric vehicles (A comprehensive review). *Renewable and Sustainable Energy Reviews*, *74*, 1210–1239. https://doi.org/10.1016/j.rser.2017.01.075

6. Butler, K. L., Ehsani, M., & Kamath, P. (1999). A matlab-based modeling and simulation package for electric and hybrid electric vehicle design. *IEEE Transactions on Vehicular Technology*, *48*(6), 1770–1778. https://doi.org/10.1109/25.806769

7. Technologies, V. (2011). *Evaluation of the 2010 Toyota Prius Hybrid Synergy Drive System 2010 Toyota Prius Hybrid Synergy*.

8. Meisel, J. (2006). An analytic foundation for the Toyota Prius THS-II powertrain with a comparison to a strong parallel hybrid-electric powertrain. *SAE Technical Papers*, *2006*(724). https://doi.org/10.4271/2006-01-0666

9. IICT Policy Update. (2016). India bharat stage VI emission standards. *The International Council on Clean Transportation*, *April*, 10.

10. *28441-hybrid-electric-vehicle-model-in-simulink @ www.mathworks.com*. (n.d.). https://www.mathworks.com/matlabcentral/fileexchange/28441-hybrid-electric-vehicle-model-in-simulink. Accessed online on 10/05/2020

11. Zhuang, Weichao, et al. "Optimal Engine Starts of an Input-Split Hybrid Electric Vehicle." *SAE International Journal of Alternative Powertrains*, vol. 4, no. 2, 2015, pp. 343–351. *JSTOR*, www.jstor.org/stable/26169093. Accessed 10 May 2020.

12. Asghar, M., Bhatti, A. I., Ahmed, Q., & Murtaza, G. (2018). Energy Management Strategy for Atkinson Cycle Engine Based Parallel Hybrid Electric Vehicle. *IEEE Access*, *6*, 28008–28018. https://doi.org/10.1109/ACCESS.2018.2835395

13. Chan, C. C., Bouscayrol, A., & Chen, K. (2010). Electric, hybrid, and fuel-cell vehicles: Architectures and modeling. *IEEE Transactions on Vehicular Technology*, *59*(2), 589–598. https://doi.org/10.1109/TVT.2009.2033605

14. Cho, S., Ahn, K., & Lee, J. M. (2006). Efficiency of the planetary gear hybrid powertrain. *Proceedings of the Institution of Mechanical Engineers, Part D: Journal of Automobile Engineering*, *220*(10), 1445–1454. https://doi.org/10.1243/09544070JAUTO176

IOP Conf. Series: Materials Science and Engineering 1080 (2021) 012002 doi:10.1088/1757-899X/1080/1/012002

CrackWeb : A modified U-Net based segmentation architecture for crack detection

Sandeep Ghosh, Subham Singh, Amit Maity and Hirak Kumar Maity
College of Engineering and Management, Kolaghat
E-mail: hmaity@cemk.ac.in

Abstract. Classical image processing methods demands heavy feature engineering, as well as they are not that precise, when it comes to manual extraction of relevant features in real life scenarios amid to various lighting conditions and other factors.Thus, detection of cracks using methods based on classical image processing techniques fails to provide satisfactory results always. Hence, we have proposed a deep convolutional neural network, that is not based on manual extraction of features as mentioned above. We proposed a modified U-Net architecture, and replaced all of its convolutional layers with residual blocks, inspired from the ResNet architecture. For evaluation of our model Dice Loss is used as our objective function and F1 score as a metric. Other than that, for better convergence and optimization, a learning rate scheduler and AMSGRAD optimizer was utilized.

1. Introduction

Sudden outbreaks or fractures that occurs through the member completely or partly due to various natural and man-made reasons results in formation of crack in concrete structure. More precisely, when the tensile stress exceeds tensile capacity formation of crack takes place. The utility of concrete can be seen in almost everywhere, from buildings, bridges, to other structures. Thus, when a crack occurs in a concrete slab, it can be an indication of some major structural defects in the whole architecture, which holds a potential risk of some serious accident. There are some existing ways to detect cracks, starting from visual inspection and monitoring to other non-destructive techniques (NDT) which uses various image processing methods to segment cracks, but due to unavoidable noise in images, the segmentation of those cracks from its background, isn't that precise.

In this paper, we tried to design a simple deep learning algorithm, that holds the potential solution to detect cracks of any thickness. As a result, it bypasses the need of manual feature extraction, by learning the essential features necessary for segmenting the crack from its background by classifying each pixel as a crack or not.

The arrangement of the paper is as follow: The scope and motivation for the work is mentioned in Section 2. A brief literature survey is mentioned in Section 3. The proposed work is discussed in Section 4 along with network architecture and mathematical modelling. In Section 5, we have mentioned the benchmark data sets that are used for this work along with the image augmentation methods.The training set-up along with the computation complexity is mentioned in section 6. The performance evaluation is done in Section 7. Finally the conclusion and scope of the future advancements is mentioned in Section 8.

2. Motivation

Formation of cracks in a structure can be termed as a significant indication of deterioration. Thus, early detection of cracks plays a crucial role in keeping the architectural integrity intact. Since, visual inspections as well as NDTs both lacks precision and objectivity in the qualitative analysis. Hence, an automated crack detection technique stands out to be a mandatory replacement. The existing literature presents a variety of techniques using classical as well as AI (Artificial Intelligence) based image processing methods. The motivation here is to compare and present an automated technique based on a Novel architecture which bypasses the need of manual feature extraction and turns out to be an improvement over the existing automated techniques too.

3. Related Works

This portion provides a brief literature review about existing methods and recent works on the segmentation of crack images in concrete structures and pavements. In the existing literature, there are two given methods: Classical methods [1, 2, 3, 4] and Deep learning methods [5, 6, 7, 8, 9]. Oliveira and Correia [1] presented a method for automatic crack detection using survey images captured at high driving speeds. They used a pre-processing technique using morphological filters for reducing pixel intensity variance and then applied dynamic thresholding to identify dark pixels in images, as compared to potential crack pixel. Ai et al. [2] mapped the probabilities of each crack pixel. Later they used a SVM(Support Vector Machine) based method and mapped the probability of crack predictions using multi scale-neighborhood information. Shi et al. [3] manually extracted the features of the annotated dataset and created crack detection network using random forest. Li et al. [4] created a model using multi variate hypothesis test for segmenting the extracted candidate cracks using a method of windowed intensity path.

Bang et al. [5] used fully connected layers as decoders and used deep residual blocks as encoders. Fan et al. [6] distributed the crack images into small patches and labelled each patch as positive or negative based on the presence of crack on the patch and then used Convolutional Neural Networks (CNN) to detect or classify those patches.As a result, the complete inference time for each testing image was very high. Zang et al. [7] used two different CNN for segmentation. Their approach was to first extract the relevant features and then feed the same into a different CNN. Yang et al. [8] used two modules namely pyramid and boosting module. The pyramid module merged two feature maps from two convolutional layers into the decoder block and at the same time, assignment of weights from easy to hard samples respectively was done using hierarchical boosting modules. Jenkins et al.[9] and Nguyen et al. [10] utilised a U-Net based architecture. Lau et al. [11] they have modified the basic U-Net architecture by replacing the encoder blocks with pre-trained ResNet-34 blocks.

4. Proposed technique

The method that we have used here is a supervised learning algorithm, where we tried to optimize an universal function approximator such that it fits as, $f : x \to y$. We mapped the crack images (x) to their respective labels, i.e, masks (y). We then optimized the parameterized neural network comprising of multiple layers of different types, having different weights and biases using a gradient descent based optimizer such that a minimised loss function can be obtained by altering the theta θ parameter upto a satisfactory level $L = g(x, y, \theta)$.

4.1. Model Architecture

We have created an U-Net based model architecture. U-Net [12] was classically used for segmenting microscopic cells with low data quantity, but when used for detection of cracks, it produced satisfactory outputs. Therefore, this architecture was ideally suited for the mentioned

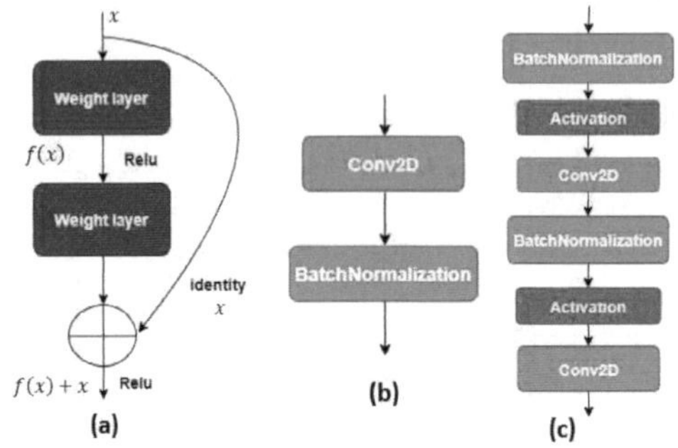

Figure 1. (a) Residual Block (b) Skip Connection(c) Full pre activation layer

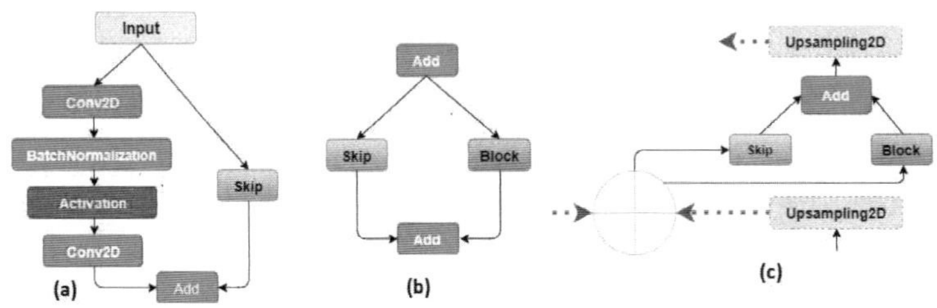

Figure 2. (a) Stem (b) Successive residual blocks and (c) Up-sampling Block

work. According to the official U-Net paper *"There is large consent that successful training of deep networks requires many thousand annotated training samples"*. U-Net represents a network with a training strategy that relies highly on image augmentation for more efficient use of the given data. The architecture involves a symmetric down sampling path that captures the context and an adjacent up sampling path that allows the precise localization of the ROI (Region of Interest). We have customised the original network by replacing all the classic convolutional layers with residual blocks, inspired by the famous ResNet [13]. The main intuition behind this alteration was "identity shortcut connection" that skips one or more layers as shown in Figure 1(a). By virtue of which the vanishing gradient problem that comes with deep neural networks was overcame. Along with that, a full pre-activation layer (shown in Figure 1(c)) with skip connections (Figure 1(b)) were introduced that provides an alternative easier path for the gradients to back propagate.

The proposed network receives a three channel (RGB) image of dimension (256,256,3) and generates an output of dimension (256,256,3). Encoder loads with a stem (Figure 2(a)), that consists of a convolutional layer with 16 kernels of size (3×3) each with stride of 1 and padding of zero, followed by a Batch Normalization, a ReLu activation layer and another convolution layer identical to the above. On the other hand, an identical convolutional layer with 16 kernels of size (1×1) followed by a batch normalization layer without activation, is fed to the successive layers below. A full pre-activated layer along with a skip connection constitutes a successive block as shown in Figure 2(b). The presented architecture used the following numbers of kernels, $l : [16_{l=1}, 32_{l=2}, 64_{l=3}, 128_{l=4} \& 256_{l=5}]$, respectively. With the kernels of size (3×3) and stride 2 at each level (l=level), followed by a final bottleneck layer with 256 filters and stride 1. The output from this layer is fed to the decoder network, where a single decoder block (as shown in Figure 2(c)) consists of concatenated activations along the channel of dimensions equivalent to its respective encoder block along with up-scaled activations of its previous convolutional

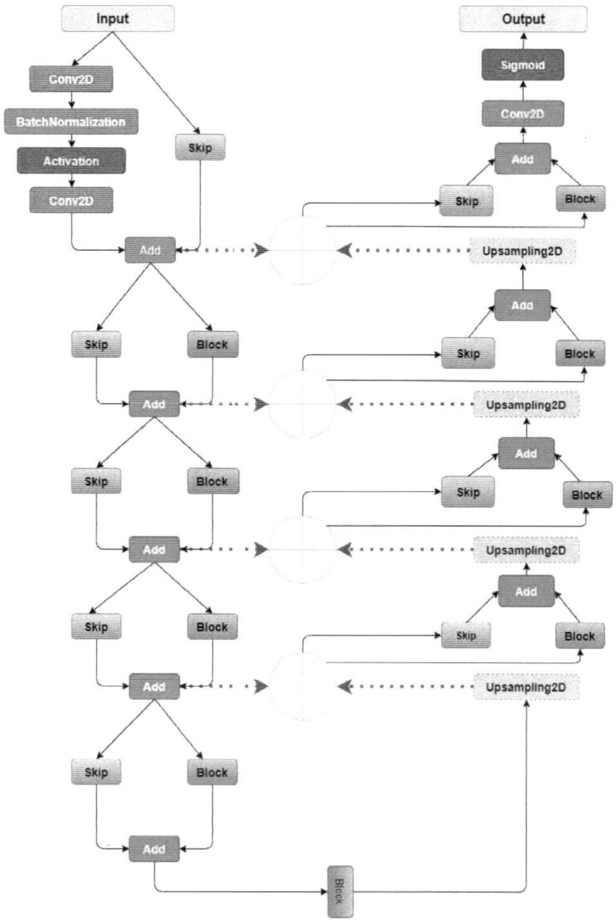

Figure 3. Proposed network architecture.

blocks. Each decoder block consist of a UpSampling2D layer with filter size (2×2) followed by a residual block. The complete network architecture is shown in Figure 3.

4.2. Loss Function

The network's output layer is connected to a sigmoid function such that each pixel in the output layer ranges from zero to one. Which determines the probability of crack in each pixel. As a result, it was noticed that the non-crack pixels outnumbered the crack pixels by a huge ratio. Hence,a huge class imbalance was observed. Thus, to deal with this issue, dice coefficient loss [14] was selected as it directly optimizes the dice score. Moreover, its equivalent to the F1 score too. Dice coefficient is actually twice the area of overlap divided by the total number of pixels in the image. Thus, the dice loss stands out to be :

$$L = \frac{1}{N} \sum_{y=1}^{N} 1 - \frac{2|y' \odot y|}{|y'| + |y|} \tag{1}$$

where, prediction and ground truth is depicted as $y' \in \Re^{h \times w}$ and $y \in \Re^{h \times w}$ respectively.The operator $\odot$ denotes the element-wise multiplication operation. The loss function produces a scalar value between '0' and '1' such that when the output matches the ground truth it becomes zero, else one.

4.3. Parameter Optimization
The optimizer that we have used here to optimize our model is ADAM and Beyond [15]. Which uses a new exponential moving average AMSGRAD. The AMSGRAD uses a smaller learning rate in comparison to ADAM [16]. In case of ADAM the decrement or decay of learning rate is not guaranteed where as AMSGRAD uses smaller learning rates, thus it maintains the maximum of all the learning rates until the present time step and uses that maximum value for normalizing the running average of the gradient unlike the learning rates in ADAM or RMSPROP [17]. Thus, it converges better than ADAM or RMSPROP.

The gradient computation is done by using

$$g_t = \frac{1}{m} \sum_{i=1}^{m} \Delta_\theta \mathcal{L}\left(x^i, y^i, \theta_t\right) \tag{2}$$

Then the moment estimation is computed as

$$m_t = \rho_1 m_{t-1} + (1 - \rho_1) g_t, v_t = \rho_2 v_{t-1} + (1 - \rho_2) g_t^2 \tag{3}$$

where, $\hat{v}_t = max(\hat{v}_{t-1}, v_t)$ and $\theta_{t+1} = \theta_t - \eta . \frac{m_t}{\sqrt{\hat{v}_t}}$. We kept the information of the maximum value of all v_t until current time step and later used that in the update rule compared to just the current v_t in ADAM algorithm.

4.4. Learning Rate
The learning rate used here, was not kept constant throughout the training of the data, instead we used a learning rate schedular, that increases/decreases the learning rate gradually after every fixed set of epochs such that the function can attain the optimum convergence by the end of the training.

5. Dataset and Preprocessing
The performance of the proposed model is evaluated with two different datasets as mentioned below:

- **Crack500** [18] The dataset consists of 500 images and corresponding masks of size (3264 × 2448). 250 images were used for training,50 for validation and 200 images were used for testing.
- **DeepCrack** [19] The dataset consists of 537 manual annotated images. Divided into 300 training samples and 237 test samples.

Two sample data are shown in Figure 4(a to d) and Figure 5(a to d) from Crack500 and DeepCrack datasets respectively.

5.1. Image Augmentation
All the data are augmented on the fly using the albumentation [20] library. Random flips and rotations are applied with random changes in lighting by increasing/decreasing contrast, gamma & brightness. Random distortions are also applied like elastic distortion, grid distortion and optical distortion.

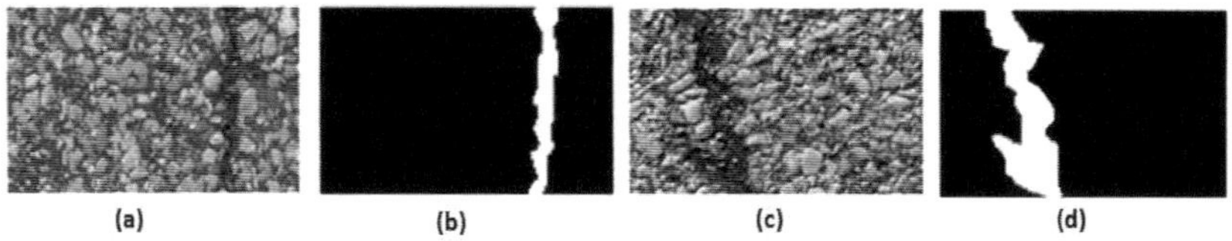

Figure 4. Sample images and their corresponding masks from Crack500 dataset

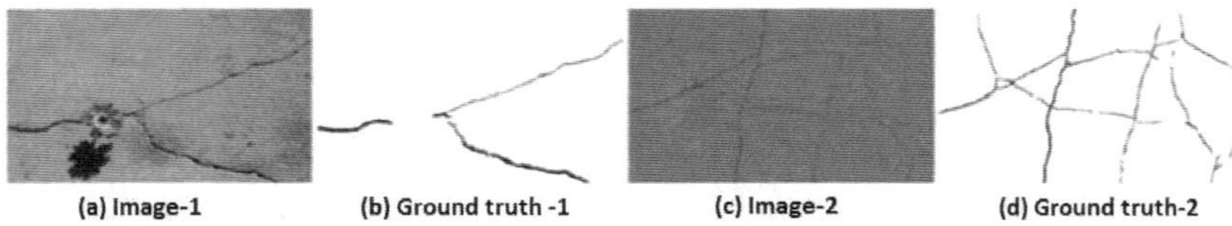

Figure 5. Sample images and their corresponding masks from DeepCrack dataset

6. Training

The estimated time taken by our architecture to train on Crack500 and DeepCrack dataset were 46 mins and 10 mins respectively. We trained our model on a Google Colab instance with a 16GB Tesla P100 paired up with an Intel Xeon processor and 12GB of RAM. The model was trained end-to-end with a batch size of 10 and input image size of(256,256,3) to compensate for the hardware limitations. The deep learning framework used here is a high level API known as Keras along with Tensorflow 2.2 as backend. Keras is a fast, easy and optimized API for prototyping research works like the one presented in this paper.

7. Results and Discussion

The proposed network is trained on Crack500 and DeepCrack datasets. F1 score is the harmonic mean of precision and recall. The required expressions are shown below:

$$Precision = \frac{TP}{(TP + FP)}; Recall = \frac{TP}{(TP + FN)}; F1 - Score = \frac{(2PrecRec)}{(Prec + Rec)} \tag{4}$$

where TP = True positive, FP = Flase Positives, FN = False Negatives. The experimental results are shown in Figure 6 and Figure 7 for the above mentioned datasets, respectively. A comparative study of our network with other well known existing methods on the mentioned datasets is shown below:

Table 1. Comparison of various methods on Crack500 dataset.

Method	Precision	Recall	F1
U-Net by Jenkins et al. [9]	0.6811	0.6629	0.6788
U-Net by Nguyen et al. [10]	0.6954	0.6744	0.6895
CNN by Fan et al. [6]	0.7123	0.6955	0.7056
Split-Attention Network [21]	0.7368	0.7165	0.7295
Lau et al. [11]	0.7426	0.7285	0.7327
Our method	**0.8095**	**0.7409**	**0.7737**

Table 2. Comparison of various methods on DeepCrack dataset.

Method	Precision	Recall	F1
Y. Liu et al. [19]	**0.861**	0.869	0.865
Our method	0.8535	**0.905**	**0.878**

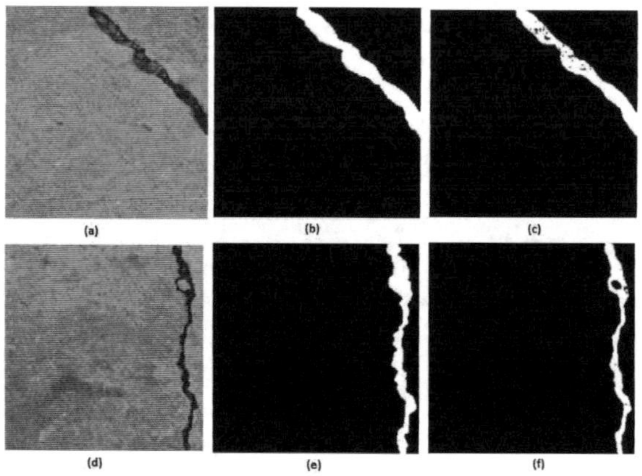

Figure 6. Original Image, Ground truth and predicted mask are shown in (a - c) and (d -f), respectively from Crack500 dataset

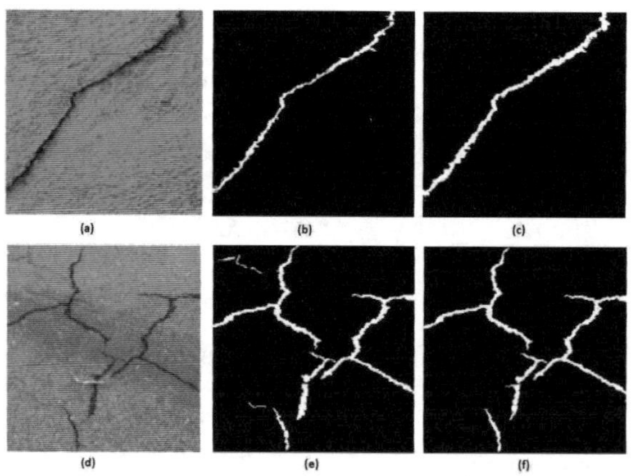

Figure 7. Original Image, Ground truth and predicted mask are shown in (a - c) and (d -f), respectively from DeepCrack dataset

8. Conclusion

The proposed algorithm requires minimal feature engineering unlike any other machine learning algorithms and it achieves state-of-the-art F1 score of about 88 on the DeepCrack data set and 77 on the Crack500 data set. Also the network uses a U-Net like architecture with residual blocks to automate the process of crack segmentation. Though achieving state-of-the-art performances on various datasets, the only drawback of our method is manual annotations of data, which makes the whole process quite expensive. Unsupervised learning holds the potential to resolve this issue as it doesn't demands labelled data.

This is also noticed that the images have uneven illumination that could be addressed with techniques like Histogram Equalization (HE) or Contrast Limited Adaptive Histogram Equalization (CLAHE).Further advancements can be done using better quality images along with more precise labeling of the data. Even more data augmentation techniques may be added to resolve the issue of small dataset.

References

[1] Oliveira H and Correia P 2009 Automatic road crack segmentation using entropy and image dynamic thresholding EUSIPCO 622-626.
[2] Ai D, Jiang G, Kei L S and Li C 2018 Automatic pixel level pavement crack detection using information of multiscale neighborhoods IEEE Access 6:24452-24463.
[3] Shi Y, Cui L, Qi Z, Meng F and Chen Z 2016 Automatic road crack detection using random structured forests IEEE Trans. Intell. Transp. Syst. 17(12) 3434-3445.
[4] Li H, Song D, Liu Y and Li B 2019 Automatic pavement crack detection by multi scale image fusion IEEE Trans. Intell. Transp. Syst. 20(6) 2025-2036.
[5] Bang S, Park S, Kim H, Yoon Y S and Kim H 2018 A deep residual network with transfer learning for pixel-level road crack detection ISARC 35: 1-4.
[6] Fan Z, Wu Y, Lu J and Li W 2018 Automatic pavement crack detection based on structured prediction with the convolutional neural network arXiv:1802.02208.
[7] Zhang A, Wang K C P, Li B, Yang E, Dai X, Peng Y, Fei Y, Liu Y, Li J Q and Chen C 2017 Automated pixel-level pavement crack detection on 3D asphalt surfaces using a deep-learning network Computer-Aided Civil Infrastructure Engg. 32(10) 805-819.
[8] Yang F, Zhang L, Yu S, Prokhorov D, Mei X and Ling H 2020 Feature pyramid and hierarchical boosting network for pavement crack detection IEEE Trans. Intell. Transp. Syst. 21(4) 1525-1535.
[9] Jenkins M D, Carr T A, Iglesias M I, Buggy T and Morison G 2018 A deep convolutional neural network for semantic pixel-wise segmentation of road and pavement surface cracks EUSIPCO 2120-2124.
[10] Nguyen N T H, Le T H, Perry S and Nguyen T T 2018 Pavement crack detection using convolutional neural network, SoICT 251256.
[11] Lau S L H et al. 2020 Automated Pavement Crack Segmentation Using U-Net-Based Convolutional Neural Network IEEE Access.
[12] Ronneberger O, Fischer P and Brox T 2015 U-Net: Convolutional Networks for Biomedical Image Segmentation ICMICCAI 234-241.
[13] He K, Zhang X, Ren S and Sun J 2016 Deep Residual Learning for Image Recognition CVPR 770-778.
[14] Sudre C H et al 2017 Generalised Dice Overlap as a Deep Learning Loss Function for Highly Unbalanced Segmentations Lecture Notes in Computer Science 240248.
[15] Reddi S J, Kale S, Kumar S 2018 On the Convergence of Adam and Beyond.
[16] Kingma D P, Jimmy B 2015 Adam: A Method for Stochastic Optimization. CoRR.
[17] Hinton G Neural Networks for machine learning online course https://www.coursera.org /learn/neural-networks/home/welcome
[18] Yang F, Zhang L, Yu S, Prokhorov D, Mei X and Ling H 2020 Feature pyramid and hierarchical boosting network for pavement crack detection IEEE Trans. Intell. Transp. Syst. 21(4) 15251535.
[19] Liu Y, Yao J, Lu X, Xie R and Li L 2019 DeepCrack: A deep hierarchical feature learning architecture for crack segmentation Neurocomputing 338 139-153.
[20] Buslaev A, Iglovikov V I, Khvedchenya E, Parinov A, Druzhinin M and Kalinin A A 2020 Albumentations: Fast and Flexible Image Augmentations Information 11 125.
[21] Zhang H, Wu C, Zhang Z, Zhu Y, Zhang Z, Lin H, Sun Y, He T, Mueller J, Manmatha R, Li M and Smola A 2020 ResNeSt: Split-attention networks arXiv:2004.08955.

Study of the Reality on Electric Vehicle in Indian Scenario

Bappaditya Kar[1], **Tarun Kanti Pal**[1*], **Suvanjan Bhattacharyya**[2],
[1*] Department of Mechanical Engineering
College of Engineering and Management, Kolaghat - 721 171, India
[2]Department of Mechanical Engineering
Birla Institute of Technology and Science (BITS)
Pilani - 333031, (Rajasthan), India
([1*]Email: tarunkantipal@gmail.com)

Abstract: In this work, the estimation of the actual CO_2 emission of a conventional IC engine vehicle and an electric vehicle have been studied in different phases and those vehicles are in same segmented also. How much time it would be taken in order to convert an electric vehicle in to a Green Vehicle (**GV**) also been studied. In our work, we compare two similar SUV's in there same segment one is Electric Vehicle (**EV**) and another is conventional IC Engine Vehicle (**ICV**). Both the cars have very similar power output; torque and design in there segment. Although there are many options of fast charging, long refuelling or recharging time is a major problem of the **EV** but this is not our concern study. In science each and every innovation has one good and a bad side, and there is no exception about the **EVs** also.

Keywords: CO_2 emission, Lithium ion battery, Conventional IC engine vehicle, Electric Vehicle, Green vehicle

1. Introduction

India emits around 3 Giga tonnes of CO_2 and other greenhouse gases per year. Nearly 25% of this emission occurs due to each and every types of road transport vehicles [1]. In order to reduce emitted greenhouse gases slowly then it should be moved in to the Electric vehicles (**EV**). That's why many vehicle manufacturers are presently manufacturing different types of **EVs**. Many of peoples are appreciated that every **EV** is a Green Vehicle (**GV**) that means all **EV** are environment friendly then the conventional IC Engine Vehicles (**ICV**). But in reality there be a short of attention.

1.1. Introduction of both cars

In this work, we consider two similar SUV segmented cars; one is conventional IC engine vehicle having 1.5 L turbo charged diesel engine which has the mileage of around 17 to 22 km pl, company claims that. That car has a power of 108 bhp and maximum 260 Nm torque. Another is electric vehicle is a fully electric suv car and it's reach upto 312 km maximum range during a single charge. The battery pack used in this is a 30.2 kwh li-ion battery. The car has a power of 127 bhp and maximum torque is around 245 Nm [2].

1.2. Dark side of Lithium- ion Battery production:

The Lithium Triangle is a region of the Andes rich in lithium reserves around the borders of Argentina, Bolivia and chilli. The area is thought to hold around 54% of world's lithium reserves. The production of a battery cell requires sourcing of as much as 20 different materials from around the worlds, which will pass through several refining stages, for those refining stages depending on the different energy sources, and those energy sources all are not renewable. Another think is for several material purification processes is not environment friendly [3].
The Car manufacturers are basically used lithium ion battery because li-ion batteries power density is much high than any type of other conventional battery's and li-ion batteries has another advantages that li-ion batteries are weight/ power ratio is high and those are less dangerous than lithium polymer batteries. Emission of high amount of CO_2 during mining and production phase is the main dark side of li-ion batteries. In li-ion batteries there is a small amount of lithium is used around 6% only, another component of this batteries is Cobalt. Blasting and electricity consumption in Cobalt mining is damaging the environment. Carbon dioxide and nitrogen dioxide emission are much larger during cobalt mining. CO_2 emission in battery production is depends upon the capacity of the battery. The maximum charge density of this battery is around 240 wh per kg. The average value of CO_2 emission is 177 kg CO_2/kwh power, the lowest value is 121 kg CO_2/kwh and highest value is 250 kg CO_2/kwh [4].

1.3. Power Electronics

Figure 1 shows power electronics employed in electric vehicles. In an EV power electronics plays the various power conversion processes from the plug to wheel. The different power electronics converters working in EVs like a) AC-DC converter b) DC-DC converter c) DC-AC converter d) AC-AC converter [1].

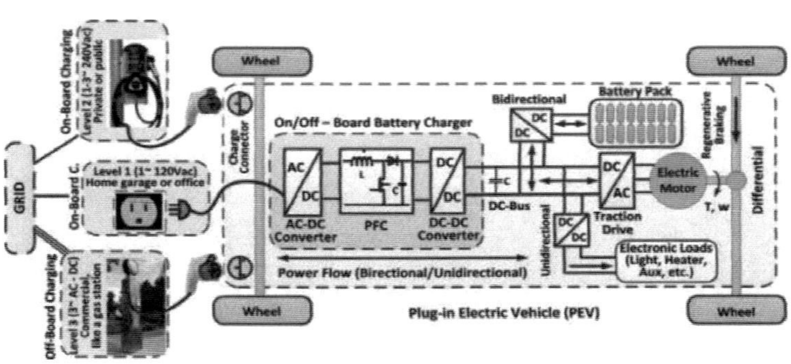

Figure 1: Power Electronics employed in an EV [1]

1.4. Different ways of Electricity produce in India

There are different ways of grid electricity production in India; most of the electricity is produced by fossil fuels simply from coal. And the coal mining is also not environment friendly.

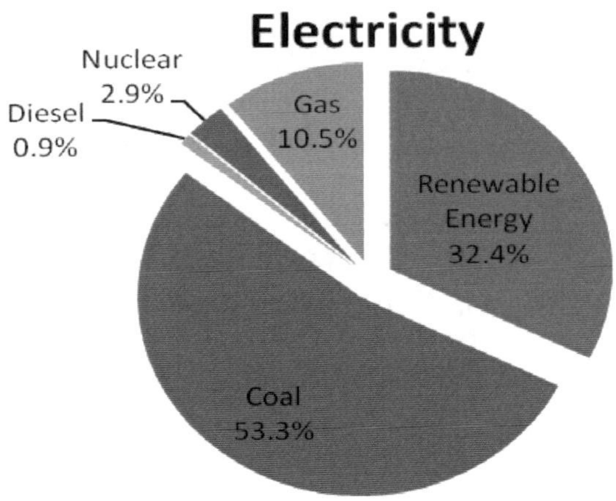

Figure 2: Resource for production of electricity in India

This is simply understandable from Fig.2 is that the maximum energy of grid is produced from burning the coal and few from natural gas is around 57%. Now if the power station with an efficiency of 34% burns coal, it emits **1.0 kg** CO_2 for generating of **one kwh** of electricity (i.e **1kwh electricity = 1 kg of CO_2.**). And approx **57%**of electricity is generated by burning of fossil fuels. This data is very vital for farther calculating the CO_2 emission [5].

2. Comparison of both cars in different phases

This work does not compare the production or manufacturing phase, because the amount of CO_2 emits for both the cases is more or less same. The reasons behind this the CO_2 emission due chassis production is same for both cases and CO_2 emission due to engine and transmission is equal to the electric motor and inverter production. Mainly we compare three phases those are:

- Fuel production for **ICV**.
- Battery production for **EV**.
- Performance and efficiency of **ICV**.
- Performance and efficiency of **EV**.
- CO_2 emits per year of running those vehicles.

And finally it has been studied how many years it will take in order to convert an **EV** in to **GV**.

2.1. Fuel production for ICV:

1.06kwh power is required to purification or refining of 1 litre of gasoline. The approximate CO_2 emission per litre of diesel fuel burning is **2.68kg**, and for petrol it is around 2.31kg**.** LPG produces around 1.51kg per litre.

2.2. Battery production phase:

The average value of CO_2 emission is **177 kg co2/kwh** power. So for the **30.2kwh** battery pack total emission is around **5.3** metric tonne of CO_2. More bigger battery means it will emits more CO_2, for 60kwh battery emits more than 10 metric tonnes of CO_2 and for 100kwh emits 17 metric tonnes of CO_2 [6].

2.3. Performance and efficiency of ICV:

The average efficiency of an IC engine is more than 34%

⇩

15% of total energy is lost due Transmission system

⇩

The overall efficiency of an ICV is more of less 25%.

The overall efficiency of an ICV is around 25%. That means we have only 25 kwh of useful power out of 100 kwh power. This is a huge power loss in the **ICV**s. This huge power loss occurs due to a lot of mechanical parts sliding or moving inside the engine. The engines mechanical loss is around 20% of total energy [7].

2.4. Performance and efficiency of EV:

AC grid transmission efficiency is 90%

⇩

AC to DC conversion + battery charge efficiency is 85%

⇩

Power electronic motor is 90%efficient.

⇩

Overall efficiency is around 69%

So the overall efficiency of an electric car is mostly around **70%**. That means if we have 100 kwh power on the electricity then we can use 70 kwh on the wheels.

This efficiency is much higher than the conventional **ICV** but gasoline has much more energy density than the batteries. Gasoline contains more than **12 kwh** power per kg of fuel burning. And the batteries have mostly 240 wh power per kg of lithium that is the battery packs are very large and weighty. For bigger battery we add more weight on EV [8].

2.5. CO_2 emission per year for both the cars

For an Indian car annually runs average around 12000km, mostly runs around **15000km**. Let's consider the **15000km** and this value is really helpful for us during compare the both cars.

ICV: For running of 15000 km that car needs to burn 833.34 lit of diesel (mileage consider as 18 kmpl). Fuel Production phase for 833.34 lit diesel purification we need 883.34 kw power that generates 503.5 kg CO_2 (833.34*0.57*1.06). During running: for burning 833.34 lit diesel CO_2 emits around 2233.35 kg CO2 per year (833.34*2.68).

TOTAL CO_2 EMISSION= [total CO_2 emission per year is =fuel production +during running the car]

=503.5***y**+2233.35***Y** [**y** is the time constant, represent the number of year]

=2736.85***Y** kg of CO_2 emits per year running the car.

EV: For running of 15000 km that car needs to charge 48 times (considering 312 km range of this car for a single charge) [15000/312=48].

Charging the battery is 30.2 kwh battery for 48 times recharging needs total power of around 1452 kwh. During charge the total amount of CO_2 emission is around 1076 kg (1452*0.57*1.3).

During Battery production: For production 30.2kwh battery emits 5345.4 kg CO_2 (30.2*177).

Total CO_2 emission= [during charge per year + battery production]

=1076***Y**+5345.4 kg CO_2.

In order to compare the CO_2 emission for a single year simply put **Y =1** and we get that the CO_2 emission for EV is much higher than ICV because the high CO_2 emission of production of battery.

But when you put **Y**=more than 3 then we get the EV emits less CO_2 then ICV that is notable point. Above equations simply said that after 3 years of using, your EV emits less CO_2 then ICV, because the CO_2 emission during battery production is very high.

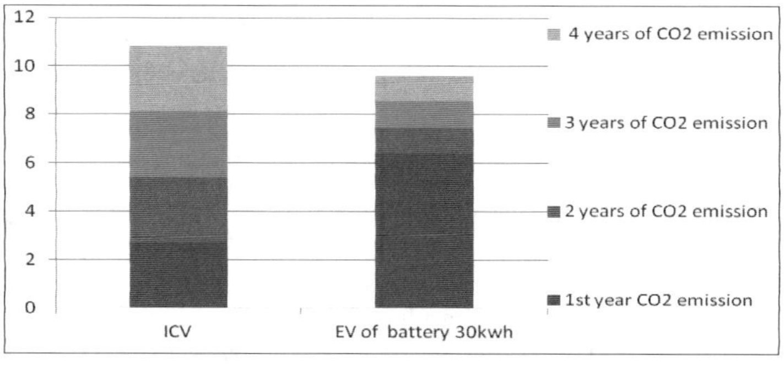

Figure 3: CO2 emission for ICV and EV

Form Fig.3 it has been observed that the total 4 years of using both the cars the EV is starts to move on to the **GV**, for more larger battery it takes more than **6** to **7** years to make it eco-friendly. But ultimately the life cars are not more than **8** years.

3. Conclusion

In this study it has been concludes that the actual CO_2 emission of a conventional **ICV** and an **EV** through its production of life time using, CO_2 for an **EV** is depends upon not only the power sources but also the battery production. Perceptibly the source of power generation played a big role. If we move on to the renewable energy sources the CO_2 emission is reduces and it will take lesser time to make it eco-friendly.

References:

[1] Kumar, A. Rakesh, Padmanaban, S., (Apr-Jun 2019), "Electric Vehicles for India: Overview and Chalanges", *IEEE India Info*. vol. 14, no.2.
[2] Sustainability Report by Tata Motors, (sustainability-report-2016-17.pdf), Online: "*https://www.tatamotors.com*"/wp-content/uploads/2015/10/06100728/tml.
[3] Sharma, R.; Manzie, C.; Bessede, M.; Crawford, R.H.; Brear, M.J., (2013), "Conventional, hybrid and electric vehicles for Australian driving conditions. Part 2: Life cycle CO2 -e emissions". *Transport. Res. C Emerg. Technol.*, vol.28, pp. 63–73.
[4] Peters, J.F.; Baumann, M.; Zimmermann, B.; Braun, J.; Weil, M., (2016), "The environmental impact of Li-Ion batteries and the role of key parameters—A review". *Renew. Sustain. Energy Rev.* vol. 67, pp. 491–506.
[5] Samundeswari , R., Vanitha, S., (November 2019), "A STUDY ON INDIAN POWER SECTOR", *Conference: Trends and Perspectives in Management Research.*
[6] "Vehicle-Grid Integration A global overview of opportunities and issues" – *NREL*
[7] Alghatgi GT., (2014), "The outlook for fuels for internal combus-tion engines". *Int J Engine Res,* vol. 15, pp. 383–198
[8] Gulhane; Vidyadhar , Tarambale; M. R., Nerkar ;Y P., (2006), "A Scope for the Research and Development Activities on Electric Vehicle Technology in Pune City", *IEEE Conference on Electric and Hybrid Vehicles.*

The future has arrived, are we ready for EV?

Probhas Bose[1] and Dipak Kumar Mandal[2]

[1,2] Dept. of Mechanical Engg., College of Engg. & Management, Kolaghat, P.O: K.T.P.P. Township, Midnapore (E) - 721 171, West Bengal, India,
Corresponding author, probhas1@yahoo.co.in[1]

Abstract. With the invention of petroleum in 1859, R & D of internal combustion (IC) engine powered vehicle commenced and the first gasoline powered automobiles hit the market of Germany in 1876. Two major problems are associated with petroleum powered automobiles; these are chemical and noise pollution. These are causing tremendous damage to the nature and its habitat. Also the petroleum reserve is fast depleting. Hence all the eyes were focused towards non-petroleum bases automobile. Automobile with alternate fuels, engine with improved efficiency were tried but keeping the health of our planet in mind, maximum thrust of research is aim to run the automobile using green energy. Solar energy, the most popular green energy, found suitable in light transport like rickshaw and three wheelers because of low energy density. EV may be the solution of future transportation system. It offers zero emission and reduces noise pollution. Cost of purchase of EV is more, though its less running cost. The technology, especially the energy storing technology is yet to be matured. Overall cost of EV will come down within affordability with the technology. EV will be popular over conventional vehicle and it appears to be the viable solution in near future.

Keywords: Automobiles, EV, HEV, Pollution, Charging

1.0 Introduction

One of the main differences between the plants and animals is that the animals can move from one place to another. The speed of move may vary from species to species; like snail's pace and the pace of leopard. The speed of walk of a young adult person is 4.5 to 6.5 km/hr [1] which may be much faster than the snail but inadequate to meet the requirements. This reason necessitated man to walk day in and day out to move from place to place. Another associated problem of human is its load carrying capacity, which is limited to 20 to 25 kg for a person weighing around 50 kg [2]. This capacity decreases with the increase of the person's weight, for self weight above 70 kg, the load carrying capacity falls below 20 kg. Use of animals for the purpose of faster movement was being used by us since the time immemorial. It was with the invention of wheel around 3500 BC [3] that the revolutionary change to move faster with load took birth in the form of animal cart. However, in order to enhance mobility and load carrying capacity, the society remained indebted to surface transport; both rail and road transport. Consequent to invention of steam engine by Thomas Newcomen in 1712 and its subsequent up-gradation by a Scottish inventor, mechanical engineer and chemist James Watt, more than the transportation system, the world witnessed the industrial revolution [4]. With the invention of petroleum, it opened a new horizon to think about internal combustion engine and the automobile industry was about to commence its journey. The age old problems of slow movement and limited load carrying capacity of mankind got an answer when the first automobile rolled out of the factory in USA in 1886 [5]. It gain huge popularity and by 1920, automobile became part of life, with additional advantage of creating job opportunity in auto industry and its ancillary sectors. These industries played an important role to improve the economy of USA [6].

1.1 Early automobiles

Though the first oil exploration was credited to Colonel Edwin Drake [7] an American businessman, who successfully drilled oil well in 1859, the cars of early 1900 were electrical motor driven. Earlier in 1769, some of them were powered by bulky and low efficiency steam engines. This is because, the initial use of petroleum was to light lamps. The situation changed with the invention of electric bulb by Thomas Alfa Edition and the first gasoline powered automobiles hit the market of Germany in

1876. Ten years later, [8] commercial production with IC engine fitted in it and gained popularity due to easy availability of refilling stations. Production of diesel car started in France in 1933 with the introduction famous brand Citroën's Rosalie. The Mercedes-Benz and the Hanomag Rekord were available in the market in 1936.

1.2 Problems with Conventional Automobiles

Two major problems associated with petroleum fuelled automobiles are chemical pollution due to emission of toxic gases through the engine exhaust and noise pollution caused by the engine and other associated moving parts.

1.2.1 Chemical Pollution: The engine exhaust emits volatile organic compounds, oxides of nitrogen, lead oxides and carbon monoxides. These harmful chemicals mix with ambient air and the atmosphere gets polluted leading to much short term and long term adverse effects on natural ecosystems. Formation of smog, depletion of ozone layer, greenhouse effect, global warming etc are some of the unwanted effect on the atmosphere. The resultant condition adversely affects health of natural habitats, agricultural productivity and changes the climatic conditions [9]. Vehicle types and pollutant for emissions from common petrol engines [10] and diesel engines [11] with fuel combinations is given in Table 1.

Table 1 Emissions of pollutants of different vehicles

Engine type	Fuel Type	Vehicle type	Major emissions
Petrol Engine	Petrol	Cars, buses (some trucks)	HC, CO, NOx, Pb & particulate
Diesel Engine	Diesel	Lorries, trucks tractors (some buses & cars)	NOx, SOx, HC, CO, Soot and particulates
Two-stroke Engine	Petrol	Motor cycles	HC, CO, NOx, Pb & particulates

1.2.2 Noise Pollution: It is caused by automobiles is primarily because of running of IC engine, frictional noise due to rolling of whets on the road surface. Another source of noise is uncontrolled honking of horn. Automobile noise is a growing public health concern for children, young and old people. This is more acute in urban areas where the traffic density is very high. Prolonged exposure to noise affect concentration, slip disturbance, high blood pressure, increased risk of hypertension and heart disease [12]. Depending upon the level and duration noise, its effect on our health can be broadly categorized into four group as given in Table 2 [13]

Table 2 Physical effect of noise

Category	Effect	Health Problem
I	Physical effects	Hearing defects
II	Physiological effects	High blood pressure, irregularity of heart rhythm and ulcers
III	Psychological effects	Sleeplessness and going to sleep late, irritability and stress
IV	Effects on work performance	Reduction of productivity and misunderstanding what is heard

For average human beings, noise level of 85 dB is considered as a safe level by hearing health professionals. [14]. Permissible noise level in residential areas as per WHO specification is 50 dB(A) [15]. According to Govt. norms the noise range for horns has been fixed between 93 dB(A) and 112 dB(A). [16]. In spite of strict pollution norms, it has not been possible to control the adverse effects due to huge vehicle density.

1.3 Consideration of Alternate Source of Power

Alternate to conventional automobiles Research for alternative to conventional automobiles are on which would not threat this planet with chemical and noise pollution. Another reason to look for alternatives is that the oil reserve is fast depleting. It is estimated that annual global consumption rate of crude oil is excess of 11 billion tones. At this rate the global oil reserve will last for another 50 years or so [17] Automobile with alternate fuels, improves engine have received some attention as potential options [18] However, keeping the health of our planet in mind, maximum thrust of research is aim to run the automobile using green energy with particular reference to solar energy. But there is a mismatch between solar energy available *vis-a-vis* energy required to run the vehicle; former being lesser than the later, solar cars are limited to light transport like rickshaw and three wheelers. Another option gaining popularity is electric vehicle (EV), probably the best answer against chemical and noise pollutions and also achievable solution against limited oil reserve.

It is observed that that EV is generally noiseless while it is running at a speed below 25 km/hr. Since we are not used to zero noise in actual road condition, it becomes very dangerous for the visually impaired pedestrians and slow movers like bicycles and rickshaws because they fail to recognize that an EV is behind them. This has resulted to add some noise while the EV is moving at slow speed to alert these movers. Toyota Corporation Japan has incorporated the audible warning system which may increase with the increase speed of EV. It was suggested that the noise level should be alarming but below the annoying level [19].

2.0 The Electric Vehicle

In IC Engine, the energy is developed using petroleum only. However, electricity is generated by generator/alternator which can run from a diversified source of energy. Moreover, limited oil reserve has forced every country to move beyond conventional IC engine driven vehicle and many countries have set time frame [20]. The UK has announced that it will stop selling of conventional cars in the year 2040 in order to achieve the target of zero emission road environments by 2050. Paris, however wants to achieve the both the targets ten years before UK. Scotland plans to stop selling IC engine vehicles by 2032. The Scandinavian countries have introduces green tax system [21] to encourage the citizens to go for environment friendly transport in place of fossil fueled system. Austria and Netherlands planned to impose ban against IC engine cars by 2025 and 2030 respectively. China is only the major country yet to set any deadline against banning of conventional cars. However, they are working to set a time line and it may be declare the dead line any time soon. In USA, sale of EV is growing at rapid pace as the tax credit ranging between $ 2500 - $ 7500 is being give to the customers against the purchase of each new [22] EV. In India, the Ministry of Heavy Industries and the [23] NITI Aayog are working together to ensure that by the year 2030, only EVs are available for the Indian roads. The sale of EV has picked up especially for the two wheels segments. However, with present demand of EV which is only one out of 400 conventional vehicles, India has a long way to go to achieve its deadline of 2030.

The overall scenario is of course quite encouraging and continuously tilting towards EV because of awareness program raised by all the countries for a greener planet. The data in Table 3 [24] shows that the demand of EV is increasing exponentially with every passing year.

Table 3 Data for EV demand over the years

Year	2012	2013	2014	2015	2016	2017	2018
No. of EV in Million	0.11	0.22	0.4	0.72	1.18	1.93	3.27

By extrapolating data beyond the year 2018, as shown in the plot in Fig. 1, the number of EV in the years 2019 and 2020 is showing tremendous increase up to 5.57 and 8.12 million respectively.

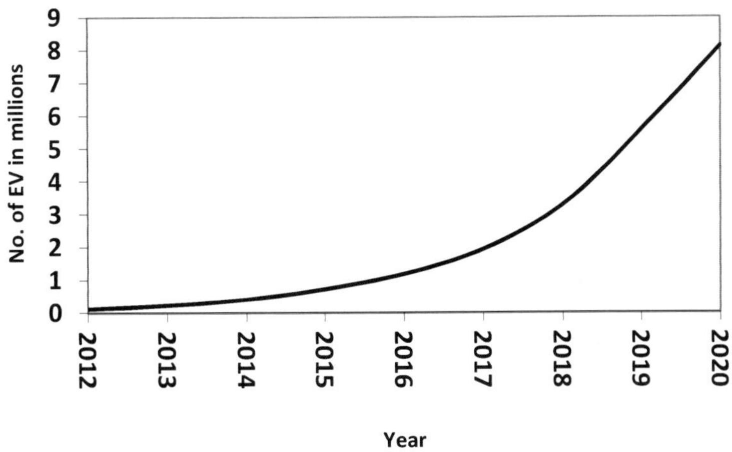

Fig.1 Global growth of EV

3.0 Variants of EV

In its simplest form EV contains a motor which is the power pack to drive it. The motor may be AC or DC motor, which run using electricity from a battery bank. The motor speed is controlled by a controller. The motor together with the battery bank and controller replaces the IC engine and fuel system. In pure form of EV, called Battery Electric Vehicle (BEV), the battery bank is charged from external power source, called charging station. The second variant is Hybrid Electric Vehicle (HEV). As the name suggests, it has twin facility of running in electricity as well as using IC Engine [25]. The battery in this case is charged through two sources of energy. First source of energy is generated by the onboard IC Engine coupled with a generator. The second method is by capturing the energy, which otherwise would have been lost due to application of brake. In this case the electric motor is made to function as a generator and charge the system. This is called regenerative braking. There is another variant of EV called Plug-in Hybrid Electric Vehicle (PHEV) which provides twin facility of charging the battery from a charging station [26], as well as by its on-board IC engine coupled with a generator.

4.0 Additional Advantages of EV

In addition to zero emission and noiseless operation, EV offers some more additional advantages which are much desired requirements of users. These are very high pick up, very low running cost and maintenance cost. Technically, it has simple power train with forward and reverse gear eliminating the gear box of a conventional car. It is estimated that for a range of 600 km, running cost of EV is only 30 to 33% of that of gasoline vehicle [27]. Energy conversion from chemical energy to mechanical work for IC Engine vehicle is 29 - 32 % for gasoline engine and 34 – 36 % for diesel engine [28] remaining energy is lost due to heat loss, noise, friction etc. For EV 85 to 90% energy of the battery is converted to mechanical work at the wheel [29].

5.0 Challenges with EV

All types of EV have higher initial cost as compare to its conventional counterpart. This is the one of the reasons that EV is unable to climb the preference chart of potential customers. It is observed that EV is more cost effective if the complete lifecycle of the vehicle is considered instead of initial cost. Comparing Tata Nexon EV and Maruti Suzuki Swift DZire (Diesel) it is noted price parity comes between the two in the 7th or 8th year of life. Maintenance and service costs of EV being lower, further years of running, EV works out to be cheaper [30]. To popularize EV, various tax exemptions is given by the Governments for initial purchase.

5.1 Charging Station: Very few charging stations are available at the road side. Average range of electric car being around 300 to 400 km, users are more worried about its next recharge rather than enjoying the drive. Few states of US and Canada has established alternative fuel data centre [31] which provides the locations, hours of operation, and type of charging equipment available etc.

5.2 Charging time: Generally considerable time is required to charge the battery resulting long idle time for the users. It is observed that for an electric car having a battery capacity of 60 kWh, the charging time from zero to full is eight hours by a 7kW charging station [32]. Penn State University of USA has developed EV battery with high energy storing capacity but low charging time. It is reported that the battery which provides a [33] range of 480 km need only 10 minutes to achieve full charge. However, using fast charger the charging time can be reduced considerably. One of the solutions of long waiting time can be use of 'cooking gas cylinder refilling principle'. This is to deposit the exhaust battery and in exchange pick up a charged battery. For this there is a requirement of common specifications by all the battery manufacturers for same type of EV.

5.3 Range: Current generation of EVs offer appreciable range with fully charged battery. Hyundai Kona offers a range of around 450 km with its 39 kWh lithium-ion battery [34]. MG ZS EV and Tata Nexon EVs have the single charge range of 340 km and 310 km respectively.

6.0 Conclusion

There is a paradigm shift of technology from conventional (petroleum fuelled) automobile to EV. Potential customers will take time to absorb this shift of technology and accept the new driving system. In one hand, EV offers zero emission, zero noise pollution and low operational cost which satisfy the environmentalists, on the other hand to delight the car lovers, in addition to all essential features; it offers a very high pick up with smooth ride. For example, popular electric car of SUV category MG ZS has been designed to accelerate from 0-100 km/h in 8.5s. However, high procurement cost, not so good support system like inadequate charging stations, long charging time, limited range of operation etc are the present challenges faced by the owners. Hence there is an urgent need to consider these aspects and provide solutions. It is accepted that the human mind is not open to welcome any new technology quickly without comparing the support system with the existing one. Hence, instead of looking toward additional features, the mind looks into the impending problems it may have to surmount as new owner. So the designers have a challenging task to make the support system owner friendly. EV has the scope to be controlled by AI, and the work is on in this regard. The technology, especially the energy storing technology is yet to be matured. As the technology is absorbed at the commercial production level, overall cost of EV will come down within affordability. There is no denying that EV is here to stay and this is the solution of future transportation system. Hence, the future has arrived to migrate to EV from the conventional vehicle and presently it appears to be the viable solution to the existing transportation system.

7.0 References

[1] *https://www.healthline.com/.*
[2] *http://www.thedoctorwillseeyounow.com/.*
[3] *https://www.livescience.com/18808-invention-wheel.html.*
[4] *https://www.livescience.com/.*
[5] *www.daimler.com/.*
[6] *www.angelfire.com/.*
[7] *https://enviroliteracy.org/.*
[8] *www.autolife.umd.umich.edu.*

[9] Bolaji B O and Adejuyigbe*et* S B 2006 Vehicle emissions and their effects on natural environment – a review, *Journal of the Ghana Institution of Engineers* 4(1):35–41.
[10] Shamil AFA 2019 The gasoline fuel quality impact on fuel consumption, air-fuel ratio (AFR), lambda (λ) and exhaust emissions of gasoline-fueled vehicles, *Journal Cogent Engineering* 6(1) 1616866.
[11] Ibrahim AS, Altinişik K and Keskin A 2015 The pollutant emissions from diesel-engine vehicles and exhaust after treatment systems, *Clean Technologies and Environmental Policy* 17:15–27
[12] Lee E Y, Jerrett M, Ross Z, Coogan P F and Seto E Y W 2014 Assessment of traffic-related noise in three cities in the United States *Environment Research* 132:182-189.
[13] Pachiappan Y et al, 2013 Regression modeling of traffic noise pollution, *Građevinar*, 65 (12):1089-1096.
[14] Fink DJ 2017 What Is a Safe Noise Level for the Public? *American Journal of Public Health* 107(1):44–45.
[15] Menkiti N U and Agunwamba J C 2015 Assessment of noise pollution from electricity generators in a high-density residential area *African Journal of Science, Technology, Innovation and Development*. 7 (4): 306–312.
[16] The Noise Pollution (Regulation And Control) Rules, 2000.
[17] *https://www.ecotricity.co.uk/our-green-energy.*
[18] Ambuj DS 1995 Automobiles and global warming: Alternative fuels and other options for carbon dioxide emissions reduction *Environmental Impact Assessment Reviewl* 15(3):241-274.
[19] *Toyota Global News Room Report* 24 Aug 2010.
[20] *The Economic Times* 20 Nov 2017.
[21] Brannlund R and Gren I M 1999 Green Taxes: Economic Theory and Empirical Evidence From Scandinavia . Cheltenham, UK: Edward Elgar, xvi, 157.
[22] Gonzalez C.M. 2019 Ready for Electric Car Revolution? *Mechanical Engineering* 42 – 45
[23] *NITI Report.*
[24] *http://www.statista.com.*
[25] *Indiana Office of Energy Development*
[26] *www.greentechmedis.com.*
[27] Reddy R 2019 Why EVs are the next big market? Is India ready for EV revolution? *IEI ATV MCDB IV* : 98-102
[28]*https://www.researchgate.net/post/what_is_an_efficiency_of_modern_average_car_IC_engines/56e13d97615e27e85804db21/citation/download*
[29] *https://www.energycouncil.com.au/*
[30] *https://inc42.com/datalab*
[31] *https://afdc.energy.gov/fuels/electricity_locations.html#/find/nearest?fuel=ELEC*
[32] *pod-point.com › Driver Guides Overview.*
[33] *https://www.financialexpress.com/auto/car-news.*
[34] *https://www.indiacarnews.com/news.*

A model of inductive charging technique in battery electric vehicles(BEVs)

Partha Mishra[1], Arpita Bhakta[2], Rittick Sarkar[3] and Farzana Jasmin[4]

[1] Assistant Professor, Department of Electrical Engineering, College of Engineering and Management, Kolaghat, West Bengal–721171, India
e-mail:pmishra@cemk.ac.in

[2,3,4] Students (batch, 2016-20), Department of Electrical Engineering, College of Engineering and Management, Kolaghat, West Bengal–721171, India

Abstract. In this paper a prototype model has been designed. In this study a theoretical concept of a simple but effective method for inductive charging in battery electric vehicles(BEV) has been proposed. It has been observed in experimental study that power can be transferred wirelessly from transmitter to receiver using inductive coupling principles. In this article a prototype model has been designed to observe the inductive charging and also the inductive resonant coupling. It has also been observed experimentally that the power output of the wireless charging model significantly based on the resonant condition between the transmitting and receiving coils. In this article a proposed model has been presented to focus on the inductive coupling power transfer technique which is used for charging of a battery electric vehicles(BEVs). The proposed model and its experimental results successfully describes the inductive charging technique and the maximum power transfer occurs when resonant condition achieved by the two coils in very smaller distance. This model is very effective to established the concept of wireless power transfer in very smaller coil size. The experimental implementation of the design is made with digital multimeter, milliammeter, power-MOSFETS, voltage regulator IC, oscillator and diode rectifier circuit.in the transmitter and receiver coils.
Keywords. Battery Electric Vehicles(BEVs), Inductive charging, Magnetic Coupling, Oscillator circuit, Rectifier.

1. Introduction

It has been observed that economic growth and development of any country highly influenced by transportation sector because it contributes the national GDP, employment generation as it meets the needs of logistics and the transportation industry and also improving the lives of millions of people [1]. In India, 22% of GDP depends on auto industry [1]. It is expected that demands of passenger vehicles, commercial vehicles and two wheelers in India will be 10 million,2.7 million and 34 million respectively [1]. Indian transportation sector largely depends on oil import and a large amount of money has been spending for this purposes. So, India's future energy sector is facing challenges to meet the high demands of energy in future. In this situation, globally automotive industry has been shifted from Internal Combustion Engine(ICE) to Electric Vehicles on account of fast depletion of fossil fuels, impact on environment and rapid increase of cost of fossil fuels [2].

The battery electric vehicles or frequently called as BEVs are fully battery based vehicles. It has high capacity rechargeable battery which provides energy to electric motor. There is no internal combustion engine or any fuel chambers. Therefore, it does not emit any harmful emission to the environment and any carbon particle or hazards to the environment [5]. It is clean sources of energy. The battery packs are charged by any external sources. The charger circuit are classified in respect of charging speed or time in which it charged the battery. It has high starting torque compare to gasoline fuel motor and also noise free [2-5]. It requires less maintenance compare to ICE vehicles [4]. The drawback is the battery technology or charging time of the battery packs [6-7]. In this proposed study a prototype model has designed to study the wireless charging technique or inductive charging with resonant condition. This proposed method is very much significant for understanding the EVs charging technique with smaller coil size. In this article the proposed study contributes significantly that, in spite of very low voltage output and very small air gap distance the proposed model successfully indicates the resonant condition for maximum output [7].

2. Principle of operation of inductive charging

Inductive charging is based on the concept of electromagnetism. When a current carrying conductor creates some magnetic field surroundings it, then a portion of the magnetic flux may link to other coil and induced voltage and corresponding current flowing through the second coil. In this situation it is said that this two coils are magnetically coupled and energy are transferred from one circuit or coil to other circuit or coil. The energy transferred is performed by the transferred of the magnetic field which is common to both of the coils. This phenomenon also called inductive coupling. The strength of the coupling depends on the mutual inductance of the coupling circuit. Power or energy transfer of the coupling circuit can be increased by increasing the number of turns, cross-sectional area, strength of the radial magnetic field and amount of current passing through the circuit.
Inductive charging is nothing but a short distance energy transfers of two coils in a charging station. In the charging station or in a base station an inductive coil creates alternating magnetic field which sets up a magnetic coupling to the coil which is kept in a portable device of the vehicle and the energy is transferred to the device which again convert it to current and charge the battery packs.

3. Working principle of the whole circuit with its schematic block diagram

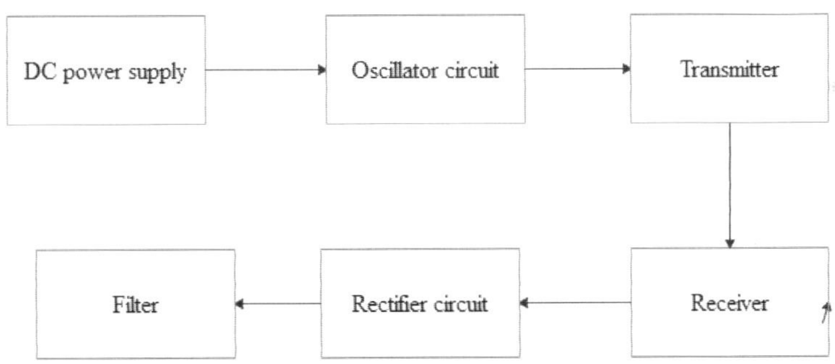

Figure 1. Schematic block diagram of the circuit.

The transmitter module is made of by a DC source, an oscillator commonly known as an inverter and a transmitter coil. First a DC power has been provided to the oscillator circuit where the oscillator circuit convert the DC power to high frequency AC current which is then fed to the transmitter coil to produce a high frequency alternating magnetic field. The input D.C. Power comes from a simple step down transformer and a rectifier circuit. The transformer steps down the voltage to a desired level and the rectifier circuit convert the A.C. voltage to D.C voltage.

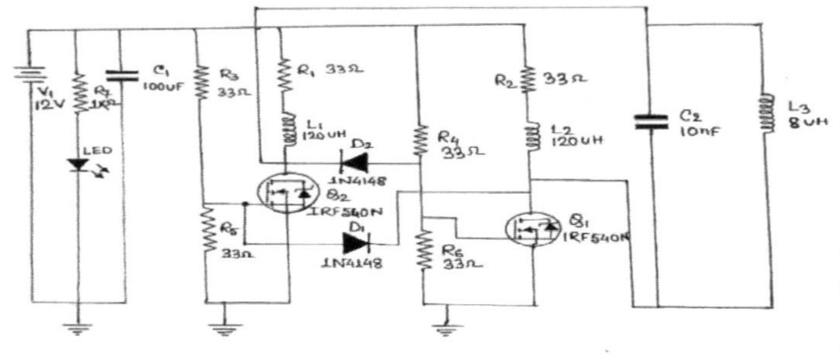

Figure 2. Schematic diagram of the transmitter circuit.

The receiver circuit consist of receiver coil, rectifier circuit and a voltage regulator IC. Furthermore, additional bulk converter has been incorporated to increase more current. An A.C. voltage is induced in the receiver coil. The rectifier circuit converts it to D.C. and the voltage regulator IC helps to maintain a constant limited voltage at the load. Bulk converter has been incorporate with the system to increase more current.

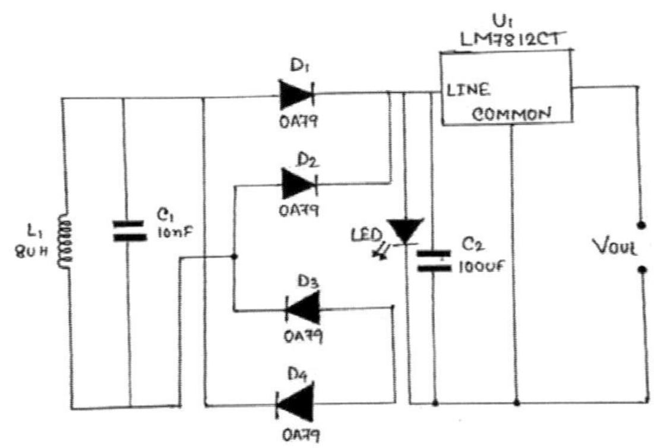

Figure 3. Schematic diagram of the receiver circuit.

A voltage regulator is used to maintain constant voltage level of the circuit. It is generally a electronic component. Depending on the design, it may be used to regulate one or more AC or DC voltages.

4. Results and Discussions

The present work successfully implemented in a prototype model and it has been a good model to study the inductive charging technology. This paper helps to design inductive resonance circuit also.

This hardware model gives output voltage in a smaller value. The model of wireless charging includes the design of transmitting coils, receiving coils, oscillator circuit and rectifier circuit. The model output power was measured and enlighten different LED indicator lamp to show the efficiency and maximum power output. The 5mm LED indicator lamps (3.6V,20mA) were connected on the receiving coils and the power output was measured by changing the distance between transmitting and receiving coils. The experimental setup is shown in the Figure 5. The source voltage is 12V DC. The receiving and transmitting coils were designed identically to obtained maximum efficiency [7]. Table 1 shows the different coils parameters.

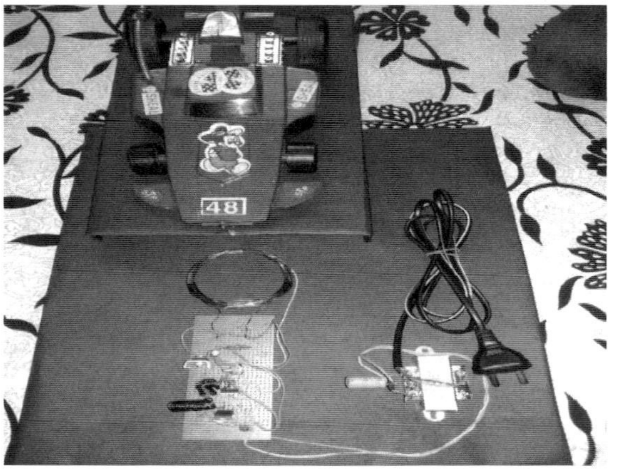

Figure 4. Complete photograph of the model.

Table 1. Coil parameter.

S. no.	Parameter	Unit
1	Wire	SWG 20 no
2	Numbers of turn	15
3	Coil diameter	6mm
4	Length	20cm

The graph of output power and different distance is shown in Figure 6. During the calculation of the power output, some assumption was made for better result for example the working current and voltage of LED with measured output power. It has been observed that the maximum power output obtained in certain distance not in minimum distance between transmitting and receiving coils. It may conclude that when the coupling between two coils is strong or resonance may have achieved between two coils then the maximum output can be obtained.

Table 2. Output power versus different distance between two coils.

Distance between transmitting and receiving coils(cm)	Output power(mW)
2	50.4
3	57.6
4	79.2
5	68.4

The graphical representation of output power with the change of distance is shown in Figure 5.

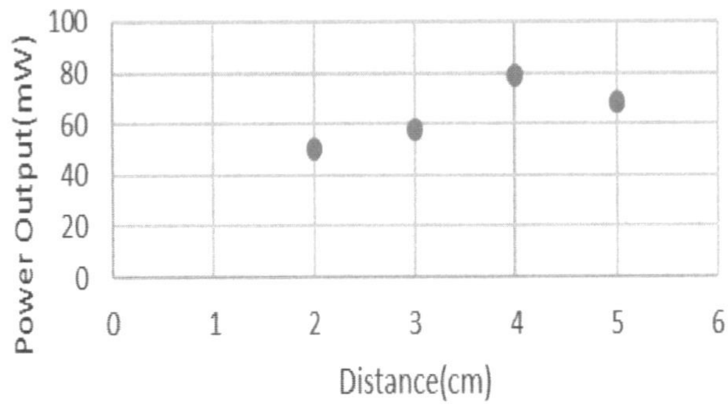

Figure 5. Graphical representation of output power with the change of distance

5. Conclusions

In this article a small model of inductive charging has been presented and a study have been carried out to understanding the resonant inductive coupling which gives maximum output power to the load. In this study a simple experimental setup has been proposed and a charging model developed successfully in smaller voltage level. The result shows that in this proposed technique power can be transferred wirelessly in a smaller coil with smaller distance. Future work may include in more distance and better charging time with more output power.

References

[1] *National Electric Mobility Mission Plan 2020* , Department of Heavy Industry, Govt. of India.

[2] Hongjun Chen, Fei Lu, Fujuan Guo, "Power Management System Design for Small Size Solar-Electric Vehicle", 2012 *IEEE 7th International Power Electronics and Motion Control Conference* - ECCE Asia, 2012, pp. 2658- 2662.

[3] B. Ganji and A. Z. Kouzani, "A study on look-ahead control and energy management strategies in hybrid electric vehicles," 2010 *8th IEEE International Conference on Control and Automation (ICCA),* 2010, pp. 388-392.

[4] Jacob Millman and Christos C. Halkias, —Integrated Electronics: Analog and Digital Circuits and Systems

[5] F. R. Salmasi, "Control Strategies for Hybrid Electric Vehicles: Evolution, Classification, Comparison, and Future Trends," *IEEE Transactions on Vehicular Technology*, **vol. 56**, 2007, pp. 2393-2404.

[6] Akar, F., Tavlasoglu, Y. and Vural, B. 2016. "An Energy Management Strategy for a Concept Battery/Ultracapacitor Electric Vehicle with Improved Battery Life". *IEEE Transactions on Transportation Electrification,* **vol 3** : 191-200.

[7] A. Elahi, A. Ahmed Amin, U. Tabraiz Shami, M. Tayyab Usman, M. Sajid Iqbal. " Efficient Wireless Charging System for Supercapacitor-Based Electric Vehicle using Inductive Coupling Power Transfer Technique" *Advances in Mechanical Engineering* 2019, **vol.11(11)**, 1-10, 2019.

Medical Image Encryption Algorithm based on Dynamic DNA Sequencing and 2D-3D Chaotic Equations

Anwesha Kar[1] and Dilip Kumar Gayen[1]

[1] Department of Computer Science & Engineering, College of Engineering & Management, Kolaghat, KTPP Township, West Bengal,721171, India
e-mail: *dilipgayen@yahoo.com*

Abstract. Digital images are now widely used in modern clinic diagnosis. The diagnostic images with confidential information related to patients' privacy are stored and transmitted via public networks. Secured schemes to guarantee confidentiality of patients' privacy are becoming more and more vital. Medical images are considered as one of the most important and sensitive data in information systems. Sending medical images over the network requires a strong encryption algorithm such that it is resistant against cryptographic attacks. Among the three security objectives for the security of information systems namely confidentiality, integrity and availability, confidentiality is the most important aspect that need to be taken much care for the secure storage and transmission of medical images. This paper proposes an adaptive medical image encryption algorithm based on Dynamic DNA Sequencing and 2D-3D Chaotic Equations that encrypts any medical image with same uniformity and efficiency.

1. Introduction

In this era of Digitalisation, we exchange a lot of personal information through the Internet. A medical image contains sensitive data [1]. To transfer any medical image securely, it must be encrypted and stored which requires developing a strong, secure and efficient encryption algorithm is necessary [2]. The three main features of any security service is confidentiality, integrity and availability. Confidentiality being the most essential criteria for exchanging medical images has to be given utmost priority. Generally medical image encryption algorithm is chaoses based or are based on information entropy [3]. DNA Sequencing has also been gaining popularity. The paper combines the image encryption, chaotic system along with DNA Sequencing to improve the security. This way, it can increase the security of key space and image in a much more efficient way.

2. Chaotic Systems

With the ever increasing demand of requirement of more and more reliable, trustable and strong security for images/videos with confidential information, chaos-based systems have is an important method in process of image encryption because of their excellent random properties and encryption performance. Chaos theory is used in various fields of physics, engineering, biology and economy in recent years. Wang et al. have used an extension of fractal Fourier transformation and a digital holographic scheme to propose a cryptographic algorithm [4]. Due to the parameters used in this scheme and security enhancement of the encryption process is very high. A type of complex nonlinear system, chaotic systems have initial value sensitivity, pseudo randomness and non-periodicity, which are consistent with the requirements for cryptography. That can be used to generate random keys, which will have equal encryption effect for all samples. This is not capable of being broken. Thus, chaotic encryption method is being used in the field of information security, more precisely in the field of image encryption. We introduce a chaotic map, the intertwining logistic map or 3D chaotic equations, which is defined as follows:

$$x_{n+1} = [\mu \times k1 \times y_n \times (1 - x_n) + z] \bmod 1$$

$$y_{n+1} = [(\mu \times k2 \times y_n + z \times 1)/(1 + (x_n + 1)^2))])] \bmod 1$$

$$z_{n+1} = [\mu \times (x_{n+1} + y_{n+1} + k3) \times \sin(z_n] \bmod 1$$

where μ, ki are the system parameters. The 2D Chaotic Equations are defined as:

$$x_{n+1} = 1 - ax_n^2 + y_n$$

$$y_{n+1} = bx_n$$

where a, b are constants.

3. DNA Sequence and Dynamic DNA Sequencing

DNA is a carrier of the biological genetic information that is stored in the body. It has a high degree of parallelism, massive storage capacity, and least energy consumption with a unique molecular structure producing a unique molecular recognition mechanism, which determines its excellent information storage and information processing ability enhancing its potential and suitability in the field of information security, information hiding, and authentication, paving a new way for the development of modern cryptography through DNA Sequencing. DNA consists of a linear string of nucleotides, or bases, for simplicity referred to by the first letters of their chemical names-A(Adenosine), T(Thymine), C(Cytosine) and G(Guanine). The process of deducing the order of nucleotides in DNA is called DNA sequencing. A RGB image encryption algorithm based on DNA encoding and chaotic map by Liu et al. proposed [5, 6]. Chai et al proposed an image encryption algorithm based on chaos combined with DNA operations [7]. A type of digital image encryption technology based on hyperchaos mapping and DNA sequence library arithmetic to realize a scrambling position transformation of image pixels and the spread of the pixel values was proposed by Niu et al. [8].

Most algorithms only chose fixed DNA encoding rules to encode with only a maximum of eight experiments can break it. In this paper, the DNA encoding rules are dynamically selected according to the random keys generated through chaotic sequences.

Algorithm:

1. Read the Medical Image
2. Convert Image to Gray scale
3. Form Image Matrix I (n*n)
4. Generate the initial values of x,y and z for 3D chaotic equation.
   - 4.1 Take random 72 characters.
   - 4.2 Convert it into binary sequence (288 bits) .
   - 4.3 Divide into three group (each group= 96 bits)
   - 4.4 Repeat for group 1 to 3
     - 4.4.1 Generate four blocks - q1, q2, q3, q4 (each block =24 bits)
     - 4.4.2 Convert blocks to decimal.
     - 4.4.3 Perform the bitwise XOR operation('X'):
       q5=(((q1 X q2) X q3) X q4)
     - 4.4.4 X(group) =0.5000+mod(($q5/2^{32}$),1)
   - 4.5 X(group{1 to 3}) = x,y,z respectively.
5. Generate n*n set of values for xn+1,yn+1,zn+1 as x, y, z respectively based on 3D chaotic equations:

   x_{n+1}= [$\mu \times k1 \times y_n \times (1 - x_n)$+z] mod 1

IOP Conf. Series: Materials Science and Engineering 1080 (2021) 012006 doi:10.1088/1757-899X/1080/1/012006

$y_{n+1} = [\ (\ \mu \times k2 \times y_n + z \times 1\ )/(1 + (x_n + 1)^2)\ ]\ ]$ mod 1
$z_{n+1} = [\mu \times (x_{n+1} + y_{n+1} + k3) \times \sin(z_n)]$ mod 1
k1, k2, k3 values are constant

6. Mix columns of I
7. Mix rows of I
8. Generate **X** : n*n set of randomized values in range of 1 to 4 for DNA operations based on equation 1 of 2D chaotic equation:
   x_{n+1}=1-a*$x_n{}^2$+y_n ------ Equation (1) a, x_n, y_n are constants.
9. Generate Y : n*n set of randomized values in range of 1 to 8 for Rule Conversions based on equation 2 of 2D chaotic equation:
   y_{n+1}=b*x_n ------ Equation (2) b and $x_n$are constants
10. Repeat for n*n elements
    10.1 Take each value of I and convert it to DNA sequence based on corresponding rule of Y. Store in P
    10.2 Take each value of z and convert it to DNA sequence based on corresponding rule of Y. Store in B
        10.2.1 repeat for each character in DNA sequence from P and B
            10.2.1.1 *Perform DNA operations*(add,sub,xor,nor) based on corresponding value from X
            10.2.1.2 Store the binary value after *reversing rule conversion* based on Y
        10.2.2 Store resultant in decimal in matrix J
11. Repeat step 4,5 for random key generation matrix.
12. Perform XOR with J and randomized matrix
13. Display the encrypted image matrix.

4. Results and Discussions

The simulation is done with the help of Matlab using different input images. We have taken different types of input images such as X-Ray of Human Teeth, CT scan for Lung Cancer, Radiology of Human Hand, Brain Imaging of Multiple Sclerosis, X-Ray detecting Pulmonary Fibrosis and Ultra-sonography Of Renal Cyst for our algorithm. In this paper we describe only one input image as shown in the Figure 1. After applying different operations as describe in the algorithm on input image, the resultants different stages are shown in Figure 2 to Figure 10.

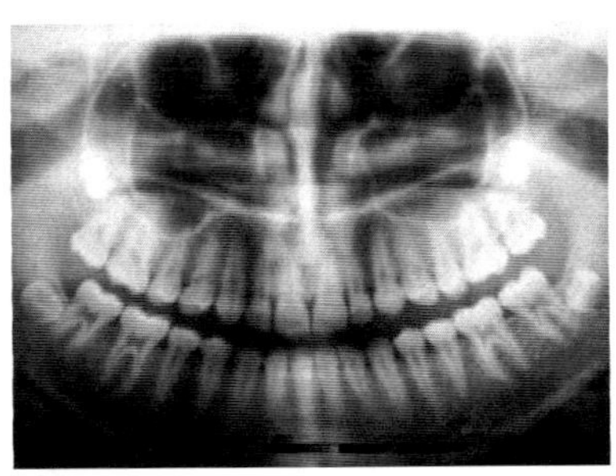

Figure 1. X-Ray of Human Teeth

Command Window

0.9907	0.9963	0.6301	0.0573	0.4282	0.6310	0.3648	0.0957	0.6428	0.3581	0.5274	0.7513
0.0097	0.5850	0.3653	0.9215	0.0802	0.8180	0.9888	0.6483	0.5580	0.0225	0.3246	0.8215
0.2186	0.5712	0.5524	0.5100	0.7695	0.0532	0.0858	0.0668	0.9490	0.6431	0.8938	0.4491
0.6100	0.8362	0.4900	0.6996	0.2028	0.8116	0.8776	0.4215	0.6363	0.3562	0.6423	0.6673
0.0034	0.2031	0.7584	0.3813	0.1429	0.5265	0.8894	0.4199	0.9428	0.8046	0.6672	0.6718
0.6390	0.2921	0.9683	0.1392	0.1761	0.2686	0.9890	0.0207	0.2338	0.5515	0.4983	0.2900
0.7232	0.4837	0.2027	0.0769	0.4774	0.4043	0.6238	0.2005	0.8449	0.9947	0.6929	0.8146
0.9114	0.1676	0.7429	0.0398	0.3516	0.8552	0.0722	0.2920	0.9449	0.4183	0.3257	0.9648
0.4623	0.3412	0.8026	0.9204	0.0733	0.3316	0.4312	0.1928	0.3563	0.6055	0.1173	0.8814
0.7543	0.8059	0.3751	0.8739	0.0128	0.7631	0.3747	0.8095	0.1481	0.6644	0.4465	0.3608
0.5705	0.9521	0.9251	0.3549	0.4649	0.7024	0.8202	0.4116	0.2246	0.1446	0.6332	0.1294
0.6181	0.6496	0.0913	0.3675	0.3985	0.1751	0.3204	0.9869	0.5071	0.6866	0.9035	0.1415
0.7420	0.7090	0.2098	0.3505	0.3885	0.6571	0.7437	0.6016	0.1638	0.7658	0.9476	0.6541
0.8853	0.5663	0.0148	0.8734	0.5585	0.4336	0.8962	0.4121	0.4636	0.2305	0.4837	0.5806
0.9013	0.1491	0.6996	0.2784	0.0203	0.2087	0.1330	0.8321	0.0551	0.2769	0.4343	0.8891
0.1111	0.4960	0.4326	0.9990	0.8883	0.0716	0.2142	0.3638	0.9551	0.7331	0.4948	0.7600
0.4048	0.0811	0.8769	0.7400	0.5540	0.2247	0.9571	0.4836	0.9853	0.1026	0.1202	0.0340
0.7132	0.5813	0.9253	0.0780	0.5910	0.8413	0.5055	0.3157	0.2065	0.0793	0.6141	0.3060
0.3507	0.2040	0.8817	0.3049	0.8347	0.1158	0.7410	0.8832	0.4481	0.2943	0.8328	0.8254
0.8728	0.9836	0.8216	0.1408	0.3329	0.2742	0.0563	0.3108	0.4074	0.5402	0.2113	0.4635
0.4972	0.2977	0.2695	0.9850	0.7067	0.1650	0.7324	0.1470	0.8199	0.6258	0.4549	0.7000
0.7377	0.6177	0.8623	0.5558	0.1249	0.4103	0.2059	0.1482	0.7721	0.1883	0.0546	0.2949
0.3537	0.2603	0.2583	0.1249	0.3668	0.4241	0.5553	0.4462	0.9676	0.7697	0.8154	0.1130
0.1692	0.8133	0.3376	0.9498	0.9588	0.1024	0.1195	0.1166	0.9483	0.7675	0.3984	0.1419

Figure 2. The input image is read, converted it to gray scale and then resized to desired size (nXn) .

Showing the pixel values of the input image based on 3D chaotic equations whose initial values are taken from 2D equations. The 2D equations are in turn generated from an arbitrary set of 72 character set.

Command Window

20	10	165	147	47	80	93	22								
28	11	141	143	49	73	104	23								
34	15	104	136	46	76	123	23								
40	27	82	123	72	87	136	33								
41	32	74	113	90	81	140	44								
46	39	80	88	101	81	147	58								
49	46	94	85	108	78	145	69								
51	61	111	89	118	78	141	69								
49	90	115	91	132	71	135	66								
53	104	114	79	141	70	133	70								
56	113	114	72	144	72	123	74								
58	116	120	69	141	72	110	76								
64	129	116	62	137	75	100	79								
61	135	122	60	136	80	87	80								
72	136	117	65	142	72	83	81								
76	149	114	68	149	82	89	83								
84	144	120	64	157	81	84	89								
85	152	107	57	168	101	77	96								
90	159	102	49	176	112	69	95								
89	158	112	46	179	105	57	94								
93	149	116	50	181	95	46	99								
88	150	114	55	174	89	47	100								
95	149	114	49	170	77	43	97								
96	145	104	44	163	75	44	96								
92	137	108	44	159	57	34	91	79	156	167	27	145	90	58	58
93	140	116	44	151	54	31	91	70	158	162	27	151	86	61	44

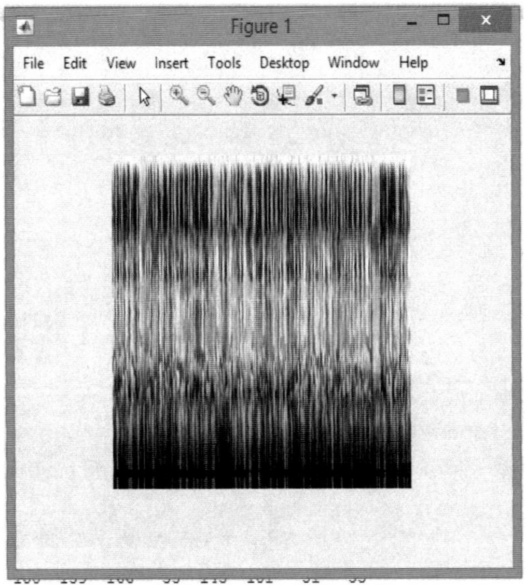

Figure 3. Showing the pixel values and image of the input image after exchanging columns of Figure 2 based on corresponding one to one indexing of sorted random values of x and y from 3D equations .

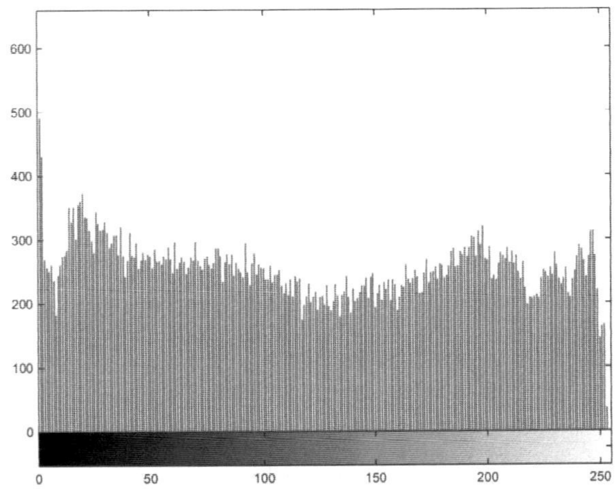

Figure 4. Histogram of the image after exchanging columns.

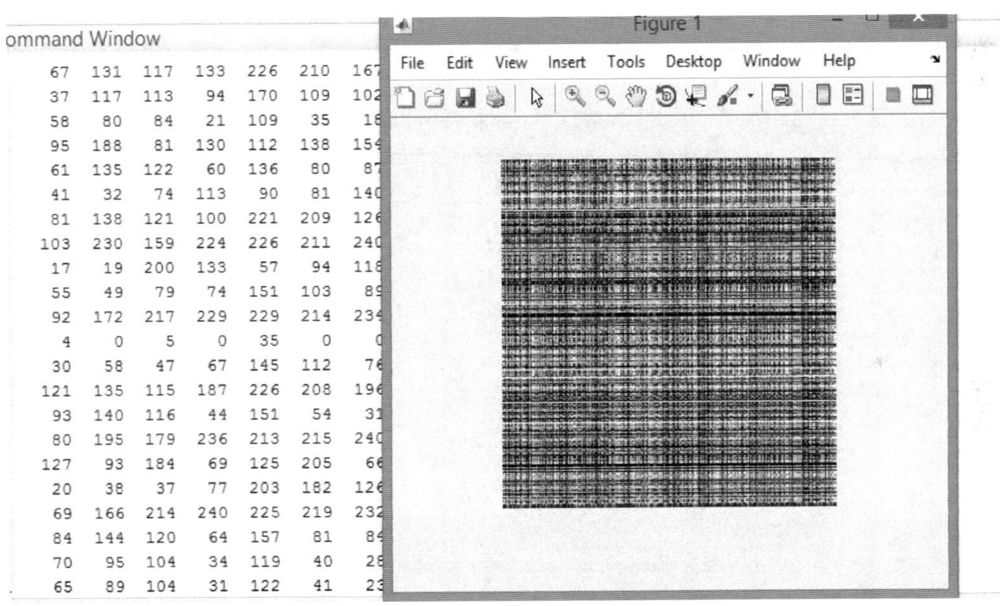

Figure 5. Showing the pixel values and image of the input image after exchanging rows of Figure 3 based on corresponding one to one indexing of sorted random values of x and y from 3D equations.

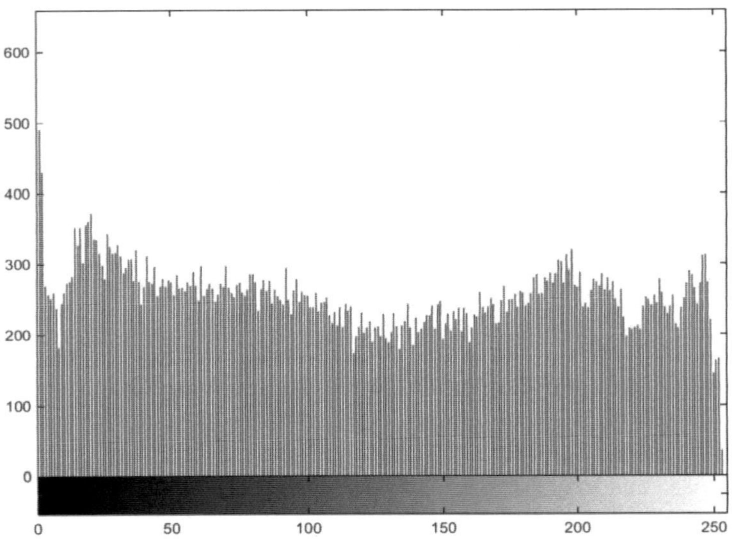

Figure 6. Histogram of the image after exchanging rows.

Command Window

166	48	213	62	182	249	87	102	181	131	89	245	252	152	95	249	174	135	160	148	168	93	249	237
242	120	141	112	183	6	235	159	116	35	214	217	122	29	191	75	169	83	23	225	27	159	253	63
92	155	88	188	11	112	79	121	122	8	90	12	252	229	76	199	91	247	135	84	130	122	255	176
144	102	55	176	8	147	46	77	209	244	58	186	211	179	34	69	250	247	15	188	225	57	239	205
80	107	90	94	154	190	162	230	246	75	195	233	111	229	3	197	138	79	210	248	200	92	171	89
214	41	231	31	141	225	153	247	189	86	191	47	69	11	143	127	164	171	126	219	81	21	206	149
151	8	241	183	219	63	74	172	126	111	1	0	240	176	188	103	75	145	53	50	251	189	150	39
154	133	96	147	170	6	135	41	176	76	133	167	215	174	97	184	215	87	135	8	223	213	255	178
236	217	121	59	63	160	69	225	198	54	121	77	230	72	81	178	155	184	209	216	238	230	83	192
38	34	148	253	84	90	159	197	28	110	136	75	27	190	87	40	94	188	191	153	0	255	37	44
153	239	219	35	88	43	246	7	106	149	98	86	153	50	50	149	224	183	169	190	177	202	11	109
48	138	23	96	73	208	131	223	180	201	192	252	16	160	240	173	201	185	110	110	216	244	0	80
111	126	73	178	85	31	126	238	255	211	77	27	187	26	12	9	64	246	176	2	253	127	232	123
208	126	215	138	212	25	121	127	114	83	200	86	244	8	212	245	7	250	16	12	1	81	147	181
201	221	252	213	35	85	134	172	214	57	191	107	75	140	36	46	228	251	93	250	106	253	151	139
69	3	228	242	49	73	26	5	246	90	56	97	111	141	168	101	161	153	65	152	251	33	126	34
57	149	245	5	12	230	239	78	155	125	218	255	91	238	239	118	173	248	51	48	47	153	177	13
149	164	104	187	48	222	137	26	249	255	215	217	247	203	52	229	218	123	157	146	184	45	18	181
21	106	82	47	0	96	177	254	174	133	142	235	182	253	6	9	204	112	185	58	229	54	175	227
170	149	127	164	112	176	68	68	131	110	107	88	49	251	230	185	162	106	14	235	18	28	134	214
7	252	166	35	212	95	131	143	147	89	3	102	218	119	172	59	4	73	173	114	51	174	238	50
154	156	34	45	157	47	224	92	111	93	102	191	14	220	229	73	60	117	118	22	37	170	56	252
237	226	56	124	236	235	103	106	211	202	47	75	255	72	156	232	223	203	157	55	242	46	254	149

Figure 7. Generate **X** : 256×256 set of randomized values in range of 1 to 4 for DNA operations based on equation 1 of 2D chaotic equation .

Generate **Y**: 256X256 set of randomized values in range of 1 to 8 for Rule Conversions based on equation 2 of 2D chaotic equations. Resultant pixel values after perform DNA operations(ADD, SUB,

XOR and XNOR) based on corresponding value from X and storing the binary value after reversing rule conversion based on Y.

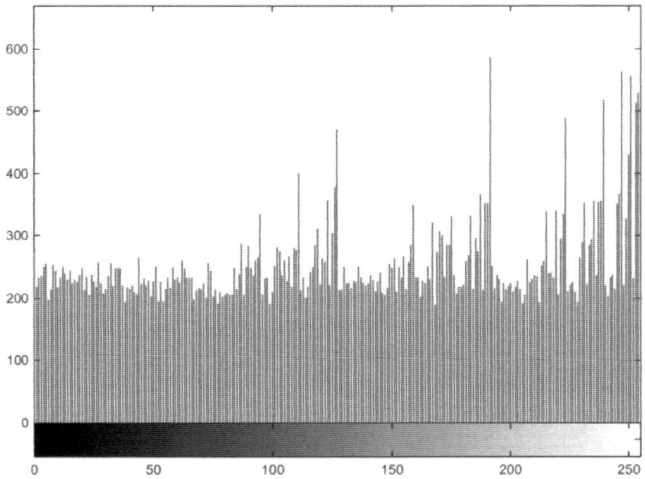

Figure 8. Histogram of resultant image after perform DNA operations(ADD, SUB, XOR and XNOR) based on corresponding value from X and storing the binary value after reversing rule conversion based on Y.

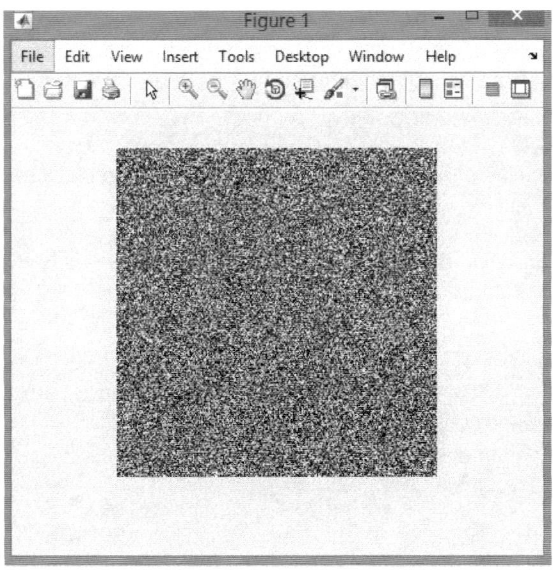

Figure 9. Encrypted image after performing the operations.

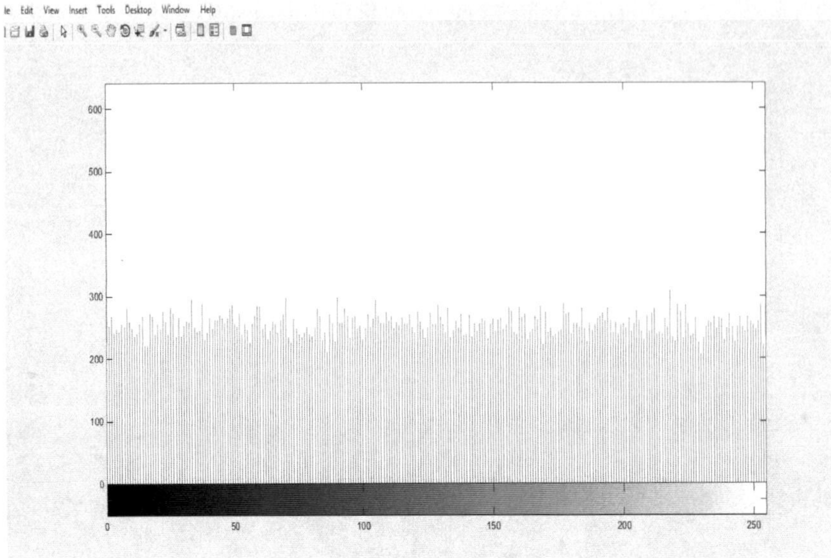

Figure 10. Histogram of encrypted image.

From the above results, clearly we can see that inspite of the original images being very tilted with non uniform histograms, the encrypted image histograms becomes very flat and uniform after encryption. We see a uniformly-distributed histogram for a encrypted image. This can be witnessed from the final encrypted medical images and its histograms.

5. Conclusions

A novel medical image encryption algorithm based on dynamic DNA sequencing and chaotic systems is proposed in this paper. A detailed process of the novel encryption algorithm is given. Based on the experimental results of our algorithm, we conclude that the proposed chaos and dynamic DNA sequencing based image encryption technique is perfectly suitable for the practical image encryption. The pseudo number sequences generated from the two-dimensional and three-dimensional logistic map equations for image encryption are very much random and complicated. The comparison of original medical image with the encrypted image and the histogram comparison between the same clearly reflect the efficiency of the algorithm. All the encrypted medical images have same histogram uniformity which makes it efficient.

References

[1] Nyeem H, Boles W and Boyd C 2013 A Review of Medical Image Watermarking Requirements for Teleradiology *Journal of Digital Imaging* **26** 326–343.

[2] Akkasaligar P T and Biradar S 2016 Secure medical image encryption based on intensity level using Chao's theory and DNA cryptography *IEEE International Conference on Computational Intelligence and Computing Research (ICCIC), Chennai* 1-6.

[3] Belazi A, Talha M, Kharbech S and Xiang W 2019 Novel Medical Image Encryption Scheme Based on Chaos and DNA Encoding *IEEE Access* **7** 36667-36681.

[4] Wang X Y, Zhang Y Q and Zhao Y Y 2015 A novel image encryption scheme based on 2-D logistic map and DNA sequence operations *Nonlinear Dynamics* **82** 1269–1280.

[5] Liu L, Zhang Q and Wei X 2012 A RGB image encryption algorithm based on DNA encoding and chaos map *Computers & Electrical Engineering* **38** 1240–1248.

[6] Liu Y, Tang J and Xie T 2014 Cryptanalyzing a RGB image encryption algorithm based on DNA encoding and chaos map *Optics and Laser Technology* **60** 111–115.

[7] Chai X, Chen Y and Broyde L 2017 A novel chaos-based image encryption algorithm using DNA sequence operations *Optics Lasers in Engineering* **88** 197–213.

[8] Niu Y, Zhang X and Han F 2017 Image encryption algorithm based on hyperchaotic maps and nucleotide sequences database *Computational Intelligence and Neuroscience* **2017** 1– 9.

IOP Conf. Series: Materials Science and Engineering 1080 (2021) 012007 doi:10.1088/1757-899X/1080/1/012007

Comparison of spectral responses of $Cs_2TiI_{6-x}Br_x$ based Perovskite device with CdS and TiO_2 Electron transport layer

Kunal Chakraborty[1] and Samrat Paul[1]

[1]Advanced Materials Research and Energy Applications Laboratory, Energy Engineering Department, North Eastern Hill University Shillong, Meghalaya, India

E-mail: paulsamrat17@gmail.com

Abstract. This article represents the spectral responses of a planar FTO/CdS or $TiO_2/Cs_2TiI_{6-x}Br_x$/CuSCN/Ag based structure where CdS and $TiO_2$is used as Electron transport layer (ETL) and CuSCN as Hole transport layer (HTL) under 300 K temperature condition with the SCAPS-1D simulator. The simulation was performed at 750 nm thickness of $Cs_2TiI_{6-x}Br_x$ absorbing material where x = 0 to 6 to understand the quantum efficiency (QE) using two different ETL (CdS and TiO_2). This article shows that TiO_2as ETL has better spectral responses from 300 nm to 900 nm wavelength compare to CdS.

1. Introduction

Perovskite solar cell (PSC) have undergone through a tremendous development for past few years due to its photovoltaic properties. It has created a huge interest for researchers towards its photovoltaic application. Recent reports suggest a power conversion efficiency above 22%. Moreover, researchers have proposed different Perovskite materials having different cell architecture to study the performance of solar cells. Despite of these progress in new generation solar cells, still stability remains a key issue for the Industrial Application & Commercialization. In early days when thin film based third generation Perovskite solar cell was made it was consists of an organic-inorganic lead compound with ABX_3 chemical formula X is an anion which could be a halide element either Br, I, Cl or F and A and B are cations [1]. If we consider a most common organic-inorganic lead compound methyl ammonium lead Iodide Perovskite which has $(CH_3NH_3)+$ cation and it is enclosed by octahedra of PbI_6 to form the crystal structure [2]. The position of anion is play important role towards formation of energy. Such organic-inorganic halide hybrid PSCs is most suitable to work as photo active layer as it has several advantages like greater light absorptivity, charge carrier mobility, higher exciton diffusion length, lower exciton binding energy and tunable band gap. As a result the power conversion efficiency (PCE, %) is also higher. Such hybrid halide was first introduced by Kojima et al. [1] with $CH_3NH_3PbI_3$ and $CH_3NH_3PbBr_3$ active material for solar cell and PCE was recorded up to 3.8%. Later he further improved the conversion efficiency by using the quantum dots technique with TiO_2 and spin coating as fabrication method recorded up to 6.5% [1, 2, 3]. Then the researcher used organic spiro-MeOTAD as the HTM (Hole transfer material) in the solar cell which has further improved the PCE upto 9.7% [4]. Further extensive research on CH3NH3PbI3 solar cell recorded highest conversion efficiency >20% which is higher than any traditional DSSCs [4, 5]. The present article, an alternative structure of $Cs2TiI_{6-X}Br_X$ based absorbing material (Cs_2TiBr_6, $Cs_2TiI_1Br_5$, $Cs_2TiI_2Br_4$, $Cs_2TiI_3Br_3$, $Cs_2TiI_4Br_2$, $Cs_2TiI_5Br_1$ and Cs_2TiI_6) has been studied [6, 7] with two different ETL (CdS and TiO_2) to find the quantum efficiency (QE, %) of the device. This result will help researcher to find the suitable ETL with this active layer for photovoltaic applications.

2. Simulation Process

The simulation work has been executed under 300 K ambient temperature, 1.5 AM light illumination with 1000 W/m^2 power which can be reset through the SCAPS-1D simulator software screen [4]. Here, Fig. (1) shows the schematic view of SCAPS tool screen-

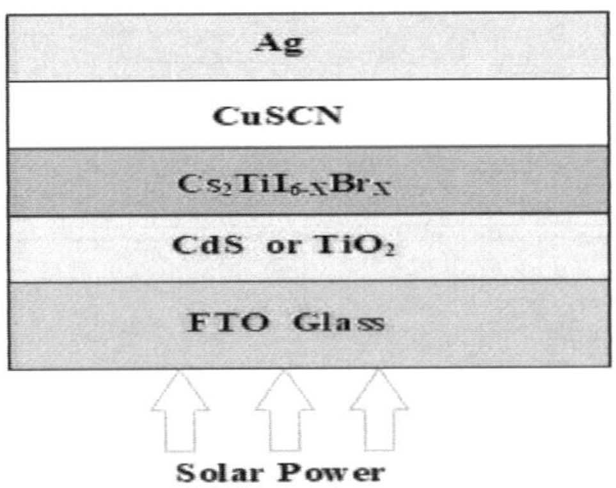

Figure 1. Schematic view of proposed solar cell model

Here, both ETL has higher band gap which is 2.6 eV for CdS and 3.26 eV for TiO_2. The above Fig. (1) shows a 5-layer structure and further it can be extended up to 7 layers as per the device structure. The active layer thickness was taken as 0.75 μm for Cs_2TiBr_6, $Cs_2TiI_1Br_5$, $Cs_2TiI_2Br_4$, $Cs_2TiI_3Br_3$, $Cs_2TiI_4Br_2$, $Cs_2TiI_5Br_1$, Cs_2TiI_6 and 0.14 μm for both CdS and TiO_2 ETL [6, 7, 8]. Here, Table 1 indicates the properties of basic material and Table 2 denoted the properties of absorbing material [9, 10].

TABLE 1. Basic Material Properties

Properties	CuSCN	CdS	TiO_2	FTO
Thickness (μm)	0.35	0.14	0.14	0.1
Band gap (eV)	3.4	2.4	3.26	3.6
Electron affinity (eV)	1.9	4.18	3.70	4.0
Permittivity	9.0	10	55.0	9.0
N_D (1/cm^3)	0	1×10^{15}	4×10^{14}	2.4×10^{18}
N_A (1/cm^3)	$1x10^{18}$	0	0	1×10^{5}
Electron mobility (cm^2/V$_S$)	$2x10^4$	100	100	100
Hole mobility (cm^2/V$_S$)	$1x10^{-2}$	25	25	25

TABLE 2. Absorbing Material Properties

Properties	Cs_2TiBr_6	$Cs_2TiI_1Br_5$	$Cs_2TiI_2Br_4$	$Cs_2TiI_3Br_3$	$Cs_2TiI_4Br_2$	$Cs_2TiI_5Br_1$	Cs_2TiI_6
Thickness (nm)	750	750	750	750	750	750	750
Band gap, E_g (eV)	1.78	1.58	1.38	1.26	1.15	1.07	1.02
Electron affinity, E_a (eV)	3.22	3.42	3.62	3.74	3.85	3.93	3.98
Relative Permittivity, ε_r	3.75-5.06	3.75-5.06	3.75-5.06	3.75-5.06	3.75-5.06	3.75-5.06	3.75-5.06
Donor density, N_D (1/cm^3)	1×10^{19}	1×10^{19}	1×10^{19}	1×10^{19}	1×10^{19}	1×10^{19}	1×10^{19}
Acceptor density, N_A (1/cm^3)	1×10^{19}	1×10^{19}	1×10^{19}	1×10^{19}	1×10^{19}	1×10^{19}	1×10^{19}
Electron mobility, μ_n (cm^2/V$_S$)	4.4	4.4	4.4	4.4	4.4	4.4	4.4
Hole mobility, μ_p (cm^2/V$_S$)	2.5	2.5	2.5	2.5	2.5	2.5	2.5

3. Result and Discussion

In that section, the effect of illumination on Perovskite solar cell device for FTO/CdS or $TiO_2/Cs_2TiI_{6-X}Br_X$/CuSCN/Ag structure has been studied at 300 nm to 900 nm wavelength light sources [11,12]. Quantum efficiency (QE) analysis at optimised thickness for the Perovskite material with TiO_2 and CdS electron transport layer (ETL) will suggests the conversion of part of wavelength of the absorbing material into the electron-hole pair and which will have collected in terms of current. Here, Fig. 2shows the quantum efficiency comparison between two different ETL CdS and TiO_2at the 0.75 μm thickness of Cs_2TiBr_6, $Cs_2TiI_1Br_5$, $Cs_2TiI_2Br_4$, $Cs_2TiI_3Br_3$, $Cs_2TiI_4Br_2$, $Cs_2TiI_5Br_1$, Cs_2TiI_6 absorbing layer. The mathematical equation of the quantum efficiency is defined as [11, 12]-

$$QE\,(\%) = \frac{h\,\gamma\,I_{ph}}{e\,P} \text{ ------------------------- } (1)$$

Where, e is charge of electron (1.6×10^{-19} C), I_{ph} is the electrons (holes) collected per second, h is the Planck's constant (6.62×10^{-34}) and P is the incident optical power (Watt).

Fig. 2(a) shows the QE comparison of Cs_2TiBr_6 absorbing layer with CdS and TiO_2 layer where highest 98 % quantum efficiency is visible at the wavelength of 360-450 nm for CdS layer and 100 % QE is visible at the 360-460 nm wavelength for TiO_2 layer. After that wavelength the QE is started to fall gradually, it means collection of charge carrier is also started to decrease after that wavelength. So, for the Cs_2TiBr_6 absorbing layer the collection of electron hole pair is higher at the 360-460 nm wavelength.

Similarly, Fig. 2(b) indicates up to 98% QE possible for $Cs_2TiI_1Br_5$ absorbing layer between 350-650 nm wavelength with CdS layer and 100 % QE is achievable at the 360-650 nm wavelength for TiO_2 layer. After that wavelength the QE is started to fall. On the other hand, Fig. 2(c) shows $Cs_2TiI_2Br_4$ absorbing layer has 100% of QE at the wave length of 330-750 nm with TiO_2 and 97% QE achievable between the 320-740 nm with CdS layer. Fig. 2(d) shows $Cs_2TiI_3Br_3$ absorbing layer with 0.75 μm thickness has 98% of QE with CdS ETL and 100% QE with TiO_2 ETL at the wave length of 350-760 nm. Similarly, the Fig. 2(e) shows 100% QE for $Cs_2TiI_4Br_2$ absorbing layer with TiO_2 ETL and 96% QE with CdS ETL at the 350-800 nm. Other two materials, $Cs_2TiI_5Br_1$, Cs_2TiI_6 has witnessed highest 100% QE with TiO_2 ETL and 95% QE with CdS ETL at the 350-830 nm wavelength which can be visible in the Fig. 2(f) and Fig. 2(g) respectively.

IOP Conf. Series: Materials Science and Engineering 1080 (2021) 012007 doi:10.1088/1757-899X/1080/1/012007

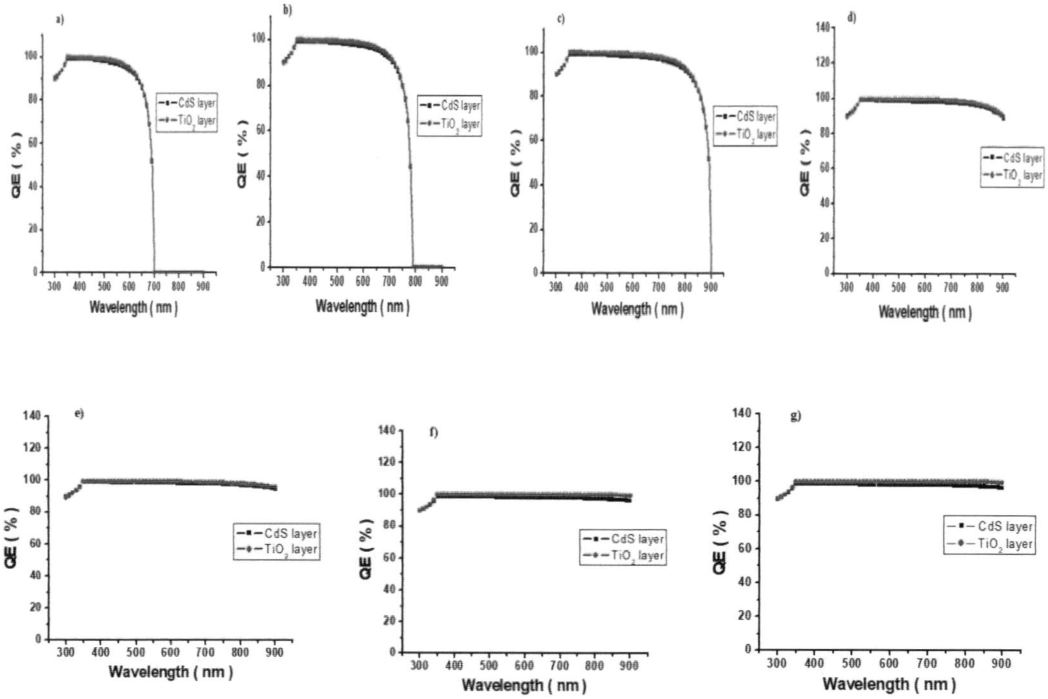

Figure 2. Spectral responses with CdS and TiO_2 ETL for different absorbing layer a) Cs_2TiBr_6, b) $Cs_2TiI_1Br_5$, c) $Cs_2TiI_2Br_4$, d) $Cs_2TiI_3Br_3$, e) $Cs_2TiI_4Br_2$, f) $Cs_2TiI_5Br_1$ and g) Cs_2TiI_6.

The above analysis shows that after 330 nm wavelength the QE started to increase, it means higher amount of solar irradiance is converted into the charge carrier is possible at the 330 nm wavelength and which will remain constant up to 650 nm wavelength for $Cs_2TiI_1Br_5$, 750 nm wavelength for $Cs_2TiI_2Br_4$, 760 nm wavelength for $Cs_2TiI_3Br_3$, 800 nm wavelength for $Cs_2TiI_4Br_2$, 830 nm wavelength for $Cs_2TiI_5Br_1$ and Cs_2TiI_6 absorbing layer. Due to the higher collection of electron- hole pair between that wavelength, the power conversion efficiency and maximum power generation is also enhanced. After that wavelength the conversion efficiency falls due to the less collection of charge carrier. The above comparison shows that TiO_2 will be most suitable as ETL compared to CdS for $Cs_2TiI_{6\text{-}X}Br_X$ Perovskite absorbing layer as at the same wavelength interval the $TiO_2/Cs_2TiI_{6\text{-}X}Br_X$ has 100 % QE.

4. Conclusion

This research article investigated the solar radiation intensity and the power conversion capability of the device with two different ETL for a $Cs_2TiI_{6\text{-}X}Br_X$ based Perovskite active layer. The spectral analysis shows that TiO_2 ETL has better conversion capability compare to CdS ETL as the opto-electronics properties of TiO_2 is superior compare to CdS ETL.

Acknowledgment

The authors are thankful to Prof. Burgelman, University of Gent, Department of Electronics and Information systems, Belgium and Department of Science and Technology (DST), Government of India, for their support (EMR/2016/002430) to complete this research work.

5. References

[1] Kojima A., Teshima K., Shirai Y., Miyasaka T. 2009 *J. Am. Chem. Soc.* **131** 6050-6051.

[2] Giustino F., Snaith H. J. 2016 *ACS Energy Lett.* **1** 1233–1240.

[3] Stranks S. D., Eperon G. E., Grancini G., Menelaou C., Alcocer M. J., Leijtens T., Herz L. M., Petrozza A., Snaith H. J. 2013 *Sci.* **342** 341-344.

[4] Yang W. S., Noh J. H., Jeon N. J., Kim Y. C., Ryu S., Seo J. 2015 *Sci.* **348** 1234-1237.

[5] Yin W. J., Yang J. H., Kang J., Yan Y., Wei S. H. 2015 *J. Mat. Chem. A* **3** 8926-8942.

[6] Ke W., Kanatzidis G. M. 2019 *Nat. Comm.* **10** 1-4.

[7] Chaudhary N., Chaudhary R., Kesari J. P., Patra A., Chand S. 2015 *J. of Mat. Chem. C* **3** 1-19.

[8] Yang Y., You J. 2017 *Nature* **544** 155–156.

[9] Ju M. G., Chen M., Zhou Y., Garces H. F., Dai J., Ma L., Padture N. P., Zeng X. C. 2018 for photovoltaic *ACS Energy letters* **3** 297-304.

[10] Chen M., Ju M. G., Carl A.D., Zong Y., Grimm R. L., Gu J., Zeng X. C., Zhou Y., Padture N. P. 2018 *Joule* **23** 558-570.

[11] Bishnoi S., Pandey S. K. 2018 *IET Optoelectronics* **12** 185-190.

[12] Heriche H, Rouabah Z, Bouarissa N 2017 *Int. J. of Hydrogen Energy* **42** 9524-9532.

Feeble ferromagnetism and double relaxation mechanism in $Bi_{0.42}Ho_{0.28}Ba_{0.24}Ca_{0.06}Fe_{0.7}Ti_{0.3}O_3$ ceramics

Chiranjib Chakrabarti[a‡], Songliu Yuan[a‡], Qingshan Fu[a], and Anirban Das[b‡]

[a] School of Physics, Huazhong University of Science and Technology, Wuhan 430074, People's Republic of China
[b] Department of physical and material Sciences, College of engineering and management, Kolaghat, Purba Medinipur, West Bengal 721171, India
e-mail: yuansl@hust.edu.cn (Songliu Yuan)
anirban_das@cemk.ac.in (Anirban Das)
chiranjib1987@yahoo.com (Chiranjib Chakrabarti)

Abstract. The structural, magnetic, and dielectric studies on $Bi_{0.42}Ho_{0.28}Ba_{0.24}Ca_{0.06}Fe_{0.7}Ti_{0.3}O_3$ ceramics present some exciting outcomes. An XRD based structural study shows that the ceramics hold an ABO3-type perovskite network. Room-temperature MH-hysteresis study detected the trace of ferromagnetism in the ceramics. The temperature-dependent dielectric study exhibits that a double relaxation mechanism is engaged for a wide range of temperature and frequency.

1. Introduction

Recently, $BiFeO_3$ has emerged as a novel material to the researchers because of its excellent multiferroic properties at room temperature that exhibits numerous commercial application possibilities in spintronics, memory elements, sensors, actuators, etc.[9] Unfortunately, some stumbling blocks, such as poor magnetic properties and defect-related conductions, creates a severe challenge for the application possibilities of this class of material.[5, 7] A substantial amount of effort was made to overcome this problem, such as the substitution of A- with Rare earth or B-site sites with a transition metal or creating binary or trinary networks with other suitable perovskite materials.[1, 7]. In our previous work,[5] the presence of Ho-holding a perovskite-ferrite form is demonstrated in a series of BFO based solid-solutions. Here, the phase-structural, magnetic, and dielectric studies on $Bi_{0.42}Ho_{0.28}Ba_{0.24}Ca_{0.06}Fe_{0.7}Ti_{0.3}O_3$ ceramics are discussed, and possible relaxation mechanism in this material is portrayed.

2. Synthesis and characterizations

Polycrystalline samples of $Bi_{0.42}Ho_{0.28}Ba_{0.24}Ca_{0.06}Fe_{0.7}Ti_{0.3}O_3$ were prepared by a solid-state reaction process. The details of the preparation procedure are elaborated elsewhere. [5] The phase structural analysis of the sample was performed by an X-ray powder diffractometer (Model no: **MXP18AHF, MARK**, Japan). The temperature dielectric properties at different frequencies were investigated using a precision impedance analyzer (Model no: **Wayne Kerr 6500B**).

3. Results and discussions

Fig. 1(a) shows the refined X-ray diffraction pattern (XRD) of $Bi_{0.42}Ho_{0.28}Ba_{0.24}Ca_{0.06}Fe_{0.7}Ti_{0.3}O_3$ ceramics at room temperature. The crystalline phases and structural parameters of $Bi_{0.42}Ho_{0.28}Ba_{0.24}Ca_{0.06}Fe_{0.7}Ti_{0.3}O_3$ were obtained from XRD pattern analysis followed by a Reitveld refinement using X'pert pro and Full prof software, respectively. The phase analysis indicated the existence of *R3c* (~23.2%), *P4mm* (~60.6%), and *Pbnm* (~16.2%) phase in the ceramics network, which identifies the presence of $BiFeO_3$, $BaTiO_3$, and $HoFeO_3$, respectively. Therefore the ceramic network contains a complete perovskite structure. Moreover, the Rietveld refinement result exhibits an excellent χ^2 value (~1.8) with reliability factors: R_p=48.5 and R_{wp} = 31.3.

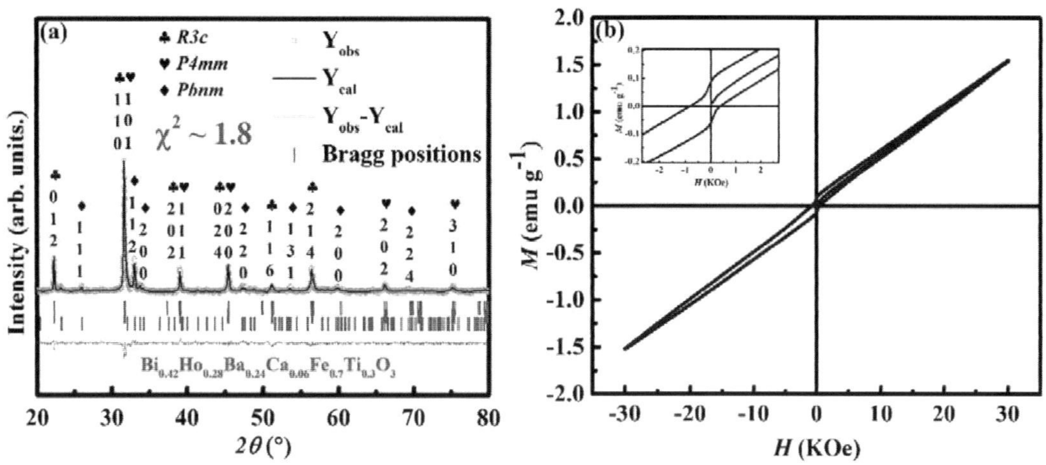

Fig. 1 Refined X-Ray Diffraction patterns measured at room temperature (a) and room temperature MH-hysteresis curve (b) with magnetified part of the loop (inset) for $Bi_{0.42}Ho_{0.28}Ba_{0.24}Ca_{0.06}Fe_{0.7}Ti_{0.3}O_3$ ceramics.

As shown in Fig. 1(b), the room temperature magnetization-magnetic field hysteresis (M-*H*) in the $Bi_{0.42}Ho_{0.28}Ba_{0.24}Ca_{0.06}Fe_{0.7}Ti_{0.3}O_3$ sample indicating a ferromagnetic behavior. The enlarged part of the hysteresis curve estimates the M_r (0.075 emu g^{-1}) and H_c (~0.0601 kOe) value of the sample. The obtained M_r value is comparable to the previously reported $0.7BiFeO_3$-$0.3Ba_{0.75}$ $Ca_{0.25}TiO_3$ (~0.07 emu g^{-1}) ceramics[3] and considerably improved than that in the BSFO (~0.0017 emu g^{-1}) ceramics.[2] Such origin of the weak ferromagnetism in the $Bi_{0.42}Ho_{0.28}Ba_{0.24}Ca_{0.06}Fe_{0.7}Ti_{0.3}O_3$ sample can be attributed to the combination of two different mechanisms (1) ferromagnetic exchange interaction between Fe and Ho ions, and (2) canting mechanism in Fe^{+3} spins due to the substitution. [5, 6]

The temperature-dependent relative permittivity in $Bi_{0.42}Ho_{0.28}Ba_{0.24}Ca_{0.06}Fe_{0.7}Ti_{0.3}O_3$ ceramics under a wide range of frequencies (10kHz-1MHz) are shown in Fig. 2(a). Careful observation reveals that two different anomalies are present; one at a lower temperature region (~407 K) and the other one at a relatively higher temperature region (~ 600 K), namely A_1 and A_2 regions, respectively. Such dual anomalous behavior also can be observed in the temperature variation of the *tan* δ curve (inset of Fig. 2(a)). A similar anomalous response was previously observed in (1-*x*) $BiFeO_3$-x($0.5CaTiO_3$-$0.5SmFeO_3$) ceramics, which was attributed to the different relaxation processes in the ceramics.[8] It was demonstrated that A_2 is probably due to the grain boundary effect, and A_1 is related to the charge ordering in mechanism between the Fe ions.

Fig. 2(b) presents the temperature-dependent AC conductivity (σ_{AC}) measurement in $Bi_{0.42}Ho_{0.28}Ba_{0.24}Ca_{0.06}Fe_{0.7}Ti_{0.3}O_3$ sample, which is obtained from the following relationship:[4]

$$\sigma_{AC}=2\pi f\,\varepsilon_0\varepsilon_r\,tan\,\delta \qquad (1)$$

Besides, the change of slopes, as indicated by the arrows, indicates different relaxation mechanism is active inside the material. Therefore, the entire temperature range is divided into three regions: (i) The frequency-independent high-temperature region I, where σ_{AC} increases sharply, partially frequency-dependent region II, where the defect conductivity is a crucial factor for conduction process, and (iii) a frequency-dependent low-temperature region III, σ_{AC} is nearly constant over the frequency range. In this region, the relaxation mechanism is attributed to the carrier hopping.[4] Interesting that, the slope of the curve primarily changes at two junction points, which can be recognized as A_2 and A_1 at lower and higher temperature region, respectively. Hence, the anomaly at A_1 can be identified as the charge ordering due to the carrier hopping between Fe^{2+}-Fe^{3+} in the perovskite network, while the anomaly at A_2 is probably due to the grain boundary defect based conductivity, such as Bismuth or oxygen vacancy triggered conduction mechanism.

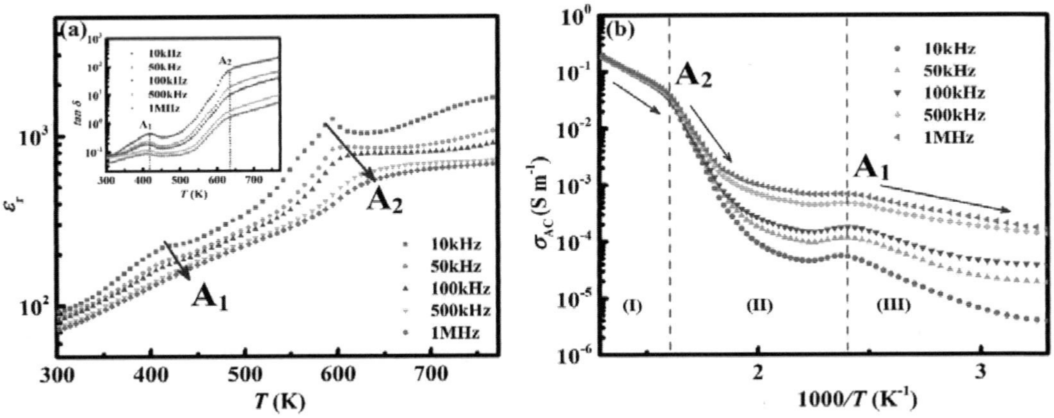

Fig. 2 Temperature-dependent relative permittivity (a) and tangent loss (inset) under different frequencies in $Bi_{0.42}Ho_{0.28}Ba_{0.24}Ca_{0.06}Fe_{0.7}Ti_{0.3}O_3$ ceramics. Variation of A.C conductivity with 1000/T (b) in the same ceramics.

4. Conclusion

The phase-structural, room-temperature magnetic, and temperature-dependent dielectric studies are performed for the $Bi_{0.42}Ho_{0.28}Ba_{0.24}Ca_{0.06}Fe_{0.7}Ti_{0.3}O_3$ ceramics. XRD analysis, along with Rietveld refinement, indicated that the ceramic holds a perovskite network containing three different (ABO_3) perovskites. Room-temperature MH-hysteresis study indicated weak ferromagnetism in the sample. Besides, a detailed high-temperature dielectric analysis revealed that two different relaxation mechanisms exist in the sample, the one at a relatively lower temperature occurs from the carrier hopping and the other one, at a relatively higher temperature, emerges from the Bismuth or oxygen vacancy related defects inside the crystal.

Acknowledgment

This work was aided by the National Natural Science Foundation of China (Grant No. 11474111). We are thankful to the staff of the Analysis Center of Huazhong University of Science and Technology (HUST) for their support in various measurements.

References

[1] Sugato Hajra, Manisha Sahu, Varsa Purohit, R. N. P. Choudhary, 2019 Dielectric, conductivity and ferroelectric properties of lead-free electronic ceramic: $0.6Bi(Fe_{0.98}Ga_{0.02})O_3$-$0.4BaTiO_3$, Heliyon **5** e01654.

[2] T. Durga Rao, Asmitha Kumari, Manish K. Niranjan, Saket Asthana, 2014 Enhancement of magnetic and electrical properties in Sc substituted $BiFeO_3$ multiferroic, Physica B Condens. Matter. **448** 267-272.

[3] X. N. Zhu, M. S. Rahman, Y. J. Wu, X. Q. Liu, Y. H. Huang, G. Liu, R. Guo, X. M. Chen, A. S. Bhalla, 2016 Enhanced ferroelectricity, piezoelectricity, and ferromagnetism in $(Ba_{0.75}Ca_{0.25})TiO_3$ modified $BiFeO_3$ multiferroic ceramics, J. Alloys Compd. **658** 973-980.

[4] Nitin Kumar, Alok Shukla, R. N. P. Choudhary, Structural,2017, Electrical and Magnetic Properties of (Cd, Ti) Modified $BiFeO_3$, Phys. Lett. A **381** 2721-2730.

[5] Chiranjib Chakrabarti, Qingshan Fu, Xinghan Chen, Yang Qiu, Songliu Yuan Canglong Li, 2020,Modulation of magnetic, ferroelectric and leakage properties by $HoFeO_3$ substitution in multiferroic $0.7BiFeO_3$-$0.3Ba_{0.8}Ca_{0.2}TiO_3$ solid-solutions, Ceram. Int. **46** 212-217.

[6] N. Wang, A. Jain, Y. Li, F. L. Wang, Y. L. Lu, H. Zhen, Y.G. Wang, F. G. Chen, 2020,Investigation of structural, ferroelectric, and magnetic properties of Ca modified $BiFeO_3$-$BaTiO_3$ ceramics, Ceram. Int. **46** 3855-3860.

[7] Chiranjib Chakrabarti, Qingshan Fu, Xinghan Chen, Canglong Li, Biao Meng, Yang Qiu, Songliu Yuan,2020, Substitution driven enhancement of ferromagnetic, ferroelectric and leakage properties in multiferroic $0.7Bi_{1-x}Er_xFeO_3$-$0.3Bi_{0.5}Na_{0.5}TiO_3$ solid-solutions, J. Sol-Gel Sci. Technol. **93** 587-95.

[8] Juan Liu, Xiao Qiang Liu, Xiang Ming Chen,2017, Ferroelectric and magnetic properties in (1-x) $BiFeO_{3\text{-}x}(0.5CaTiO_{3\text{-}0.5}SmFeO_3)$ ceramics, J. Am. Ceram. Soc. **100** 4045-4057.

[9] N. A. Spaldin, R. Ramesh, 2019,Advances in magnetoelectric multiferroics, Nature. Mater. **18** 203-212.

Nomenclature

σ_{AC}	AC conductivity ($S\ m^{-1}$)	ε_0	absolute permittivity
f	frequency of ac current (Hz)	ε_r	relative permittivity
tan δ	tangent loss		

A review of NiO based composite electrode materials for electrochemical energy storage

Sujit Chatterjee and Swapan Kumar Bhattacharya
Department of Chemistry, Jadavpur University, Kolkata 700032, India
e-mail: *skbhatt7@yahoo.co.in*

Abstract. At the present era supercapacitor is widely recognized as the most promising energy-storage device to meet the challenges of high power-supply, long cycle life and reliability. From literature it has well established that the electrode material is one of the significant components to determine their performance. In this regard, the three–dimensional hierarchical mesoporous unique NiO nanocomposites with large surface area, high electrical conductivity, low equivalent series resistance and possibility of advantageous proportionate mixing of suitable components, demonstrate substantial promise for high performance supercapacitor applications, because of synergistic effect. This review summarizes the NiO based nanocomposites for supercapacitor electrode materials as well as a brief overview regarding supercapacitors active electrode materials.

Keywords: Nanocomposite, Pseudocapacitor, NiO, BET surface area, Porous network

1. Introduction

Supercapacitor, also known as electrochemical capacitor or ultra-capacitor can store the energy at the electrode electrolyte interface. Electrchemical capacitor (EC) was introduced at first by Becker as patent on 1957, by using carbon as electrode material [1]. Later, during 1990–2000 it became highly interesting due to its potential application in the field of electric vehicles and hybrid electric vehicles to provide the necessary power. It has now become much important in the context of decreasing the availability of fossil fuels since it can store intermittent renewable energy for sustainable development of human society [2]. Recently, several companies like Panasonic, Maxwell Technologies invest for development of high performance supercapacitors. It is to be noted that EC fills in the gap between batteries and conventional capacitors i.e., it has greater energy density than conventional capacitors with improved power density and much increased cycle life than batteries. Figure 1 depicts the Ragone profile for a few valuable useful energy devices. One of the most significant components for performance of supercapacitors is active electrode materials.

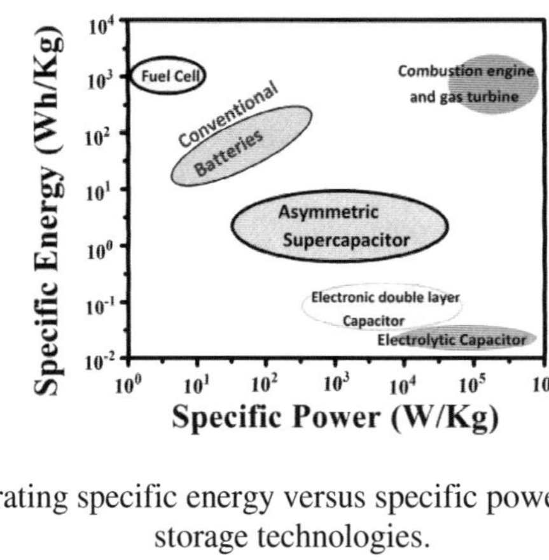

Figure 1. Ragone plot illustrating specific energy versus specific power of different electrical energy-storage technologies.

Based on energy storage mechanism EC may be distinguished as electric-double layer capacitors (EDLC), pseudocapacitors and hybrid supercapacitors (such as asymmetric, composite, and battery-type). In EDLC the charge storage is achieved by electrochemical reaction-less non–faradic process i.e., physical accumulation and depletion of ions with generation of basic stern double layer at the electrode–solution junction. In case of pseudocapacitors, the electroactive reagent undergoes faradic redox reaction i.e., the charge storage is achieved by transfer of electron from electrolyte to the electrode material via change in oxidation state of active electrode materials according to the equation 1 [3]:

$$O_{ad} + ne \leftrightarrow R_{ad} \qquad (1),$$

where, 'O_{ad}' and 'R_{ad}' are the adsorbed oxidants and reductants respectively and 'n' indicates the number of electrons. On other hand, hybrid supercapacitors use both faradic and non-faradic processes i.e., charge storage is achieved by both electrostatically and electrochemically. Generally, in asymmetric hybrid supercapacitors carbon rod serves as the negative electrode while a metal or metal oxide (or composite of metal oxide) as positive electrode. Asymmetric hybrid supercapacitors exhibit higher energy density and comparable cycling stability to supercapacitors. The active electrode materials of EDLC are activated carbon, porous carbon, carbon nanotubes, carbon fiber etc., whereas for pseudocapacitors are conducting polymers (polythiophine, polypyrrole, polyaniline and their composites), metal oxides (RuO_2, MnO_2, Mn_2O_3, Co_3O_4, SnO_2, Fe_2O_3, V_2O_5, V_2O_3, VO_2, WO_3, MoO_3, CuO, Ni_2O_3 and NiO etc.) and transition metal nitrides such as VN, TiN, WN and MoN etc. [4, 5]. In 25 years ago, it was found that RuO_2 exhibited very high specific capacitance of up to 750 F g^{-1} even at very low temperature [5]. RuO_2 is highly conductive and it has three distinct oxidation states. Moreover, it is accessible within large potential window (1.2 V). During fast reversible redox reaction the oxidation state of Ru had changed from (II) to (IV) as in equation 2 [6]:

$$RuO_2 + xH^+ + xe^- \leftrightarrow RuO_{2-x}(OH)_x \qquad (2),$$

where, $0 \leq x \leq 2$. Since, it is highly toxic and expensive so fabrication of low cost electrode materials with high specific capacitance became naturally highly interesting and desirable. Thus several attempts were undertaken since last two decades to manufacture suitable high performance of electrode materials based on nanomaterials for supercapacitors application [7, 8]. Other less expensive transition metal oxides, such as iron oxide, vanadium oxide, nickel oxide, cobalt oxide, molybdenum oxide, manganese oxide were extensively used as active materials for pseudocapacitors. In case of MnO_2 the charge storage mechanism occurs by adsorption of cations (e.g., Na^+ from aqueous Na_2SO_4) and proton (H^+) incorporation according to the following reaction (Eq 3):

$$MnO_2 + xM^+ + yH^+ + (x+y)e^- \leftrightarrow MnOOM_xH_y \qquad (3),$$

where, x+y =1 in terms of mole and M^+ may be Li^+ or Na^+, or K^+ depending upon the system. Although MnO_2 is cheap, it possesses large natural abundant but due to poor electronic and ionic conductivity result in poor materials utilization. Generally three types of electrolyte are used in EC are aqueous electrolyte (aqueous solution of KOH, Na_2SO_4 etc.), organic electrolyte (such as acetonitrile, propylene carbonate etc.), and ionic liquids. The most important factor behind the high capacitance active electrode material is that it has high specific surface area and highly electronically conducting substance [8]. Fabrication of composites is one of the significant approach to synthesize high performance electrode materials due to the fact that it exhibits synergistic effect through inducing porous networks, enhancing specific surface area (SSA), reduced particle size, preventing agglomeration, facilitating electrical conductivity, extending the working voltage, improving life span and extra pseudocapacitance [9]. Over the last two decades, there is continuation of demand to invent better and suitable composite electrode materials of NiO for supercpactors application due to its high theoretical capacitance value (2573 F g^{-1} at a potential window of 0.5 V), easily synthesized, cost effectiveness, pseudocapacitive behavior and good thermal and chemical stability [10]. The surface faradic reaction of NiO in alkaline electrolytes (e.g. KOH) can be expressed as Eq (4):

$$NiO + OH^- \leftrightarrow NiOOH + e^- \qquad (4)$$

2. Application of NiO and NiO based composites as an electrode material

X. Zhang et al. synthesized porous NiO nanostructures with different morphologies (nanoslices, nanoplates, and nanocolumns) with control SSA by simple hydrothermal methods followed by annealing at 400 °C for 2 h [11]. In addition X. Zhang et al. reported that porous nanoparticle of NiO with column morpholology exhibited higher specific capacitance (390 F g^{-1}) than that of slices (176 F g^{-1}) and plates (285 F g^{-1}) when studied at a current density of 5 A g^{-1} using 1 M aqueous KOH solution as electrolyte. This group initially synthesized β-$Ni(OH)_2$ of different morphologies by hydrothermal method using alkaline solution of $NiCl_2{\cdot}6H_2O$. The morphology of $Ni(OH)_2$ may be cotrolled by altering the pH of the medium. The nanocolumn morphology of stacked $Ni(OH)_2$ nanoplates were obtained by maintaining pH 13 of precursor solution used in the hydrothermal process. The large specific capacitance of NiO with mesoporous nanocolumn morphology can be attributed by their high SSA (102.4 m^2 g^{-1}) and smaller pore dimension (pore size 2 nm) which facilitates the ion diffusion and provides large number of electrochemically active sites than others. S.I. Kim et al. prepared porous NiO nanostructures with shapes of flower, slice, and particle by hydrothermal decomposition of $Ni(OH)_2$ ancestor [12]. The specific capacitance of nanoflower, nanoslice, and nanoparticles electrode materials was found to be 480, 116, and 260 F g^{-1} respectively at mass normalized current of 0.5 A g^{-1}, although the measured SSA of these active electrode materials are 159, 193, and 233 m^2 g^{-1} respectively. Aqueous KOH solution (2 M) was used as an electrolyte in this study. The authors suggested that unexpectedly high specific capacitance of porous nanoflower NiO with largest pore size (16.7 nm) facilitated maximized contact and fast diffusion of OH^- ions through the three-dimensional nanochannels in NiO structures and thus provided large electrochemical active sites. Through this example we can know that the behavior of supercapacitors is not only related to the SSA of active electrode materials but also to the morphology and the pore sizes of the active electrode materials. B. Zhao et al. prepared mono layer graphene/NiO composite with nanosheet morphology as electrodes for supercapacitor application by precipitation method followed by pyrolysis of the precursor in N_2 atmosphere at 400 °C. The electrode exhibited a specific capacitance of 525 F g^{-1} at 200 mA g^{-1} current density, which is ample greater than that of pure NiO (307 F g^{-1}). This is however illustrated by the large SSA of the graphene/NiO composite (134.5 m^2 g^{-1}) than that of pure NiO (42 m^2 g^{-1}) [13]. According to B. Zhao et al, the observed capacity of this nanocomposite is mainly due to Faradic reaction (Eq 4) which is conformed from two redox peaks observed in CV plots registered for 6 M KOH aqueous solution as the electrolyte using potential speed of 5 mV s^{-1} in the potential domain between –0.1 and 0.35 V. Moreover, this composite electrode exhibited 95.4% capacity retention after 1000 cycles, revealing that the composite materials are promising for supercapacitor performance. M. Huang et al. investigated a supercapacitor using a hierarchical NiO/CuO nanocomposite electrode [14]. The as-synthesized hybrid nanocomposites showed a high specific capacitance of 280 F g^{-1} at the mass normalized current of 1 A g^{-1}, and retained about 0.914 fraction of its first capacitance afterward 3000 cycles. X. Zheng and Co-workers prepared ZnO–Au–NiO hybrid composite [15]. The composite exhibited maximum specific areal capacitance of 4.1 F/cm^2 by application of a current density of 5 mA/cm^2 and high cycle life with 80.3% retention after 4000 cycles. D. Yao et al. reported that NiO@$NiCo_2O_4$ core–shell nanosheet prepared by solvothermal method followed by a postannealing process exhibited excellent specific capacitance of 1623.6 F g^{-1} at a mass normalized current of 2 A g^{-1}, and the capacitance retaining of 90% after 10000 cycles at a current/mass of 20 A g^{-1} [16]. A. Ray et al. prepared various spherical mesoporous Ni-Mn-Oxide nanocomposites with Ni:Mn molar ratio as 1:1, 1:2, 1:3, and 1:4 by precipitation method at high temperature followed by calcinations. Among the different as synthesized products, the Ni-Mn-Oxide (Ni:Mn = 1:3) with large Brunauer–Emmett–Teller (BET) surface area 127.09 m^2 g^{-1}, mean diameter of 16 nm and average pore dimension 8.2 nm exhibits excellent specific capacitance of 1215.5 F g^{-1} at a potential speed of 2 mV s^{-1}, and the capacity retained is 87.2% after 5000 cycles at a current density of 10 A g^{-1} [17]. Furthermore, A. Roy et al. prepared porous spinel $NiMn_2O_4$ nanocomposite with average diameter of 8 nm by simple sol–gel method which had a specific capacitance of 875 F g^{-1} at a scan rate of 2 mV s^{-1}. 1 M aqueous Na_2SO_4 solution was used as electrolytic solution, and demonstrated columbic efficiency retention of 97.6% after 2000 cycles at current/mass of 2.5 A g^{-1}

[18]. X. Zhao et al. prepared mesoporous MnO_2@NiO nanocomposite with nanosheets@nanowires hierarchical morphology and BET surface area of 47.8 m^2 g^{-1} by hydrothermal techniques. The composite can exhibit specific capacitance of 374.6 F g^{-1} at a current/mass of 0.25 A g^{-1} as well as the capacitance retaining of 92.7% after 5000 cycles [19]. Recently, J. Zhang et al. reported that $NiCo_2O_4/MnO_2$ core-shell nanosheet electrodes were prepared for asymmetric supercapacitor by two-step cathodic electrodeposition method. The best one exhibited specific capacitance of 3.81 F cm^{-2} at current density of 2 mA cm^{-2} and demonstrated the capacitance retaining of 86.1% after long 10000 rounds at 1 mA cm^{-2} [20]. To provide some insights for future research on NiO and NiO composites based supercapacitors we summarize the specific capacitance and cycling stability of some efficient supercapacitor electrodes in Table 1. It reveals that more work has been performed with NiO based electrode material in alkali rather than acid plausibly because of corrosion and much decreased capacity retention with increased cyclic operation. The Table 1 will benefit researchers to fabricate proper active electrode ingredients for supercapacitor application.

Table 1. Comparison of the supercapacitor performance of NiO based nanocomposite electrode materials reported in the literature.

Nanocomposites	Morphology	Potential Window (V)	Electrolyte	Maximum Specific capacitance	Capacitive retention	Ref.
NiO/C	Nanosheets	– 0.1 to 0.65	1 M NaOH	1012 F g^{-1} @ 5 mV s^{-1}	102% after 1000 cycles	[21]
NiO/Co_3O_4	Core/shell	– 0.1 to 0.5	6 M KOH	510 F g^{-1} @ 5 mA cm^{-2}	92% after 1000 cycles	[22]
NiO@$NiMoO_4$@PPy	Porous spherical	0 to 0.6	2 M KOH	941.6 F g^{-1} @ 20 A g^{-1}	77.1% after 30000 cycles	[23]
$NiO/Ni(OH)_2$/PEDOT	3D Porous Nanoflowers	0 to 1.45	6 M KOH	404.1 mF cm^{-2} @ 4 mA cm^{-2}	82.2% after 1000 cycles	[24]
$ZnCo_2O_4$/RGO/NiO (ZCGNO)	Nanowires	– 0.2 to 0.4	6 M KOH	1256 F g^{-1} @ 3 A g^{-1}	80% after 3000 cycles	[25]
Polyaniline–nickel oxide (PANI-NiO-3)	Nanosheets on nanofibre of PANI	– 0.2 to 0.8	1 M H_2SO_4	514 F g^{-1} @ 1 mV s^{-1}	64% after 1000 cycles	[26]
$NiO/Ni_3V_2O_8$ (PVK1)	nanorods	0 to 0.6	1 M NaOH	653 F g^{-1} @ 1 A g^{-1}	96% after 1000 cycles	[27]
CoS/NiO	Nanoflakes/ nanosheets	0 to 1.6	1 M KOH	1620 F g^{-1} @ 1.0 A g^{-1}	93.5% after 3000 cycles	[28]
$NiO-Co_3O_4$	Nanosheet	– 0.5 to 0.5	3 M KOH	324 F g^{-1} @ 10 mV s^{-1}	99.15% after 5000 cycles	[29]
NiO/CNT (1:1)	Nanoflakes	0 to 0.5	2 M KOH	258 F g^{-1} @ 1 A g^{-1}	86% after 2500 cycles	[30]

3. Outlook

Now the major challenges for NiO based nanocomposites for supercapacitor applications include enhancing of energy density, power density, energy efficiency, self-life, and shelf cyclic stability. Specific capacitance larger than the theoretical capacitance of NiO has also been reported. The possibility of such improvement should be critically reviewed. The improvement is possible in fabrication of suitable NiO based nanocomposites with large electrochemical surface area, suitable morphology, optimum proportion of components, high electrical conductivity, possibility of surface functionality, porous network, suitable porosity (shape, size, pore dimension distribution), squat internal impedance, low maintenance requirement and better electrochemical and mechanical stability [31, 32]. For the design of high-performance supercapacitors, electrolyte ions along with active electrode-materials play very crucial role especially in case of porous nanomaterials, but unfortunately majority of researchers omitted the variation of electrolytes for a particular electrodes.

4. Conclusions

Synthesis of NiO based composite is one of the significant approach to construct high performance electrode materials for supercapacitors, since it exhibits synergistic effect in presence of other suitable materials through inducing porous networks, enhancing specific surface area, reduced particle size, preventing agglomeration, facilitating electrical conductivity, extending the working voltage, improving life span, as well as it provides extra pseudocapacitance. In this short review, we have chosen only adequate number of important NiO based asymmetric hybrid supercapacitors to be analyzed in terms of their active electrode materials along with their fabrication techniques. It is hoped that this should be reinforced of our reader for designing NiO based electrochemically active electrode materials with high capacity, cycle stability, operated in a large working voltage.

References

[1] H.E. Becker, Low voltage electrolytic capacitor, US Patent 2 800 616 (to General Electric) 1957
[2] Kötz R and Carlen M 2000 *Electrochim. Acta.***45** 2483–2498
[3] Najib S and Erdem E 2019*Nanoscale Adv.* **1** 2817
[4] Shi J, Jiang B, Li C, Yan F, Wang D, Yang C and Wan J 2020 *Materials Chemistry and Physics* **245** 122533
[5] Zheng J P, Cygan P J and Jow T R 1995 *J. Electrochem. Soc.***142(8)** 2699–2703
[6] Simon P and Gogosti Y 2010 Materials for electrochemical capacitors *Nanoscience and technology*: a collection of reviews from Nature journals (pp 320–329)
[7] Ray A, Roy A, Sadhukhan P, Roy-Chowdhury S, Maji P, Bhattachrya S K and Das S 2018 *Appl. Surf. Sci.***443** 581–591
[8] Majumdar D and Bhattacharya S K 2017 *J. Appl. Electrochem.***47** 789–801
[9] Teng Y, Yu D, Li Y, Meng Y, Wu Y, Feng Y, Hua Y, Wang C, Zhao X and Liu X 2020 *J. Electrochem. Soc.* **167** 020510
[10] Kate R S, Khalate S A and Deokate R J 2018 *J. Alloys Compd.***734** 89–111
[11] Zhang X, Shi W, Zhu J, Zhao W, Ma J, Mhaisalkar S, Maria T- L, Yang Y, Zhang H, Hng H-H and Yan Q 2010*Nano Res.***3(9)** 643–652
[12] Kim S-I, Lee J-S, Ahn H-J, Song H-K, Jang J-H 2013 *ACS Appl. Mater. Interfaces* **5** 1596–1603
[13] Zhao B, Song J, Liu P, Xu W, Fang T, Jiao Z, Zhang H and Jiang Y 2011 *J. Mater. Chem.* **21** 18792–18798
[14] Huang M, Li F, Zhang Y X, Li B and Gao X 2014 *Ceram. Int.***40** 5533–5538
[15] Zheng X, Yan X, Sun Y, Bai Z, Zhang G, Shen Y, Liang Q and Zhang Y 2015 *ACS Appl. Mater. Interfaces* **7** 2480–2485
[16] Yao D, Ouyang Y, Jiao X, Ye H, Lei W, Xia X, Lu L and Hao Q 2018 *Ind. Eng. Chem. Res.* **57** 6246–6256
[17] Ray A, Roy A, Saha S, Ghosh M, Roychowdhury S, Maiyalagan T, Bhattacharya S K and Das S 2019 *Langmuir* **35** 8257–8267
[18] Ray A, Roy A, Ghosh M, Ramos-Ramón J A, Saha S, Pal U, Bhattacharya S K and Das S 2019 *Appl. Surf. Sci.* **463** 513–525
[19] Zhao X, Liu X, Li F and Huang M 2020 *J. Mater. Sci.* **55** 2482–2491
[20] Zhang J, Wang Y, Yu C, Zhu T, Li Y, Cui J, Wu j, Shu X, Qin Y, Sun J, Yan J, Zhang Y and Wu Y 2020 *J. Mater. Sci.* **55** 688–700
[21] Leontyeva D V, Leontyev I N, Avramenko M V, Yuzyuk Y I, Kukushkina Y A and Smirnova N V 2013 *Electrochim. Acta.***114** 356-362
[22] Fan M, Ren B, Yu L, Song D, Liu Q, Liu J, Wang J, Jing X and Liu L 2015 *Electrochim. Acta.***166** 168–173
[23] Yi T-F, Qiu L-Y, Mei J, Qi S-Y, Cui P, Luo S, Zhu Y-R, Xie Y and He Y-B 2020 *Sci. Bull.* **65** 546–556
[24] Yang H, Xu H, Li M, Zhang L, Huang Y and Hu X 2016 *ACS Appl. Mater. Interfaces* **8** 1774–1779
[25] Sahoo S and Shim J J 2017 *ACS Sustain. Chem. Eng.* **5** 241-251
[26] Singu B S, Palaniappan S and Yoon K R 2016 *J. Appl. Electrochem.* **46** 1039–1047
[27] Vishnukumar P, Saravanakumar B, Ravi G, Ganesh V, Guduru R K and Yuvakkumar R 2018. *Mater. Lett.* **219** 114-118
[28] Yang Y, Li S, Liu F, Wen J, Zhang N, Wang s and Liu K 2018 *Nanotechnology* **29** 455401
[29] Saykar N G, Pilania R K, Banergee I and Mahapatra S K 2018 *J. Phys. D: Appl. Phys.***51** 475501
[30] Sannasi V, Maheswari K U, Karthikeyan C and Karuppuchamy S 2020 *Ionics*
[31] Edward B.O. Sihite*et al* 2020 *IOP Conf. Ser.: Mater. Sci. Eng.***725** 012042
[32] Majumdar D, Mandal M and Bhattacharya S K 2020 *emergent mater.* **3** 347–367.

Catalytic reduction of 4-nitrophenol to 4-aminophenol using an efficient Pd nanoparticles

Sujit Chatterjee, Malay Chakraborty, Kamal Kanti Bera, Ankita Mahajan, Senjuti Banik, Partha Sarathi Roy and Swapan Kumar Bhattacharya

Department of Chemistry, Jadavpur University, Kolkata–700032, India
e-mail: *skbhatt7@yahoo.co.in*

Abstract. Highly stable catalytically active polyvinyl alcohol supported Pd nanoparticles (Pd NPs) are synthesized and used as catalysts in reduction of 4-nitrophenol in aqueous phase by sodium borohydride as reducing agent. The as–synthesized Pd NPs are characterized with the help of UV–vis. absorption spectroscopy, X–ray diffraction (XRD), transmission electron microscopy (TEM). The TEM and high resolution TEM images reveal that the synthesized Pd NPs are mostly global shaped with mean radius of c.a. 2.1 nm. The progress of the reaction is conveniently monitored by UV–vis. spectroscopy. The rate constant of the pseudo–first–order reaction increases with temperature up to 318 K. The rate constant and the activation energy for the reduction of 4-nitrophenol by using Pd nanoparticles were found as 0.013 s^{-1} and 14.12 kJ mol^{-1} respectively at 318 K. The as-prepared Pd NPs are stable, easy to prepare, cost effective and have significant efficiency for reduction of 4–nitrophenol in green protocol.

1. Introduction

Metallic nanoparticles (NPs) especially noble metal nanoparticles (MNPs) have recently attracted considerable attention in the fields of fuel cells, sensors, catalysis etc. due to their unique properties compare to bulk materials [1]. Scientists were highly interested for this reason to fabricate catalysts involving MNPs for catalytic and electrocatalytic applications, in the last few decades [2]. The catalytic efficiency of the MNPs in heterogeneous catalytic reactions depends on the particle shape, dimension, morphology etc. As a result a universal model reaction is necessary to investigate the efficacy of metal and metallic oxide NPs in a heterogeneous catalytic environment. Recently, the hydrogenation reaction of 4-nitrophenol (4-NP) by $NaBH_4$ in aqueous media to 4-aminophenol (4-AP) have been used as a model reaction to study the efficacy of various metal and metal oxide nanocatalysts [3]. NPs of a variety of metals like Ni, Ag, Pt, Au, Ru, and Pd can be employed as catalysts for this reaction. For the synthesis of a large number of pesticides, herbicides, and synthetic dyes etc. 4-NP and its derivatives are obtained as by–products. Moreover, these are common hazardous organic pollutants in agricultural wastewaters. 4-NP has harmful effects on human health, since it damages the central nervous system, liver, kidney etc. Hence its removal from the environment is a crucial task. In contrary, 4-AP is a significant chemical intermediate in the preparation of antipyretic drugs such as paracetamol, dye stuffs, corrosion inhibitors etc. The important method for the removal of hazardous 4-NP and preparation of 4-AP is the direct hydrogenation of 4-NP catalyzed by Pt, Pd, Au, Ni etc. Although, Raney Ni has been used successfully for catalytic hydrogenation of nitro group, but one of the major drawback of its use is that it catalyzes the hydrogenation of both the nitro group and aromatic ring of 4-NP. J. Chen et al. have reported the hydrogenation of 4-NP catalysed by nitroreductase in presence of nicotinamide adenine dinucleotide under green protocol [4].

Furthermore, J. Chen et al. reported that 4-AP and 4-hydroxylaminophenol were found as the reductive products. Moreover, the catalytic performance of metal nanocatalysts largely depends on the morphologies of nanoparticles because of their extremely high surface energy. So, there is possibility of self-aggregation to minimise the extremely high surface energy especially in absence of stabilizer causing decrease in the catalytic performance of such nanocatalysts. For this reason, polymer matrix, dendrimers, metal organic frame works and micelle encapsulated metal nanoparticles have demonstrated excellent catalytic efficacy towards the reduction of 4-NP to 4-AP [5]. He and co-workers synthesized immobilized silver nanoparticles on leaky bio–scaffolds like a membrane of natural egg shell, at room temperature. Procyanidine, a polyphenol was used as a reductant and a stabilizer during synthesis [6]. C. Na et al. prepared multifunctional and reusable carbon nanotube ponytails by integrating CNT's into micrometer-sized colloidal particles. These carbon nanotube ponytails act as catalysts for the hydrogenation of 4-nitrophenol [7]. Ramaraj groups synthesized gold nanoparticles at room temperature and reduced with sodium citrate and $NaBH_4$. They were then supported on original gel of phloroglucinol carboxylic acid-formaldehyde gel [8]. This Au nanocatalyst exhibits good catalytic activity towards the reduction of 4-NP. Zero-valent Pd is a well-known catalyst and widely used in several reactions like C-C coupling reactions, hydrogenation reactions etc. [1]. In the present work, we have investigated the catalytic activity of polyvinyl alcohol (PVA) stabilized Pd nanoparticle, for the reduction of 4-NP to 4-AP with $NaBH_4$ in aqueous medium. The catalytic capability in terms of the rate constants and different thermodynamics parameters which are still relatively rare in literature, are fully discussed. Our as-synthesized PVA embedded Pd nanocatalyst appear to exhibit remarkable performance towards the reduction of 4-NP in aqueous medium.

2. Experimental Section

2.1. Materials

Palladium chloride ($PdCl_2$) was supplied from Arora-Matthey Limited, Millipore water (Synergy) was used through-out the experiment. 4-nitrophenol, $NaBH_4$ were purchased from sigma Aldrich, potassium chloride and polyvinyl alcohol (number average relative molar mass: 150000) were purchased from Merck. The all other materials were supplied from Merck, India.

2.2. Instrumental characterization

UV–visible spectra were captured using UV–1800 spectrophotometer (Shimadzu) (source lamp: deuterium (D2); tungsten-halogen). X–ray diffraction pattern was obtained by a Bruker D8 Advance diffractometer using Ni–filtered Cu$K\alpha$ source of radiation (λ = 1.5418 Å) operating at 40 kV and 40 mA, using a scan rate of 0.2° / second. TEM and HRTEM measurements were obtained by H-7600 transmission electron microscope (Hitachi, Japan) instrument.

2.3. Synthesis of PVA embedded Palladium nanoparticles

For the synthesis of PVA embedded Pd NPs, initially 0.5 mL 0.056 (M) aqueous K_2PdCl_4 (2.81×10^{-5} mol) solution was added drop-wise into the solution of 24.5 mL 1% (W/V) polyvinyl alcohol in a 100 mL beaker under stirring. Then 1.07 mg of $NaBH_4$ was added to this solution and stirred for 30 min at room temperature (30 °C) [3]. The yellow colour of K_2PdCl_4 solution changed to greyish black, manifesting the formation of Pd(0) NPs [scheme S1 in Supporting Information].

2.4. Reduction of 4-nitrophenol by using palladium nanocatalyst

The catalytic hydrogenation of 4-nitrophenol by using sodium borohydride as reductant was carried out in a standard quartz cuvette by using as-synthesized Pd NPs as catalyst. For catalytic reduction of 4-NP 0.03 M aqueous solution of $NaBH_4$ was prepared in ice-cold water 10 min before of each set of kinetic study. About 2.14 mL of water, 0.5 mL of 8.0×10^{-4} M solution of 4-NP, and 0.35 mL of 0.03 M $NaBH_4$ were mixed in a standard quartz cuvette (path length = 1 cm). That led to a colour change immediately from pale yellow to bright yellow. To this reaction mixture 11 μL 1.127×10^{-3} M Pd NPs solution was added and the time-dependent UV–vis. absoption spectra were recorded at a temperature of 318 K. The background correction was performed with Millipore water as reference. The rate constants of the reduction reaction were determined at an ambient temperature by measuring the decrease of absorbance of 4-nitrophenolate ion (at 400 nm) as a function of time. For evaluation of the thermodynamic data, the temperature was varied from 298 K to 318 K.

2.5. *Calculation of rate constant*

The kinetic parameters were analyzed by using pseudo first order kinetic equations. The absorbance of electromagnetic radiation of wavelength 400 nm (λ_{max} of 4-nitrophenolate ion in aqueous media) was utilized to determine the concentration of 4-NP. The natural logarithm of the absorbance of 4-nitrophenolate at 400 nm was plotted against time, and the steepest part of the curve was fit with a line, the negative slope of this line was taken as the apparent rate constant (k_{app}).

3. Results and Discussions

The as–synthesized Pd nanoparticles exhibited typical UV–visible spectra of Pd. Surface Plasmon resonance band was completely absent and the peak for Pd^{II}–PVA complex disappears indicating complete reduction of Pd(II) to Pd(0) (Figure S1a in Supplementary Information). The X-ray diffractogram is shown in Figure S1b which depicts diffraction peaks at 2θ values of 40.01°, 79.80° corresponding to the lattice planes (111) and (311) of face-centred cubic Pd [JCPDS No. 87-0637]. Moreover, the XRD profile shows a peak at 2θ = 44.52° marked by (101) plane of hexagonal primitive lattices of graphite carbon [JCPDS Card No. 89-8487]. Morphological studies were performed by using TEM analysis of as–synthesized Pd nanocatalysts. Figure 1a depicts the TEM image of Pd NPs and corresponding histogram is shown in Fig 1b. Figure 1b clearly reveals that the average diameter of as-prepared Pd NPs is 4.2 nm. Figure 1a clearly indicates that the as-synthesized Pd materials are predominantly globular shaped having nano-dot structure revealing unique three-dimensional quantum confinement of electron.

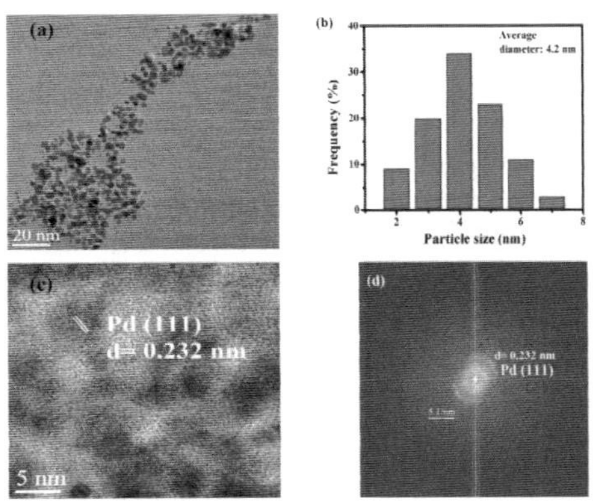

Figure 1. TEM image of as–prepared Pd NPs (a), and their corresponding histogram (b), HR-TEM image (c), and SAED pattern (d).

The HR-TEM micrograph and the selected area electron diffraction pattern of the Pd nanodots are shown in Figs. 1c and 1d, respectively. Figure 1c exhibits the lattice spacing is about 0.232 nm which correspond to the (111) plane of face-centred cubic Pd. The SAED pattern is also revealed the (111) plane of Pd with the spacing of 0.232 nm (Figure 1d). Therefore, HR-TEM image and SAED pattern further confirmed the XRD results. To test the efficacy of as–synthesized Pd nanodots for catalytic conversion of 4-NP to 4-AP, $NaBH_4$ was chosen as reducing agent. The main advantage is that the absorbance of 4-nitrophenolate ion obtained from 4-NP in presence of $NaBH_4$, can be easily monitored by UV–vis. spectrophotometrically. Fig. 2a shows the successive decrease of absorbance with time at 400 nm (λ_{max} of 4-nitrophenolate ion) with the concomitant evolution of a peak at 300 nm (λ_{max} of 4-AP) at 318 K, clearly revealing the catalytic reduction of 4-NP to 4-AP. In all the experiments we used excess $NaBH_4$ so that the reduction of 4-NP in presence of Pd NPs was zero order with respect to $NaBH_4$ and first-order with respect to 4-NP. To describe the reaction kinetics quantitatively typical plot of ln (A_t/A_0) versus the reaction time is shown in Fig. 2b, where A_t and A_0 are the absorbance of 4-nitrophenolate ion at any time t, and at the initial stage respectively at about 400 nm. The linear relationship between the ln (A_t/A_0) versus the reaction time suggests that the reduction of 4-NP follows pseudo-first-order kinetics. The apparent rate constant of the reduction of 4-NP in presence of Pd NPs, as calculated from the slope of a linear fit of ln (A_t/A_0) versus reaction time, is 1.3×10^{-2} s^{-1} at 318 K. It demonstrates that the catalytic efficiency of as–prepared Pd NPs is significantly higher than other synthesized Pd NPs found in the literature. The rate constant (k) calculated from the Eq (1) in here significantly greater than that of others synthesized nanocatalysts presented in the literature. The catalytic reduction of 4-NP in presence of as–prepared Pd NPs was studied at different temperatures of 298 K, 308 K, 313 K, and 318 K to evaluate the activation energy (E_a) of the catalyst (Fig. 3a). Table T1 (in Supplementary Information) shows the value of rate constants for reaction of 4-NP at different temperatures.

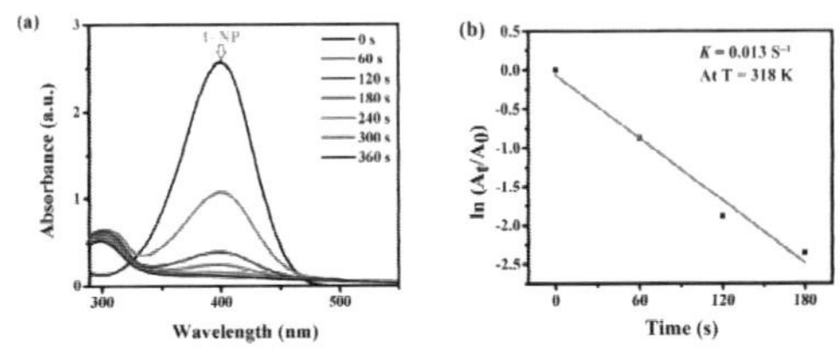

Figure 2. (a) UV–visible spectra that monitor the time-bound reduction of 4-nitrophenol to 4-aminophenol over Pd NPs at 318 K, and (b) the corresponding plot of ln (A_t/A_0) against the reaction time at λ_{max} = 400 nm.

$$ln\left(\frac{A_t}{A_0}\right) = -k \times t \qquad (1)$$

$$\ln k = lnA - \left(\frac{E_a}{RT}\right) \qquad (2)$$

As depicted in Fig. 3a, the catalytic reduction of 4-NP by using of as-prepared Pd NPs increases with an increase of the temperature. The E_a obtained by linear fitting of ln k against 1/T, (Fig. 3b) and by using Arrhenius equation (Eq 2).

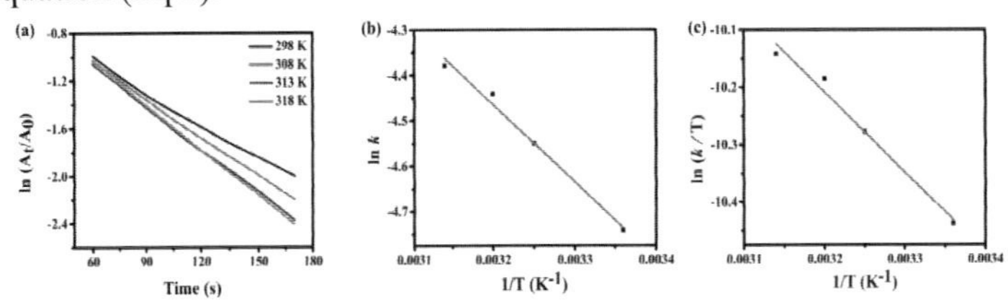

Figure 3. (a) Plots of ln (A_t/A_0) versus reaction time at λ_{max} = 400 nm at different temperatures, (b) plot of ln k versus 1/T, and (c) plot of ln (k/T) versus 1/T for the hydrogenation of 4-NP by $NaBH_4$ with Pd NPs as catalyst.

Moreover, the calculated activation energy was 14.12 kJ mol^{-1} which was much lower than that of several recently reported nanocatalysts. Table 1 depicts the comparison of the catalytic performance of as–synthesized Pd NPs in the reduction of 4-NP in aqueous medium with some catalysts presented in the literature. On the basis of above analysis, our synthesized Pd NPs exhibit excellent catalytic performance towards 4-nitrophenol reduction. The reusability of the Pd NPs was examined by catalyzing the reaction with the used catalyst for four times more. Figure S2 exhibits this recycling activity of the catalyst. It is found that the reaction is almost completed within 6 min for the 1st run but takes 10.4 min for the 4th run. In the study, the value the activation enthalpy ($\Delta H^{\#}$) and activation entropy ($\Delta S^{\#}$) for the reduction of 4-NP were determined by using Eyring equation. These are 11.6 kJ mol^{-1} and – 0.25 J mol^{-1} respectively as obtained by Fig. 3c. The later supports the proposition that the adsorption of 4-nitrophenol is the rate determined step in Langmuir-Hinshelwood mechanism [9]. Initially, 4-NPs and the borohydride ions (BH_4^-) are adsorbed on the surface of Pd NPs, then by the transfer of electron from the borohydride ions to the 4-NP via the metal surface converted it 4-AP, which then desorbed from the surface of Pd NPs. The slim size distribution of spherical Pd NPs with smaller dimensions, elevated surface region, and unique three-dimensional quantum confinement increase the diffusion of the reactant molecules and thus exhibit excellent catalytic activity towards 4-NP reduction reaction.

Table 1. Comparison of catalytic performances of various nanocatalysts for reduction of 4-NP in aqueous media.

Entry	Catalysts	Average diameter (nm)	Rate constant (s^{-1}) at 25 °C	Activation energy (kJ mol^{-1})	Ref.
1	Pd	4.2	8.7×10^{-3}	14.12	This work
2	Au@[C4C16Im]Br	10	1.1×10^{-4}	99.4	[9]
3	Pt@Ag NPs	185.47	5.9×10^{-3}	–	[10]
4	Pd13-DENs	1.33 ± 0.15	–	30.6	[5]

4. Conclusions

In summary, PVA embedded stable Pd NPs with average diameter of 4.2 nm are easily prepared. The particles exhibit remarkable catalytic capability for catalytic conversion of 4–nitrophenol by $NaBH_4$ in aqueous media. It might be applicable in wastewater treatment. The low poly-disperse characteristic and large adsorbent area make the Pd NPs an efficient catalyst for the reduction of 4–nitrophenol. Our catalyst exhibits greater rate constant and lower activation energy than Au, Pt and Ag catalysts available in the literature.

References

[1] Chatterjee S and Bhattacharya S K 2018 Size-Dependent Catalytic Activity and Fate of Palladium Nanoparticles in Suzuki–Miyaura Coupling Reactions *ACS Omega* **3** 12905–12913
[2] Kumar B S, Amali A J and Pitchumani K 2015 Fabrication of Pd Nanoparticles Embedded C@Fe3O4 Core–Shell Hybrid Nanospheres: An Efficient Catalyst for Cyanation in Aryl Halides *ACS Appl. Mater. Interfaces* **7** 22907–22917
[3] Aditya T, Pal A and Pal T 2015 Nitroarene reduction a trusted model reaction to test nanoparticle catalysts *Chem. Commun.* **51** 9410–9431
[4] Chen J, Dai R J, Tong B, Xiao S Y and Meng W 2007 Reduction of 4-nitrophenol catalyzed by nitroreductase *Chin. Chem. Lett.* **18** 10–12
[5] Bingwa N and Meijboom R 2014 Kinetic Evaluation of Dendrimer-Encapsulated Palladium Nanoparticles in the 4-Nitrophenol Reduction Reaction *J. Phys. Chem. C* **118** 19849–19858
[6] Liang M, Su R, Qi W, Yu Y, Wang L and He Z 2014 Synthesis of well-dispersed Ag nanoparticles on eggshell membrane for catalytic reduction of 4-nitrophenol *J Mater Sci.* **49** 1639–1647
[7] Wang H, Ma H, Zheng W, An D and Na C 2014 Multifunctional and Recollectable Carbon Nanotube Ponytails for Water Purification *ACS Appl. Mater. Interfaces* **6** 9426–9434
[8] Sundarapandi M, Shanmugam S and Ramaraj R 2019 Synthesis and Catalytic Activities of Metal Shells (Monolayer, Bilayer, and Alloy Layer)-Coated Gold Octahedra toward Catalytic Reduction of Nitroaromatics *J. Phys. Chem. C* **123** 21066–21075
[9] Thawarkar S R, Thombare B, Mundec B S and Khupse N D 2018 Kinetic investigation for the catalytic reduction of nitrophenol using ionic liquid stabilized gold nanoparticles *RSC Adv.* **8** 38384
[10] Lv Z-S, Zhu X-Y, Meng H-B, Feng J-J and Wang A-J 2019 One-pot synthesis of highly branched Pt@Ag core-shell nanoparticles as a recyclable catalyst with dramatically boosting the catalytic performance for 4-nitrophenol reduction *J. Colloid Interface Sci.* **538** 349–356

IOP Conf. Series: Materials Science and Engineering 1080 (2021) 012010 doi:10.1088/1757-899X/1080/1/012010

Supplementary Information

Catalytic reduction of 4-nitrophenol to 4-aminophenol using an efficient Pd nanoparticles

Sujit Chatterjee, Malay Chakraborty, Kamal Kanti Bera, Ankita Mahajan, Senjuti Banik, Partha Sarathi Roy and Swapan Kumar Bhattacharya

Department of Chemistry, Jadavpur University, Kolkata–700032, India
e-mail: *skbhatt7@yahoo.co.in*

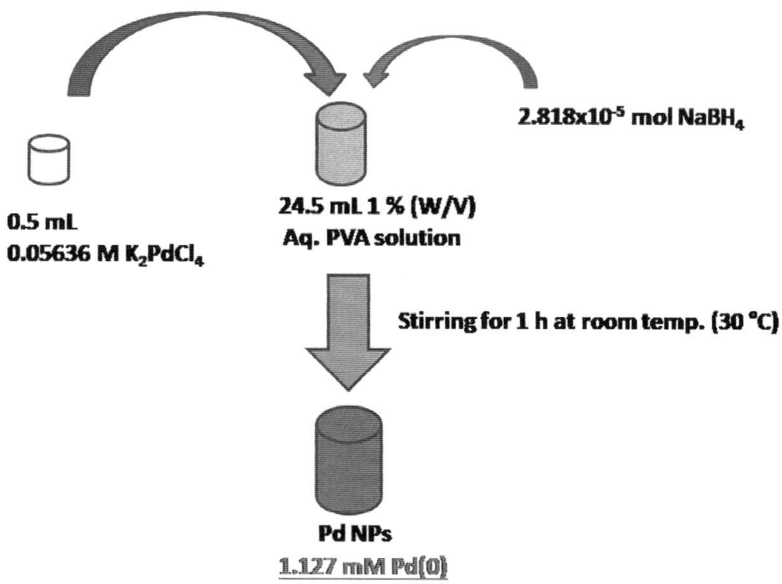

Scheme S1. A schematic illustration of the synthesis of PVA embedded Pd NPs.

Supplementary Figures

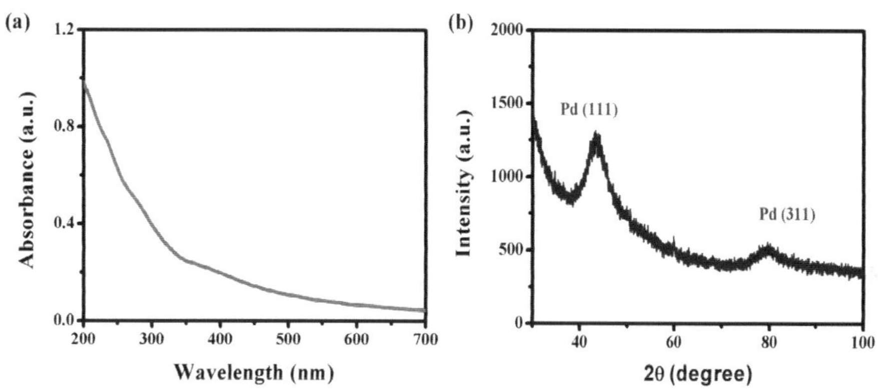

Figure S1. (a) UV–visible absorption spectrum of ten times diluted aqueous solution Pd NPs. (b) the XRD pattern of Pd NPs (in presence of activated charcoal).

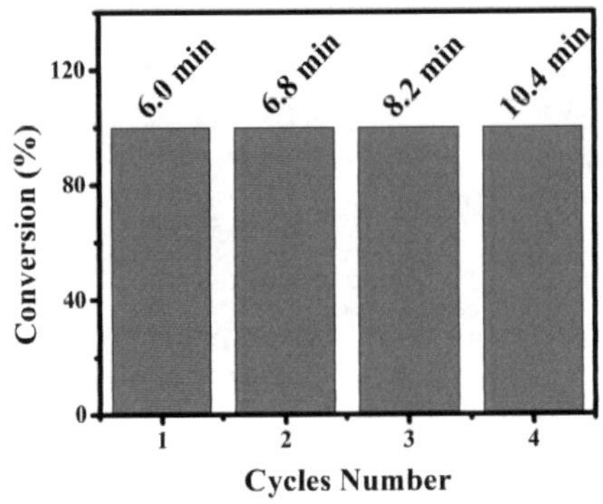

Figure S2. Reusability of Pd nanocatalysts for the hydrogenation of 4-NP at 45 °C in aqueous media.

Supplementary Table

Table T1. Influence of temperature in rate constants for catalytic reduction of 4-NP.

Temperature (K)	$10^3 \times$ Rate constant (s^{-1})
298	8.7
308	10.6
313	11.8
318	12.5

AI SHOPPING SOLUTION - The Smartest All-in-One Shopping Solution

Soumyojit Paul[1], Pritam Mahapatra[2], Maheswar Banerjee[3] and Dr.Tapas Kumar Nandi[4]
Mechanical Engineering, 3rd Year, Techno International New Town, Kolkata-700156, India
[1]e-mail:paul.soumyojit1998@gmail.com, [2]e-mail:pritammahaptra2050@gmail.com
[3]e-mail:maheswar252000@gmail.com, [4]e-mail:tapasnanditamal@gmail.com

Abstract: In today's world, purchasing and shopping at supermarkets and malls are becoming part and parcel of urban life. We notice the huge rush in the markets and malls on holiday as well as in the weekends. The problem occurs at the customer end to find their likely products and to pay the bills standing at long queue; both of these are time consuming and tiring. In this paper, we discuss about an all new complete AI (Artificial Intelligence) SHOPPING SOLUTION which will make our marketing experience completely smart and hassle free. Especially during and after the COVID-19 situation, social distancing and human health will become the foremost priority of the modern civilization. This is the ideal situation to make use of this all new technology in which the customer can purchase his/her needful products at one's fingertips by visiting the nearby market or getting the things delivered comfortably to his/her place by self-sanitizing drones. This entire shopping system will be controlled by our own indigenous developed mobile app or website.

Keywords: Social Distancing, Mobile App, Website, Smart trolley, RFID Tag, RFID Reader, NODE MCU, Online Payment, Self- Sanitizing Drone, Cash back

1. Introduction

Innovation and technology is changing the world in leaps and bounds. In today's hectic schedule, a large gathering in the shopping malls creates a more annoying picture in the shopper's mind and then the long queue in the billing section adds some extra bit of exhaustion to their minds. To make life more comfortable **AI SHOPPING SOLUTION** is here to serve a delicious recipe in front of shoppers' taste buds. A customer can just open our app [10] or log into one's account to create the shopping list and can enjoy a hassle free shopping experience with stunning rewards and offers. This paper presents not only a shopping element but a total digital platform of each and every shopping brands ranging from the food products-fashion wares-grocery-electronic goods food court items-baby products to every shopping items guided by smart shopping trolley[4] (if one visits the mall) or at his/her doorsteps within a couple of hours.

The issues addressed or faced along with our solutions are as follows: -

- Customers' dissatisfaction because of hours of waiting for checkout or billing process[2] and [7]- our smart shopping solution is designed in a way to accept payment in all forms that is Cash, E-wallet or the Smart Card depending upon user's convenience as soon as it is through with its shopping.

- Involvement of huge man-power, which is expensive- we have come with a design or a system that automates the checkout procedure by billing the products simultaneously as soon as we add the product in the Smart shopping trolley.
- Dishonest behaviour of customers- we all know that everything has its own advantages as well as disadvantages. Automation has its own worries. Absence of human operators can lead to forgery. We propose a solution that handles all the cases of deception, if any so that it is an attractive option for both the buyers as well as the sellers too.

2. Problem Statement and Working Procedure

The AI SHOPPING SOLUTION is based on four major systems which are as follows:-

- Smart billing system**: -** To develop RFID based billing system for supermarkets in order to expedite billing process and avoid crowd [2].
- Smart management:-Getting product information is easy and no extra time needed to update entry or sell.
- Hassle-free payment:-Use RFID TAG for billing instead of barcode scan, need of any staff for billing [6] and [7].
- Easy product handling and tracking:-User and admin of manager can easily track the product position and quantity.

2.1 Working Procedure

- Customer sign in to his or her account of our app [10]
- Customer create his or her shopping list as required
- The list is saved into the account and the availability is checked through the database
- In case of unavailability. Alternate suggestions are made
- Then after saving the final list the customer gets option for home delivery or can visit the mall
- on entering the mall a small trolley can be unlocked through the shopping kart ID and the listed product will transfer to it
- The AI trolley will guide the customer to the required stores and the eventually to their Precise locations in the shelf
- Every product has an RFID tag which contains a Unique ID. These ID's are fed in the database assigned to the corresponding products [1].
- There will be another option provided to get the location of required product.
- If there needs to be a purchase done, then that product can be dropped in the cart where the RFID reader reads the tag [1]. The information of the product is extracted and displayed on the Web Application. At the same time billing information is also updated.
- When a customer wants to remove any product from the trolley, then that product needs to be scanned again.
- At the same time the billing information is updated. The total amount of purchases is also displayed on Application.
- These steps are repeated until the end of shopping button or send bill button is pressed. This generated bill is sent to billing side computer to get the computerized bill [7].
- The customer can straight away pay the bill and leave.
- Inventory status of the products is also updated at the end of shopping. Simultaneously the temporary data present in microcontroller is reset, so that it can be reused.
- If the customer has registered user card, the payment can be done.
- If the customer opts for home-delivery option then the package will be delivered to the required location by our Logistic drone.

- At the time of home delivery, a programmed call will be made to the customer informing about the approach of the delivery item exact 10 minutes prior to the arrival of the drone using GPS location of the drone.
- The customer needs not to wait for the delivery. On arrival of the logistic drone an OTP will be sent to the customer's registered mobile number.
- The customer has to enter this OTP into the drone's number panel using registered number and the product will get unlocked and the customer can collect his/her item.

The total working is shown in the flowchart given in the Figure 1.

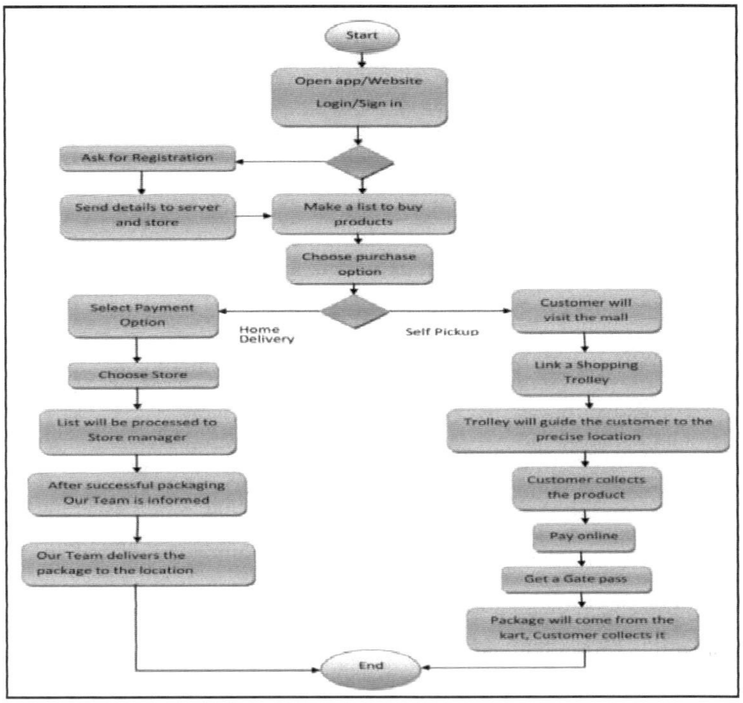

Figure 1: Flowchart of the entire AI SHOPPING SOLUTION

2.2. Working of the App: MechITEasy

The working of the app is shown in the Figures 2(App Icon), 3(Customer Account Details), 4(Navigation), 5(Browsing through the shopping items), 6(Adding to cart before proceeding for payment), 7(Various sections of App)

Figure 2: App Icon

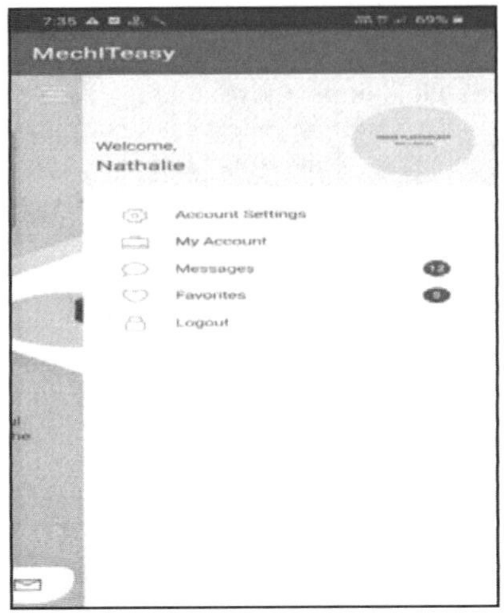

Figure 3: Customer Account Details

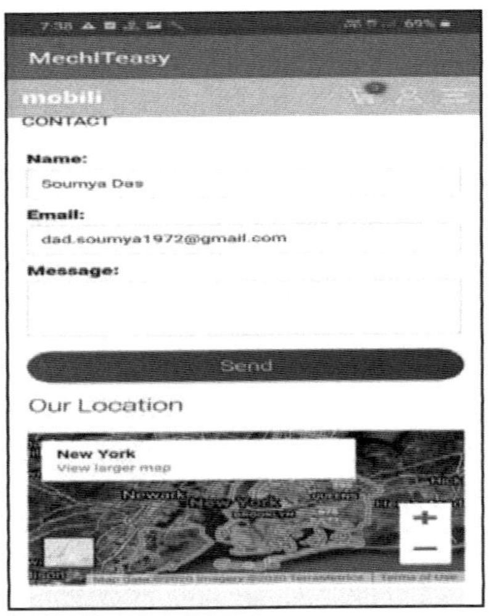

Figure 4: Navigation

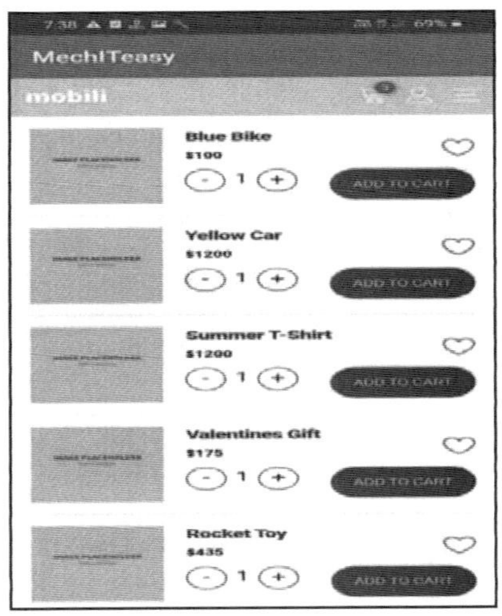

Figure 5: Browsing through the shopping items

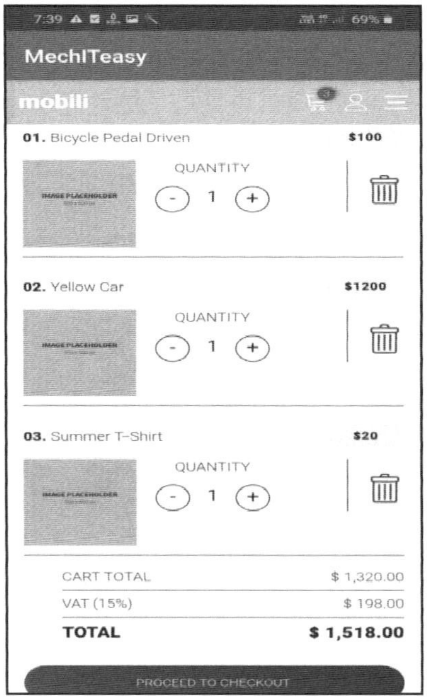

Figure 6: Adding to cart before proceeding for payment

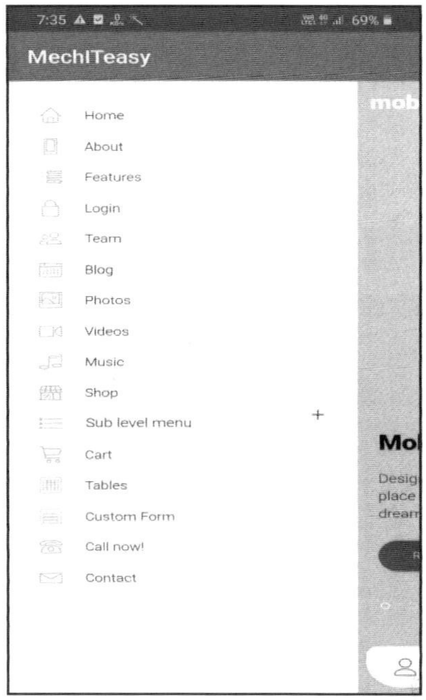

Figure 7: Various sections of App

We are using Node MCU as an IoT (Internet of Things) device. The Node MCU is always connected to the given Wi-Fi network SSID and password defined in code. Then the he connected RFID reader or BARCODE scanner scan the product. The product ID will be processed by Node MCU (IoT Device).

Node MCU send all the data to a given link. When the link is triggered, the product will be added to cart of the mobile app. Hence, we repeat this process to continue the shopping.

2.3. The Application of the Softwares:-

- **Arduino IDE, Python-** To program Node MCU
- **Android IDE-** To develop the android app
- **HTML5, CSS3, JavaScript-** To design and develop the application
- **PHP-** To make the application dynamic
- **Json-** To send the value from Node MCU to server
- **MySQL-** To store the details in database
- **jQuery-** For the development of the application

The hardware interface components are listed in Table -1

A. Hardware Interface

Sl. No	Components	Quantity
1	Shopping Trolley	1
2	Node MCU	1
3	Arduino UNO	1
4	RFID Reader	1
5	Barcode Scanner	1
6	RFID tag	3
7	Buzzer	1
8	LED	2
9	DC Motor (for rack and pinion mechanism) and DC Motor driver	1
10	Logistic Drone	1
11	Battery (12V)	1

Table 1: Hardware Components

B. Software Interface

- **Programming Language**
  1.Python
  2.PHP
  3.HTML5, CSS3
  4.JavaScript
  5.jQuery
  6.Json
  7.Arduino IDE
  8.Android IDE
- **Database**
  9.MySQL

App link: - https://webingoinfotechsolutions.in/spice.apk

3. Result and Discussion

This all new AI Shopping Solution is developed to improve the quality of services to the customer and save time and man power thereby reducing maintenance cost and maximizes profit. The proposed system provides customer a choice whether he/she wants to shop by going to a smart mall or chooses to get the required items through home delivery.

The customer opens the app/website; he/she is required to login/signup. Then one has to make a list to buy products and choose the store and then one has to decide the purchase option (home delivery/self-pickup). If one chose for home delivery then the purchase list will be processed to the store manager who will do the needful. Once the items are ready then the information will pass on to our team who will do the required packaging and will deliver to the given address. If the distance is within a short range then the package will be delivered by the self-sanitizing logistic drone as shown in Figure 11. If one chooses for the self-pickup option, then one has to visit the mall/store, and then has to link a smart trolley to his account. The smart trolley will guide the customer to recognised row and column and the customer has to just collect the product and the trolley will get locked. After the shopping is over, the customer has to pay the billing amount online (customers will get added discounts and rewards, if they use our payment gateway system).Lastly a gate pass will be generated and the package will come from the kart which ends the system.

The system will make use of a shopping trolley, node MCU, Arduino UNO, RFID Reader, Barcode Scanner, RFID tag [1] ,Buzzer, LED, DC Motor (for rack and pinion mechanism), DC Motor driver, Battery. The softwares which we are using are Arduino IDE, C, Python, Json, PHP and HTML. The mechanism and architecture of the Trolley is shown in Figure 8 and the working of the entire solution is depicted in Figure 9.

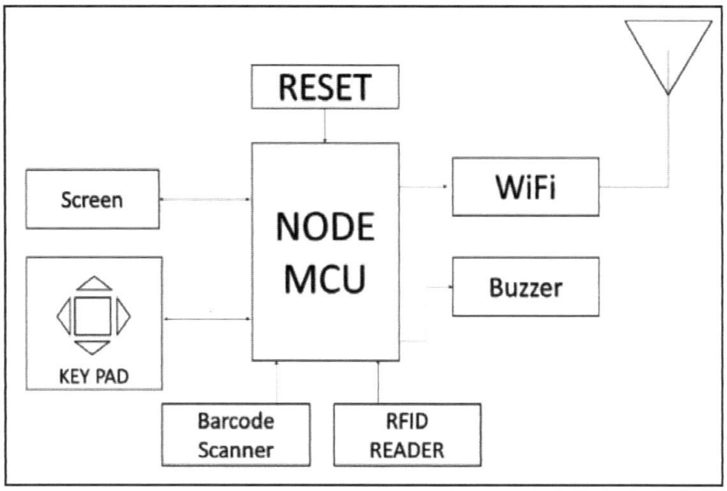

Figure 8: System Architecture of Trolley

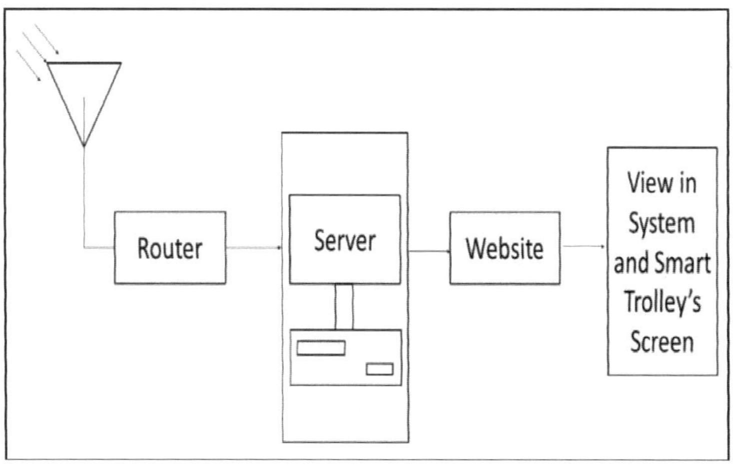

Figure 9: Block Diagram of the AI SHOPPING SOLUTION

The experimental set-up is tested for various test cases, with various test products. The system gives the customer an option to make his/her choice for shopping (home-delivery/self-pickup).The system guides the customer through his/her choice to get the required outcome. This makes the shoppers life easy going avoiding long queues or getting everything in ones homely comfort. The entire shopping system is maintained by our team with full proof security of the customers and their payment gateway. On paying through our gateway the customer will receive attractive offers and discounts and also cash back offers. This entire system is server based and also reduces man power and related problems. The system is cost-effective for large scale implementation and strengthens the idea of smart mall and smart shopping solution in developing countries like India.

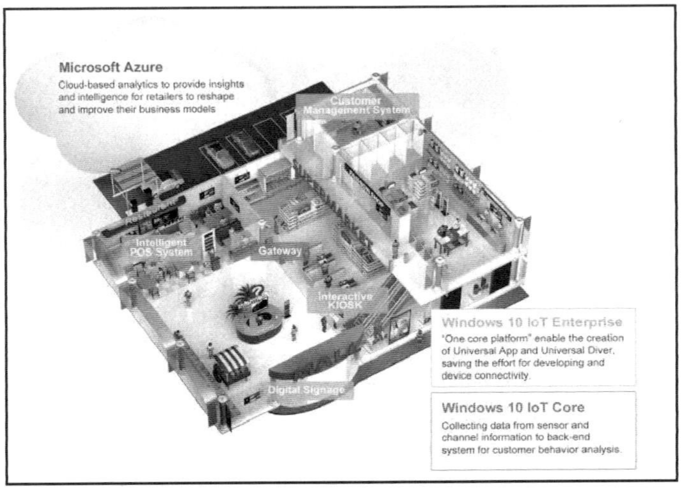

Figure 10: Graphic Representation of Smart Shopping Mall

Figure 11: The Logistic Drone

3.1. Cost of the project

HARDWARE REQUIREMENT

1. Shopping Trolley	Rs. 2500
2. Node MCU	Rs. 450
3. Arduino UNO	Rs. 400
4. RFID Reader	Rs. 300
5. Barcode scanner	Rs. 1500
6. RFID tag	Rs. 100
7. Buzzer	Rs. 30
8. LED	Rs. 10
9. DC Motor (for rack and pinion mechanism)	Rs. 100
10. DC Motor driver	Rs. 250
11. Battery (12)	Rs. 700
Total-	Rs. 5984

Cost of logistic drone (Maximum Capacity-1Kg, Average Runtime-20mins)	Rs. 19500
Total cost of the project	Rs. 25484

3.2. Advantages of MECHITEASY Shopping Solution

The success of MECHITEASY Shopping Solution lies in the satisfaction of customer. The all new technology and simple working grabs customer attention. The following advantages of this shopping solution which keeps us ahead of the currently existing shopping websites:-

- This project introduces the idea of smart shopping mall which is still not being initialized in India as shown in Figure 10.
- The brightest feature of this MECHITEASY Shopping Solution is a new feature has been added which will serve people during the time of need to save one's life. A feature is all set to be introduced which will play a vital role in the time of medical emergency. If a customer orders for some first aid or supporting medicines even at the remote areas, our drone or delivery team will reach the location within 20 minutes to serve.
- This project provides a golden opportunity for the employment of younger generation in software, hardware and management sectors.
- Generation of high revenue which will strengthen the associated business chain.
- This AI Shopping Solution utilises highly automated infrastructure with easy working which provides a hassle free shopping experience.

- Boundless offers, discounts and rewards have a magnetic effect on the customers.

4. Conclusion

This project successfully demonstrates the idea of Smart Shopping Solution which makes the life of a shopper more exciting. The system which is developed by MechITEasy is highly reliable, secured and cost effective. The all new AI Shopping Solution gives a complete new vision to the global direct investment in India. The smart shopping solution will brighten up the prospect of Smart Mall. The concept of digital shopping will no longer be an obscure dream but reality in the developing countries like India. The implementation of the AI Shopping Solution will provide huge employment which will brighten up the future of our young generation. This concept can be hired or implemented by any world class global online shopping platform which will enhance their impact over the competitors and will also increase the revenue. This solution has its brightest of future even in the darkest post COVID-19 phase. The mixture of Indigenous idea along with world class facility will strengthen the foundation of ATMANIRBHAR BHARAT.

References

[1] Kamble S, Meshram S, Thokal R, Gakre R 2014 Developing a Multitasking Shopping Trolley Based On RFID Technology: *International Journal of Soft Computing and Engineering (IJSCE)* ISSN: 2231-2307, Volume-3, Issue-6

[2] S. Sainath, K. Surender, V. , Arvind V 2014 Automated Shopping Trolley for Super Market Billing System*: International Journal of Computer Applications (0975 – 8887)* International Conference on Communication, Computing and Information Technology (ICCCMIT)

[3] Shaw R, Akhil R, Radhakrishnan A, Senthil R 2017 Smart Kart: *Advances in Computational Sciences and Technology,* © Research India Publications http://www.ripublication.com ISSN 0973-6107 Volume 10, pp 2273-2288

[4] Ng YenLeng, Lim Cheng Siong, Danapalasingam K A, Tan M L P, Tan C W: Automatic Human Guided Shopping Trolley with Smart Shopping System

[5] Shopping Trolley Underpayment Cases to Proceed. 2013, September 1. Australian Payroll Association. Retrieved September 16, 2014.

[6] Dawkhar K, Dhomase S, Mahabaleshwarkar 2015 S Electronic Shopping Cart For Effective Shopping based on RFID: *International Journal OF Innovative Research in Electrical, Electronics, INSTRUMENTATION and Control Engineering* Vol. 3, Issue 1

[7] Aryan P, Pise P, Tamhane S Smart Shopping Cart with Automatic Billing System through RFID and Bluetooth *International Journal of Emerging Technology and Computer Science*

[8] Shih C, Liang B and Lin C 2011 "An Automatic Smart Shopping Cart Deployment Framework based on Pattern Design", *IEEE 15th International Symposium on Consumer Electronics*

[9] Dhokte S.S., Patere B.S. ,Magar M.T, Kulkarni V.S., Patil P.S. ,Patil R 2015 "Smart Shopping Trolley Using Rechargeable Smart Card " *in International Journal of Emerging Technology and Advanced Engineering*, Volume 5, Issue 5

[10] Bedi H.S., Goyal N, Kumar S and Gupta A 2017 Smart Trolley using Smart Phone and Arduino: *Journal of Electrical & Electronic Systems*

Exciton and Bi-exciton Binding Energy Calculation in a Core Shell Quantum Dot

Shivani Rana[1], Sanjib Kabi[2], Kamakhya Prakash Misra[3], Saikat Chattopadhyay[3]*

[1] Department of Physics & Electronics, Indian Maritime University, Navi Mumbai Campus, Mumbai 400706, India
[2] Department Physics, Sikkim Manipal Institute of Technology, Sikkim Manipal University, Majitar, Sikkim 737136, India
[3] Department of Physics, School of Basic Sciences, Manipal University Jaipur, Jaipur 303007, Rajasthan, India
*e-mail: *the_saikat@yahoo.com*

Abstract. We report the effect of shell thickness on exciton and biexciton binding energy calculation for a CdSe/ZnSe core shell quantum dot. Wentzel Kramers Brillouin approximation help us to create proper wave function for the system. It incorporates core shell boundary effect correctly. For excitons, the binding energy is affected by the Columbic confinement potentials and during biexciton binding energy calculation, the exchange and correlation effects were taken into account. It is observed that the excitonic binding energy shows an initial increase with increasing shell thickness but saturates for thick shell while it always gives a negative value for bi-excitonic binding energy. This confirms the existence of antibonding states for biexcitons.

Keywords: Core Shell quantum dot, Band gap, Exciton, Bi-exciton, Binding energy

1. Introduction:

The rapid progress in semiconductor nanoscience and technology can be envisaged from its reach technological applications and its incredible growth. The semiconductor quantum dots (QDs), due to its great demand for encouraging the miniaturization of modern technological devices, are studied well. Depending on their tunable optical properties, QDs have been widely used for numerous optoelectronic devices from detectors to QD Lasers source for different wavelength ranges or in nano-spintronic devices [1-8]. A new type of nanostructures, commonly recognized as core shell quantum dots (CSQDs), where the carrier particles (i.e. electrons and holes) are confined within the size of the core material having lower band gap and surrounded by another higher band gap shell material. It has been reported, introducing a shell layer externally on the surface of a bare QD enhances its optical properties. This external shell layer enhances the localization probability of the charge carriers along with the passivation of surface nonradiative recombination sites. These interesting materials are under intense investigation due to its high photoluminescence (PL) efficiency, optical amplification, spectral tunability etc. Some well-known examples of CSQDs are CdSe/ZnS, ZnS/CdS, CdSe/CdS, ZnSe/ZnS, etc. [4, 7, 9]. CdSe QDs are considered as one of the most proficient material in visible spectral range because of its emission wavelength can precisely be adjusted from blue to red region. Moreover, the ZnSe shell in CdSe/ZnSe CSQD, can be excited by UV light, to use it as fluorescent biological labelling [10]. The possibility to tunnel out from the core for the carriers is always high as the built-in potential well is not high enough to block carriers (electrons and holes) inside CdSe core in a CdSe/ZnSe CSQDs. Due to that, in presence of

ZnSe shell CdSe nanocrystals exhibit very high PL efficiency. Recent results by Shen et al. [11] and Moon et al. [12] reports that the PL efficiency in these CSQDs are enhancing around 20% (approximate) due to the presence of shell which makes them a correct choice for promising light source development or to utilize them in color display applications.

In the present paper, our aim is to investigate the eff ect of shell creation, on exciton, bi-exciton binding energy and on the transition energy for ground-to-exciton state (ω_{0e}) in a CdSe/ZnSe CSQD. Theoretical formulations interpreted in terms of a model, which includes Wentzel-Kramers-Brillouin (WKB) wave functions in such a way that, the core shell boundary eff ect may incorporate properly during calculations. The final result shows that with increasing shell thickness, the exciton binding energy increases while the calculated energy shows negative values which confirms the presence of antibonding states and the ground-to-exciton state transition energy ($\sim \omega_{0e}$) shows a decreasing trend in a CdSe/ZnSe CSQD, as the external ZnSe shell continues to grow on CdSe core.

2. Theoretical Formulation:

The conduction and valence band (CB and VB) alignments for core and shell materials classifies CSQDs into 2 types. Type-I CSQD, where both career particles are strongly confined inside the core region only; and Type-II CSQD are those in which one of the particle (electron/hole) is restricted in the core region and the other may extend to the shell [3, 4, 9]. Fig. 1 shows the geometry of CdSe/ZnSe CSQD used for calculation. Here a and b are the radii of the core and total CSQD, respectively. Thus, the shell thickness is $|b - a| = d$. The confinement potential of the charge carriers is considered as parabolic in nature and given as [3]

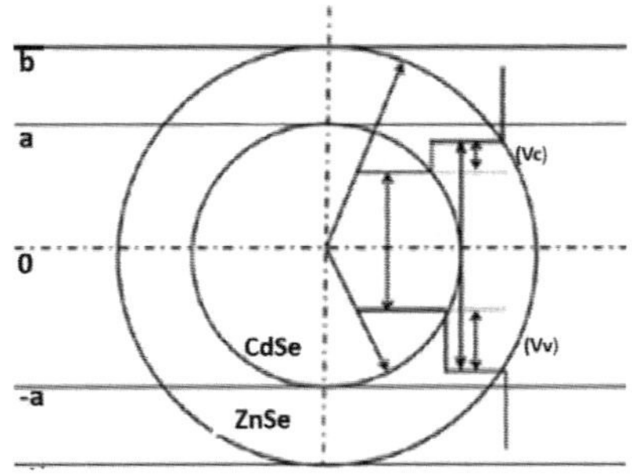

Figure:1. Schematic illustration of the of a core shell quantum dot.

$$V_e(r) = \frac{V_c}{a^2}(r^2 - a^2) \quad (1)$$

and

$$V_h(r) = \frac{V_v}{a^2}(r^2 - a^2) \quad (2)$$

V_c and V_v denote CB and VB offsets, respectively. A strong confinement potential experienced by carriers, inside the core due to the existence of ZnS shell layer, as detailed in Ref. [3, 4]. The WKB wave functions for single particles, under such condition are [3, 4]

$$\phi_j(a,b) = \begin{cases} \frac{A_S}{\kappa_j}\exp\left(\int_{-b}^{-a}\kappa_j dr\right) & for -b < r < -a \\ \frac{A_C}{k_j}\exp\left(\int_{-a}^{a}k_j dr + \frac{\pi}{4}\right) & for -a < r < a \\ \frac{A_S}{\kappa_j}\exp\left(-\int_{a}^{b}\kappa_j dr\right) & for\ a < r < b \end{cases} \quad (3)$$

Here, the subscript j represents individual electron (e) or hole (h) particles. Details of normalization constant A_C and A_S are already explained in Ref [3, 4]. The parameter k_j can be explained mathematically as [3]

$$k_j = \sqrt[4]{\frac{mV_{e,h}(r^2 - a^2)}{\hbar^2 a^4}} \quad (4)$$

The running parameter r for bare QD is in the range $-a \leq r \leq a$. The effect for inclusion of shell in the structure, taken into account for calculations by considering r as $-b \leq r \leq -a$ and $a \leq r \leq b$. Eq. (3) represents the WKB wave functions which are the single particle envelope functions with proper consideration of the energy and effective mass of the carriers. The total wave functions for electron (ψ_e) and for hole (ψ_h) may described as [3]

$$\psi_e(a, b, r_e) = \phi_e(a, b) \cdot u_e(r_e) \quad (5a)$$

and

$$\psi_h(a, b, r_h) = \phi_h(a, b) \cdot u_h(r_h) \quad (5b)$$

Here we are considering the concept of effective mass approximation (EMA). The product of the envelope function, as described above, with the Bloch functions (u_e, u_h) helps to complete the total wave functions in such nanostructures under EMA approximation. r_e and r_h represents the positions of different career particles like electron and hole of the core shell system, respectively. The optical properties of semiconductor QDs are determined by the paired electron-hole (e-h) states popular as exciton. Coulombic influence comes into the picture for any small QD to form e-h pair or exciton. Complete exciton wave function (ψ_X) in ground state with Hartree approximation is [13]

$$\psi_X(a, b, r_e, r_h) = \phi_e(a, b)u_e(r_e) \cdot \phi_h(a, b)u_h(r_h), \quad (6)$$

and the exciton binding energy Δ_X is [3, 4]

$$\Delta_X(a,b) = \left\langle \psi_X(a,b,r_e,r_h) \right| \left(V_e + \frac{e^2}{\varepsilon_0 r}\right) \left| \psi_X(a,b,r_e,r_h) \right\rangle + \left\langle \psi_X(a,b,r_e,r_h) \right| \left(V_h + \frac{e^2}{\varepsilon_0 r}\right) \left| \psi_X(a,b,r_e,r_h) \right\rangle \quad (7)$$

3. Results and Discussions:

The band gap energy for CdSe and ZnSe are 1.716 eV [14] and 2.8 eV [15], respectively. In CdSe/ZnSe CSQD, ZnSe offers strong confinements to the carriers due to its higher band gap energy. The CB offset for electrons and VB offset for holes in CdSe/ZnSe is 0.735 eV and 0.135 eV, respectively [16]. We are using these parameters for our calculation of binding energy.

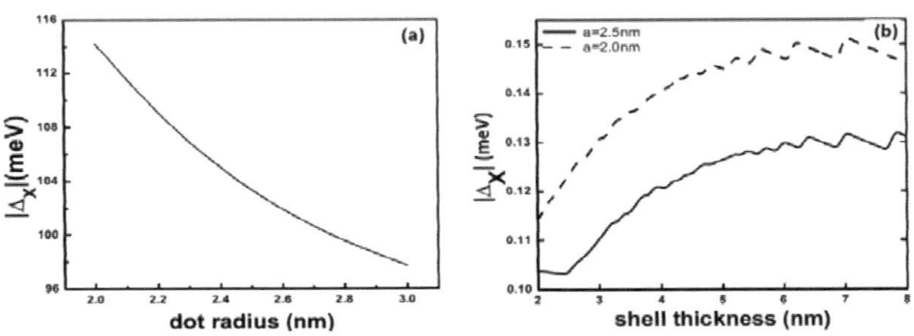

Figure 2: (a) Exciton binding energy (Δ_X) as a function of CdSe QD radius (b) Change in exciton binding energy ($|\Delta_X|$) with increasing shell thickness in a CdSe/ZnSe CSQD.

Fig. 2 (a) depicts the effect of increasing dot dimension on the exciton binding energy for a bare CdSe QD. The theoretical results suggest bigger QDs leads to a decreasing exciton binding energy. Fig. 2 (b) shows the eff ect of the shell structure on the binding energy of excitons (Δ_X).

The calculation is applied to two diff erent CSQDs of core radii 2.5 and 2.0 nm. For each core, the shell thickness is varied up to 8nm. It is evident from the figures that the exciton binding energy increases up to a thickness of the shell. The exciton binding energy starts saturating after a certain thickness for the shell for CdSe/ZnSe CSQD. According to Talapin et al. [10] there is a remarkable improvement in PL quantum efficiency ZnSe shell around CdSe core structure. They also confirmed a clear red shift in absorption wavelength spectra in such structures. A comparative study clearly confirms the external shell is the reason of such observation in any CSQD. The observed red shift confirms the fact that ZnSe shell fails to support with enough strong barrier potential to avoid the outflow of the carrier particles from core to the shell. So, we can conclude that the increment in the exciton binding energy attributed to the spreading of single particle wave function into the shell region. The observed red shift in transition energy can explain using the Eq. (8) below,

$$\hbar\omega_{0e} = \left[\epsilon_g - |\Delta_X(a,b)|\right] \tag{8}$$

Fig. 3, shows the change in ground ($|0\rangle$) to exciton ($|e\rangle$) state transition energy with increasing shell thickness. As the exciton binding energy i.e.$|\Delta_X(a,b)|$ increases with increasing shell thickness (Fig.3), from Eq.(8) it is confirmed that the ground-to-exciton level transition frequency ω_{0e} reduces, causing a red-shift in transition energy with increasing shell thickness. We further investigated the effect of shell formation on bi-exciton binding energy in the same CdSe/ZnSe CSQD. A well-established mathematical formula to calculate bi-exciton binding energy $\Delta_{XX}(a,b)$ is [3, 4]

$$\begin{aligned}\Delta_{XX}(a,b) =& \\ &\left\langle\psi_e(a,b,r_e)\right|\left(V_e + \frac{e^2}{\varepsilon_0 r}\right)\left|\psi_e(a,b,r_e)\right\rangle - \left\langle\psi_h(a,b,r_h)\right|\left(V_h + \frac{e^2}{\varepsilon_0 r}\right)\left|\psi_h(a,b,r_h)\right\rangle + \\ &\left\langle\psi_e(a,b,r_e)\right|\left(V_h + \frac{e^2}{\varepsilon_0 r}\right)\left|\psi_e(a,b,r_e)\right\rangle - \left\langle\psi_h(a,b,r_h)\right|\left(V_e + \frac{e^2}{\varepsilon_0 r}\right)\left|\psi_h(a,b,r_h)\right\rangle\end{aligned} \tag{9}$$

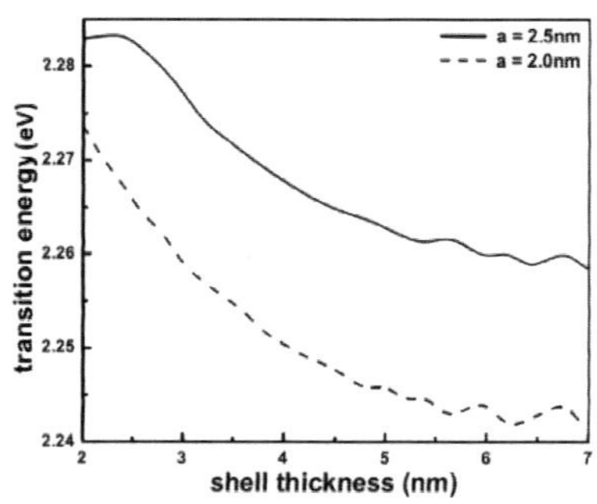

Figure 3: Ground-to-exciton state $|0\rangle \rightarrow |e\rangle$ transition energy $\hbar\omega_{0e}$, as a function of shell thickness in a CdSe/ZnSe core shell quantum dot.

Fig. 4 (a) shows increase in the bi-exciton binding energy ($\Delta_{XX}(a,b)$) for varying dot radius in a bare CdSe QD. Fig. 4 (b) illustrates the same due to the growth of ZnSe shell on CdSe core. The similar experimental findings by Sarkar et al. [17] report about the bi-exciton binding energy and on the fine structure splitting in case of a single self-assembled InAs/AlAs QD, further validate our theoretical prediction for CdSe/ZnSe CSQD. The increase in the binding energy indicates the possibility to attain the antibonding state. The biexciton binding energy (plotted as $\Delta_{XX}(a,b)$ with shell thickness in Fig. 4(b)) drops down suddenly with growing shell thickness. The negative value for bi-exciton binding energy indicates the presence of antibonding states which appears due to the stronger carrier confinement from external ZnSe shell in CdSe/ZnSe CSQD. These leads to an enhancement in the interaction between the carriers represented in Eq. (9). The strong interaction reveals an escalation in the repulsive force activating antibonding of bi-excitons.

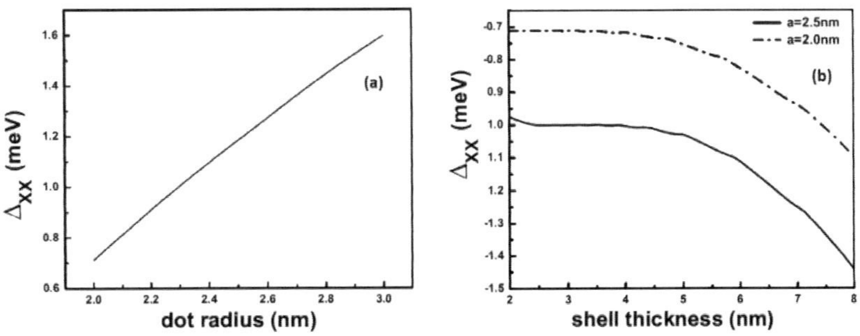

Figure 4: (a) bi-exciton binding energy (Δ_{XX}) with dot radius in a bare CdSe QD. (b) Variation in Δ_{XX} with growing shell thickness in a CdSe/ZnSe core shell quantum dot.

4. Conclusion:

In conclusion, theoretical analysis of the exciton, bi-exciton binding energy and transition frequency of a CdSe/ZnSe CSQD has been presented in this paper using WKB wave function to include the exact surface effect near core-shell boundary during calculation. The influence of QD radius on binding energy and the impact of externally grown ZnSe shell on CdSe core has also been calculated and compared with bare CdSe QD. The study is helpful to understand the experimentally observed red shifted nature in absorption and PL spectra. This can be explained using Eq. (7) and Eq. (8), which shows that as the exciton binding energy increases, the transition energy reduces, and this leads to the red shift during the investigation of optical properties. The theory effectively explains the tendency to attain the antibonding states for biexcitons in a CSQD.

5. References:

[1] Ramya E, Rao V M, Rao N D 2019 Third-order nonlinear optical properties of CdSe/ZnS/CdSe core-shell-shell quantum dots Physica E **107** 24.

[2] Yano S, Goto T, Itoh T, Kasuya A 1997 Dynamics of excitons and biexcitons in CuCl nanocrystals embedded in NaCl at 2 K, *Phys. Rev. B* **55** 1667.

[3] Sen P, Chattopadhyay S, Andrews J T, Sen P K 2010 Impact of shell thickness on exciton and biexciton binding energies of a ZnSe/ZnS core-shell quantum dot *J. Phys. Chem. Solids* **71** 1201.

[4] Chattopadhyay S, Sen P, Andrews J T, Sen P K 2012 Effect of Shell and Shell Thickness on Photoluminescence (PL) of a CdSe/ZnS Core-Shell Quantum Dot J. of Phys: Conf. Series **365** 012037.

[5] Das Gupta P, Chattopadhyay S, Choudhary R J, Phase D M, Sen P 2011 Band offset in $Zn_{0.965}Cd_{0.035}O$/ZnO bilayer films *Mat. Lett.* **65** 2073.

[6] Dutta S, Das B, Electronic analog of the electro-optic modulator 1990 *Appl. Phys. Lett.* **56** 665.

[7] Chattopadhyay S, Sen P, Andrews J T, Sen 2014 P, Low temperature nano-spin filtering using a diluted magnetic semiconductor core-shell quantum dot *Physica E* **61** 198.

[8] Gibbs H M 1985 Optical Bistability: Controlling Light with Light, Academic Press, Orlando.

[9] Lad A, Mahamuni S 2008 Effect of ZnS shell formation on the confined energy levels of ZnSe quantum dots *Phys. Rev. B* **78** 125421.

[10] Talapin D V, Mekis I, Goltzinger S, Kornowski A, Benson O, Weller H 2004 CdSe/CdS/ZnS and CdSe/ZnSe/ZnS core-shell-shell Nanocrystals, *J. Phys. Chem. B* **108** 18826.

[11] Moon H. Lee C, Lee W, Kim J, Chae H 2019 Stability of Quantum Dots, Quantum Dot Films, and Quantum Dot Light - Emitting Diodes for Display Applications Adv. Mater. **31** (34) 1804294.

[12] Shen H, Gao Q, Zhang Y, Lin Y, Lin Q, Li Z, Chen L, Zeng Z, Li X, Jia Y, Wang S, Du Z, Song L, Zhang Z 2019 Visible quantum dot light-emitting diodes with simultaneous high brightness and efficiency, Nat. Photonics **13** 192.

[13] Madelung O 1996 *Introduction to Solid State Theory*, 3rd ed., vol. 2, Berlin.

[14] Shan W, Walukiewicz W, Ager J W, Yu K M, Wu J, Haller E E 2004 Pressure dependence of the fundamental band-gap energy of CdSe, *Appl. Phys. Lett.* **84** 67.

[15] Chadi D J 1994 Doping in ZnSe, ZnTe, MgSe and MgTe wide-band-gap semiconductors, *Phys. Rev. Lett.* **72** 534.

[16] Puls J, Rabe M, Wunsche H J, Henneberger F 1999 Magneto-optical study of the exciton fine structure in self-assembled CdSe quantum dots, Phys. Rev. B **60** R-16303.
[17] Sarkar D, Van der Meulen H P, Calleja J M, Becker J M, Haug R J, Pierz K 2006 Fine structure splitting and biexciton binding energy in single self-assembled InAs/AlAs quantum dots *Physica E* **32** 111.

Anodic oxidation of sodium formate and methanol on Pd and Pt electrodes in alkali: a comparative study

Kamal Kanti Bera, Malay Chakraborty and Swapan Kumar Bhattacharya*
Physical Chemistry Section, Department of Chemistry, Jadavpur University, Kolkata – 700032, India.
*Email – skbhatt7@yahoo.co.in Tel.: +919831699643, Fax: +913324146584

Abstract. Formate and methanol oxidation reaction (FOR and MOR)s are compared at room temperature on Pd and Pt electrodes in alkali. The onset potential and tafel slope for FOR on Pd are less by 0.203 V and 0.481 $Vdec^{-1}$ than these on Pt electrode. The peak current densities on Pd are 15.8 and 144 times greater than that on Pt in MOR and FOR respectively. The greater catalytic activity of Pd in FOR is supported by the greater formation of carbonate as obtained from IR spectroscopic and chromatographic investigation of products. The equilibrium exchange current density on Pd for FOR and MOR are 52.3 and 2.3 times of the corresponding data on Pt electrode. Despite methanol provides greater steady current than formate, the latter is better fuel particularly on Pd electrode in respect to formation of carbonate and power development.

1. Introduction

The use of formate as a fuel is an alternate to methanol in fuel cell reaction. Formate fuel cell produces an overall theoretical E^0_{cell} of 1.45 V which is 0.24 V greater than the alkaline methanol fuel cell [1] (DMFC) respectively. In addition, the formate salts have low toxicity, high solubility in water to form liquid fuel, food additive and air plane road de-icing agents [2], non-flammable property, renewable fuel. The anodic oxidation of formate and methanol in alkali is a two and six electron transfer reaction respectively on complete oxidation as given by

$HCOO^- + 3\,OH^- \rightarrow CO_3^{2-} + 2H_2O + 2e^-$ (1)

$CH_3OH + 8\,OH^- \rightarrow CO_3^{2-} + 6H_2O + 6e^-$ (2)

The differences in mechanisms for Pt and Pd anodes in both MOR and FOR are not completely established. Moreover, most of the works of such oxidation have been done on modified Pt and Pd electrodes [1,3-5], the study on pure Pt and Pd electrodes is very rare particularly as a comparative study, involving product analysis, order determination and mechanistic information, etc. of the reactions.

2. Experimental

2.1. Electrochemical measurement

All electrochemical measurements were carried out using AUTOLAB electrochemical workstation. The working electrodes were platinum (99.99%), palladium (99.97%) wires (Arora Matthey) with a geometrical area of 0.457 and 0.112 cm^2 respectively. A platinum foil (3.00 cm^2 area) and $Hg/HgO/OH^-$ (1M) electrode (MMO) were used as counter and reference electrodes respectively. Following our previous studies [1,3,5], FTIR study (Perkin Elmer, SN 74514 Spectrum RX1, Resolution 4 cm^{-1}) and Chromatographic determination of products of FOR and MOR was executed using a high performance liquid chromatography (HPLC) (Shimadzu Corporation, Japan).

3. Results and Discussion

3.1. Cyclic Voltammetry study

The steady cycle of formate oxidation on Pt and Pd electrode (Fig.1a) reveal that the observed peak-current which occurs mainly at lower potential, is much greater on Pd than that on Pt electrode and the adsorption of formate on Pd electrode is too large to form Pd-OH or PdO. Table 1 reveals that the peak current densities on Pd are 15.8 (22.23/1.41) and 144 time (10.57 / 0.073) greater than that of Pt in MOR and FOR respectively. The reduction of the onset potential of 200 mV (-0.6027 V + 0.8052 V) indicates significant improvement of catalytic activity of Pd electrode (than Pt) for direct formate fuel cells. Figs. 1b and 1c exhibit the forward peak current density of methanol (1.41 mA cm^{-2} at ca -0.175 V) is ~20 times higher than the anodic peak current density of formate (0.073 mA cm^{-2} at ca -0.073 V) for Pt electrode. It indicates that methanol is a better current providing fuel than formate on Pt electrode.

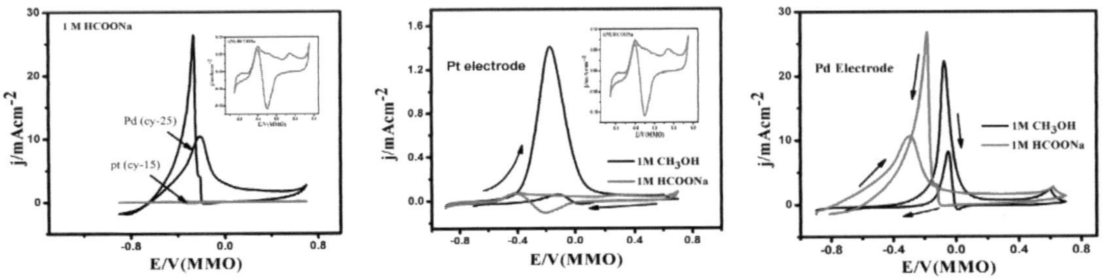

(a) (b) (c)

Figure 1. (a) Cyclic voltammograms of steady cycle for Pt and Pd electrodes in 1M HCOONa in 1M NaOH at scan rate 50 mV s^{-1}. Top inset represents steady cycle of HCOONa on Pt in a different scale. Comparative study of cyclic voltammograms of 1M HCOONa and 1M CH_3OH in 1M NaOH solution for (b) Pt and (c) Pd electrodes.

In the forward scan of potential, Pd electrode exhibits current density of 22.23 mA cm^{-2} at -0.073 V for MOR with which is only ~2 times higher than the corresponding peak of formate. Significant improvement of peak potential is found in selecting HCOONa as fuel instead of CH_3OH for both Pt and Pd electrodes. These are -0.224 V (-0.399 V + 0.175 V) and -0.222V (-0.295 V + 0.073 V) for Pt and Pd electrodes respectively. So, for the same change of fuel from methanol to formate, the changes in apparent power density ($\Delta(I_f \times E_f)$ in mW cm^{-2}) at the respective peak potentials are 1.50 (3.118-1.623) for Pd and -0.218 (0.029-0.247) for Pt.

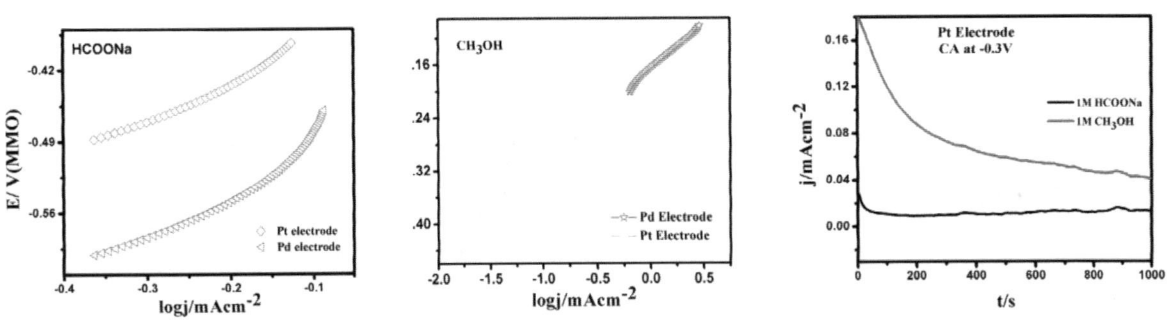

(a) (b) (c)

Figure 2. Tafel plots (*E* or V vs log of current density) for Pt and Pd electrodes in aqueous alkaline (1M) solution of 1M (a) sodium formate (b) methanol (**c**) Comparative studies of chronoamperometry of 1M HCOONa and 1M CH_3OH in 1M NaOH on Pt electrode at -0.3 V.

It means that power density increases 92% for Pd and decreases 88% for Pt for changing the fuel from methanol to formate. So, the choice of formate instead of methanol as fuel is beneficial in alkaline fuel cell particularly when Pd is used as anode.
Tafel plots were obtained using potentiodynamic measurement by application of 0.001 Vs^{-1} scan rate of Pt and Pd in alkaline methanol and formate (Figs. 2a and 2b) Equilibrium exchange current density i_0 is calculated [3] and found to be greater on Pd (1.151 mA cm^{-2}) than that on Pt (0.022 mA cm^{-2}) indicating that the former is a better electrode catalyst than the latter in the near equilibrium region.

3.2. Chronoamperometric studies

Fuel	Electrode	Onset potential (V)	Forward peak potential E_f (V)	Backward peak potential E_b (V)	Forward peak current density I_f (mA cm^{-2})	Backward peak current density I_b (mA cm^{-2})	Tafel slope (V dec^{-1})	Exchange current density I_0 (mA cm^{-2})
Methanol	Pt	-0.5028	-0.175	-0.117	1.41	0.058	0.061	1.16×10^{-9}
	Pd	-0.4326	-0.073	-0.047	22.23	8.816	0.057	2.63×10^{-9}
Formate	Pt	-0.6027	-0.399	-0.409	0.073	0.064	.9750	0.022
	Pd	-0.8052	-0.295	-0.181	10.57	26.96	0.4940	1.151

The analysis of Figs. 2c and 3a reveal that the steady current density for methanol oxidation is 3 times greater than that for oxidation of formate on Pt electrode but ~5 times on Pd electrodes. The steady current density on the Pd electrode (alike corresponding cyclic-voltammetric peak current density) is greater using methanol signifying relatively less blocking of electrode surface during MOR than FOR.

Table 1. Comparison of electrochemical- performances of formate and methanol as fuel on Pd and Pt electrodes immersed in alkali. Hg/HgO/OH-(MMO) is used as reference electrode.

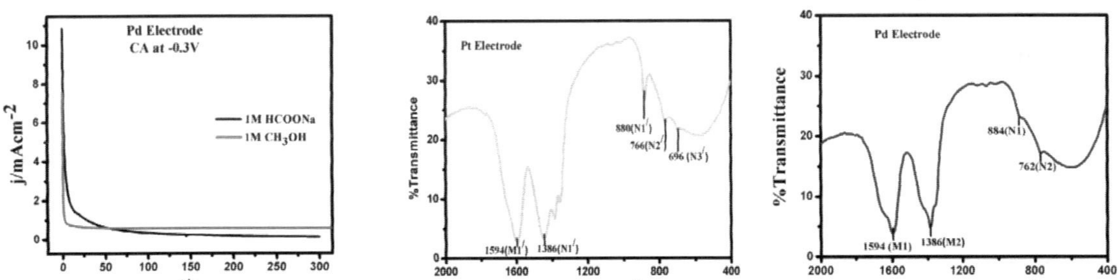

(a) (b) (c)

Figure 3. (a) Comparative studies of chronoamperometry of 1M HCOONa and 1M CH_3OH in 1M NaOH on Pd electrode at -0.3 V. FTIR spectra of the concentrated residue after electro-catalytic oxidation of 1M HCOONa in 1M NaOH using (b) Pt and (c) Pd anodes respectively.

3.3. FTIR study of the products

Figs. 3b and 3c indicate the presence of the bifurcated bands (M1 and M2) at 1594 and 1386 cm^{-1} due to C-O stretching for formate and weak peaks (N1 and N2) around 884 and 762 cm^{-1} for carbonate on analysis of solution of anode- chamber where FOR occurs on Pd electrode. It is observed that absorbance ratio corresponding to carbonate : formate (N1: M1 and N2: M2) are 0.587 and 0.462 for Pd electrode and similar ratios are 0.435 and 0.372 for Pt electrode respectively. These indicate that formation of carbonate occurs more on Pd electrode compare to that on Pt electrode.

3.4. Order with respect to formate in FOR

The order with respect to any species k is determined at various potential using the following equation:[5, 6]

$$\rho_k = \left(\frac{\mathrm{dlog}i}{\mathrm{dlog}C_k}\right)_{\Delta\phi, C_{OH^-}, T}$$

The order with respect to formate ion in FOR (Fig. 4a) is greater for Pd than Pt indicating adsorbed formate ion is more susceptible to oxidation on Pd surface. The decrease of order for Pd is due to the lack of surface-sites for formate and more adsorption of OH^-ions at higher potential. Above a certain potential, the increase of order for Pt is due to removal of poisonous intermediate at higher potential.

3.5. Chromatography Study

The HPLC profiles (Figs. 4b and 4c) exhibits retention time of 5.01 min, 6.62 min and 9.44 min correspond to the evolution of peaks and detection of pure formaldehyde, sodium carbonate and sodium formate solutions respectively [3]. The ratio of absorbances of the peaks of carbonates obtained as product in FOR using Pd and Pt electrodes is found to be 1.63 (12945 / 7905) indicating Pd is better electrode than Pt in respect to FOR. The a small peak of carbonate is observed for the solution of the products of MOR on Pt electrode reveals that methanol oxidation does not pass through formate as an intermediate on Pt electrode and Pd electrode is unable to oxidise methanol completely.

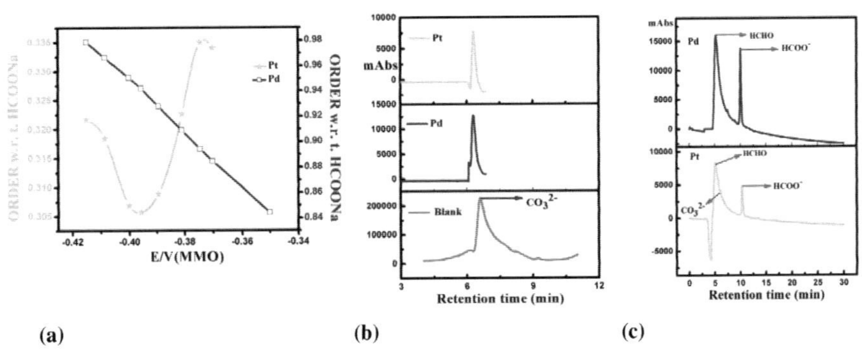

Figure 4. (a) Plots of order with respect to sodium formate, versus potential for oxidation of formate on Pd and Pt electrodes. (b) HPLC profiles of known concentration of sodium carbonate and the product of formate oxidation in NaOH on Pd and Pt electrode. (c) HPLC profiles of products of methanol oxidation in NaOH on Pd and Pt electrodes.

3.6. Mechanism of FOR and MOR

The oxidation of formate ion follows three paths on Pd and Pt electrodes as shown in Scheme 1.

Scheme 1

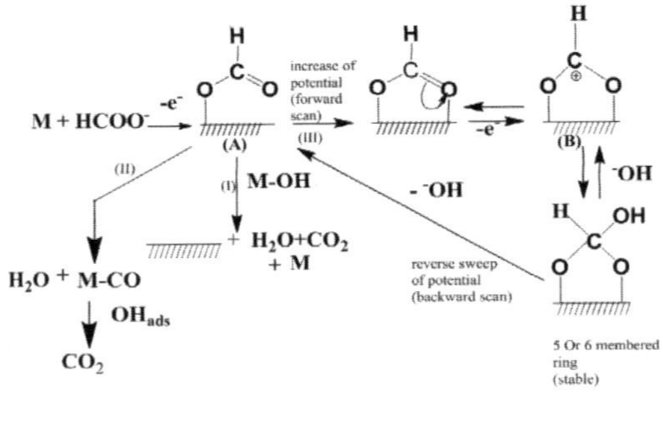

Scheme 1. Mechanism of Formate.

At higher potential path II is followed particularly for Pt and path III is for Pd. For methanol oxidation (scheme 2), two different paths: path (1) may be dominated at lower potential with the formation of adsorbed intermediate, (a). At higher potential MOR passes through the second path.

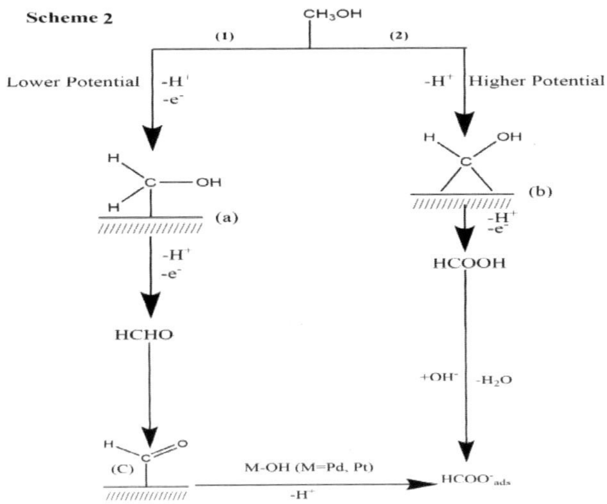

Scheme 2. Mechanism of methanol oxidation.

4. Conclusions

Pd is more beneficial than Pt as anode for oxidation of both formate and methanol in alkali. The equilibrium exchange current density increases 52.3 times and power density 107 times at peak potential of CV, for replacing Pt by Pd in FOR. The power density at the peak potential of CV is increased by ~6.57 times for replacing Pt by Pd as anode catalyst for oxidation of methanol. For MOR the ratio of exchange current density is 184 times higher with Pt compare to Pd. The experimental results conform to the probable mechanism of the reaction as presented by scheme 1. Formate is easily oxidised at lower potential but the reaction is retarded at higher potential due to blockage of the electrode surface. Methanol is not easily and not at all oxidised to CO_3^{2-} on the bulk Pt and Pd electrode because the product formate is not oxidised to carbonate at higher potential. Adsorption and oxidation of formate at lower potential is substantially high on Pd electrode than these on Pt in FOR.

Reference

[1] Roy Chowdhury S, Ghosh S and Bhattacharya S K 2017 Improved catalysis of green-synthesized Pd-Ag alloy-nanoparticles for anodic oxidation of methanol in alkali *Electrochim Acta* **225** 310-321

[2] Hellesten P P, Salminen J M, Jorgensen K S and Nysten T H 2005 Use of potassium formate in road winter deicing can reduce groundwater deterioration *Environ. Sci. Technol.* **39** 5095-5100

[3] Bera K K, Chakraborty M, Roy Chowdhury S, Ray A, Das S and Bhattacharya S K 2019 Significantly improved and synergistic effect of Pt–ZnO–Bi_2O_3 ternary hetero-junctions toward anode-catalytic oxidation of methanol in alkali *Electrochim Acta* **322** 134775-134791

[4] Gralee B and Lewera A 2016 Catalytic activity of unsupported Pd-Pt nanoalloys with low Pt content towards formic acid oxidation *Appl Cat B Environ* **192** 304-310

[5] Mahajan A, Banik S, Roy P S, Roy Chowdhury S and Bhattacharya S K 2017 Kinetic parameters of anodic oxidation of methanol in alkali: Effect of diameter of Pd nano-catalyst, composition of electrode and solution and mechanism of the reaction *Int J Hydrogen Energy* **42** 21263-21278

[6] Mukherjee P and Bhattacharya S K 2015 Anodic oxidation of butan-1-ol on Pd and Pt electrodes in alkaline medium *J Appl Electrochem* **44** 857-866

Quality Analysis of Tribal Casting Products by TOPSIS for Different Gating System

Soumyajit Roy[1], Akshay Kr. Pramanick[2] and Prasanta Kr. Datta[2]

[1] PhD Student, Met. And Mat. Engg. Department, Jadavpur University, Kolkata.

[2] Professor, Met. And Mat. Engg. Department, Jadavpur University, Kolkata.

e-mail: soumyajit.roy009@gmail.com

Abstract. Indian Tribal Products has high demand in the International market but the quality of the Tribal Products are not been analyzed by scientific technique, as no R&D is involved with the sector. A Lost-Wax Investment cast metal craft traditions were adopted for the project which has been in practice from the ancient age. An experiment was designed and done to understand the effect of gating design which controls the quality of the casting. The quality of castings was analyzed based on Eccentricity, Uniformity of Thickness, Surface Quality and Surface Roughness. The preferable gating system was chosen using TOPSIS.

1. Introduction

Indian sub-continent has a very sophisticated Non-Ferrous Investment Casting tradition from Cooperage [1] named “Dhokra” or “Dokra” which is cultured by the Tribal population of West Bengal, Orissa, Jharkhand and Chattishgarh. Bikna Shilpa Danga at Bankura District (23°15.3”N 87°5.9”E) and Dariapur, Guskara at Burdwan District (23°29’1”N 87°44’6”E) [2] are the most popular site in West Bengal.

The Lost Wax casting process is used in Dhokra Casting. They produce Complex-shaped items and most of the products are Hollow and Thin-walled (up to 0.4 mm thin). Thin-walled hollow products are very difficult to cast as liquid metal experienced high friction force due to its high viscosity[3], and also resistance is created due to capillary action caused by narrow mold cavity[4]. Hot Clay Mold has been used from an early age to overcome this problem [5].

Gating system controls the quality of a cast product [6] by controls the kinetics of liquid metal flow [7, 8]. Different Gating System was experimented adopted from the Tribal Casting Technique. The Eccentricity, Uniformity of Thickness, Surface Quality and Surface Roughness of all samples was analyzed and better gating system was determined using a very popular MCDM (Multi-criteria decision making) method- TOPSIS (Technique for Order of Preference by Similarity to Ideal Solution) [8].

2. Experimental Methodology

Wax Patterns of Vermillion Case and Cap of the same design and same dimension were made. The procedure of production is described pictorially (Fig.-1). Four Gating System was chosen for the experiment (Fig.-2) (i) Central Axial Top Gating System (TYPE-I), (ii) Peripheral Axial Top Gating System (TYPE-II), (iii) Peripheral Radial Top Gating System (TYPE-III), (iv) Peripheral Radial Bottom Gating System –(TYPE-IV). Few samples of each product were cast for each of the gating systems. Cast Product Quality was analyzed based on (i) Uniformity of Thickness, (ii) Eccentricity, (iii) Surface Quality and (iv) Roughness.

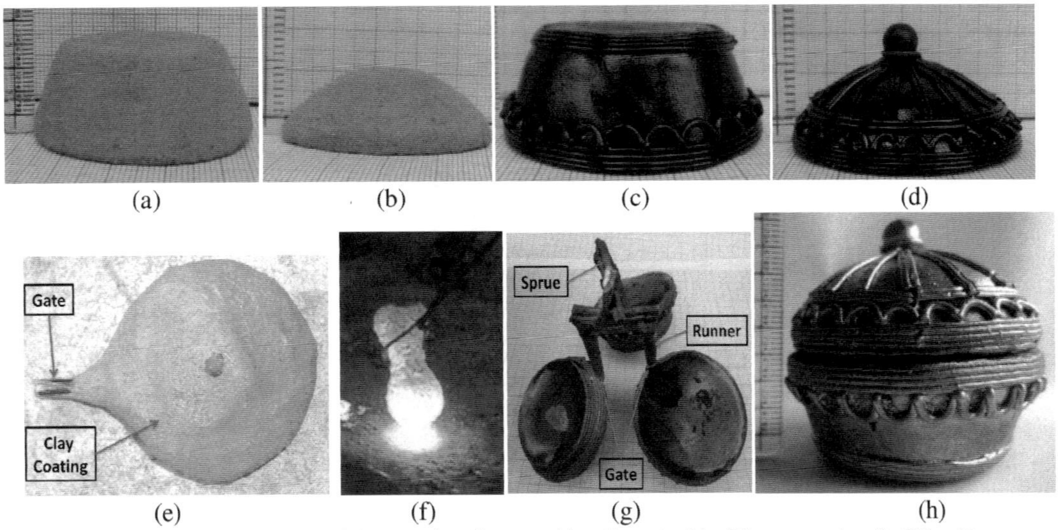

Figure 1. Production Steps of Cosmetics Case and its Cap: (a, b)- Clay core, (c, d)- Wax Pattern, (e) Clay Coating with Gating, (f) Fired mold- pouring of liquid metal also done, (g) Samples with Gating System (Details in Fig.-2), (h) Final product.

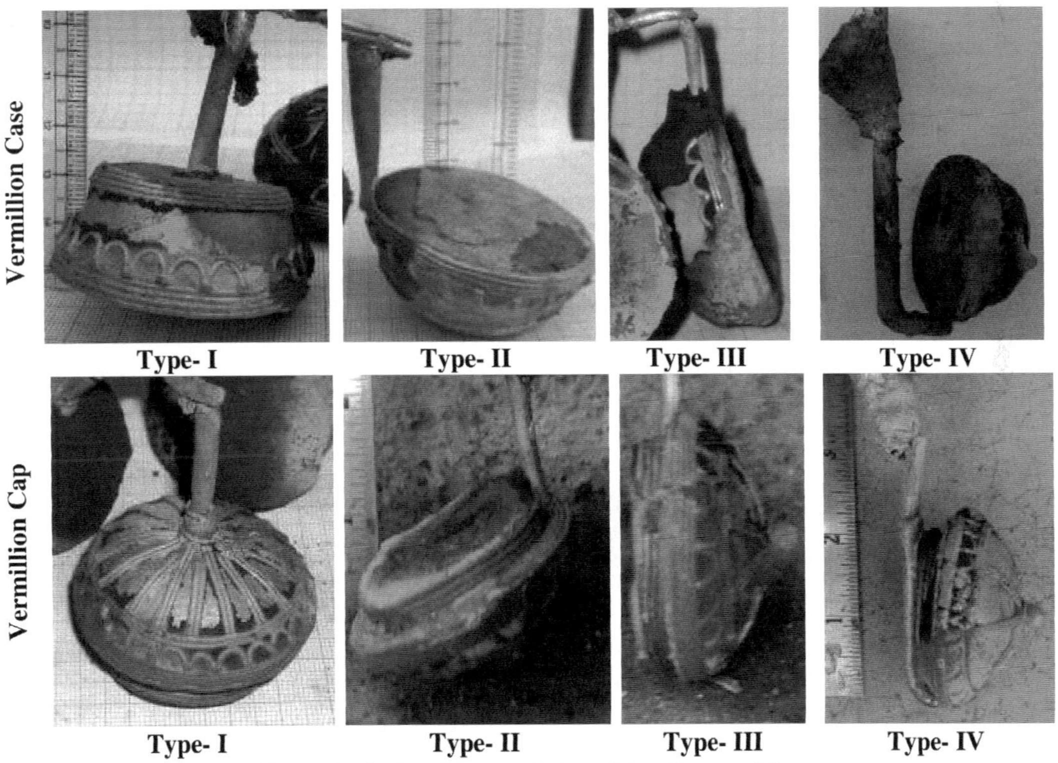

Figure 2. Gating System of Vermillion Case and Cap

3. Calculation Procedure of TOPSIS to determine the Rank of the Product

TOPSIS [9] (Technique for Order Preference by Similarity to Ideal Solution) is one of the popular and scientific numerical methods of the multi-criteria decision making (MCDM) which is helpful to measure the quality of products with non-commensurable parameters. Solution Method of a problem, using TOPSIS [10] is discussed below.

Step-1: Consider m alternatives A_1, A_2, ……., A_m. Each alternative A_i respects n criteria C_1, C_2, ………., C_n which are expressed with positive numbers C_{ij}. The criteria C_1, C_2, …… C_k are benefit (monotonically increasing preference), and criteria are non-benefit (monotonically decreasing preference). Weights w_j of the criteria C_j are given so that $\sum_{j=1}^{n} w_j = 1$.

The Most optimal alternative, criteria and its weights are placed in the table- 1.

Step-2: The pay off matrix must be balanced, as the numbers c_{ij} of different criteria having different measuring units. The given weights w_j of the criteria must be taken into account. First, the measuring numbers c_{ij} of the criteria c_j are replaced with the normalized or relative numbers (r_{ij}) are followed by: $r_{ij} = \frac{c_{ij}}{\sqrt{\sum_{i=1}^{m} x_{ij}^2}}$

The Normalized Payoff Matrix has been shown in Table- 1

Table-1: Payoff Matrix and Normalized Payoff Matrix [11]

		Payoff Matrix					Normalized Payoff Matrix			
Criteria		C_1	C_2	…	C_n		C_1	C_2		
Weights		w_1	w_2	…	w_n					
Alternatives	A_1	c_{11}	c_{12}	…	c_{1n}		r_{11}	r_{12}	….	r_{1n}
	A_2	c_{21}	c_{22}	……	c_{2n}		r_{21}	r_{22}	….	r_{2n}
	: :	: :	: :	: :	: :		: :	: :	: :	: :
	A_m	c_{m1}	c_{m2}	…	c_{mn}		r_{m1}	r_{m2}	….	r_{mn}

Step-3: The relative numbers (r_{ij}) belonging to the open interval <0, 1>. Then, according to the share w_ic_j of the criteria c_j, the normalized numbers (r_{ij}) are replaced with the weighted normalized numbers given by:

$$a_{ij} = w_j r_{ij} = w_j \frac{c_{ij}}{\sqrt{\sum_{i=1}^{m} x_{ij}^2}} \quad (1)$$

Determine ideal (f_i*) value (maximum) and negative ideal (f_i**) value (Minimum) for each criterion j. The Weighted Normalized Payoff Matrix, ideal (f_i*), negative ideal (f_i**) values are tabulated in Table-2.

Table-2: Weighted Normalized Payoff Matrix [11]

Criteria	C_1	C_2	……	C_{n-1}	C_n
A_1	a_{11}	a_{12}	……	……	a_{1n}
:	:	:	:	:	:
C_m	a_{m1}	a_{m2}	……	……	a_{mn}

Step-4: Parameters are required for the computation of separation measure are f_j^*, f_j^{**} value for each criterion j. Calculate separation measure D_a^+, D_a^-, in J- dimension Euclideam distance for each alternative.

(i) The separation measure of each alternative 'a' from the positive and negetive ideal solution is the distance of each criterion from its ideal value, and summing these for all criterion for given alternative 'a', i.e.,

$$D_a^+ = \sqrt{\sum_{j=1}^{j} \left(f_j(a) - f_j^*\right)^2} \quad (2) \qquad D_a^- = \sqrt{\sum_{j=1}^{j} \left(f_j(a) - f_j^{**}\right)^2} \quad (3)$$

(ii) Computation of relative closeness of each alternative 'a' with reference to negative ideal measure D_a^- is :

$$C_a = \frac{D_a^-}{D_a^+ + D_a^-} \quad (4)$$

Rank of the alternative base on the C_a value The Lower the C_a value, the better the alternative.

4. Results and Discussions

The Quality of Vermillion Case and Cap was tabulated in Table-3. The measurement was done by the following parameters:

- Uniformity of Thickness: Root Mean Square (RMS) technique was followed.
- Eccentricity: Inner diameter, Outer diameter of the case for both the lip and base was measured averaged for calculation. For the Cap, the only base diameter was measured.
- Surface Quality: Percentage of the good quality surface was measured.
- Surface Roughness: (Ra): Surface roughness [Arithmetic mean of departures (Ra)] of the Samples was measured with the sampling length (L_C) of 0.8 mm.

The Criterions taken to determine the Rank of the Products is as discussed above and also
(i) All the criteria chosen such a way, that lowest values are the best for each one,
(ii) Equal Weight (0.25) was chosen to determine the rank of the castings.

Based on the calculation of TOPSIS technique, the products have been ranked, with preferable and rejected gating system with explaination in Table-4. Preferable Gating System for different geometrical shape was given. The Rejected Gating System was mentioned with the reason and defects. Another important observation was found that "Bottom Gating" always be avoided for thin- walled investment casting production.

Table-3: Quality Analysis of Vermillion Case & Cap with Individual Ranking (R)

	Gating Type	Thickness Uniformity (RMS)	R	Eccentricity	R	Surface Quality (%)	R	Surface Roughness (Ra)	R
Case	Type-I	0.381	7	0.1255	5	0	1	1.16	2
	Type-II	0.378	6	0.1085	3	1	2	1.96	6
	Type-III	0.248	2	0.07	2	2	3	2.84	7
	Type-IV	0.276	3	0.2315	8	25	6	1.46	3
Cap	Type-I	0.198	1	0.1413	6	0	1	1.12	1
	Type-II	0.304	4	0.02995	1	3	4	1.48	4
	Type-III	0.336	5	0.1205	4	15	5	1.76	5
	Type-IV	0.579	8	0.16	7	45	7	2.11	8

Table-4: Preferable and Rejected Gating System with Ranking

Product	Gating Type	**Product Ranking**		**Preferable Gating System**	Rejection (%) Gating System with Defects
		Individual	Oveall		
Cosmetics Case	**I**	**1**	**4**	1. Type-I 2. Type-II 3. Type-III	**Type-IV:** (100% Rejected, Defects: Misrun)
	II	**2**	**3**		
	III	**3**	**5**		
	IV	**4**	**7**		
Cosmetics Cap	**I**	**2**	**2**	1. Type-II 2. Type-I	**Type-IV:** (100% Rejected; Defects: Misrun) **Type-III:** (40% Rejected; Defects: Misrun)
	II	**1**	**1**		
	III	**3**	**6**		
	IV	**4**	**8**		

5. Conclusions

The experiment proved that the quality of Indian Tribal Cast Products is better enough for mass production also. The experiment showed the following conclusion.

- The overall Surface Roughness irrespective of Gating System was better for Cosmetics Cap which is Hemispherical Geometrical Shape. The reason was the flow liquid metal was smoother than other geometrical shapes.
- The maximum unfilled section was found at opposite to the gate for the bottom gating system. Also, the dimension variation was large for the bottom gating system.
- Gating system should be chosen according to the shape, size and thickness of the product. Gating type and position should be chosen such a way that the liquid metal flow should be smoother and also the position of the gate was important for good quality economic casting production.

Acknowledgment

The author is thankful to the Department of Metallurgical and material engineering of Jadavpur University. Also special thanks to Mr Gopan Karmakar- Artisan of Bikna for his joyful co-operation.

References

[1] Mandal Barnali and Datta Prasanta Kumar, May 2020, Hot mold casting process of ancient East India and Bangladesh, *China Foundry*, Vol- 7.2, 171-177.

[2] Roy S et.al. August-2019 The Technology Inherent in Indian Traditional Craft Metal Casting Process, *Indian Foundry Journal*, Volume-65, Issue-8, 27-37.

[3] Poirier D R et. al., 1973, Transport Phenomena in Mat. Processing, Addison-Wesley, 13-22

[4] Stefanescu D M, 2002, Science and Engineering of Casting Solidification, Springer Sc., 65-66.

[5] Roy S et.al., 2017, Kinetics of liquid metal flow in gating design of investment casting production, *Slevarenstvi*, Vol- 5– 6, 149-154

[6] Masoumi M, et.al., 2005, Effect of Gating Design on Mold Filling, *AFS Transactions.*

[7] Kalpakjian S et. al., 2001, Manufacturing Engineering and technology, Pearson, 266-269.

[8] Flinn R A, 1963, Fundamental of Metal Casting, Addision-Wesley Pub. Co. inc., 11-96.

[9] Jyoti Vimal, Vedansh Chaturvedi, Ashutosh Kumar Dubey, Application Of Topsis Method For Supplier Selection In Manufacturing Industry, *IJREAS* Volume 2, Issue 5 (May 2012), 25-35.

[10] Pavić Z, et. al., 2013, Notes on TOPSIS Method, *Int. Journal of Research in Engineering and Science,* Vol 1 Issue 2, pp-05-12.

[11] Srinavasa K Raju, et. al., 2010, Multicriterion analysis in Engineering and management, PHI Learning Pvt. Ltd., New Delhi pp- 80-81.

IOP Conf. Series: Materials Science and Engineering 1080 (2021) 012015 doi:10.1088/1757-899X/1080/1/012015

Creating Edge Element from Four Node Quadrilateral Element

Durgarao Kamireddy[1] **and Arup Nandy**[2]

[1,2] Department of Mechanical Engineering, Indian Institute of Technology Guwahati, Guwahati, 781039, India.

e-mail: *durga176103010@iitg.ac.in, arupn@iitg.ac.in*

Abstract. When nodal based FEM is applied to vector field problems, some spurious eigen values appears. This can be avoided by adopting edge based FEM in the formulation. A separate conversion algorithm is used which takes the nodal information of quadrilateral element as input. Edge element is created from four node quadrilateral element to solve the 2D eigen value problem.

1. Introduction

Finite element method (FEM) has been extensively used in Electromagnetic application [1-3]. Nodal based FEM have following drawbacks [4-13]:

1. In Eigen value problems presence of spurious modes can not be avoided due to lack of proper continuity condition at material interfaces.
2. Nodal based elements can not capture singular eigen values due to sharp corners and edges.
3. Due to nodal continuity requirement, these elements cannot give direct solutions in terms of electric and magnetic field variables. We have to follow potential formulations.
4. Vector fields require special elements due to their tangential continuity and normal discontinuity across material interfaces.

These drawbacks are alleviated by adopting Edge element in FEM implementation. Whitney introduced these elements in the field of FEM in electromagnetics. These are constructed by curl and divergence conforming spaces called Whitney spaces. In this element [4-17] electric fields are along the edge of the element satisfying required tangential continuity.

2. Coversion of node to edge

In edge based FEM, element information are required to be supplied in edge data structure. In order to attain this, a separate conversion algorithm is required which uses the nodal data as input and gives the edge information as output. Each edge of the element is formed by connecting two nodes of the element. Here we have used a conversion algorithm which converts the nodal element data of a quadrilateral element into edge element data. Such algorithm is essential as most of the mesh generator create nodal element data structure.

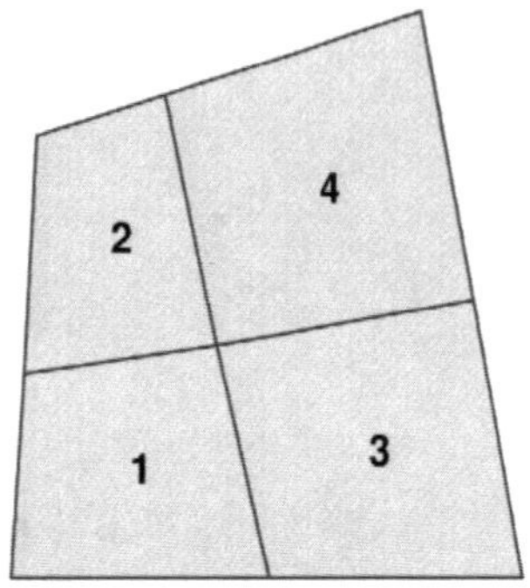

Figure 1. Global elements of 4-node quadrilateral element of domain

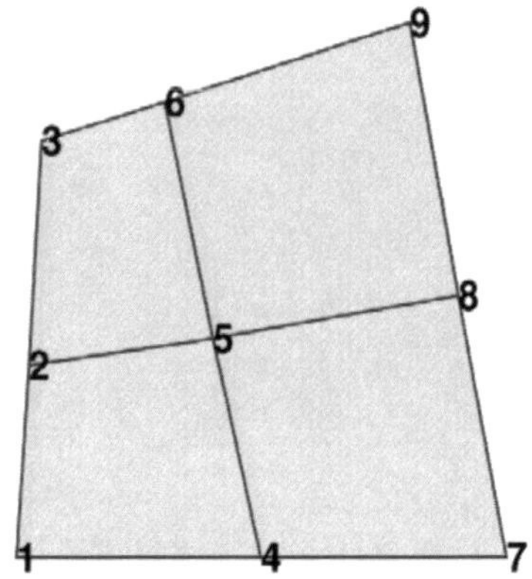

Figure 2. Global nodes of 4-node quadrilateral element of domain

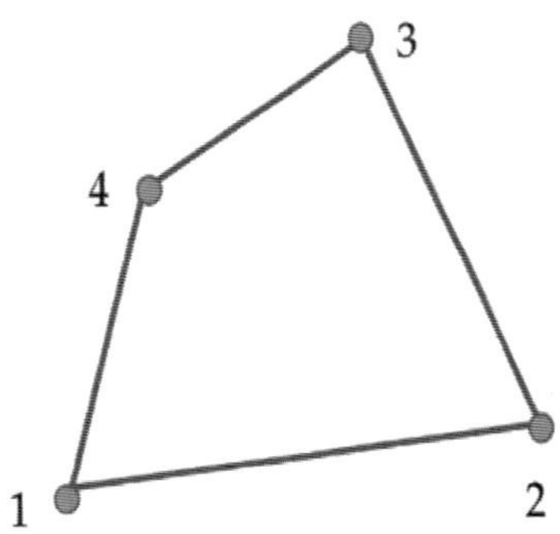

Figure 3. Local nodal connectivity of 4-node quadrilateral elements

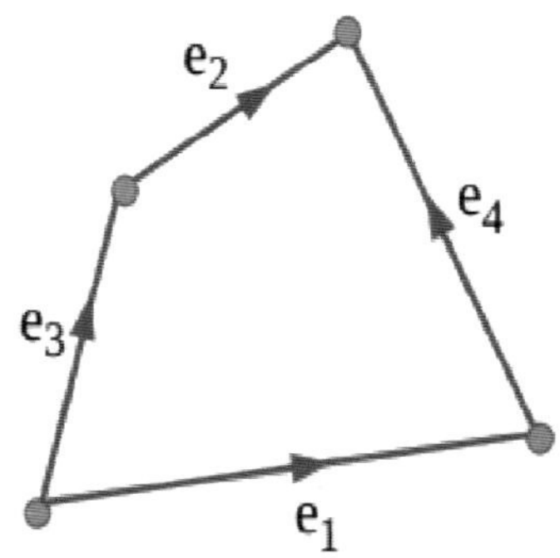

Figure 4. Local edge connectivity of 4-edge quadrilateral elements

In this conversion algorithm, a simple finite element quadrilateral mesh as shown in Fig. 1 is used. Global nodal connectivity of discretized domain is shown in Fig. 2. Fig. 3 shows the local nodal connectivity of quadrilateral element and Fig. 4 is the desired local edge connectivity of such element. Four edges $\mathbf{e}_1$, $\mathbf{e}_2$ $\mathbf{e}_3$ and $\mathbf{e}_4$ are formed by connecting local node sets **(1, 2)**, **(4, 3)**, **(1, 4)** and **(2, 3)** respectively. In this algorithm, the outer loop runs over the total number of discretized quadrilateral elements. The inner most loop runs over the total number of edges (*nlocedge*) of each element. Inside the local edge loop with the help of the local nodal connectivity as shown in Fig. 3 and local edge connectivity as shown in Fig. 4 two end nodes of the edge of quadrilateral element are returned.

Global Node	Total no. of connected edges (nodeedgenum)	Connecting edge and corresponding other node(nodeedge(1:8))							
		1st edge	Other node	2nd edge	Other node	3rd edge	Other node	4th edge	Other node
1	2	1	4	3	2	-	-	-	-
2	3	-3	1	2	5	6	3	-	-
3	2	5	6	-6	2	-	-	-	-
4	3	-1	1	4	5	8	7	-	-
5	4	-2	2	-4	4	7	6	9	8
6	3	-5	3	-7	5	11	9	-	-
7	2	-8	4	10	8	-	-	-	-
8	3	-9	5	-10	7	12	9	-	-
9	2	-11	6	-12	8	-	-	-	-

Table 1. Nodeedge array of nodes of the discretized domain.

2.1 Formation of edge connectivity array

The information of the two end nodes (starting and end nodes) of each local edge, last assigned global edge and existing edge connectivity array are supplied to another subroutine. Here in this subroutine checking is performed whether any edge is already existing between the current starting and end nodes with the help of *nodeedge* array. It stores the information of other edges connected with the current starting node in the odd columns and the corresponding end nodes in the even columns of each row. This array also have the direction information of the edge. Table. 1 shows the information about the *nodeedge* array for all the global nodes of the discretized domain with the quadrilateral element. If there is no current edge is found in between the two end nodes then existing global edge number is incremented to digit 1. This is assigned to the current edge and this information is updated in the existing edge connectivity array (*edgearr*). Table. 2 is the edge connectivity of quadrilateral elements of the discretized domain as shown in Fig. 1. Fig. 5 shows the edge connectivity of each element, arrow shows the direction of the edge pointed. For each global node total number of edges shared are counted by one global counter and this information is stored in *nodeedgenum* array. After successful updation of various arrays local edge loop runs for next local edge. This process repeats for all the edges of the element.

Element number	Edge connectivity (Global edge no.)
1	1, 2, 3, 4
2	2, 5, 6, 7
3	8, 9, 4, 10
4	9, 11, 7, 12

Table 2. Edge connectivity array of elements of the discretized domain.

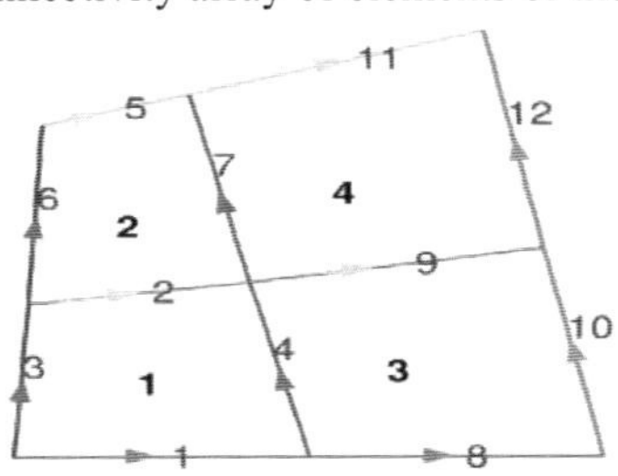

Figure 5. Edge connectivity of elements

3. Numerical Analysis

The governing differential equation under harmonic excitation can be given as [3]

$$\nabla \times \left(\frac{1}{\mu_r}\nabla \times \boldsymbol{E}\right) - k_0^2\epsilon_r \boldsymbol{E} = -i\omega\mu_0 \boldsymbol{j}, \quad (1)$$

where $i = \sqrt{-1}$, $k_0 = \omega^2/c^2$ is the wave number, ω is the excitation frequency, c is the speed of light in vacuum, $\boldsymbol{E}$ is the electric field, $\boldsymbol{j}$ is current density, $\mu_r = \mu/\mu_0$, $\epsilon_r = \epsilon/\epsilon_0$ are the relative permeability and relative permittivity, and ϵ_0 and μ_0 are the permittivity and permeability for vacuum. Assuming no external load condition, $\boldsymbol{j}$ is to be zero, the above equation reduces to

$$\nabla \times \left(\frac{1}{\mu_r}\nabla \times \boldsymbol{E}\right) - k_0^2\epsilon_r \boldsymbol{E} = 0. \quad (2)$$

This equation is used to solve the eigen value problems

To validate the edge element implemented, standard 2D eigen value problem in computational field as reported in [18] is considered. For the problem considered we have assumed $\epsilon_r = \mu_r = 1.0$. In order to find the square of the eigen values equation (2) is considered.

3.1 Square domain with perfectly conducting sides

Square domain having side length of π is considered. All the sides/boundaries of square domain are perfectly conducting. $\boldsymbol{E} \times \boldsymbol{n} = 0$ is considered as the boundary condition where $\boldsymbol{n}$ is the normal of the sides of the square. Here the domain is discretized with the 256 quadrilateral elements (Q4) to perform numerical analysis. In Table. 3 square of eigen values are presented. These values are compared with analytical results reported in [18] and and numerical results(nodal based) from [19]. Here, B9 is the conventional 9-node quadrilateral element. The perecentage of error with analytical results against the numerical results of edge element are calculated as shown in Table. 3. It is observed that error lies below 0.8%. Proposed element(mesh) results are showing good match with the analytical and numerical results of nodal elements. It also gave correct multiplicity of eigen values. For all the elements first non zero eigen value occurred after stating the number of zeros generated at machine precision level. These zeros signifies the approximation of null space.

Analytical (Benchmark)	Nodal element (B9)	Edge element (Q4)	% Error between analytical results and edge element results
1	1.000033	1.000803	0.0803
1	1.000033	1.000803	0.0803
2	2.000066	2.001607	0.0803
4	4.002049	4.012868	0.3217
4	4.002049	4.012868	0.3217
5	5.002081	5.013671	0.2734
5	5.002081	5.013671	0.2374
8	8.004097	8.025735	0.3217
9	9.022487	9.065245	0.7249
9	9.022487	9.065245	0.7249
10	10.022520	10.066048	0.6605
10	10.022520	10.066048	0.6605
13	13.024535	13.078112	0.6009
13	13.024535	13.078112	0.6009
Number of computed zeros			
-	961	220	-

Table 3: k_0^2 on the square domain for different elements.

4. Conclusion

Edge element based FEM will nullify the presence of spurious modes and singularities in modal analysis and harmonic analysis. In implementation of FEM information of the edges of the element are to be supplied. With the conversion algorithm 4-node quadrilateral element data is converted to 4-edge quadrilateral element data. Square domain of dimension π is discretized with the quadrilateral edge element to perform eigen analysis and is compared with the analytical values.

References

[1] P. P. Silvester and R. L. Ferrari 1996, *Finite elements for electrical engineers*, Cambridge: Third Edition, Cambridge University Press.
[2] P. P. Silvseter and G. Pelosi, eds. 1994, *Finite elements for wave electromagnetics*, IEEE Press.
[3] Jian-Miing Jin, 2018, *The finite element method in electromagnetics*, Third Edition, Wiley-IEEE Press.
[4] J C Nedelec 1980 Mixed finite elements in R3, *Numer.Math. 35*, 315–341.
[5] Alain Bossavit et Jean-Claude Verite 1982 A mixed FEM-BIEM method to solve 3-D Eddy-Current Problem, *IEEE Transactions of Magnetics* **18(2)**: 431–435.
[6] M.L.Barton, Z.J.Cendes 1987 New vector finite elements for three-dimensional magnetic field computation, *Journal of Applied Physics* **61**, 3919.
[7] Alain Bossavit 1988 A rationale For "Edge-Elements" in 3-D fields computations, *IEEE Transactions of Magnetics* **24(1)**:74-79.
[8] A. Bossavit and I. Hiayergoyz 1989 Edge elements for scattering problems, *IEEE Transactions of Magnetics* **25(4)**: 2816–2821.
[9] Jin-Fa Lee, Din-Kow Sun, Zoltan J. Cendes 1991 Full-wave analysis of dielectric waveguides using tangential vector finite elements, *IEEE Transactions of Microwave Theory and Techniques* **39(8)**: 1262–1271.
[10] Jin-Fa Lee 1991 Tangential vector finite elements for electromagnetic field computation, *IEEE Transactions of Magnetics* **27(5)**: 4032–4035.
[11] A. Chatterjee, J. M. Jin and J. L. Volakis 1992 Computation of cavity resonances using edge-based finite elements, *IEEE Transactions of Microwave Theory and Techniques* **40(11)**: 2106–2108.
[12] J. P. Webb 1993 Edge elements and what they can do for you, *IEEE Transactions of Magnetics* **29(2)**:1460–1465.
[13] A. Nandy and C. S. Jog 2016 An amplitude finite element formulation for electromagnetic radiation and scattering, *Computer and Mathematics with Application*; **71**: 1364-1391.
[14] T. V. Yioultsis and T. D. Tsiboukis 1996 Multiparametric vector finite elements: a systematic approach to the construction of three-dimensional higher order tangential vector shape functions, *IEEE Transactions of Magnetics* **32(3)**: 1389–1392.
[15] Roberto D. Graglia, Donald R. Wilton, Andrew F. Peterson 1997 Higher order interpolatory vector bases for computational electromagnetics, *IEEE Transactions of Antennas and Propagation* **45(3)**: 329–342.
[16] J. P. Webb 1999 Hierarchal vector basis functions of arbitrary order for triangular and tetrahedral finite elements, *IEEE Transactions of Antennas and Propagation* **45(8)**: 1244–1253.
[17] Seung-Cheol Lee, Jin-Fa Lee, Robert Lee 1999 Hierarchical vector finite elements for analyzing waveguiding structures, *IEEE Transactions of Microwave Theory and Techniques* **51(8)**: 1897–1905.
[18] D. Boffi, M. Farina, L. Gastaldi 2001 On the approximation of Maxwell's eigen problem in general 2D domains, *Comput. Struct.* **79**:1089-1096.
[19] C. S. Jog and Arup Nandy 2014 Mixed finite elements for electromagnetic analysis, *Computers and Mathematics with Applications* **68**:887-902.

IOP Conf. Series: Materials Science and Engineering 1080 (2021) 012016 doi:10.1088/1757-899X/1080/1/012016

Design and kinematics of a 3-DOF compliant triglide micromanipulator

Suraj Kumar Mishra[1] **and C S Kumar**[2]

[1]Research Scholar, Robotics and Intelligent Systems Laboratory, Mechanical Engineering Department, Indian Institute of Technology Kharagpur, Kharagpur, India
[2]Professor, Robotics and Intelligent Systems Laboratory, Mechanical Engineering Department, Indian Institute of Technology Kharagpur, Kharagpur, India

E-mail: [1]surajkmishra@iitkgp.ac.in

Abstract. This paper proposes a novel 3-degrees of freedom (DOF) compliant parallel micromanipulator. The design of the mechanism is based on triglide parallel manipulator. The compliant version of the triglide is prepared by making use of flexure joints. Displacement amplifiers are incorporated in the design to increase range of motions. The pseudo-rigid-body modeling technique is then used to formulate and solve the inverse kinematic problem for the design. To check the correctness and accuracy of the analytical kinematic model FEA simulations are performed in Abaqus-6.14 environment. Simulation results reveal that the proposed micromanipulator can perform different motions very effectively. Additionally, it is found that results predicted by the analytical model conform very closely to the simulation outcomes. The errors are very low, and the motions involve negligible parasitic motions.

1. Introduction

A micromanipulator is a device that usually operates under a microscopic field with the objective of manipulation of tools precisely [1]. Based on application they are also referred to as a micro-positioning stage. Micromanipulators find applications in a variety of fields. Biological cell manipulation, optical alignment, micro-surgery, micromachining, scanning electron microscopy, and micro-assembly are some of the applications.

A mechanism that gets its mobility from the deformation of some of its members is called a compliant mechanism [2]. Due to some of their inherent characteristics, compliant mechanisms are one of the most popular solutions for micromanipulation. Compliant mechanisms with the parallel structure are called compliant parallel micromanipulators or CPMs.

Most of the micromanipulation activities require 3-DOF; hence the demand for 3-DOF CPMs is highest among all. In their paper Koseki *et al.* [3] presented a translational 3-DOF micromanipulator. The authors used matrix method for kinematic modeling of the mechanism. Stiffness modeling of a 3-DOF flexure parallel mechanism was discussed by Pham and Chen [4]. A novel design of a 3-DOF translational CPM was proposed by Li and Wu [5]. The authors used pseudo-rigid-body modeling for the analysis of the mechanism. For surface grinding, Zhang *et al.* developed a 3-DOF micro-positioning stage. The design has a high stiffness to give better performance at high frequencies. Some other works in the relevant area can be found in [6], [7], and [8]. Piezo-actuators are generally considered as the ideal choice for the actuation of CPMs. However, they suffer from the drawback of low range. In this paper, a novel design of CPM is proposed. To increase the range of motions, the proposed design incorporates displacement amplifiers. In one of their earlier works [9], the authors successfully employed a similar amplifier to increase the range of motion of their 6-DOF compliant Stewart micromanipulator. In each of the three displacement amplifiers, there are two similar parallelograms mirroring each other. This arrangement gives the mechanism more stiffness, and hence better dynamic performance. The double parallelogram structure also eliminates unwanted parasitic input motions.

The organization of the rest of the paper is as follows. Section 2 presents the design of the proposed micromanipulator. Kinematic modeling of the mechanism is formulated in Section 3. In Section 4, the results of the kinematic analysis are summarized. The FEA validation of the results is also discussed in this section. Finally, concluding remarks are outlined in Section 5.

(a) Revolute flexure or compliant revolute joint (b) Compliant spherical joint

Figure 1: Compliant flexure joints.

2. Design of compliant triglide micromanipulator

The design of the micromanipulator is based on the 3-DOF triglide parallel manipulator. A triglide is a spatial parallel manipulator and is capable of producing 3-DOF xyz motions. For the preparation of the compliant version of the triglide, flexure joints are used. The current design involves use of revolute and spherical flexure joints (see Fig. 1). Figure 2 shows the final

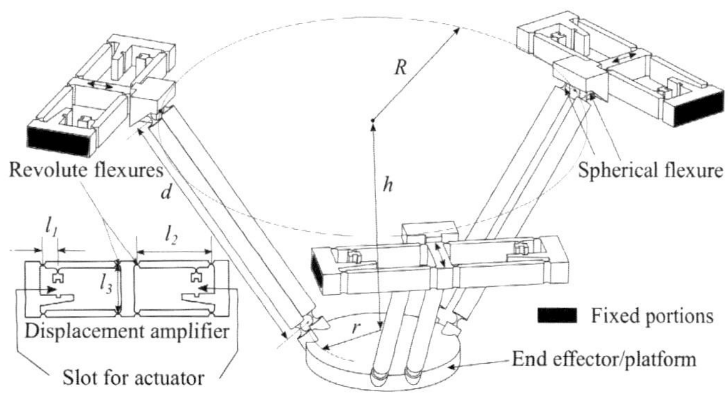

Figure 2: CAD model of the compliant triglide.

compliant version of the triglide. In the design, three displacement amplifiers are introduced to improve the range of motion of the mechanism. The displacement amplifier consists of two mirroring parallelograms made with revolute flexures. Each parallelogram receives input from an actuator, and then works as a lever to amplify the input. The output of the displacement amplifier connects to the platform by a passive parallelogram made by four spherical joints and two cylindrical links as represented in Fig. 2.

3. Kinematics

Pseudo-rigid-body (PRB) modeling is a technique in which the flexible members or flexure joints of a compliant mechanism are replaced by suitable conventional kinematic joints. The resulting mechanism can then be analyzed through already available methods of robot kinematics or theory of mechanisms.

3.1. Inverse kinematics

The inverse kinematics problem for a mechanism concerns with finding the inputs for a desired position of the end effector. For the present case, the inverse kinematics can be solved by considering the loop-closure equation for any of the three limbs. With reference to Fig. 3, the loop-closure equation can be written as

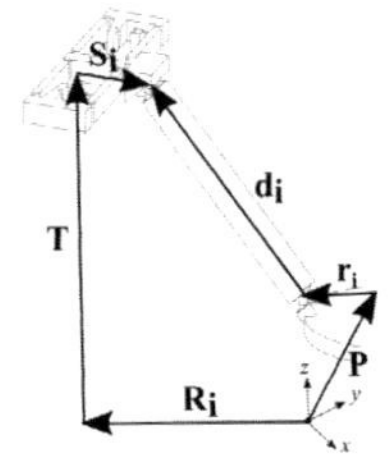

Figure 3: Loop closure for one limb of the micromanipulator.

$$\mathbf{P} + \mathbf{r_i} + \mathbf{d_i} = \mathbf{R_i} + \mathbf{T} + \mathbf{S_i} \tag{1}$$

The previous equation can be written in the matrix form as

$$\begin{bmatrix} x \\ y \\ z \end{bmatrix} + \begin{bmatrix} r\,cos(\alpha_i) \\ r\,sin(\alpha_i) \\ 0 \end{bmatrix} + \begin{bmatrix} d_x \\ d_y \\ d_z \end{bmatrix} = \begin{bmatrix} R\,cos(\alpha_i) \\ R\,sin(\alpha_i) \\ 0 \end{bmatrix} + \begin{bmatrix} 0 \\ 0 \\ h \end{bmatrix} + \begin{bmatrix} -S_i\,cos(\alpha_i) \\ -S_i\,sin(\alpha_i) \\ 0 \end{bmatrix} \tag{2}$$

$$\begin{bmatrix} d_x \\ d_y \\ d_z \end{bmatrix} = \begin{bmatrix} cos(\alpha_i)\,(R - r - S_i) - x \\ sin(\alpha_i)\,(R - r - S_i) - y \\ h - z \end{bmatrix} \tag{3}$$

$$d_x^2 + d_y^2 + d_z^2 = (cos(\alpha_i)\,(R - r - S_i) - x)^2 + (sin(\alpha_i)\,(R - r - S_i) - y)^2 + (h - z)^2 \tag{4}$$

This finally yields the following quadratic equation in S_i.

$$S_i^2 + 2(x\,cos(\alpha_i) + y\,sin(\alpha_i) - K)\,S_i + x^2 + y^2 + (z-h)^2 + K^2 - d^2 - 2K(x\,cos(\alpha_i) + y\,sin(\alpha_i)) = 0 \tag{5}$$

where, $R - r = K$ and $d_x^2 + d_y^2 + d_z^2 = d^2$. Out of the two results yielded by the Equation 5, one corresponds to the case wherein the actuations would be in the order of the size of the mechanism. Due to limits in joint motions, this magnitude of actuation is not permitted in the flexure-based micromanipulators. The other feasible solution of this equation is presented below.

$$S_i = K - x\,cos(\alpha_i) - y\,sin(\alpha_i) - \sqrt{d^2 - h^2 - x^2\,sin^2(\alpha_i) - y^2\,cos^2(\alpha_i) - z^2 + xy\,sin(2\alpha_i) + 2zh} \tag{6}$$

This equation provides the actuations needed for any position reached by the end effector.

Table 1: Values of major structural parameters of the design (in mm).

b_r	d	h	l_1	l_2	l_3	r	R	r_r	r_s	t_r	t_s
30	351	270	25	125	80	113	337	5	8	1	2

4. Results and FEA validation

The analytical model for inverse kinematics was formulated in the previous section. In this section, the results from the analytical model are obtained, and then they are validated with the help of FEA simulations. With reference to Fig. 1 and Fig. 2, the values of the major structural parameters of the micromanipulator are presented in Table 1.

4.1. Results from analytical inverse kinematics model

From Eq. 6, one can obtain the actuation values for a desired position of the end effector or platform. Table 2 presents a compilation of actuations needed for four different types of movements of the platform.

Table 2: Inverse kinematics results obtained from analytical model.

Output motion	Required actuation (in μm)		
(in μm)	S_1	S_2	S_3
40, 0, 0	-4.000	8.000	-4.000
0, 40, 0	-6.928	0.001	6.928
0, 0, 40	-9.598	-9.598	-9.598
19, 19, 19	-9.751	-0.759	-3.168

4.2. Validation of the results using FEA simulations

In this section, the actuations obtained from the analytical model will be given to the FEA model of the micromanipulator. Furthermore, it will be checked whether the mechanism performs the expected motions accurately or not. The FEA model of the compliant triglide is prepared in the Abaqus-6.14 environment. For meshing C3D10 element is used. Moreover, as the deformations are expected to be concentrated mainly at flexure joint areas, they are provided with a fine mesh. Table 3 summarizes a comparison between the expected analytical motions and actual motions performed by the FEA model. Deformations contours for the FEA results are shown in Fig. 4. The results presented in Table 3 can be interpreted as follows. The maximum error in the exploration was found to be less than 7%. This error can be attributed to the out of axis deformation of the flexures, and deformation of the connecting links which were assumed to rigid in the kinematic model. On the other hand, the undesired parasitic motions for the mechanism were found to be below 0.18%. This shows that the output of the proposed design involves negligible parasitic motion.

Table 3: Comparison of results of analytical model and FEA simulations.

Expected motion as per analytical model (in μm)	Motion performed by FEA model (in μm)	Error/Parasitic motion (%)
40, 0, 0	37.844, 0.037, -0.024	5.390, 0.098, -0.063
0, 40, 0	0.065, 37.803, 0.040	0.172, 5.492, 0.105
0, 0, 40	0.128, -0.064, 37.394	0.342, -0.171, 6.515
19, 19, 19	17.997, 18.021, 18.034	5.279, 5.153, 5.084

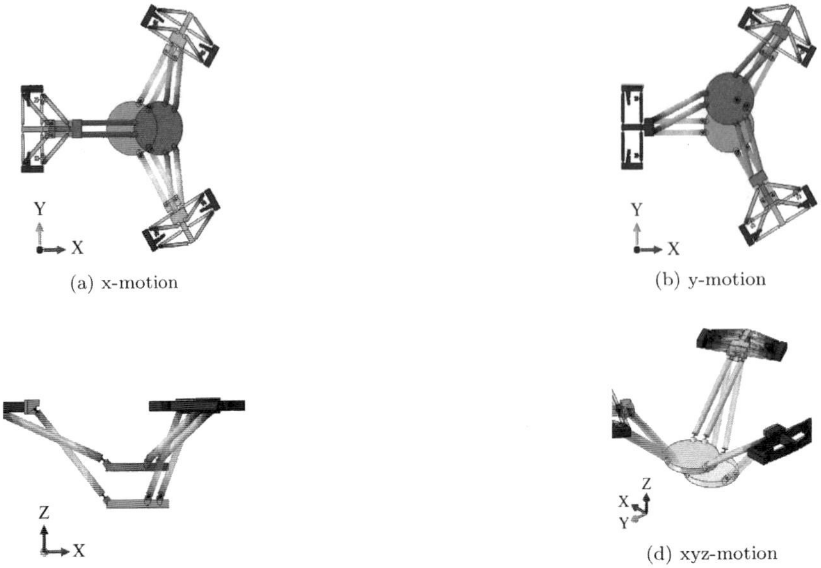

(a) x-motion (b) y-motion (c) z-motion (d) xyz-motion

Figure 4: Deformation contours representing the motions performed by FEA model of triglide CPM (deformations have been scaled for the sake of visibility).

5. Conclusions

A novel design of a compliant 3-DOF micromanipulator was presented in this paper. The compliant version of the triglide parallel manipulator was successfully prepared by making use of flexure joints. It was shown that the pseudo-rigid-body modeling could be very effectively employed to formulate the inverse kinematics model of the proposed design. The results indicated that the errors were within the acceptable limit, and the micromanipulator performed the desired motions with negligible parasitic motions.

The effect of payload and gravity was not considered in the present work. A stiffness model of the micromanipulator would give insights on the effect of payload on end-effector motion. Moreover, the proposed design in this paper is an initial version of the micromanipulator. To find the dimensions for optimal performance more elaborate analysis and testing are needed.

References

[1] El-Badry H M 1963 *Micromanipulators and Micromanipulation* (New York: Springer Science & Business Media)

[2] Howell L L 2001 *Compliant Mechanisms* (New York: John Wiley & Sons)

[3] Koseki Y, Tanikawa T, Koyachi N and Arai T 2002 Kinematic analysis of a translational 3-dof micro-parallel mechanism using the matrix method *Ad. Rob.* **16** 3:251-264

[4] Pham H H and Chen I M 2005 Stiffness modeling of flexure parallel mechanism *Pre. Eng.* **29** 4:467-478

[5] Li Y and Wu Z 2016 Design, analysis and simulation of a novel 3-dof translational micromanipulator based on the PRB model *Mech. and Mach. Th.* **100** 235-258

[6] Yi B J, Chung G B, Na H Y, Kim W K and Suh I H 2003 Design and experiment of a 3-dof parallel micromechanism utilizing flexure hinges *IEEE Tr. on Rob. and Aut.* **19** 4:604-612

[7] Jensen K A, Lusk C P and Howell L L 2006 An xyz micromanipulator with three translational degrees of freedom *Robotica* **24** 3:305-314

[8] Lu T F, Handley D C, Yong Y K and Eales C 2004 A three-dof compliant micromotion stage with flexure hinges *Ind. Rob.: An Int. J.* **31** 4:355-361

[9] Mishra S K and Kumar C S 2018 Design and kinematics of a compliant stewart micromanipulator *Int. Conf. on Manipulation, Automation and Robotics at Small Scales* (Nagoya: IEEE) pp 1-6

Flattening analysis of plastically graded cylindrical contact under plane stress condition

Tamonash Jana, Anirban Mitra and Prasanta Sahoo

Department of Mechanical Engineering, Jadavpur University, Kolkata-700032, India

e-mail: *tamonash.jgec@gmail.com*

Abstract. The present work deals with finite element based flattening analysis of a functionally graded cylindrical contact against a rigid flat surface under plane stress condition. The yield strength of the semi-cylinder is varied radially according to an exponential function. The contact between the semi-cylinder and the rigid flat is considered to be perfect-slip type in nature. To reduce the computational time, a 2D quarter circle is used to model the cylinder and rigid flat surface is modelled using a straight line. The effect of the gradation parameter on contact behaviours e.g. contact area, contact force, contact pressure etc. of the semi-cylinder are presented and discussed.

1. Introduction

Spherical and cylindrical contact interfaces are the most basic and fundamental unit of rough surface asperity contact models. Cylindrical contact models could be useful to analyze wide range of contact problems ranging from micro scale rough surface asperity interactions to macro level surface interactions, e.g. wheel-on-rail, roller bearings etc. Additionally cylindrical contact model can be useful to analyze the contact interactions of stretched asperities which are commonly found in anisotropic and fibre-structured material surfaces [1]. Hertz first analyzed the elastic deformation of parabolic surfaces in contact [2]. This analysis was later extended by Johnson [3] to get the results of cylindrical elastic contact. Afterwards, elastic-plastic cylindrical contact analysis has been carried out by number of researchers.

When it comes to surface materials, functionally graded materials (FGM) are found to be superior as compared with isotropic or layered materials. Functionally graded surface materials can efficiently shield contact surface from damage and deformation [4]. Plastically graded material (PGM) is a special type of FGM which has its material property graded in the form of yield strength. If incorporated appropriately, plastically graded materials suppress the onset of cracking, yielding by altering stress distribution at the contact interface and in its vicinity. PGMs can be fabricated in several ways including controlling microstructural size distribution, directional composition variation, surface treatments etc. Due to the major damage resistance potential of PGM, it had been a significant domain of interest for a number of researchers [5-8].

Single-asperity contact analysis can be broadly subdivided into two types, viz. flattening and indentation contact analysis [9]. Most of the research works involving single asperity functionally graded cylindrical contact deals with indentation of functionally graded substrate with rigid cylindrical indenter. The current work involves flattening contact analysis of radially plastically graded semi-cylinder under plane stress condition.

2. Finite element modelling

A finite element based contact analysis is performed using finite element simulation tool ANSYS to investigate the flattening contact interaction between a plastically graded semi-cylinder and a rigid flat surface. In order to bring down the finite element model size, a quarter circular area is used to model half of the semí-cylinder due to its symmetry w.r.t. the vertical axis [figure 1(a)]. The longitudinal axis of the semi-cylinder is assumed to be perpendicular to the cross-sectional plane shown in figure 1(a). The flat surface, which initially just touches the summit of the semi-cylinder is modelled by a straight líne in a similar fashion. The nature of the contact between the two is assumed to be completely frictionless. 2D PLANE223 element, capable of coupled-field analysis is adopted to mesh the quarter circular area. The arc of the quarter circular area is meshed with CONTA172 elements which come in contact with rigid flat meshed with TARGE169 elements. The near-tip zone, where the semi-cylinder comes in contact is finely meshed to capture the results with accuracy. However, to optimize the solution time and accuracy, the mesh density is graded in such way that, the mesh becomes coarser towards the base of the semi-cylinder [figure 1(b)].

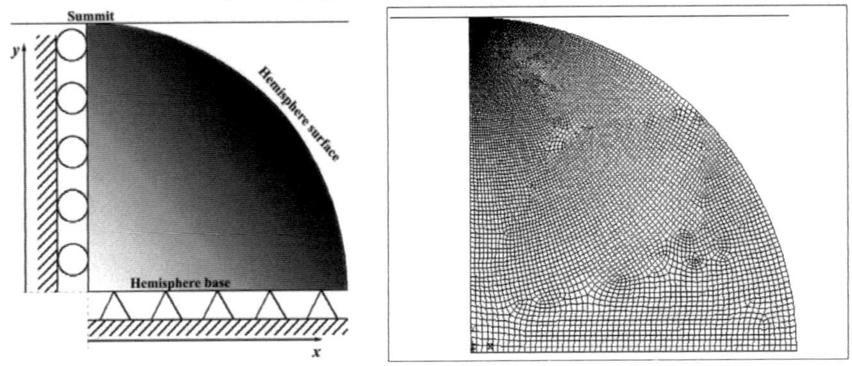

(a) (b)

Figure 1. (a) Representative finite element model and (b) the mesh configuration.

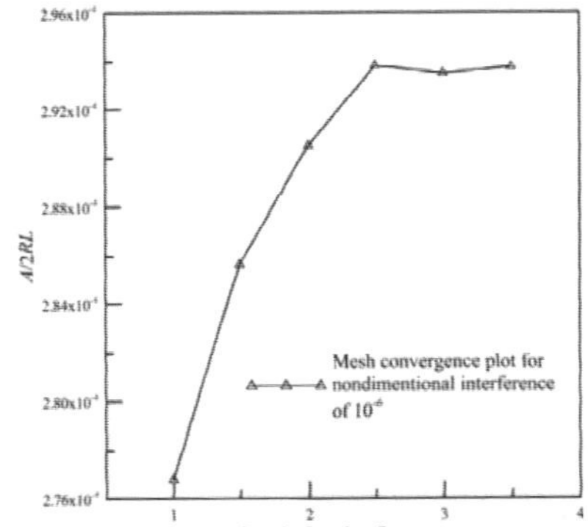

Figure 2. Mesh convergence plot ín terms of nondimensional contact area corresponding to nondimensional interference of 10^{-6}.

A proper convergence study is carried out to determine the optimum mesh configuration for the desired level of accuracy. Figure 2 presents the mesh convergence plot in terms of non-dimensional contact area (*A*/*2RL*) against level of mesh refinement level (*x*-axis of the plot). The mesh refinement level is explained as, if at level 1, an unmeshed line is divided into *n* number of divisions; there would be $n \times N$ number of divisions at level N. The final mesh configuration consists of 13524 PLANE223, 284 CONTA172 and 2 TARGE169 elements.

The semi-cylinder is constrained in such way that, every node on the vertical axis can move in vertical direction only and the nodes located at the base are set fixed. The material is assumed to follow bi-linear isotropic hardening model in the post-yield regime. Yielding of material is identified using the Von Mises criterion. The gradation is implemented along the radial direction in the form of yield strength (Y). Young's modulus (E), Poisson's ratio (ν) and tangent modulus (E_t) are kept invariant for the entire semi-cylinder. The stress-strain plot of the adopted PGM model is presented in figure 3(a). The gradation in yield strength is realized by assigning temperature to all the nodes of the semi-cylinder as per the desired gradation function of the coordinate points and subsequently by defining temperature dependent yield strength accordingly. Exponential function is one of the most common gradation functions in for FGM surfaces [10, 11]. For present analysis, the yield strength is graded according to the following exponential function

$$Y = Y_0 e^{\gamma_p (R-r)} \tag{1}$$

Here, Y is the yield strength and '0' subscript denotes the same at the outer surface. γ_p, R and r and are plastic inhomogeneity parameter, radius of the semi-cylinder and radial distance of a point from the central axis of the semi-cylinder, respectively. It can be comprehended from equation (1) that, for varying γ_p values, the yield strength (Y) at the outer surface remains invariant. The yield strength increases towards the core of the semi-cylinder for positive value γ_p and decreases for negative value. The material is isotropic, if γ_p is zero.

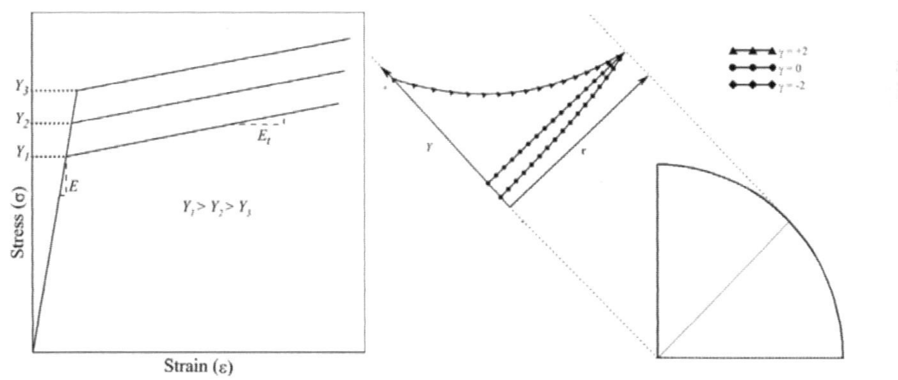

(a) (b)

Figure 3. (a) Adopted stress-strain diagram of the PGM model and (b) representative diagram of the yield strength gradation.

Yield strength of the hemisphere is varied with Y_0 as 250 MPa. Young's modulus (E) and Poisson's ratio (ν) for the entire semi-cylinder is set as 200 GPa and 0.3, respectively. The E_t/E ratio is set as 0.04 anywhere in the semi-cylinder. For the analysis, the length (L) and radius (R) of the semi-cylinder are adopted as 1 μm. Downward vertical displacement is imparted to the flat surface in steps. For each step of rigid flat surface displacement or interference, of the cylindrical contact, contact force, stresses etc. are recorded.

3. Results and Discussions

The effect of γ_p on the nondimensional contact force ($P/2ERL$), contact area ($A/2RL$) and average contact pressure (P/AY_0), w.r.t. the nondimensional interference (ω/R) are illustrated in figure 4. The figure shows that for a constant interference value, higher γ_p results in higher contact force, contact area and average contact pressure. Since higher yield strength causes less yielding of the material, semi-cylinder with higher γ_p provides more resistance against deformation, hence resulting in higher contact force and contact pressure. On the other hand, for far into elastic-plastic domain, yielding of the material located near the contact interface becomes the primary influential factor on the trend of contact area.

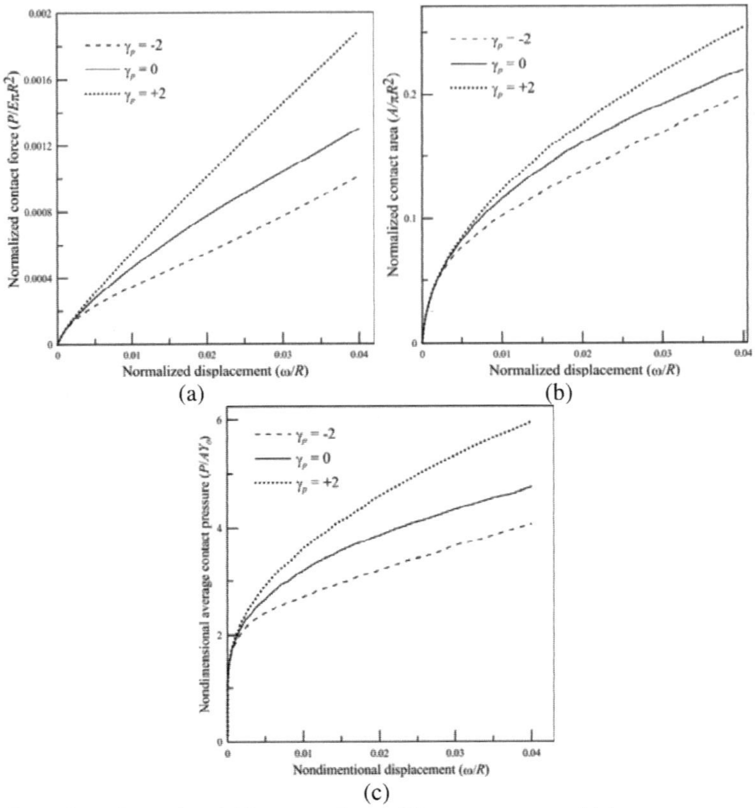

Figure 4. Plots of nondimensional (a) contact force, (b) contact area and (c) average contact pressure w.r.t. nondimensional interference corresponding to γ_p = -2, 0 and 2.

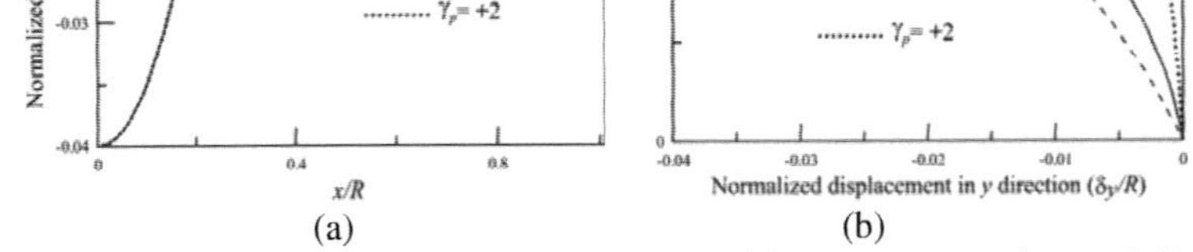

Figure 5. Nondimensional deformation in y direction (δ_y/R) of (a) top surface and (b) y axis corresponding to nondimensional interference (ω/R) of 0.04.

It can be observed from the stress-state plot (figure 7) that higher value of γ_p causes the yielded zone to be restricted at the vicinity of the contact interface, hence creating higher contact area. Figure 5 plots the nondimensional vertical displacements (δ_y/R) of the nodes located at the top surface as well as at the vertical axis, corresponding to nondimensional interference (ω/R) of 0.04. Lower value of γ_p causes higher downward displacement at the top surface as well as at the entire vertical (y) axis of the semi-cylinder. Figure 6 provides the visualization of the vertical deformation for the entire semi-cylinder through contour plots of average nondimensional displacement in y-direction (δ_y/R). The figure shows that, for lower value of γ_p, vertical displacement transmits deeper towards the bottom of the semi-cylinder.

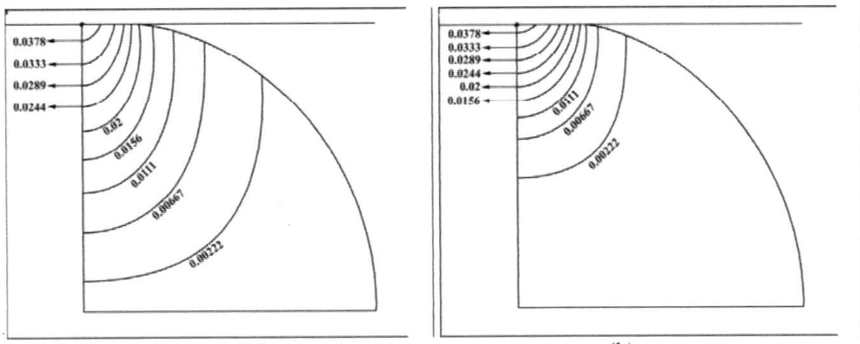

(a) (b)

Figure 6. Contour plot of δ_y/R at nondimensional interference of 0.04 corresponding to γ_p = (a) -2 and (b) +2.

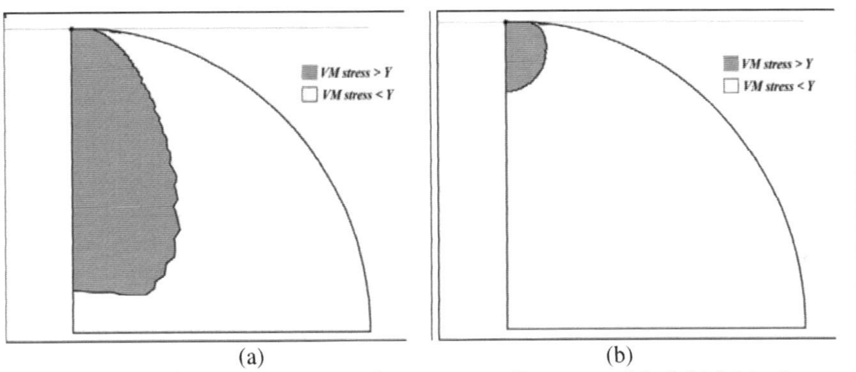

(a) (b)

Figure 7. Stress-state contour plots corresponding to γ_p = (a) -2 (b) 0 (c) +2.

As mentioned previously, one of the primary objectives of applying FGM at contact level is altering the stress distribution to suppress yielding at some location of interest. To visualize the influence of the gradation parameter in terms of altering stress distribution, stress state contour plots are presented in figure 7. These contour plots compare the Von Mises stress value at any point of the semi-cylinder with the yield strength of the material located at that point and shows whether the material is at elastic phase or in post-elastic phase. The plots show that, for lower γ_p, the yield zone gets transmitted more towards the base, whereas for low γ_p, the yield zone remains confined in the near contact zone. Moreover, the volume of the material of the semi-cylinder that gets yielded increases with the increase of the value of γ_p.

4. Conclusions

The paper presents a finite element based study of a radially functionally graded single asperity cylindrical contact with varying yield strength under plane stress condition. The yield strength is varied following an exponential function along the radial direction of the semi-cylinder, keeping the Young's modulus and tangent modulus fixed. The gradation is controlled by varying the parameter named plastic inhomogeneity parameter. From the analysis it is found out that contact force, contact area and contact pressure increases with higher inhomogeneity parameter. In addition, vertical displacement is found to transmit deeper into the semi-cylinder for low inhomogeneity parameter. In a similar way, the yielded zone is found to occupy higher volume and it tends to propagate more towards the base for lower inhomogeneity parameter value.

References

[1] Sharma A and Jackson R L 2017 A finite element study of an elasto-plastic disk or cylindrical contact against a rigid flat in plane stress with bilinear hardening *Tribol. Lett.* **65** 1–12.

[2] Hertz H 1882 Ueber die Berührung fester elastischer Körper *J. Reine Angew. Math.* **92** 156–171

[3] Johnson K L 1987 *Contact Mechanics* Cambridge UK Cambridge University Press.

[4] Suresh S 2001 Graded materials for resistance to contact deformation and damage *Science* **292** 2447-2451.

[5] Choi I S, Dao M and Suresh S 2008 Mechanics of indentation of plastically graded materials—I: Analysis *J. Mech. Phys. Solids* **56** 156-171.

[6] Choi I S, Detor A J, Schwaiger R, Dao M, Schuh C A and Suresh S 2008 Mechanics of indentation of plastically graded materials—II: Experiments on nanocrystalline alloys with grain size gradients *J. Mech Phys. Solids* **56** 172-183.

[7] Yuan F, Jiang P, Xie J and Wu X 2012 Analysis of spherical indentation of materials with plastically graded surface layer *Int. J. Solids Struct.* **49** 527–536.

[8] Prasad A, Dao M and Suresh S 2009 Steady-state frictional sliding contact on surfaces of plastically graded materials *Acta Mater.* **57** 511–524.

[9] Jackson R L and Kogut L 2006 A comparison of flattening and indentation approaches for contact mechanics modeling of single asperity contacts *J. Tribol.* **128** 209-212.

[10] Suresh S, Giannakopoulos A E and Alcala J 1997 Spherical indentation of compositionally graded materials: theory and experiments *Acta Mater.* **45** 1307-1321.

[11] Alinia Y, Beheshti A, Guler M A, El-Borgi S and Polycarpou A A 2016 Sliding contact analysis of functionally graded coating/substrate system *Mech. Mater.* **94** 142-155.

IOP Conf. Series: Materials Science and Engineering 1080 (2021) 012018 doi:10.1088/1757-899X/1080/1/012018

Evaluation of Mechanical Properties of Developed Graphite-reinforced- Aluminium Metal Based Composites

Sourav Debnath[1], Akshay Kumar Pramanick[2]

[1, 2] Metallurgical and Material Engineering Department, Jadavpur University, Kolkata- 700032, India

e-mail: *souravdebnath262@gmail.com*

Abstract. Aluminium metal matrix composites have developed for its attractive properties and performances. Graphite- reinforced- aluminium composites have developed with varying the volume percentage (vol. %) of reinforcement, typical variations of graphite have conducted from 2.28 % to 13.19 %. Presence of different phases after sintering have specified by X-ray diffraction (XRD). Surface morphology of each developed metal has studied with scanning electron microscopy (SEM). Ultimate tensile strength (UTS) has computed from the measured hardness as per ASTM A 370-68 standards. Other mechanical properties like tensile strength and young's modulus (also termed as modulus of elasticity) have found out and reported.

Keywords: Graphite- reinforced- Aluminium, Powder Metallurgy, Surface Morphology, UTS, Tensile Strength, Young's Modulus.

1. Introduction

Continuous efforts have observed among materials engineers to develop new materials, which can fulfil the requirements of various demands on modern industries. Industries are continuously searching for strong, high strength materials, which can cater the requirements, and having longer lifespan. Aluminium is lightweight, durable metal and is easily available on earth crust. Aluminium has not only good corrosion properties, excellent casting abilities but also has low thermal expansion, wear resistant [1-5]. For this reason, aluminium is being used in different industrial applications like electrical and electronics industries, transport industry including aerospace, aeronautics, automobile applications etc. [1][6-7]. Still low melting point converge the applications of aluminium in limited fields [7-9]. However, this problem can overcome with incorporating suitable second phase into it [2-6] [10-15].

Powder metallurgy is an art or technology by which one can fabricate composites cost effectively with exact size and dimensions [16-17]. Under this fabrication technique, composites can be developed without melting its constituent materials with consequent improvement of properties [16-17].

Authors are tried to develop graphite- reinforced- aluminium metal based composites and also finding out the relevance of incorporating graphite as reinforcement. Study has also tried to visualise the various mechanical properties of developed composites.

2. Experimental procedure

The Graphite- reinforced- aluminium composites have developed followed by conventional 'powder metallurgy' technique maintaining the steps of milling and mixing, compaction and sintering carried out simultaneously into hot press unit (makes by 'NascorTechnogiesPvt. Ltd,' W.B., India). Mixing and grinding was conducted into morter parcel to produce fine graphite- aluminium powder, having mesh size ~ 250 μm. Various compositions of produced powder have considered separately. Powders of each composition has poured into separate steel die and placed into hot pressing unit towards developing composites with applying 550^0C temperature for 1.5 hours maintaining 4.2 ton pressure. After sintering, each sample has cooled slowly into the same furnace unit. After removing from furnace, each composite has polished and considered for further studies.

3. Results and Discussions

Obtained results and corresponding discussions are elaborated below.

3.1. XRD Analysis

To get a thorough insight of the present problem, a transient study is conducted first for a set of typical values of flow parameters (Gr = 10^5, Re = 200, Bi = 1 and Ma = 1000) considering both downward and upward motion of the sidewalls. The corresponding results on the temporal evolutions of flow and thermal fields are analyzed in Figs. 2–3. Non-dimensional time instants are marked by different values of τ ranging from 0.01 to 10. First, the development of flow pattern in terms of XRD pattern for graphite- reinforced- aluminium metal based composite has plotted in fig. 1. Aluminium has detected at 2θ angle ~ 38.493^0, 44.739^0, 65.012^0, 78.019^0, 82.07^0 and graphite is found out at 2θ angle of ~ 26.7^0 with low abundance. Noted that XRD analysis is carried out from 2θ angle of ~ 20^0 to 80^0 with scan 5^0/minute utilising 'RigakuUltima III analytical diffractometer'.

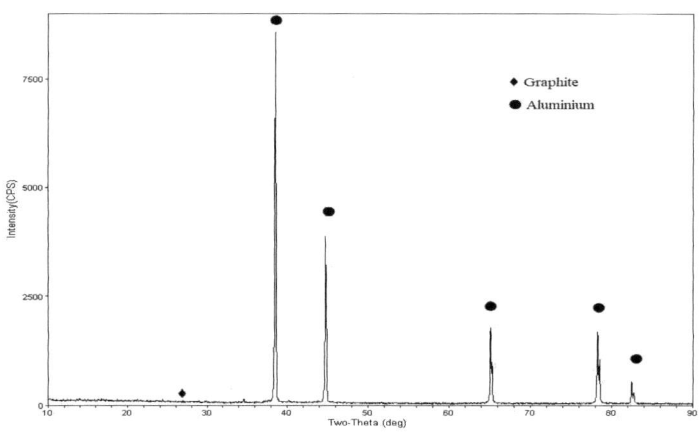

Figure 1.XRD pattern for graphite- reinforced aluminium metal based composite.

3.2. Surface morphology

Surface morphology based on SEM has shown in fig. 2. Graphite has spread in random way into aluminium matrix. The tendency of forming pores has observed with graphite reinforcement. Mentioned that maximum porosity has found out for 13.19 vol. % graphite- reinforced- aluminium composite, value only 0.52 %.

3.3. Hardness and UTS

Hardness survey has carried out using 'Leco Micro-hardness tester, model LM248SAT' and obtained hardness values are reported in table 1. Hardness has seen to be improved with graphite reinforcement. Ultimate tensile strength (UTS) has computed utilising theoretical conversion [18-19] and plotted in fig. 3. UTS has also improved with graphite reinforcement into graphite- reinforced- aluminium metal based composites as shown in fig. 3.

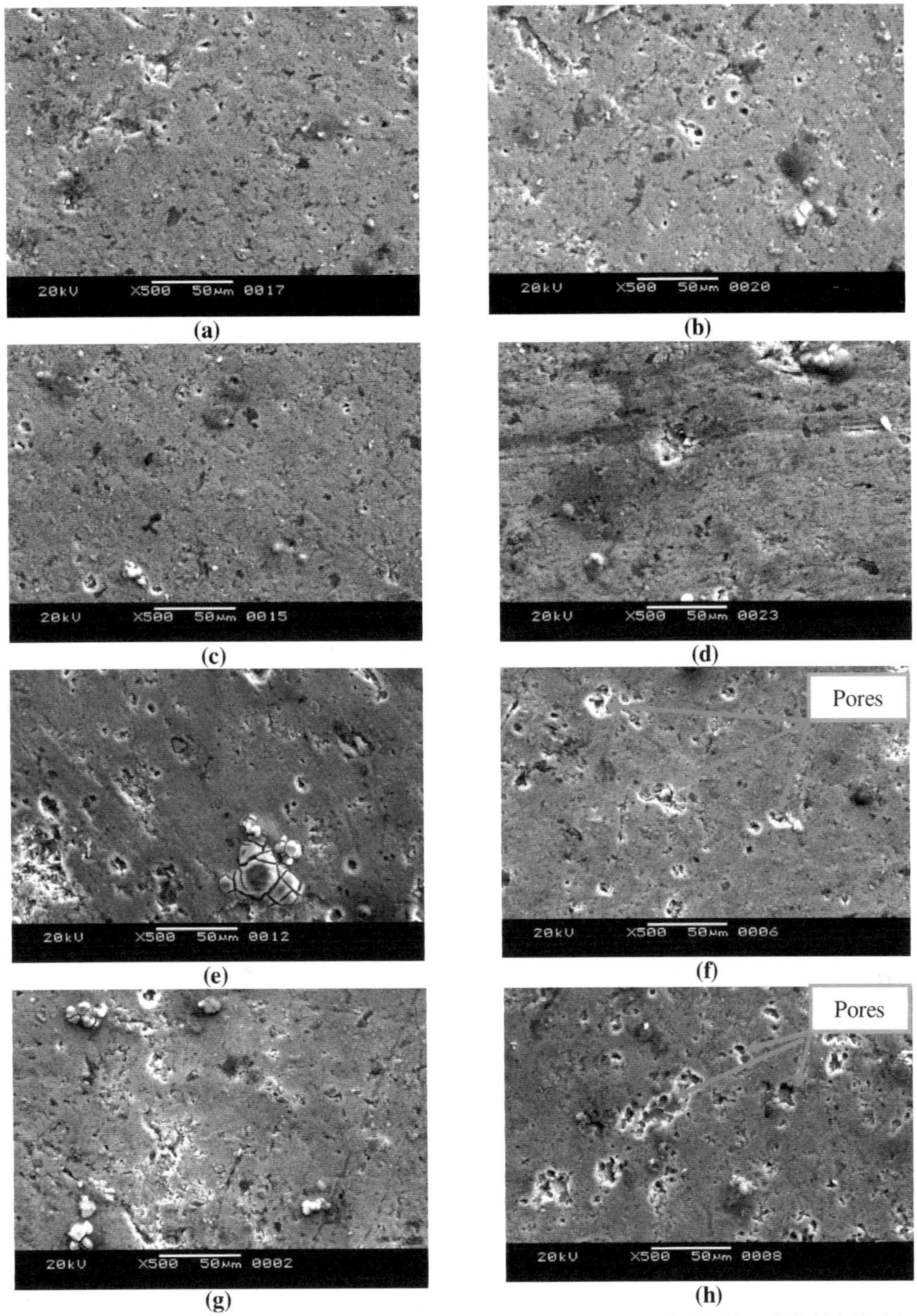

Figure 2. Surface morphology (SEM) of (a) 2.28 %, (b) 4.537 %, (c) 6.017 % , (d) 7.484 %, (e) 8.933%, (f) 10.37 %, (g) 11.789 %, (h) 13.19 % (based on vol. %) graphite- reinforced- aluminium metal based composite.

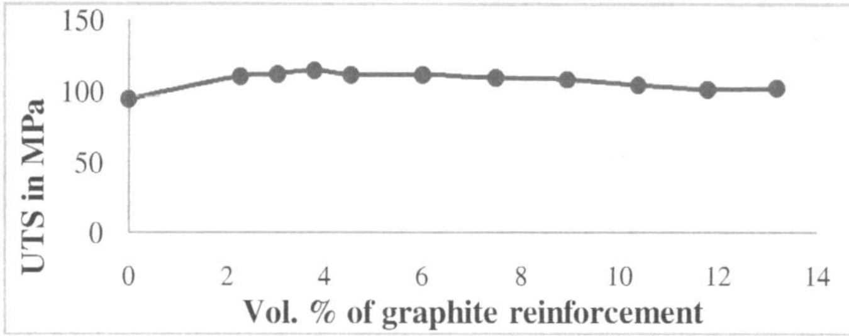

Figure 3. Variation of ultimate tensile strength (UTS) with reinforcement.

Table 1. Hardness for graphite- reinforced- aluminium composites.

Composition	Hardness in MPa
Sintered aluminium	283.3
2.28 % graphite- reinforced- aluminium composite	332.5
6.017 % graphite- reinforced- aluminium composite	334.9
8.933 % graphite- reinforced- aluminium composite	325.0
13.19 % graphite- reinforced- aluminium composite	306.6

3.4. Determination of tensile strength and modulus of elasticity

Tensile Strength and the modulus of elasticity can be calculated from the rules of mixture for composites. Modulus of elasticity is considered as 69 GPa and 9.6 GPafor aluminium matrix and graphite reinforcement respectively [20-24]. The obtained results for developed composites are demonstrated in table 2. Tensile strength has taken 90 MPa and 10 MPa for aluminium and graphite respectively from reputed articles [20-24].

Table 2. Young's modulus and tensile strength for graphite- reinforced- aluminium composites.

Composition	Young's modulus in GPa	Tensile strength in MPa
2.28 % graphite- reinforced- aluminium composite	67.784	88.365
3.04 % graphite- reinforced- aluminium composite	67.200	87.568
3.79 % graphite- reinforced- aluminium composite	66.740	86.959
4.537 % graphite- reinforced- aluminium composite	66.309	86.367
6.017 % graphite- reinforced- aluminium composite	65.428	84.642
7.484 % graphite- reinforced- aluminium composite	64.553	83.998
8.933 % graphite- reinforced- aluminium composite	63.689	82.790
10.37 % graphite- reinforced- aluminium composite	62.839	81.704
11.789 % graphite- reinforced- aluminium composite	61.988	80.559
13.19 % graphite- reinforced- aluminium composite	61.159	79.439

4. Conclusions

Significant conclusions based on this study have noted below.

- Successful development of graphite- reinforced- aluminium composites have observed cost effectively through power metallurgy technique.
- Graphite- reinforced- aluminium composites are much lighter in weight. Mentioned that these composites are weight saving even pure aluminium.
- Improvement of hardness, hence UTS, has found out with graphite reinforcement.
- Tensile strength and the modulus of elasticity for developed metals are slightly affected with graphite reinforcement.
- These developed metal based composites can be used to design the outer shell of an automobile body, fabricating various engine parts in automobiles/ aerospace industries. Mentioned that these types of metals having tolerable amount of pores may be appreciated for designing gearwheel arrangements.

References

[1] Prasad S V and Asthana R 2004. Aluminium metal matrix composites for automotive applications; Tribological Considerations, Tribology Letters 17(3), pp. 445-453.
[2] Hesabi Z R,Simchi A, Raihani S M S 2006. Structural evolution during mechanical milling of nanometric and micrometric Al2O3 reinforced Al matrix composites, Materials Science and Engineering, A428.
[3] Tong X C and Ghosh A K 2001. Fabrication of in situ TiC reinforced Aluminum matrix composites, Journal of Materials Science 36, pp. 4059-4069.
[4] Magesh M, Baruch L J, Oliver D G 2014. Microstructure and Hardness of Aluminium Alloy-Fused Silica Particulate Composite, International Journal of Innovative Research in Advanced Engineering 1(5), pp. 199-204.
[5] Hamouda A M S, Sulaiman S, Vijayaram T R, Sayuti M, Ahmad MH M 2007. Processing and characterization of particulate reinforced aluminium silicon matrix composite, Journal of Achievements in Materials and Manufacturing Engineering 25(2), pp. 11-16.
[6] Chawla N and Chawla K K 2006. Metal Matrix Composites in ground transportation, Springer-Verlag, New York, pp. 401.
[7] Dixit A 2018. Application of silica-gel-reinforced aluminium composite on the piston of internal combustion engine: Comparative study of silica-gel-reinforced aluminium composite piston with aluminium alloy piston, Composites and Advanced Materials for Industrial Applications, pp. 63-98. DOI: 10.4018/978-1-5225-5216-1.ch004
[8] Usman A M, Raji A, Waziri N H, Hassan M A 2014. Production and Characterization of Aluminium Alloy - Bagasse Ash Composites, IOSR Journal of Mechanical and Civil Engineering 11(4) Ver. III (Jul- Aug. 2014), pp. 38-44.
[9] Wakeel S 2017. Fabrication and Mechanical Properties of Aluminium Composites, Scholar-press, omniscriptum publication group- Germany, ISBN 978-3-639-71638-2.
[10] Bharath V, Nagaral M, Auradi V 2012. Preparation of 6061Al-Al2O3 metal matrix composite by stir casting and evaluation of mechanical properties, International Journal of Metallurgical & Materials Science and Engineering 2(3), pp. 22-31.
[11] Toptan F, Karaaslan A A, Cigdem M, Kerti 2009. Processing and microstructural characterization of AA 1070 and AA 6063 matrix B4Cp reinforced composites, Materials and Design 31, pp.87-91.
[12] Tong X C and Ghosh A K 2001. Fabrication of in situ TiC reinforced Aluminum matrix composites, Journal of Materials Science 36, pp. 4059-4069.
[13] Hembrom S, Roy B N, Khobragade N, Roy D 2016. Studies on Amorphous Alloy Dispersed Aluminium Matrix Composite Prepared by High Pressure Torsion, Journal of Materials Science Research 5(1), pp. 89-96.
[14] Huo H and Woo KD 2006. In situ synthesis of Al2O3 particulate reinforced Al matrix

composite by low temperature sintering, Journal of Materials Science 41, pp. 3249-3253.
[15] Shi Z 2001. The oxidation Of SiC Particles and its interfacial characteristics in Al-Matrix composite, Journal of Materials Science, 36, pp. 2441 – 2449.
[16] Debnath S and Pramanick A K 2016. Development and study on different properties of aluminium- crystalline silica ceramic matrix composites at different sintering temperatures, IJERGS 4(3), pp. 415- 423.
[17] Debnath S and Pramanick A K 2016. Development and Evaluation of Various Properties of Crystalline Silica- Aluminium Metal Based Composites, IJERGS 4(2), pp. 236-245.
[18] Zhang P and Zhang Z F 2011. General relationship between strength and hardness, Materials Science and Engineering: A, ScienceDirect, 529, pp. 62-73. https://doi.org/10.1016/j.msea.2011.08.061
[19] Gaško M and Rosenberg G 2011. Correlation between hardness and tensile properties in ultra-high strength dual phase steels – short communication, Materials Engineering - Materiálovéinžinierstvo 18, pp. 155-159. http://fstroj.uniza.sk/PDF/2011/27-2011.pdf
[20] Kumar S S, Bai V S, Rajkumar K V, Sharma G K, Jayakumar T, Rajasekharan T 2009. Elastic modulus of Al–Si/SiC metal matrix composites as a function of volume fraction, Journal of Physics d: applied physics, IOP publishing 42, pp. 1-10. doi:10.1088/0022-3727/42/17/17550
[21] Luis GuilhermeBorzaniManhania, Luiz Claudio Pardinib, Flamínio Levy Netoc 2007. Assessement of Tensile Strength of Graphites by the Iosipescu Coupon Test, Materials Research, 10(3), pp. 233-239.
[22] Eung Soo Kim and Chang H Oh 2007. Estimation of graphite density and mechanical strength of vhtr during air-ingress accident, The 12th International Topical Meeting on Nuclear Reactor Thermal Hydraulics (NURETH-12), Log Number: 185, pp. 1-13, Sheraton Station Square, Pittsburgh, Pennsylvania, U.S.A.
[23] Butt J, Mebrahtu H, Shirvani H 2016. Strength analysis of aluminium foil parts made by composite metal foil manufacturing, ProgAddit Manuf. 1,pp. 93-103. DOI 10.1007/s40964-016-0008-5
[24] Imran M and Sadik S 2016. Study of hardness and tensile strength of Aluminium-7075 percentage varying reinforced with graphite and bagasse-ash composites, Resource-Efficient Technologies, ScienceDirect, 2(2),pp. 81-88.

Acknowledgments
Authors wishing to acknowledge Mr. Rajib Gupta, an inspection engineer at Dr. Amin controllers Pvt. Ltd. for helping in every step to carry out this study.

Novel cryogenic coupling for loading LPG bullet truck

Goutam Kumar Bose[1] and Tapas Kumar Mandal[2]

[1]Mechanical Engineering Department, Haldia Institute of Technology, Haldia, Purba Medinipur, West Bengal, India. Pin 721657.
[2]Department of Process, Indian Oil Petronas Pvt. Ltd. Haldia, Purba Medinipur, West Bengal, India. Pin 721657.

E-mail: gkbose@yahoo.com, mandalt72@yahoo.com

Abstract This research work has been carried out in the Indian Oil Petronas private limited company's plant located at Haldia, West Bengal, India. The study is related to the LPG Bullet Truck Loading. Due to higher customers (Cooking gas) demand it was necessary to improve the productivity of the gantry by about 10% to face the increasing demand. The time of LPG Bullet Truck fitting with loading arm needed improvement. This work is related to develop the bullet trucks loading capacity of the terminal to face the increasing demand. After successfully implementing the improved cryogenic coupling for loading the bullet truck the productivity improvement took place. The LPG bullet truck loading is increased average from 315 bullet trucks per day to 357 bullet trucks per day or LPG loading 5512.5 MT per day to 6247.5 MT per day i.e. equivalent to increase bullet trucks loading by 13 %. This will eventually save cost.

Keywords: Bullet truck, Cryogenic coupling, Productivity improvement, Cost saving

1. Introduction

There are various bottling plants both private unit and PSU unit which can undertake to supply the extra demand of LPG day to day, but one should their existing resources to maximize the profit. Considering all possible ways and safety, in order to improve the production, the time of bullet trucks fitting with loading arm has to be reduced using modern temporary fixing method which can minimize the fixing time.

A real multi-compartment container loading problem arises in the liquefied material distribution under several practical constraints [1]. The problem for a logistics company is that it has to serve its customers by first putting the products on pallets and then loading the pallets into trucks [2]. The multi container loading problem of a company can be developed and solved through integer linear models [3]. A two-stage packing problem can be solved for a daily distribution process of a trading company [4]. Therefore a container loading problem can be dealt with frequently with the help of operations research [5]. A tree search algorithm for the three-dimensional container loading problem (3D-CLP) can be developed. The 3D-CLP is the problem of loading a subset of a given set of rectangular boxes into a rectangular container so that the packing volume is maximized [6]. The existing approaches to container loading problems are each applicable only to a narrow part of the spectrum of situations encountered in practice and that there are many scenarios for which there are no adequate methodologies [7]. There are some characteristics of major accidents in the petrochemical sector included in the major accident reporting system (MARS). Through the statistical analysis focused on the main categorization fields of the MARS short reports and additionally a refinement of the immediate causes of major accidents with focus on the organizational factors can be attempted through the details provided in the full reports [8]. A specific analysis of the accidents occurred in the 21st century can be performed by comparing them with the total set of accidents. Finally, a set of specific recommendations can be inferred from the results provided [9]. Several researchers have contributed their work related to the container loading problem, accidents related to the petroleum industry and routing problems of container vehicles. Attanasio et. al. (2007) has studied about the variants of the Three-dimensional Loading Capacitated Vehicle Routing Problem (3L-CVRP) a MILP model is proposed for the periodic 3L-CVRP. Pedruzzi et.al. (2016) has conducted a logistical study using a mathematical model based on the Three-dimensional Bin Packing Problem to optimize the arrangement of boxes inside the vehicles, leading to a better use of the vehicle's volumetric capacity to transport food. Haessler and Talbot (1990), presents a complex computer-based heuristic procedure for sizing customer orders and developing three dimensional load diagrams for rail and truck shipment of low density products. Chen et al. (1995) for the problem of loading a single container with variable height and proposed a solution for 3D-BPP based on Genetic Algorithms. Oueidat et.al. (2015) has aimed is to assess whether the prescribed safety program covers all of the system's phases; namely operations and audits. Moura and Oliveira (2009) present a framework to integrate these two problems using two different resolutions methods the Vehicle Routing Problem with Time Windows and the Container Loading Problem. Sanjeeb et.al. (2010) has studied the separation problem for single-constraint sets and show that it can be solved in polynomial time when the resulting inequality is required to be sufficiently different from the associated MIR inequalities.

This study has been carried out in the Indian Oil Petronas Private Limited company's plant located at Haldia, West Bengal, India. The current work is related to loading of LPG Bullet trucks. The average capacity of the bullet truck is about 17.5 MT. There are 16 bays in the plant and the plant runs in three shifts in seven days. The LPG Bullet Truck is loaded for different LPG Bottling plant in eastern India as per requirement of PSU (IOC & BPC) and other Private customer's demand. There are two cryogenic storage tank one for Propane and another for Butane. Beside this there are different size mounted bullet storage tank out of which three bullets capacity 1900 MT, two bullets capacity 600 MT and another two bullets capacity 150 MT. These bullets tanks are approved by concerned authority (PESO) and are tested as per norms.

On the basis of secondary data on an average 320 X 26 bullet trucks each capacity is 17.5 MT per month have been loaded in the last six months from August 2018 to January 2019. The heat exchanger capacity is 375 MT LPG / hour. But the gantry loading capacity is low i.e. 318 to 320 bullet truck per day. Now due to different government schemes and increasing population of the country, the demand is ever increasing. This resulted in loading extra 20000 MT LPG per month to cater the customer demand. In other words the present LPG loading capacity of the gantry should be increased from 145600 MT LPG to 165600 MT LPG per month. Hence the number of LPG bullet tank loading is increased from 320 bullet trucks per day to 384 bullet trucks per day or LPG loading 5617 MT per day to 6825 MT per day i.e. equivalent to increase bullet trucks loading by 22%. Here production is improved by about 1208 MT per day which will reduce production cost and also fulfill the increasing customer demand. It will also increase revenue of

the plant by about Rs.21 lakh per month (for material handling charge only). For that extra production extra consumable will be used but per unit production (bullet truck loading) use of consumable per bullet truck will be reduced. That will be given details in saving from consumable head.
The plant has only 16 bays in the bullet truck loading gantry. These16 bays are working with 85% efficiency but total production is less than the demand. Therefore, our scope is limited to the improvement in the process related to the different step involved in the bullet truck loading. In this current research the design of the bullet trucks loading line flange is to be modified and loading arm flange connect and disconnect before loading the LPG and after completion the loading of LPG in order to accommodate more bullet trucks. At first the trial on two bays has been carried out. After inspecting and checking and time study of all parameters of bullet trucks before loading, during loading and after loading, it is found that a huge time is saved when dry cryogenic coupling for LPG bullet truck loading is applied. This new attachment increases the speed of the LPG bullet truck loading. The additional loading bay will be a costly affair and it is beyond scope of this study. However, it revisits existing process parameters for improving effectiveness is considered in this project. The different steps performed before the bullet truck loading are reconsidered to maintain efficient time management accordingly some modifications in the parameters such as few steps are modified. Necessary trial and errors of the same have been conducted to make final decision.

1.1 The problem on hand

It has been said that the bullet trucks are loaded at gantry through 16 bays in the plant. There is also decantation provision of leakage or disputed bullet trucks in 2 bays out of 16 bays but normally loading operation continue at 16 bays for three shifts per day. After proper checking of bullet trucks, it enters in gantry for loading LPG/Propane/Butane as per requirement of the customer. But the bullet trucks loading line flange has to mesh with the loading arm flange of the bay at gantry. There are eight nut and bolts to be fixed manually for proper fitment. This resulted in loss of time in loading bay. Therefore, in order to eradicate this time loss in loading bays, design modifications are needed to increase production for achieve customer demand. This will help the company in planning for increasing the manufacturing capacity to fulfill the rising demand from the customers.

2. Cause and effect relationships

First cause and effect analysis is undertaken to determine the factors responsible for lower performance of Bullet Truck Loading process as shown in table 1. The aim behind this is to improve productivity of the process.

Table-1: Cause and effect relationships

Cause	Effect
1. Less number of bays runs in the gantry.	1. Lower number of bullet truck loaded.
2. Design of the fastener nut-bolt number decrease.	2. Increase chances of leakage.
3. Jump step of loading procedure.	3. Increase risk of failure.
4. The system design change.	4. Has to change maximum equipment.
5. Present capacity of the pumps and equipment change.	5. Has to change maximum pipeline.
6. Fastener design change.	6. Effective in time saving.

It is evident from the above table is that the effect is lower number of bullet truck can be loaded with the present set up and design. Therefore it is necessary to improve present setup in terms of quick coupling, pump setting and other relevant changes in order to improve productivity. This problem is critical because all steps of loading has to be maintained properly with respect of safety and also has to increase the productivity to face the increasing demand. Therefore introducing improved technology/ equipment will help improving production and utilizing available resources.

The data for the research study have been collected from the production department. The production was in shift A, B and C in total 316 to 320 bullet trucks before implementation of the project. Now the demand has increased about 12 %, this initiates the project. In the historical perspective we have furnished production for the period of 10 months. In a month the company works for 26 days average. This project intends to fulfill the shortfall in the demand by improving capacity of the bullet truck loading by 13 %.

3. LPG bullet truck loading

The LPG Bullet Truck Loading process is very similar to other liquid filling in trucks. But the difference is that it is more dangerous than diesel, kerosene or any edible oil because any leakage or release of it become more dangerous to human for gas inhalation causing nausea, breathlessness, headache and other pulmonary obstructive problems. Further LPG is highly inflammable; in open air immediately it starts to evaporate. In tank /vessel heating will cause pressurizing and chance of explosion. So excess filling / loading is prohibited, and before loading all safety parameter are checked step wise. The stepwise Standard Operating Procedure (SOP) for LPG bullet trucks loading is shown in table 2.

Table-2: Operating procedure for LPG bullet trucks loading

SL No	Job Description	Action By	Time
1	Calling bullet trucks by their registration number through announcing system.	Security	NA
2	Checking of documents /TREM Card prior to entry at Terminal.	Security	4 min
3	Prior to entry Check for availability of 2 numbers 9/10 Kg DCP type fire extinguishers, Approved type spark arrestor welded on exhaust and 300 Amp master cut-off switch available.	Security	3 min
4	Prior to entry, check for availability of prohibited items like mobile phone set, matches box, bidi/cigarettes, torch, camera etc.	Security	2 min
5	Prior to entry, check that self-starter is in good working condition and no naked electrical wire exists/visible in the bullet truck.	Security	3 min
6	Allow entry of the bullet trucks after recording.	Security	2 min
7	Weighing of empty bullet trucks at weigh-bridge and record maintained.	Officer at Weigh-Bridge	3 min
8	Marshalling and guiding of the empty bullet truck for parking behind the selected loading bay of the Gantry.	Security	4 min
9	Opening of blind flange of the loading nozzle of thebullet trucks.	Driver of TT	5 min
10	Inspection of condition of flange and its raised face portion for any unevenness, pitting, corrosion etc.	IPPL's Officer at Gantry	2 min
11	Allowing entry of the bullet truck at loading gantry after exit of previous bullet trucks.	IPPL's Officer at Gantry	1 min

12	Positioning of the bullet truck at loading bay, switching off engine, Master cut-off switch to be isolated, chock blocks to be positioned on wheels and earthing clip to be connected at designated point of the bullet trucks.	Tank Truck Driver	3 min
13	Loading arm to be connected using IPPL's supplied spiral wound metal gaskets and studs & nuts.	Tank Truck Driver under supervision of IPPL's Officer	8 min
14	Check tightness of loading arm with the flange of loading nozzle of the bullet trucks followed by leak test by crack opening the delivery valve of loading arm.	IPPL's Officer Gantry.	1 min
15	Setting the quantity of product to be loaded in BCU and open the loading arm valve slowly to start loading of the bullet trucks	IPPL's Officer at Gantry.	1 min
16	Monitoring of loading operation of the bullet trucks and the driver of the tank truck must be present near the valve box of thebullet trucks for immediate access in case of emergency.	IPPLs Officer at Gantry and TT Driver	40 min
17	Close the loading arm valve, 3" loading line valve and bullet truck's loading nozzle after completion of loading and recover the liquid from the said pipe section by opening the ½" vent valve.	Driver under supervision of IPPL's Officer	3 min
18	Disconnect the loading arm and place the same on its position. Keep the gasket and Stud-nuts at designated place.	Driver under supervision of IPPL's Officer	8 min
19	Disconnect the earthing clip and keep the same at designated place. Remove chock blocks.	Driver under supervision of IPPL's Officer	1 min
20	Inspect the loading bay and verify &record the quantity loaded for information to Weigh-bridge.	IPPL's Officer at Gantry	1 min
21	Allow the bullet trucks to move out for weighment.	IPPL's Officer at Gantry	NA
22	Weighment of loaded bullet trucks and issue weighment slip.	Officer at weigh-bridge	NA
23	Submission of weighment slip to S&D.	Driver	NA
24	Preparation of Challan / invoice.	S&D Officer	NA
25	Allow exit of the bullet trucks after verification and recording of the challan / invoice.	Security	NA

In the SOP, it is observed that a bullet truck occupies a bay from step No.11 to step No.20 and during which it takes 67 minutes. If we consider idle time 5%, as a result in a shift in 16 bays bullet truck can be loaded = 16 x 8 x 60 /(67 x 1.05) = 109 number. From the above steps No.11 to No. 20 time span is more vital because in this period the bullet trucks occupy the loading bay. It is observed that only if we can cut down time taken from step no. 11 to step No. 20 by 13 minutes then production may increase by 13 %. This time cut down or time reduction of bay occupying during loading is our main target to solution the problem or to increase productivity. It is observed that in SOP step No.13 & step No.18 refers to the time of connecting and disconnecting of loading arm to bullet trucks flange. In loading bay if we use quick coupling & decoupling (QCDC) part instead of flange joint then it may reduce time & as a result productivity will be improved. Beside this we will go through all the possible of study to check how to reduce time span in occupying the loading bay and we will implement the best ways to this project. Thus, the problem is to improve the productivity of the LPG bullet truck loading at bay in the gantry.

Fig.-1: Loading Arm fitting with LPG Bullet Truck by Dry Cryogenic Coupling

4. Dry cryogenic coupling

Dry Cryogenic Coupling consist of a "tank" unit which is a type of a non-return valve and a "hose" unit with a valve driven by an internal cam curve to open both valves at the same time. Operation is a single action using a straight forward turning motion to connect the couplings and using open flow path. An initial push and turn action on the hose unit provides engagement with the tank unit thus locking and sealing the two units together. Further through rotation the internal valves are opened, thereby allowing full flow with a minimum of pressure drop. A selective system allows the system to be used to a wide range of cryogenic liquefied gases without any risk of failure owing to "human" error. Dry Cryogenic Coupling is made with stainless steel material having different diameter as shown in figure 2. There is provision to place ERC fitting upstream it to avoid accident.

Fig.- 2: Dry Cryogenic Coupling.

The Cryogenic Coupling material can bear a wide range of temperature for that no problem arise in ambient temperature, loading or temperature of gases like Propane, Butane or LPG (range from - 42° C to + 10° C). The figure 3 shows the LPG loading arm.

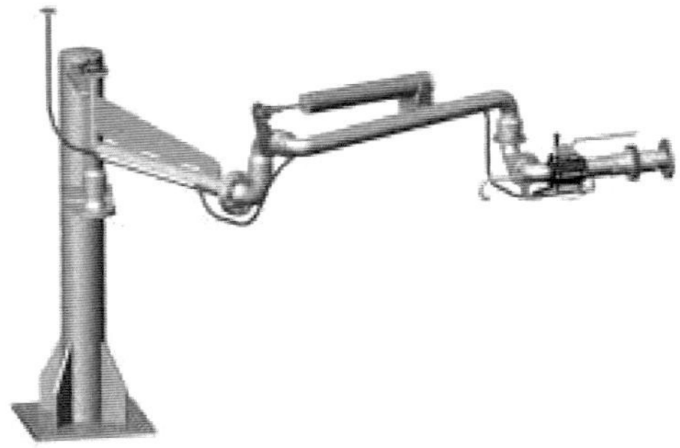

Fig.- 3: LPG Loading arm

5. Analysis

In this analysis on the truck loading for productivity improvement 4W and 1H is carried out as explained in the table 3. It is followed by Cause and Effect diagram as portrayed in the figure 4 below.

Table-3: 4W-1H analysis

Where	What	Who	When	How
In the truck loading gantry.	Fitment mechanism changed.	Plant head.	In the month of August, 2019 started.	Under close observation change one by one.
At IndianOil Petronas Pvt.Ltd., Haldia.	Use dry cryogenic coupling instead of flange joint.	Operators. Team members.	In the month of August, 2019 completed.	Observe advantage & disadvantage.
West Bengal, India.	That is more quick, reliable, and cost saving.	Mainte nance department members.	Initially face few problems. Now all are normal and stream line.	Some training has been given to all gantry operator for smooth and easy handling.

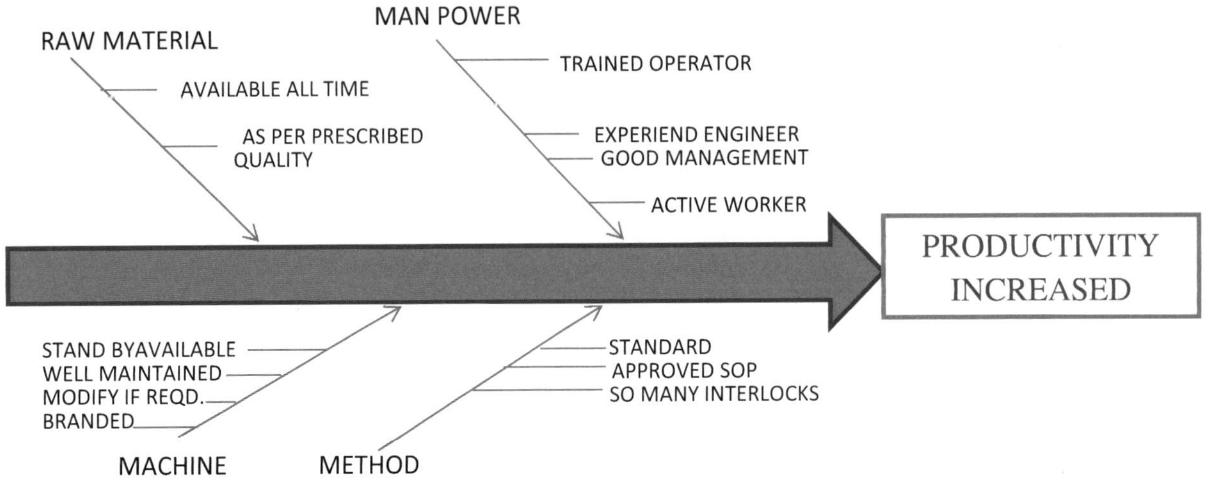

Fig.-4: Cause and effect diagram

6. Production before and after project

The production quantity before and after implementation of the project based on the number of LPG Bullet truck loading will depict the improvement. Here 12 months production data before and 3 months production data after implementation of the project is considered and it is portrayed in figure 5. However, magnitude of the improvement is calculated on the basis of average production.

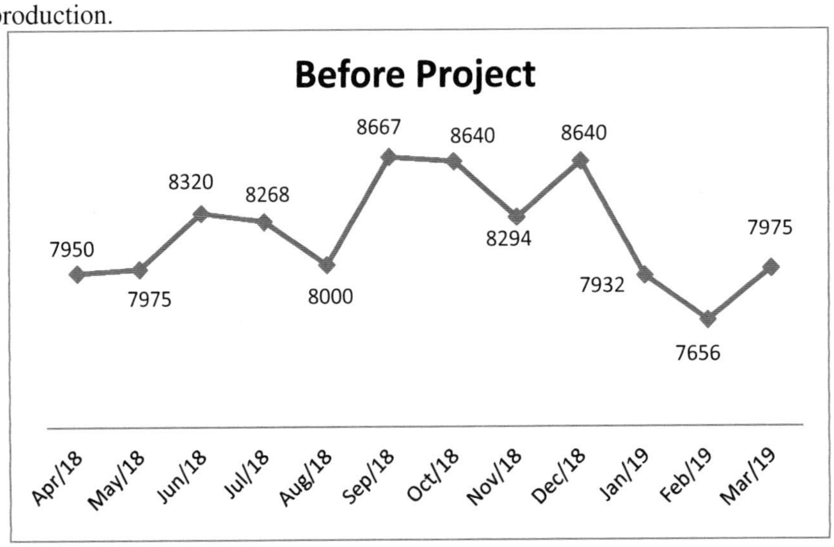

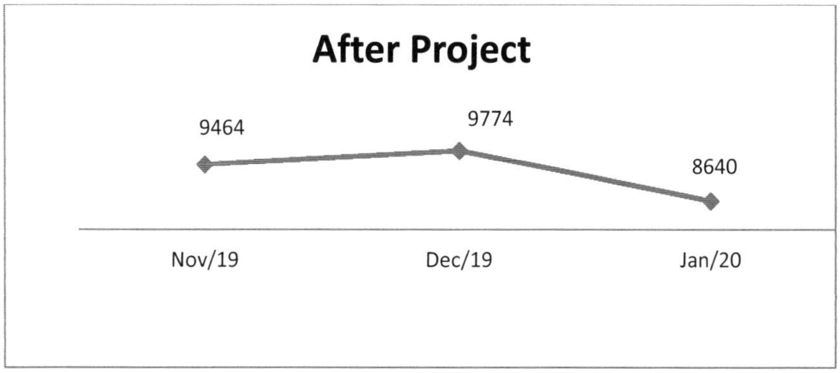

Fig.-5: Monthly production before and after project implementation

Before implementation the company used to load average 8193 bullet truck per month which has increased to about 9292. Thus there is about 1099 bullet trucks more loading in a month, which is 13% of the previous production. This production may face the aggregate demand of the clients.

Consumption of various resources for the period of 12 months from April 2018 to March 2019 is calculated and is shown in table 4.

Table-4: Consumption of various resources

Resources/ bullet truck	Before	After	Difference	Production	Rate	Unit	Amount Rs.
Electricity	150.13	149.8	0.33	357 x 26	8.0	Rs/ Kwh	24504
Steam.	406.9	405.8	1.1	357x 26	5	Rs/ Kg	51051
DM water	427.2	424.3	2.9	357 x 26	160	Rs / KL	4306
Industrial Water	543.8	539.0	4.8	357 x 26	100	Rs / KL	4455
Gasket Pc.	0.125	0.1	0.025	357 x 26	25	Rs/ Pc.	5801
Total Amount saved per month from resources considering 26 working day and average 357 bullet truck loading per day.							90117

7. Cost of the project

The estimated the cost of project implementation includes manpower, material and administrative cost at 15%. The total cost of the project implementation is Rs.55.73 Lakh. This is one time cost. It includes manpower cost of about Rs. 54934/- for decision making, training and implementation cost. The material cost is about Rs. 55.17 Lakh. The major amount required for purchase of Dry Cryogenic Coupling and its attachments. The total cost of Rs. 56 Lakh is considered for the calculation of economic feasibility.

The rates of the consumables are same before and after implementation to show true result of this project. Thus, on above mentioned five consumables Rs. 90117 per month is saved if on an average 9292 bullet trucks per month are loaded.

Table-5: Total manpower requirement

Before the Project implementation			After the Project implementation		
SL.No.	Nature of Job	No. of Employees	SL.No.	Nature of Job	No. of Employees
1	Control room officer	2	1	Control room officer	2
2	W.B. officer	1	2	W.B. officer	1
3	Gantry officer	2	3	Gantry officer	2
4	Gantry operator	2	4	Gantry operator	2

8. Economic justification

As per agreement, Indian Oil Pertonas Pvt. Ltd. receives a terminaling charge from PSU customer and few private customers as handling charge over per MT of LPG handling.

To implementation of this project the number of the manpower before and after are same.

Man-hours before = 26 days x 8 hours x 3 shift x 7 = 4368 man – hours = Man-hours after.

Output of the truck loading before was 8193 trucks per month (average) which has increased to 9292 per month (average).

Thus, the labour productivity before was = 8193 / 4368= 1.9 bullet trucks per man hour.

The labour productivity after is = 9292/ 4368 = 2.1 bullet trucks per man-hour. Thus, the labour efficiency has increased from 1.9 to 2.1bullet trucks per man hour or by 10.5 %.

Thus saving from manpower is as follows:

Cost of man power saving:
6 Operator average @ 18000 = 108000
15 Officer average @ 45000 = 675000
Total cost of manpower Rs. = 783000 for 8193 truck loading per month (average).

Therefore manpower cost for one truck loading before the project
= 783000 / 8193 = Rs. 95.56/- or Rs. 96/- per bullet truck.

Therefore manpower cost for one truck loading after the project
= 783000 / 9292 = Rs. 84.26/- or Rs.84/- per bullet truck.

Therefore manpower saving = (96 – 84) x 9292 = Rs. 111504/-

The saving from lesser consumption of Resources (5 Nos.) = Rs. 90117 per month. A bullet truck average capacity about 17.5 MT

Beside this extra terminaling charge which is collecting from our customer (9292 – 8193) x 17.5 x 105 = Rs. 2019412/- per month (if average bullet truck loading is considered before the project 8193 and after the project Rs.9292/- per month, terminaling charge Rs.105 per MT.).

The net saving is Rs.111504 + Rs. 90117+ Rs. 2019412= Rs. 22,21,033/- month.

9. Conclusion

This research study is a capacity improvement program based on customer demand. The demand is increasing day by day depending upon government policy and increasing population. Accordingly, the modification is required to fulfill the demand through improved productivity of

bullet truck loading. In order to achieve this loading arm fitting procedure with bullet truck loading flange is changed. Before implementation of the project the system was capable to load average about 8193 bullet trucks per month which was not sufficient to cater the demand of the clients.
Therefore cause and effect, why-why analysis is conducted to find out the causes behind lower productivity. Simply improving efficiency of the truck loading increase the expected level of production, which is about 13% of the previous production. The manpower productivity has increased from 1.9 to 2.1 tank trucks per man-hour. By comparing the data before and after implementation of the project it is observed that five key resources are reduced per unit production due to improvement in the productivity. The saving from Electricity, Steam, Demineralized water, Industrial water, and Gasket is about Rs.90117 per month. For additional loading of tank truck (production) about Rs.2019412/- extra income is done by the organization as terminaling charge and saving from manpower is Rs.111504. Overall the saving from this project is Rs. 2221033/- per month. The reduction in consumable is also important from the economic as well as environment viewpoint. Industrial water and Dematerialized water consumption per bullet trick reduced. Electricity consumption is also reduced by 0.2 %. Thus it is also a part of social obligations.

For further increase the production we need to change system parameter and the setting of equipment according to the pumps& pipeline capacity and also synchronizing all of them. This may further increase the production of the truck loading. Thus, there is scope for improvement till 9984 bullet truck per month. The system efficiency can be improved in such way in future if required. Thus, there is scope for improvement in the present system.

Acknowledgement
The authors are thankful to the management of the Indian Oil Petronas Private Limited Haldia, West Bengal, India. Also special thanks to the department of process for their cooperation and help.

References

[1] Ranck, R. J., Yanasse, H. H., Morabito, R. and Junqueira L. (2019) A hybrid approach for a multi compartment container loading problem, Expert Systems with Applications, 137, 471- 492.

[2] Alonso M. T., Alvarez-Valdes R, Parreño F. and Tamarit J. M. (2016) Algorithms for Pallet Building and Truck Loading in an Interdepot Transportation Problem, Mathematical Problems in Engineering, 8, 1 – 11.

[3] Alonso, M.T., Alvarez-Valdes, R., Iori, M., Parreño, F. & Tamarit, J.M. (2017) Mathematical models for multicontainer loading problems, Omega, Elsevier, 66, 106-117.

[4] Moura, A. and Bortfeldt, A. (2017) A two – stage packing problem procedure, International Transactions in Operational Research, 24, 43–58.

[5] Bortfeldt, A. and Wäscher, G. (2013) Constraints in container loading – A state-of-the-art review, European Journal of Operational Research, Elsevier, 229(1), 1-20.

[6] Fanslau, T., Bortfeldt, A., 2010 A tree search algorithm for solving the container loading problem, INFORMS Journal on Computing, 22, 222–235.

[7] Bischoff, E. E. and Ratcliff, M. S. W., (1995) Issues in the development of approaches to container loading, Omega, Elsevier, 23(4), 377-390.

[8] Nivolianitou, Z. Konstandinidou and M. Michalis, C. (2006) Statistical analysis of major accidents in petrochemical industry notified to the major accident reporting system (MARS), Journal of Hazardous Materials, 137 (1) , 1-7.

[9] Hemmatian, B., Abdolhamidzadeh, B., Darbra, R., and Casal, J. (2014) The significance of domino effect in chemical accidents, Journal of Loss Prevention in the Process Industries, 29, 30-38.

[10] Attanasio, A., Fuduli, A., Ghiani, G., & Triki, C. (2007) Integrated shipment dispatching and packing problems: a case study, Journal of Mathematical Modelling and Algorithms, 6(1), 77-85.

[11] Pedruzzi S., Paulo L., Nunes A., Rosa R. and Arpini B. P. (2016) A mathematical model to optimize the volumetric capacity of trucks utilized in the transport of food products, Gestão & Produção, 23 (2), 350 – 364.

[12] Haessler R. W. and Talbot F. B. (1990) Load planning for shipments of low density products, European Journal of Operational Research, 44 2, 289–299.

[13] Chen, C. S., Lee, S. M., and Shen, Q. S. (1995) An analytical model for the container loading problem, European Journal of Operational Research, 80(1), 68-76.

[14] Oueidat, D., Guarnieri, F., Garbolino, E., Rigaud E., (2015) Evaluating the Safety Operations Procedures of an LPG Storage and Distribution Plant with STAMP, Procedia Engineering, 128, 83-92.

[15]. Moura, A. and Oliveira, J. F., (2009) An integrated approach to the vehicle routing and container loading problems, Operations Research Spectrum, 31, 4, 775–800.

[16]. Sanjeeb D., Marcos G., Oktay G., (2010) Two-Step MIR Inequalities for Mixed Integer Programs, INFORMS Journal on Computing, 22(2), 236–249.

IOP Conf. Series: Materials Science and Engineering 1080 (2021) 012020 doi:10.1088/1757-899X/1080/1/012020

Machining alumina plates using abrasive jet of silicon carbide

D K Adak[1], P Dutta[2], S Das[3] and B Haldar[4]

[1] Department of Mechanical Engineering, College of Engineering and Management, Kolaghat, Purba Medinipur –700106, West Bengal, India.
[2&3] Department of Mechanical Engineering, Kalyani Govt. Engineering College, Kalyani, Nadia-741235, West Bengal, India.
[4] Mechanical Engineering Department, College of Engineering, Al Imam Mohammad Ibn Saud Islamic University, PO Box: 5701, Riyadh- 11432, Kingdom of Saudi Arabia.
E-mail: sdas.me@gmail.com

Abstract. The abrasive jet machining (AJM) is used in many engineering applications including brittle material machining and finishing, surface preparation of the metal for welding and thermal spray coatings, etc. In this process, fine grain abrasive particles are given high speed through carrier gas which is directed to impinge on a target surface to be machined. The high kinetic energy of the abrasive particles is responsible to erode materials from the surface. In this work, three input cutting parameters such as abrasive flow rate, abrasive grain size, and stand-off distance (SOD) are chosen to experiment on 99.7% pure alumina plate of 4 mm thickness, using silicon carbide (SiC) abrasive. Material removal rate (MRR) and nozzle wear rate are the output responses that are to be investigated using response surface methodology, keeping working pressure constant at 6 kgf/cm^2 throughout.

Keywords: Abrasive Jet Machining, Material removal rate, Abrasive grain size, Abrasive flow rate, Stand-off distance, Response surface methodology.

1. Introduction

Superior quality sintered ceramic materials are now in use for various industrial applications. Uses of alumina (Al_2O_3) parts are in biomedical implants (hip and knee joints), ballistic protection components, ceramic seals, corrosion and wear protection linings, etc. [1]. Material removal in a brittle material was proposed [2] to be due to solid particle impact. Fracture toughness and hardness of workpiece was found to be critical parameters affecting MRR in Abrasive Jet Machining (AJM).

AJM process is capable of providing higher tolerance and greater surface finish [4]. It was invented a few decades ago, but still, several research works are being continued [5] towards its advancement. Wakuda et al. [6] examined the response of using different abrasive particles like aluminum oxide, silicon carbide (of moderate hardness), and Synthetic Diamond (SD) on alumina ceramic material, whereas Xu et al. [7] investigated the Abrasive Water Jet (AWJ) cutting of alumina ceramics with controlled nozzle oscillation. They found cutting at small nozzle oscillation angles could increase the depth of cut by as much as 82%. A statistical model was developed [8] for prediction and process optimization of micro-abrasive intermittent jet machining. Similar work was also done [9] to increase the efficiency of a jet in micro-grooving process. Hot air AJM was also carried out [10] to enhance MRR. The air temperature was found responsible for higher MRR and roughening of the machined surface due to plastic deformation accompanied by brittle fracture. Srikanth et al. [11] investigated the MRR and kerf characteristics by the AJM process using tungsten carbide nozzle by varying process parameters on glass, granite, and ceramic sheets. Results obtained by Taguchi were nearly close to that of ANOVA. Wear resistance for AJM and WAJM nozzles could be increased [12] by using porous cylindrical nozzle inserts, while some others investigated [13] on AJM of alumina using a fluidized bed, SiC abrasive and response surface methodology (RSM) to perform the experiment. AJM performance on porcelain tiles, soda-lime glass and soda-lime laminated glass [14, 15] was explored on an indigenously made system using SiC and common silica sand.

AJM has its inherent advantages, but still has many challenges to overcome, such as pollution issues, homogeneous mixing of air and abrasive, maintaining a set exit flow rate of abrasive, etc. In this investigation, an indigenously made AJM setup is used to optimize MRR and nozzle wear by setting suitable mass flow rate, abrasive grain size and SOD using SiC while machining alumina plate.

2. Experimental Investigation

The experiment is carried out on 99.7 % pure alumina plate. As per the RSM matrix, experiments are carried out with three levels of three parameters i.e. stand-off distance, mass flow rate, grain size as shown in Table 1. The pressure is kept constant at 6 kgf/cm^2 during all experiments. The nozzle is made from stainless steel with 10 mm inlet diameter, 2 mm outlet diameter and 20 mm in length.

Figure 1 Close view of nozzle and alumina plate

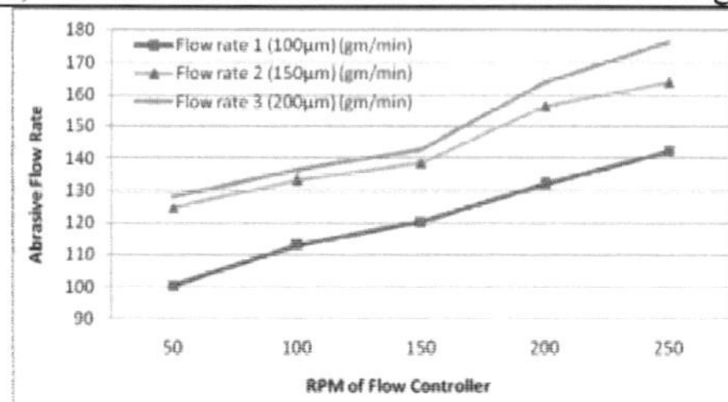

Figure 2 Flow rate calibration graph of silica sand at 6 kgf/cm^2

Table 1 Process parameter used in experiments at 6 kgf/cm^2 pressure

Parameters (Coded value)	Level 1 (-1)	Level 2 (0)	Level 3 (1)
Stand off distance (mm)	2	3	4
Grain size (μm)	100	150	200
Flow rate(gm/min)	120	130	140

2.1 Flow Rate calibration of Silicon Carbide

To calculate the abrasive flow rate, a constant mass of 50 gm SiC passed through an abrasive flow controller. Required abrasive flow rate obtained by controlling rotational RPM of the flow controller. The graph of the abrasive flow rate with the variation of rotor RPM of the flow controller is presented in Figure 2. The flow rate is directly proportional to the rotor RPM of the flow controller that is expected as per the design of the abrasive flow controller. It is observed that flow characteristic lines of 150 μm and 200 μm abrasive sizes are closer than 100 μm size of abrasive. This may be due to higher adhesive force between smaller grain sizes of abrasive grains and the flow passage wall. This plot of flow characteristic line guides to opt for any intermediate abrasive flow rate.

2.2 Procedure for Experiments

Experiments are conducted as per the design matrix as obtained from RSM and the Box-Behnken Design matrix. In these experiments, silicon carbide (SiC) with three different grain sizes are used as abrasive particles. With 20 sec of machining time, blind cavities are formed in the workpieces for calculating MRR and nozzle wear rate. Changes in nozzle diameter are measured by digital Vernier calipers. Nozzle wear is measured by taking the weight of the nozzle after each experiment.

3. Results and Discussions

3.1 Results and Discussion for MRR

Regression analysis has been done on MRR results obtained in the experiments conducted at 6 kgf/cm^2 pressure. The regression equation of MRR corresponds to linear and quadratic input parameter interaction is evaluated and shown in equation (1).

$$MRR = -5.12 + 0.00458\ D + 0.323\ \delta + 0.0689\ \dot{m} - 0.000015\ D^2 - 0.01405\ \delta^2 - 0.000250\ \dot{m}^2 - 0.000072\ D\times\delta + 0.000001\ D\times\dot{m} - 0.001713\ \delta\times\dot{m} \quad (1)$$

where, D = grain size, $\dot{m}$= mass flow rate, δ = SOD

Table 2 Analysis of Variance for MRR

Source	DF	Adj SS	Adj MS	F-Value	P- Value
Model	9	0.009645	0.001072	5.19	0.042
Linear	3	0.001268	0.000423	2.05	0.226
D	1	0.000368	0.000368	1.78	0.239
δ	1	0.000191	0.000191	0.92	0.380
$\dot{m}$	1	0.000709	0.000709	3.44	0.123
Square	3	0.007150	0.002383	11.55	0.011
$D * D$	1	0.004982	0.004982	24.14	0.004
$\delta * \delta$	1	0.000729	0.000729	3.53	0.119
$\dot{m} * \dot{m}$	1	0.002301	0.002301	11.15	0.021
2-Way Interaction	3	0.001227	0.000409	1.98	0.235
$D * \delta$	1	0.000051	0.000051	0.25	0.639
$D * \dot{m}$	1	0.000002	0.000002	0.01	0.929
$\delta * \dot{m}$	1	0.001174	0.001174	5.69	0.063
Error	5	0.001032	0.000206		
Lack-of-Fit	3	0.001032	0.000344	3684.45	0.000
Pure Error	2	0.000000	0.000000		
Total	14	0.010677			

The analysis of variance (ANOVA) shown in Table 2, is used to analyze the effect of individual and interactive parameters on MRR. Statistical analysis software, Minitab 17 is utilized to tabulate experimental results at a 95% confidence level. Small values of p for interactions of SOD and mass flow rate (p = 0.063) and the squared terms of grain size (p= 0.004) and mass flow rate (p= 0.021) correspond to a good correlation with the response. The correlation coefficient, R^2 of the regression model for MRR has a value of 90.34% which is greater than 90%, and so, the model is valid and parameters are significant. This is also supported by the small p-value of 0.042 of the model. Observed contour and surface plots of MRR with respect to interactive input parameters at 6 kgf/cm^2 pressure are shown Figure 3 and Figure 4.

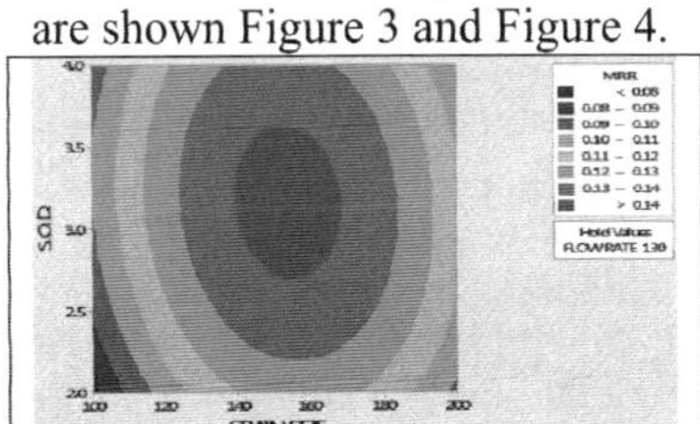

Figure 3 Contour plot of MRR with variation in grain size and SOD

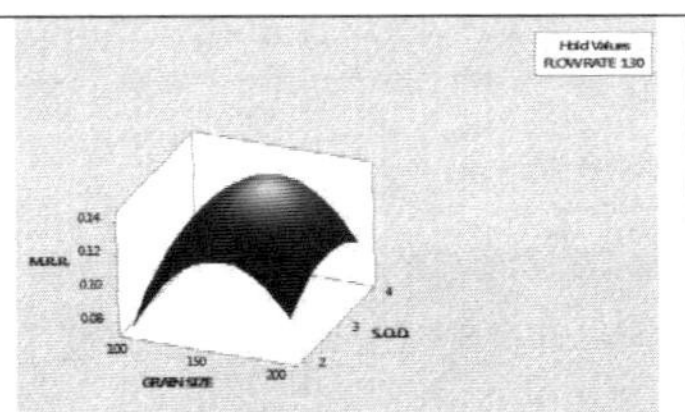

Figure 4 Surface plot of MRR with variation in grain size and SOD

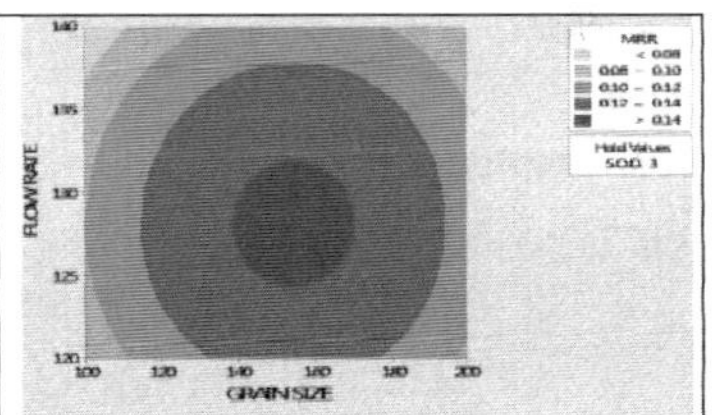

Figure 5 Contour plot of MRR with variation grain size and flow rate

IOP Conf. Series: Materials Science and Engineering 1080 (2021) 012020 doi:10.1088/1757-899X/1080/1/012020

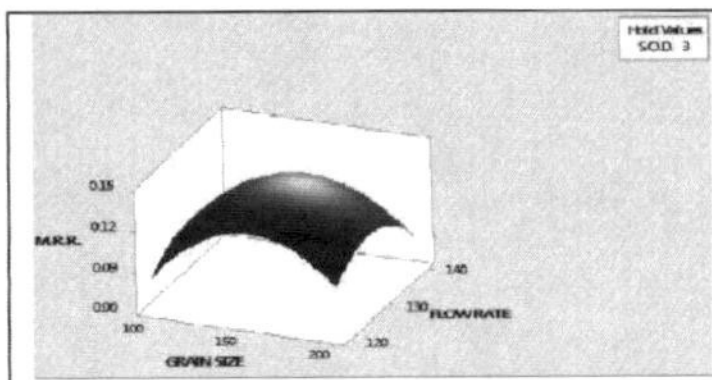

Figure 6 Surface plot of MRR with variation grain size and flow rate

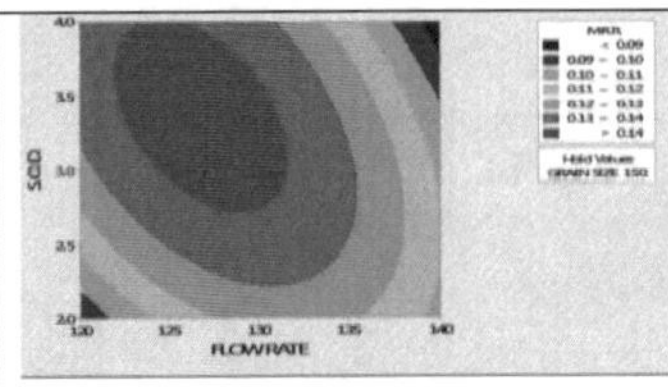

Figure 7 Contour plot of MRR with flow rate and SOD

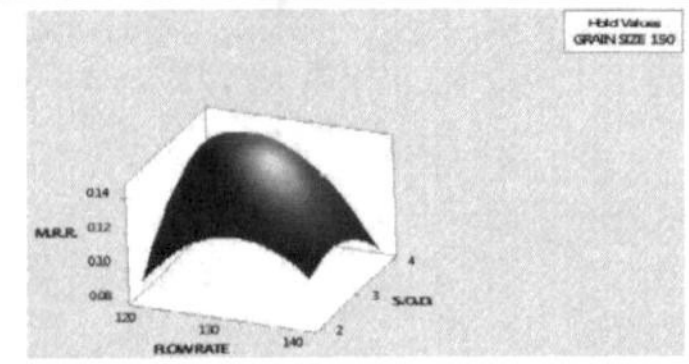

Figure 8 Surface plot of MRR with variable flow rate and SOD

In Figure 3 and Figure 4, MRR varies with SOD and grain size keeping flow rate constant at 130 gm/min. It is observed that MRR gradually increases with the use of bigger grain size and attains maximum MRR when grain sizes are within 145-165μm. The MRR then decreases with further large size particles. It may be because larger grain sizes transfer more impact energy up to a certain level at the constant air pressure, and cause more minute brittle fractures of work material thereby removing more materials from the work surface. Further increase in grain size may likely cause fewer impacts of the particles on the unit work surface area and higher intra-grain impacts thereby loosing energy of impact on to the work surface within a certain time resulting in decreasing MRR. MRR gradually increases with the increment of SOD and attains a maximum value near 3mm then MRR decreases with a further increase in SOD. Initially, with the increase of SOD, abrasive particles get sufficient displacement to be accelerated before striking the work surface, but after passing a certain SOD, velocity is reduced due to normal air resistance and flow becomes divergent. Because of these, high SOD beyond a value causes less impact on the work surface, decreasing MRR.

Surface and contour plots of MRR with the variation of flow rate and grain size at a constant SOD of 3 mm are depicted in Figure 5 and Figure 6 respectively. It is observed that MRR is maximum within 145-165μm grain sizes and decreases when it goes further left or right like before. At the flow rate of between 125 gm/min and 130 gm/min above 160μm grain sizes, it gives maximum MRR, and it gradually decreases if the flow rate increases further. For smaller grain-size particles, MRR decreases with increasing flow rate because when flow rate increases, more particles are likely to collide with each other and may lose their energy.

In Figure 7 and Figure 8, variation of MRR is shown concerning flow rate and SOD by contour and surface plots respectively keeping grain size constant at 150 μm. These figures show that MRR is maximum with a SOD in between 2.75-4 mm and a flow rate within 122-132 gm/min. It can be said that within a higher range of SOD, abrasive particles get sufficient length to accelerate to increase the kinetic energy of the particles with increasing SOD thereby increasing MRR. The above observations also show that change in MRR is more rapid towards lower abrasive flow rate and higher SOD.

3.2 Results and Discussion for Nozzle Wear

In some experiments with the same input parameters, stainless steel nozzles wears are calculated and a regression model is developed for nozzle wear.

Table 3 Analysis of Variance for Nozzle Wear

Source	DF	Adj SS	Adj MS	F-Value	P- Value
Model	9	0.000056	0.000006	7.48	0.019
Linear	3	0.000012	0.000004	4.81	0.062
D	1	0.000004	0.000000	4.47	0.088
δ	1	0.000000	0.000008	0.08	0.788
$\dot{m}$	1	0.000008	0.000009	9.89	0.026
Square	3	0.000026	0.000026	10.60	0.013

D * D	1	0.000000	0.000000	31.43	0.002
δ * δ	1	0.000001	0.000001	0.17	0.694
$\dot{m}$ * $\dot{m}$	1	0.000018	0.000006	1.05	0.353
2-Way Interaction	3	0.000015	0.000015	7.33	0.028
D * δ	1	0.000001	0.000001	18.86	0.007
D * $\dot{m}$	1	0.000002	0.000002	0.64	0.460
δ * $\dot{m}$	1	0.000004	0.000001	2.50	0.175
Error	5	0.000004	0.000001		
Lack-of-Fit	3	0.000000	0.000001	3898.95	0.000
Pure Error	2	0.000060	0.000000		
Total	14				

Analysis of variance (ANOVA) shown in Table 3, is also done at a confidence level of 95%. Small *p* values of mass flow rate (p= 0.026), the squared terms of grain size (p= 0.002), and interactions of grain size and SOD (p= 0.007) show that there is a curvature in the response surface. The correlation coefficient R-sq for nozzle wear has a value of 93.17% which reveals that the modeled equation is significant; this is also validated by the small model *p*-value of 0.019. The regression equation for nozzle wear rate (WR) is given in Equation (2).

$$WR = -0.0941 + 0.000543D - 0.00213\,\delta + 0.00105\,\dot{m} - 0.000001D^2 - 0.000197\,\delta^2 - 0.000005\,\dot{m}^2 - 0.000039\,D\times\delta - 0.000001\,D\times\dot{m} + 0.000072\,\delta\times\dot{m} \quad (2)$$

where, WR = nozzle wear rate, D = grain size, $\dot{m}$= mass flow rate, δ = SOD

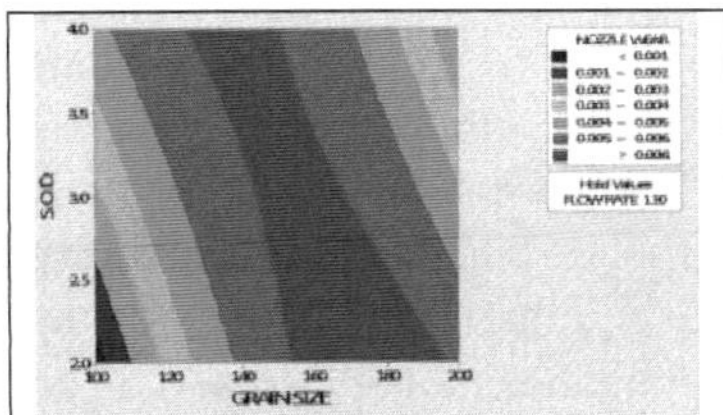

Figure 9 Contour plot of nozzle wear with variation of grain size and SOD

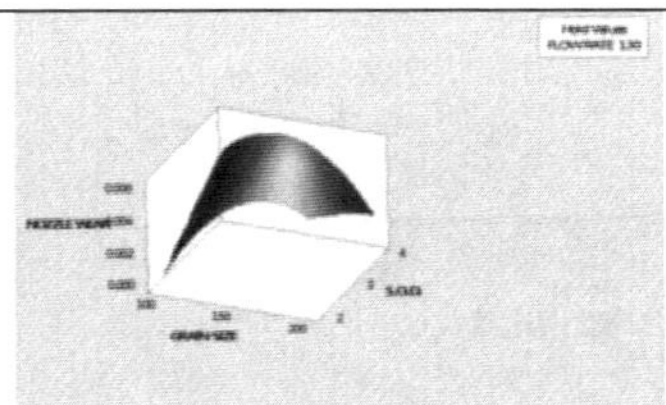

Figure 10 Surface plot of nozzle wear with variation of grain size and SOD

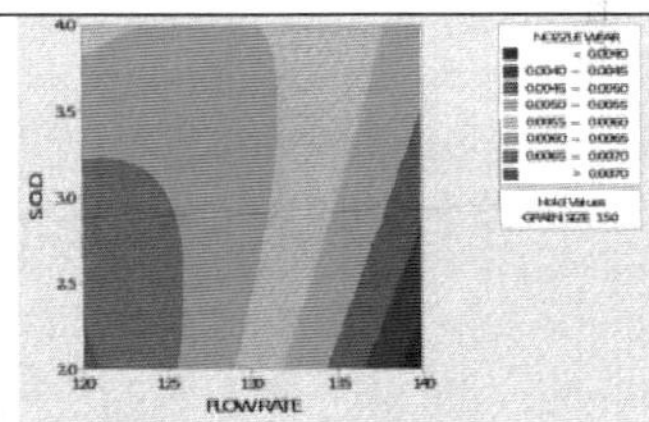

Figure 11 Contour plot of nozzle wear with variation of flow rate and SOD

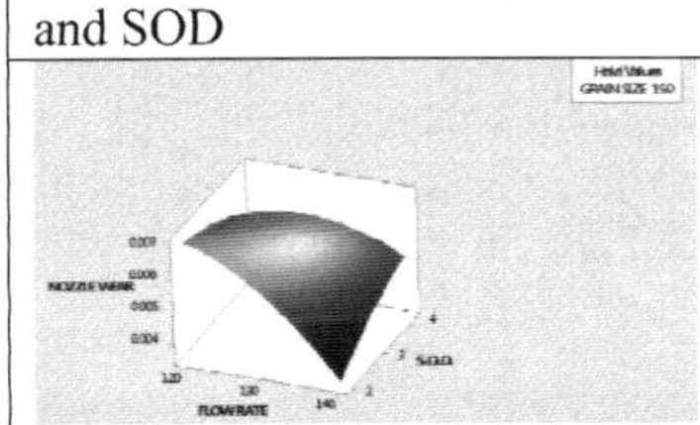

Figure 12 Surface plot of nozzle wear with variation of flow rate and SOD

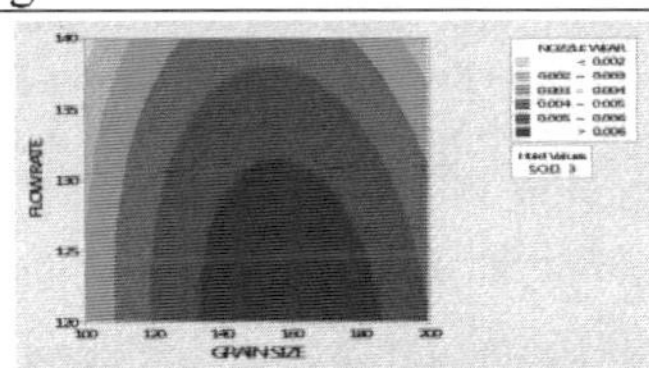

Figure 13 Contour plot of nozzle wear with variation grain size and flow rate

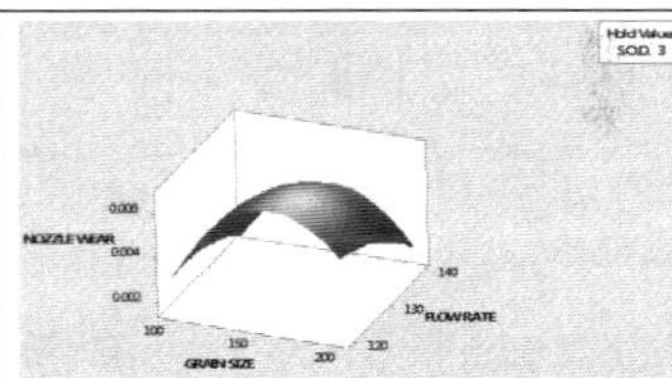

Figure 14 Surface plot of nozzle wear with variation of grain size and flow rate

In Figure 9 (contour plot) and Figure 10 (surface plot), variation of nozzle wear rate concerning grain size and SOD is shown, keeping flow rate constant at 130 gm/min. From the above graphical representation, it is observed that nozzle wear increases with increasing grain sizes and attains the maximum value between 152μm-195μm grain sizes with 2 mm SOD. With the increase of SOD, nozzle wear rate increases more rapidly with the same grain sizes and attains a maximum with 127μm-

152μm grain size and at 4 mm SOD; after that nozzle wear rate decreases. Increase of SOD increases abrasive velocity up to an extent, thus increasing nozzle wear rate even with small grain sizes.

Variation of nozzle wear concerning grain size is more significant at higher SOD. With lower grain size, nozzle wear rate increases with SOD. But with higher grain size, nozzle wear rate decreases as SOD increases. Due to less dragging effect with smaller grain size, nozzle wear rate increases with SOD may be because of velocity enhancement. Similarly, more dragging effect may have reduced the velocity of higher grain size particles, thus reducing nozzle wear rate.

In the contour and surface plots (Figure 11 and Figure 12), variation of nozzle wear rate with mass flow rate and SOD are shown when grain size of 150μm is constant. It is observed that nozzle wear rate decreases rapidly with increasing mass flow rate at lower SOD. At higher SOD, the rate of wear decreases slowly with mass flow rate. At a lower mass flow rate, nozzle wear rate decreases slowly as SOD increases. At the higher mass flow rate, nozzle wear rate increases as SOD increases. Nozzle wear rate becomes maximum when both mass flow rate and SOD are minimum (120-121 gm/min and 2-2.2 mm respectively). High mass flow rate reduces abrasive velocity and nozzle wear at low SOD.

In Figure 13 and Figure 14, contour and surface plots of nozzle wear rate with varying grain size and mass flow rate are shown keeping SOD constant at 3mm. Nozzle wear rate initially increases with increasing grain size and after a certain limit, it decreases with increasing grain size. With the same grain size, nozzle wear rate decreases as the mass flow rate increases. At a higher mass flow rate, nozzle wear rate shows a reducing trend with grain size increment. Initially nozzle wear rate increases as grain size increases due to higher plowing, but as grain size increases, may be due to dragging effect, wear rate reduces. Similarly when mass flow rate increases, nozzle wear rate also decreases.

4 Conclusions

- With increasing grain size at the constant mass flow rate, MRR and nozzle wear both are initially increased and then decreased. With increasing mass flow rate, initially, MRR increases but then decreases, but the nozzle wear rate decreases. The primary reason for nozzle wear is plowing and striking with the rebounding particle, rather than friction with nozzle surface.
- The flow becomes divergent with increasing SOD as a result the rebounding of the abrasive particle is decreased with increasing SOD. The abrasive grain size and mass flow rate are having significant effects on the nozzle wear rate as compared to SOD.

References

[1] Pertti A 1996 *Research Notes 1792, Technical Research Centre of Finland (VTT).*
[2] Wakuda M, Yamauchi Y and Kanzaki S *J. Int. Soc. Prec. Engg. Nanotech.* vol 26 p193.
[3] NPTEL 2018 *Module 9 Lesson 35 Version 2 ME IIT KGP* (accessed on 11th November 2018).
[4] Mishra P K 2014 *The Institution of Engineers (India) Text Book Series, Kolkata,* p12.
[5] Muju M K and Pathak A K 1988 *J. of Mechanical Working Technology* vol 17 No 10 p 325.
[6] Wakuda M Yamauchi Y and Kanazi S 2003 *Mat. Proc. Techno.* vol 132 p 177.
[7] Xu S and Wang J 2006 *Inter.J. of Advanced Manufacturing Technology* vol 27/7-8 p 693.
[9] Jhang L, Kuriyagawa T, Yasutomi Y and Zhao J 2005 *Int. J. Mach. Tools Manuf.* vol 45(7–8) p 873.
[10] Jagannatha N, Hiremath, Somashekhar S S, Sadashivappa K and Arun K V 2012 *Mach. Sc. Tech.: An Int. J.* vol 16(3) p 459.
[11] Jagannatha N, Hiremath S S and Sadashivappa K 2012 *Int. J. Mech. Mat. Engg.* vol 7(1) p 9.
[12] Srikanth D V and Sreenivasa Rao M 2014 *Procedia Materials Science* vol 6 p 1303.
[13] Sychuk V, Zabolotnyi O and McMillan A 2015 *Powder Met. Metal Cer.* vol 53 p 600.
[14] Nanda B K, Mishra A and Dhupal D 2017 *Int. J. Adv. Manuf. Tech.* vol 90 Issue 9-12 p 3655.
[15] Ghara T, Ansari R, Adak D K, Ahmed M, Das S and Haldar B 2018 *Proc.1st Int. Conf. Mech. Engg Jadavpur Univ. Kolkata India* Jan 4-6 p 534.
[16] Karmakar A, Ghosh D, Adak D K, Mandal B, Das S, Ahmed M and Haldar B 2019 *Adv. in unconv. Mach. and Comp* p 163.

A simple online tool condition monitoring system using artificial neural networks

A Mukherjee[1] and S Das[2]

Mechanical Engineering Department, Kalyani Government Engineering College, Kalyani, West Bengal, India

E-mail:[1]avishekmukherjee100@gmail.com, [2]sdas.me@gmail.com

Abstract. During machining occurrence of tool wear is a common phenomenon. As tool wear increases, rubbing of flank surface and workpiece also gets increased. Then, the desired quality of workpiece is not possible to achieve. To minimize this loss of quality, a cutting tool should be changed or ground after reaching certain amount of average flank wear (0.3 mm for uniform and 0.6 mm or non-uniform flank wear). For the detection of worn out state of a cutting tool, condition monitoring is required. During past decades, a lot of research works had been done on both offline and online monitoring of cutting tool. Most of the researchers used high cost setups and sensors for wear detection purpose. In this work, tool wear is detected using spindle speed as the wear detection parameter. Artificial Neural Networks is used as data analysing tool. Back propagation algorithm is used as learning algorithm. Results show that proposed methodology is capable to detect tool wear satisfactorily.

Keywords: machining, turning, online, tool condition monitoring, tool wear, spindle speed, artificial neural networks, ANN, NN.

1. Introduction

In a conventional machining process, there is a physical contact between a workpiece and a cutting tool. Due to this contact, cutting tools have to withstand rubbing action, high stresses and high temperature gradient at the work-tool interface. As a result, wearing of cutting edges occurs, or, there may be a chance of catastrophic failure of cutting tool. If cutting tool wear exceeds a limit during production of job, there will be a chance of degradation in quality of product. That is why tool condition needs be monitored. Another reason of monitoring the condition of tool is that the Risk Priority Number(RPN) of cutting tool is high comparatively to other component of a typical machining setup [1]. RPN = O x S x D,where O = Occurrence of failure, S= Severity of failure, D = Detectability. According to Failure Modes, Effects and Criticality Analysis (FMECA), tool condition should be monitored [2] continuously. Tool condition can be monitored in two ways, either online, or offline. Offline monitoring consumes more time, but in online monitoring, real time values of tool wear can be obtained. In offline monitoring,usually direct measurement of tool wear is done. Reduction indimensions of tool after being worn out is considered. For this, premature failure of cutting edge cannot be detected. To reduce these limitations, online monitoring systems, commonly using some indirect techniques,are developed.Researchers,however, reported [3] direct methods to be more accurate than indirect ones. While gathering online data or signals, there may be disturbances which can distort original signals though there were many filters used for minimizing those disturbances.

According to literature [4], some parameters like cutting force, acoustic emission, machine tool vibration, motor current and spindle power had taken a lion's share in online tool wear detection. Ghani et al. [5] used a strain gauge type dynamometer for measuring of triaxial cutting forces. This work seemed to be very effective because the error in real time value from measured value varied from 5.91 to 16 percent only. A neural networks based monitoring of coated carbide tool was tried by Daset al. [6, 7] during turning of medium carbon steel using the strategy ofsensor fusion. A three-axis cutting force dynamometer and a piezoelectric accelerometer were used for measuring of cutting forces and vibration signals coming out from machine tool while machining. Authors used back propagation neural networks to detect tool wear.Bhuiyan et al. developed [8] a monitoring system to detect plastic deformation during occurring of tool wear. For this, authors used an acoustic emission sensor and collectedsignals during machining of different frequencies to find out the characteristic signal corresponding to tool wearing. A case study [9] for detection of tool wear had been done by Shen etal. by sensing power signal of spindle and extracted the characteristic feature. Hilbert-Huanng transformation tool was used to process the data taken from experiments. On the other hand, a monitoring systemto detect tool wear in end milling cutter was developed by Karandikar et al. [10]. Cutting forces at different rotations of spindle were taken as extracted features. Raoet al. [11] reported nose radius to play a significant role during regression analysis of experimental data, and it had a correlation coefficient of 0.4581 against tool wear value. Axinte et al. developed a system using input spindle power signal. Experiments were done by both intermittent, i.e. milling, and continuous cutting operation such as drilling and turning. Das et al. used the Analytical Hierarchy Process (AHP) [12,13] to identify the states of tool condition in a simple way. According to Swain et al., adaptive system [14] is quite helpful as industries are dependent on mostly automated systems. For this type of system, authors mentioned sensors to have played a key role. Cai et al. did experiments on hybrid model [15] based on long short-term memory-based network for tool wear detection, which resulted in 'outstanding' performance, especially when operating conditions were varying. In a review, Ostasevicius [16] stated that during industrial revolution (IR 4.0), wireless sensor communication along with internet of things (IOT) made condition monitoring easy comparatively. Denkena et al. [17] reported about non sensitive feature selection for risk minimization in condition monitoring.

From literature, it can be said that, established methods for tool condition monitoring till date are quite costly in general. So, aim of this project is to explore online monitoring of tool wear by different means with a low-cost setup. Using low cost sensors anda software with a graphical user interface (GUI), a monitoring system is tried to be developed in this work. Tool wear value can be determined as real time value by sensing rotational speed of spindle as theparameter to detect wear.

2. Cause of choosing RPM as wear detection parameter

Machine tool used in this work is anNH22 HMT lathe. Main power drive of this machine tool relates to a three-phase induction motor. In this motor, slip percentage, $S= 100(N_S-N_R)/N_S$,where,Ns= Synchronous speed, N_R = Rotor speed. According to Figure1, as torque changes there is a change in slip percentage. For all rotating machines, synchronous speed is almost constant. So, slip changes because of change in rotor speed. If rotor speed increases, slip will decrease and vice-versa. It can be stated that if torque changes, there should be a change in rotational speed of the machine. If load changes, torque is also changed.During turning, when there is large value of tool wear, load on the motor will be increased. Thus, there will be an increase in torque as well as slip of motor. If slip changes, RPM of rotor fluctuates. So, using the values of fluctuated rotational speed (RPM),tool wear can be detected. That is why RPM is selected as a parameter to detect wear by online condition monitoring.

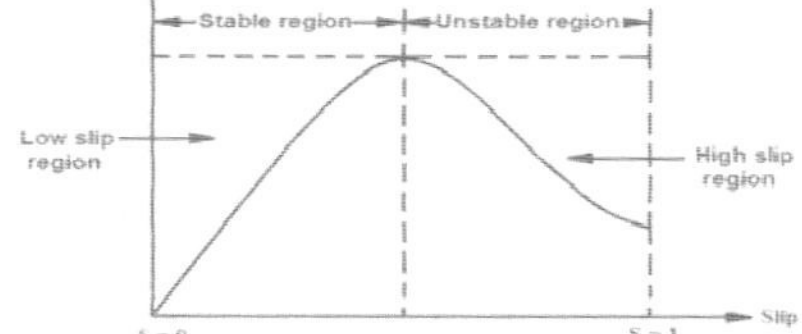

Figure 1. Torque-slip characteristics of a three-phase induction motor [18]

3. Experimental procedure

Experiments are performed on turning on AISI 1045 alloy steel rod as the workpiece and uncoated single carbide (SNUN 120408- P30) as the cutting tool. Total twenty-four experiments are performed on HMT lathe (NH22). RPM values are measured using infra red (IR) sensors and Arduino board. Also, an Arduino based data acquisition system is developed using LABVIEW software to get the real time values of RPM in excel sheet. Skewness values are extracted from each set of data. Before and after each experiment, tool wear values are taken. Gradual increment of tool wear is shown in Figure 2, 3 and 4 respectively.

During the trial tests of this work, long cylindrical bar has been taken. However, fluctuation of rpm has not been detected well. This is likely to be due to lack of stiffness along the cylindrical bar. A cylindrical bar is quite stiff at the headstock end and stiffness becomes the least in the middle portion of the bar close to the tailstock. Hence, variation of torque due to stiffness is considered to be a problem to the measurement of rpm. For this, short length of bar of 130 mm is taken for the experiment.

After getting tool wear values, those are processed through Neural Networks toolbox in MATLAB. Back propagation algorithm is used to train the data set. Feed forward algorithm is used to change the weights of the network to get the desired output. This process is continued until error between network output and original output becomes too less. At the end of this process, the final neural networksis obtained. Using the obtained neural networks, tool wear can well beestimated. After training, some inputs other than training data set are given and the neural networks gives estimation of tool wear value. Error is calculated between NN output and the measured tool wear value. In this way, the proposed method is found working satisfactorily.

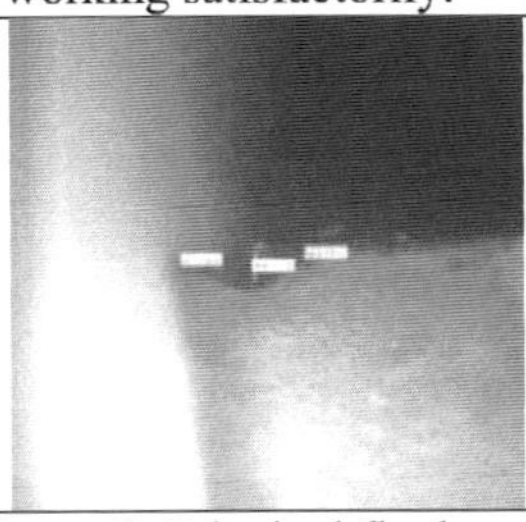		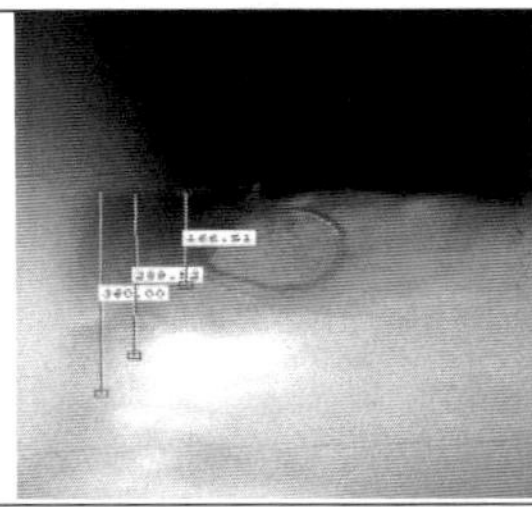
Figure 2. Principal flank wear after experiment 1	**Figure 3.** Principal flank wear after experiment 9	**Figure 4.** Principal flank wear after experiment 23

4. Experimental data and results

Input and output parameters are written in tabular form. V_c= Cutting velocity in m/min, S_o= Feed in mm/rev, t = Depth of cut in mm, W_a= Average auxiliary flank wear value in micron, W_p= Average flank wear value in micron, where V_c, S_o, t and skewness are the input parameters to the network. W_a and W_p are the output of the network whose structure is 4-32-16-2. Tan sigmoid function has been used as the activation function. Total 58 iterations were done to reach minimum error.

After this, this data is organised in three sections as low wear section, medium wear section and high wear section. Table 1 contains data for low wear section for constructing the network. After training the neural network with this set of data, NN estimated Output becomes equal to 0.99 * Target + 0.15, when Target is the measured wear value. During testing stage, NN estimated Output becomes 1.8 * Target - 44. On the whole, NN Output = 1.1 * Target - 3.7 considering both training and testing datasets.

Bar diagram (Figure 5) has been made with few experimental wear values (auxiliary flank wear (column 1 and 2) and principal flank wear (column 3 and 4) of the last two data sets) to compare them with NN output. In the legend, Series1 denotes the NN estimates and Series 2 indicates measured values of average flank wear during experiment.

Table1. Data for low wear section

V_c	95	95	92	89	89	89	86	84	84
S_o	0.04	0.04	0.04	0.04	0.04	0.04	0.04	0.04	0.04
t	2	2	2	2	2	2	2	2	2

Skewness	-.1768	-0.4544	-0.0372	-0.1035	-0.6392	-0.2267	-0.0979	-0.0005	0.0226
W_a	25.112	27.67	28.22	35.343	36.39	37.2	44.67	51.427	59.087
W_p	38.267	42.37	27	52.7167	55.429	60.4067	62	62.25	76.7

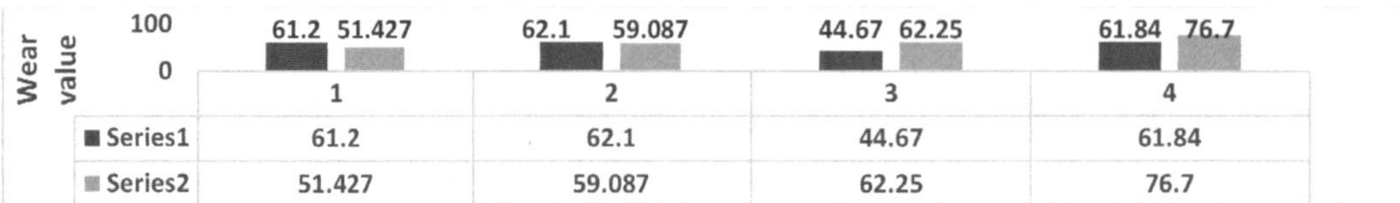

Figure 5. Bar diagram of comparison of low wear test

Table2 shows data for medium wear section. Regression equation of the neural network corresponding to all the training and testing data setsis:Output=1*Target-8.2*10^{-15}. Figure 6 shows the typical comparison between experimentally measured wear values and NN predicted values of few data (auxiliary flank wear (column 1 and 2) and principal flank wear (column 3 and 4) of the last two data sets). In the legend, Series1 denotes the predicted values by NN networks and Series2gives measured values of wear during experimentation.

Table2. Data for medium wear section

V_c	95	95	92	89	86	84
S_o	0.04	0.04	0.04	0.04	0.04	0.04
t	2	2	2	2	2	2
Skewness	-0.1801	-0.1801	0.0087	-0.0844	-0.0486	0.1485
W_a	76.15	76.15	82.467	87.34	90.52	108
W_p	86.86	86.86	100.8	129.61	131.33	132.667

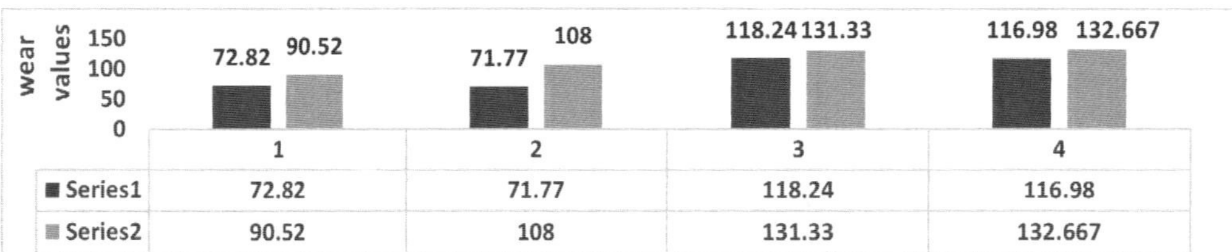

Figure 6. Bar diagram of comparison of medium wear test

Table 3 shows the data set for high wear section to form the neural networks. For training, NN estimated Output= (1* Target+ 0.66). With the testing data, the NN Output becomes (0.8*Target+1.9). Overall, with training and testing stage data, neural networks estimated Output =0.96*Target+1.7.Figure 7 shows comparison between two sets of experimental wear(auxiliary flank wear and principal flank wear of 1st and 3rd column data given in Table 3) values and NNestimated values. In the legend, Series1 denotes the NN output and Series2 denotes measured values of average flank wear during experiment.

Table 3. Data for high wear section

V_c	95	95	95	92	92	89	89	86	86	84	84
S_o	0.04	0.04	0.04	0.04	0.04	0.04	0.04	0.04	0.04	0.04	0.04
t	2	2	2	2	2	2	2	2	2	2	2
Skewness	0.0153	0.0152	0.0525	0.2726	0.3572	0.1178	-0.3382	.0741	-0.0078	0.0398	-0.1292
W_a	113.811	113.81	157.01	114.76	176	119.36	183	137	193.153	146.79	289.27
W_p	143.52	143.52	206.86	146.69	215.83	147.51	238.83	156.51	273.183	157.52	369.67

	1	2	3	4
Series1	110.84	115.71	136.43	144.12
Series2	113.811	157.01	143.52	206.86

Figure 7. Bar diagram of comparison of high wear section

As the results show, deviation in estimated value by neural networks from the measured value increases in high wear cases. This may be due to some causes to explain. As flank wear is increasing, contact of tool with work piece will get increased. So, variation of torque is not the same as occurred for low wear sections. This may be an important cause for high fluctuation at high wear region. Kaya et al. [19] reported the variation of torque and force to be higher with high tool wear. This was concluded for a milling process where an artificial neural network was employed. As shown in experimented data, depth of cut and feed are constant throughout the experiment. Cutting velocity is not given a wide variety, but one can see that skewness extracted from RPM values has a wide variation. So, this can be a key feature for wear detection, where cutting tool is in rotating condition such as grinding, milling, drilling, etc. Again, this newly designed tool condition monitoring setup costs quite less than the equipment used by previous researchers, and hence, may be employed at any small and medium scale industries.

5. Conclusion

Tool wear is successfully estimated with a low deviation with the proposed system. The setup is of quite low cost compared to that proposed by the previous researchers, and it is easy to connect the setup with a machine tool. Signals coming from the sensor used may have some disturbances that can be filtered out to reduce the error in output. In this work low-cost sensor is used, but one can also use high cost available sensor. At last, it can be stated that the proposed methodology can easily be implemented as the system developed is able to estimate flank wear with quite good accuracy.

References

[1] Mohanty AR 2014 *Machinery Condition Monitoring: Principles and Practices*.
[2] Ma L Kaneko S (Ed) 2007 *Proc. 12th Asia Pacific Vibration Conf. JSME CD Rom* p 1.
[3] Ambhore N Kamble D Chinchanikar S and Wayal V 2015 *Mat. Today Proc.* vol **2**(4-5) p3419.
[4] Dan L and Mathew J 1990 *Int. J. Mach. Tools & Manuf.* vol **30**(4) p579.
[5] Ghani J Rizal M, Nuawi M Ghazali M and Haron C 2011 *Wear* vol **271**(9-10) p2619.
[6] Das S Chattopadhyay A and Murthy A 1996 *Neural Networks* vol **9**(9) p1639.
[7] Das S Bandyopadhyay P and Chattopadhyay A 1997 *J. Mat. Proc. Tech.* vol **63**(1-3) p187.
[8] Bhuiyan M Choudhury I Dahari M, Nukman Y and Dawal S 2016 *Measure.* vol **92** p 208.
[9] Shen B Gui Y Chen B Lin Z Liu Q and Liu Q 2019 *Int. J. Adv. Manuf. Tech.* vol **106**(3-4) p1385.
[10] Karandikar J McLeay T Turner S and Schmitz T 2014 *Int. J. Adv. Manuf. Tech.* vol **77**(9-12) p1613.
[11] Venkata Rao K Murthy B and Mohan Rao N 2013 *Measure.* vol **46**(10) p4075.
[12] Das S Islam R and Chattopadhyay A 1997 *Proc. IMechE Part B: J. Engg. Manuf.* vol **211**(1) p19.
[13] Das S and Chattopadhyay A 2003 *Int. J. Mach Tools & Manuf.* vol **43**(1) p1.
[14] Swain S Panigrahi I Sahoo A and Panda A 2020 *Materials Today: Proc.* vol **23,** p474.
[15] Cai W Zhang W Hu X and Liu Y 2020 *Journal of Intelligent Manufacturing* vol **31**(6), p1497.
[16] Ostasevicius V Karpavicius P Jurenas V Cepenas M Cesnavicius R and Eidukynas D 2020 *Int. J. Adv. Manuf. Tech.* Vol **110**(3-4) p1015.
[17] Denkena B, Bergmann B and Stiehl T 2020 *Proc. CIRP* vol **88** p312.
[18] https://wwwgooglecom/search/torque-slip characteristics accessed on 8/10/2019.
[19] Kaya B, Oysu C and Ertunc H 2011 *Adv. Engg. Software* vol **42**(3), p76.

Tribological characterization of electroless Ni-P-Cu-TiO_2 coatings

Anupam Jana, Supriyo Roy*, Bal Mukund Mishra and Goutama Kumar Bose

Department of Mechanical Engineering, Haldia Institute of Technology, Haldia, India – 721657

*Email: supriyoroy11@gmail.com

Abstract
Present research work aims to describe the deposition of electroless quaternary Ni-P-Cu-TiO_2 coating on mild steel substrate and its tribological characterization. Energy dispersive x-ray analysis confirms the presence of all four elements within the coating. Tribo-testing at different sliding velocity revealed wear depth is high but stable at high sliding velocity whereas the coefficient of friction increased linearly with sliding velocity.

Keywords: Quaternary coating, Electroless, Ni-P-Cu-TiO_2, Wear, Friction

1. Introduction

The emphasis on electroless composite coatings for superior micro-hardness, corrosion, friction and wear resistance properties dates back to the notable works of few scientists with evidence [1, 2]. Payder and Araghi concentrated on the micro-hardness of the composite and so for its increase went for the incorporation of boron in the form of B_4C to form Ni-P-B_4C composite [3]. As a result, there was an increased wear resistance of AZ91D alloy substrate significantly in the presence of hard B_4C composites. Leon et al. reported excellent friction and wear resistance found in the Ni-P-BN(h) composite coatings [4]. Bozzini heat-treated the Ni-P/diamond composite layers and found out an increased hardness along with an increased Young's Modulus [5]. The incorporation of diamond particles and heat treated up to 400°C showed results of fantastic higher hardness, corrosion resistance and wear resistance. To gain improved micro-properties, uniform incorporation of particles in matrices was required and hence Sol-Gel technique was developed. The technique of very modern day over electroless coating for uniform agglomeration of particles in composites and Chen used it in his Ni-P-TiO_2 matrix [6]. The experimental results confirmed that the wear resistance and micro-hardness of the sol-gel TiO2 coatings were higher as compared to conventional TiO_2 coatings. At later stages Jappes used X-Ray Diffraction method and heat treatment technique for the measurement of wear resistance of Ni-P deposits in reference to different temperatures [7]. It was observed that at 350°C the coating showed higher wear resistance than at 500°C. His work was a confirmation of the technique used by Chen [6]. Few authors observed improvements in the tribological properties of the composites when there is a change in the orientation of the specimen (say from vertical to horizontal). Even electroless nickel with polymer matrix showed an improved result too as an effect of the addition of soft particles. The work of Mohammadi and Ghorbani on EN-PTFE-MoS_2 showed better experimental results relative to wear resistance with respect to any other incorporated soft particles to improve wear resistance, corrosion resistance [8]. Other than that, with the incorporation of certain hard particles like SiC, Ni composites showed a prospect of high micro-hardness and high Friction co-efficient. These were the works performed by Zarebidaki and Allahkaram [9] and prior to them it was the observation of Grosjean [10] as well. There was a decrease in the corrosion resistance of Ni-P-SiC as well due to the porosity and agglomeration of nano-particles. It improved when heat treated at annealing temperature of 400°C for 1 hour. With that result, Mallafati [11] observed that due to the reduction in the exposed area of metallic matrix as by the presence of SiC particles coated layer, the Ni-P alloy showed an improvement in the corrosion properties.

Above discussion ensures that, with an increased crystallinity, low porosity and decreased residual stresses, elements like Al_2O_3, B_4C, SiC, ZrO_2, TiO_2 are able to provide the degree of hardness required. Thus, the aim of this work is to develop and characterize electroless coated Ni-P-Cu-TiO_2 composite coatings where, Copper and Titania acts as a ternary and quaternary components respectively. The coatings have been deposited using hypophosphite bath. The effects of sliding speed on the tribological behavior of this composite coating have been studied at close proximity.

2. Experimental details

The chosen substrate for the purpose of coating in this paper is Mild Steel (AISI 1040). As per the size stringency, to fit in the pallet comfortably for further analytical tests the pin-type samples with specifications of 30 mm length and 6 mm diameter were used. The pin-type samples were degreased for the experiment by the process as mentioned. Three different beakers were taken containing dilute hydrochloric acid solution (50% HCL and 50% distilled water), Distilled water (deionized water), palladium solution respectively ($PdCl_2$). The sample was stringed carefully for the preliminary chemical treatment for activating the substrate for deposition. The sample was dipped in deionized water and then into dilute HCL solution, then after a few moments into palladium solution (warmed up to a few degrees above room temperature). The specimen activated for coating deposition was then immersed (dipped) into the hot electroless bath composed of Nickel sulphate ($NiSO_4$) 35 g/l, Sodium Hypophosphite ($NaPO_2H_2$) 20 g/l, Trisodium citrate ($Na_3C_6H_5O_7$) 15 g/l, Sodium acetate ($C_2H_3NaO_2$) 5 g/l, Copper sulphate ($CuSO_4$) 0.5 g/l, Sodium Dodecyl sulphate ($NaC_{12}H_{25}SO_4$) 0.2 g/l and Titanium Oxide (TiO_2) 5 g/l. The bath was kept at constant temperature of 85°C for 3 hours. To achieve higher thickness the coated samples were immediately placed into a fresh bath for next 3 hours. The baths were continuously stirred by a magnetic stirrer to keep the TiO_2 particles in suspension. After coating the samples were taken out of the bath and washed in distilled water and dried. Energy dispersive x-ray analysis was performed to confirm the presence of four elements. The tribological study of the as-deposited electroless Ni-P-Cu-TiO_2 coated pin-type samples was performed on a multi-tribotester (DUCOM-TR-20LE) with a pin-on-disc configuration. The test on the pin samples were performed in dry (non-lubricated) condition. The tests were performed at a normal room temperature under a normal load of 10 N. The speed was varied for four different samples ranging from 2 to 8 m/s for a sliding distance of 600 m. The test data was captured by associate software beside this the ware volume was also measured.

3. Results and discussion

3.1. Composition analysis

Energy dispersive x-ray analysis confirms the presence of all the four components of the coating. The EDX spectrum is shown in Figure 1. Table 1 shows the range of wt% of those four components at different test points.

Table 1: Range of elements present in the coatings

	Minimum value (wt.%)	Maximum value (wt.%)
Nickel	48.94 %	82.62 %
Phosphorus	4.99 %	12.7 %
Copper	3.96 %	13.40 %
Titanium	0.97 %	36.67 %

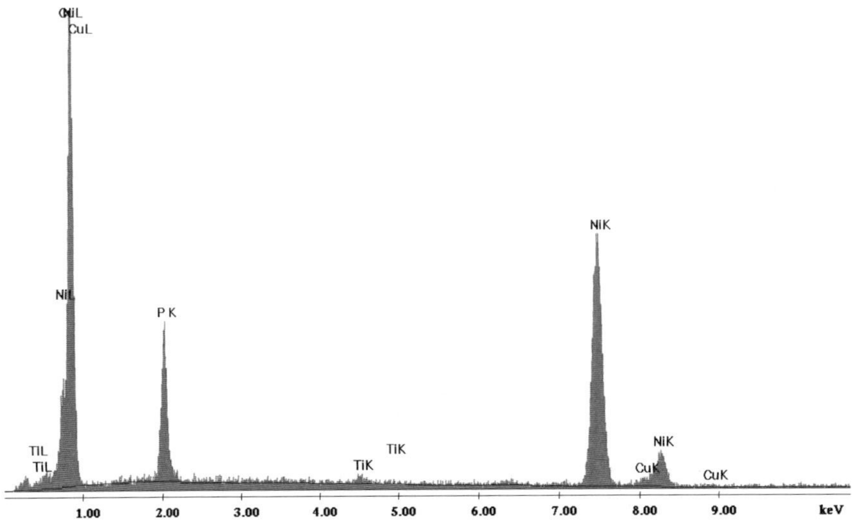

Figure 1: EDX spectrum of electroless Ni-P-Cu-TiO2 coating

3.2. *Tribological test*

Wear depth and frictional force were captured by allied software during test. Coefficient of friction was calculated from the friction force. Wear depth and coefficient of friction as a function of sliding distance for different sliding velocities are shown in Figure 2.

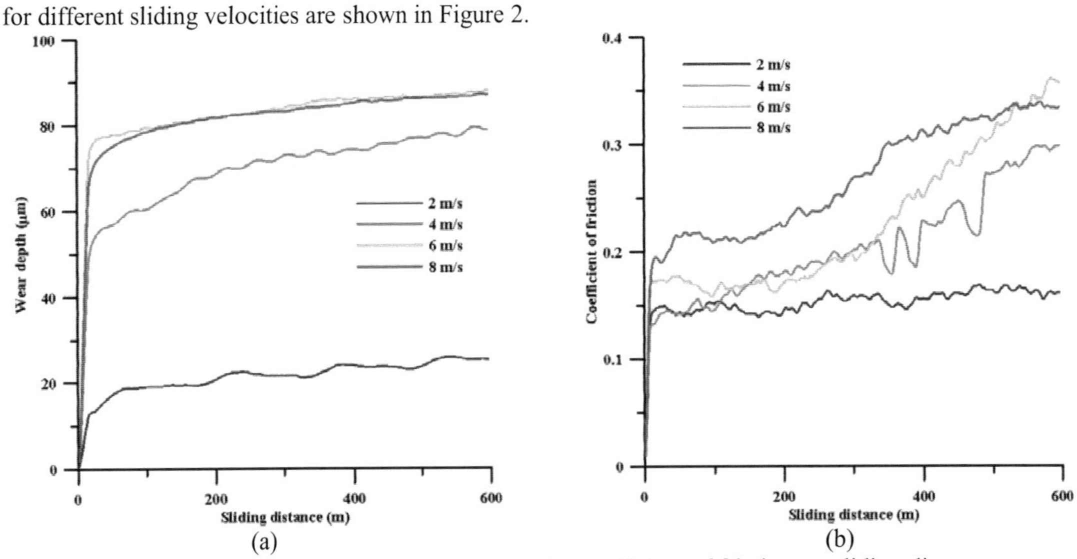

(a) (b)

Figure 2: (a) wear depth vs. sliding distance, (b) coefficient of friction vs. sliding distance

It is clear from figure 2 (a) that the wear depth jumps to a high value when, sliding velocity increased from 2 m/s to 4 m/s but a small increase in coefficient of friction was observed in this case (figure 2 (b)). However at higher speed, wear depth is high but stable with sliding distance but coefficient of friction increased with the sliding distance. In order to have a close observation on tribological behavior, a correlation between wear coefficient (Eq. 1) and frictional power (Eq. 2) was established.

$$Wear\ coefficient = \frac{V_w}{F_n \times s} \quad (1)$$

$$Friction\ power = F_f \times v \quad (2)$$

Where, V_w is the wear volume, F_n is normal force, s is sliding distance, F_f is friction force and v is sliding velocity.

The ambiguity of Figure 2(a) can be understood clearly from Figure 3. It shows that wear coefficient or wear volume follows same pattern with respect to friction power and sliding velocity respectively. Both wear parameters reach at maximum value while sliding with 6 m/s and remain almost constant or a little bit decreased while sliding with next higher speed of 8 m/s. In both cases friction coefficient is seen to be increased gradually in a linear way. Thus a stable wear can be achieved at higher sliding speed.

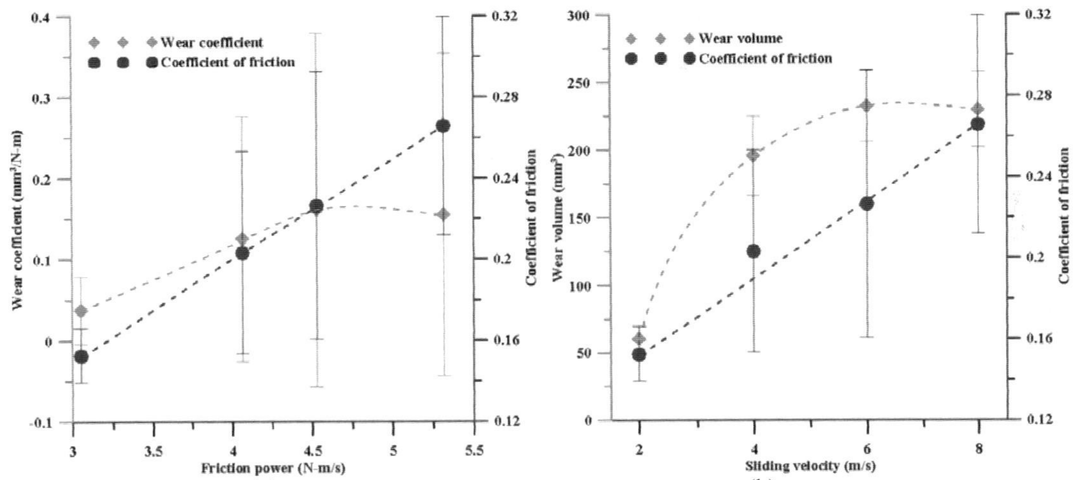

(a) (b)

Figure 3: (a) Wear coefficient and Coefficient of friction vs. Friction power, (b) Wear volume and Coefficient of friction vs. sliding velocity

4. Conclusion

The paper describes the successful deposition of electroless Ni-P-Cu-TiO2 composite quaternary coating over mild steel substrate and characterizes the effort with EDXA and tribotesting. The subject comprises with vast scope of futuristic research work relative to microstructure study, hardness, anti-corrosion properties and many more. Hence, such futuristic researches should be carried out on electroless Ni-P coatings in lieu of its existence in automobile, aerospace to alloy industries.

Acknowledgement

Authors are grateful for financial support provided by CSIR-HRDG, India, vide sanction letter no.: 22(0804)/19/EMR-II dated 25/07/2019, to carry out this work.

References

[1] Ram M, Kumar M, Ansari A, Sharma S, and Sharma A 2020 Corrosion Resistance of Electroless Ni-P-SiC/Ni-P-TiO_2-ZrO_2 Nano-Coatings in Paper Mill Bleach Plant. *Mater. Today: Proceedings*, **21**, 1200-1212.

[2] Khollari M A R, Ghorbani M, and Afshar A 2019 Fabrication and characterization of TiO_2 deposited black electroless Ni-P solar absorber. *Appl. Surf. Sci.* **496**, 143632.

[3] Araghi A, and Paydar M H 2014 Wear and corrosion characteristics of electroless Ni–W–P–B4C and Ni–P–B_4C coatings. *Tribology-Materials, Surfaces & Interfaces*, **8**(3), 146-153.

[4] Leon O A, Staia M H, and Hintermann H E 2005 Wear mechanism of Ni–P–BN (h) composite autocatalytic coatings. *Surf. Coat. Tech.* **200**(5-6), 1825-1829.

[5] Bozzini B, Boniardi M, Fanigliulo A, and Bogani F 2001 Tribological properties of electroless Ni-P/diamond composite films. *Mater. Res. Bull.* **36**(11), 1889-1902

[6] Chen W, Gao W, and He Y 2010 A novel electroless plating of Ni–P–TiO_2 nano-composite coatings. *Surf. Coat. Tech.* **204**(15), 2493-2498.

[7] Jappes J W, Ramamoorthy B, and Nair P K 2005 A study on the influence of process parameters on efficiency and crystallinity of electroless Ni–P deposits. *J. Mater. Process. Tech.* **169**(2), 308-313.

[8] Mohammadi M, and Ghorbani M 2011 Wear and corrosion properties of electroless nickel composite coatings with PTFE and/or MoS_2 particles. *J. coat. Technol. Res.* **8**(4), 527-533.

[9] Zarebidaki A, and Allahkaram S R 2011 Corrosion assessment of electroless nickel–phosphorous/nanosilicon carbide composite coatings. *Micro Nano Lett.* **6**(11), 937-940.

[10] Grosjean A, Rezrazi M, Takadoum J, and Bercot P 2001 Hardness, friction and wear characteristics of nickel-SiC electroless composite deposits. *Surf. Coat. Tech.* **137**(1), 92-96.

[11] Malfatti C D F, Ferreira J Z, Santos C B D, Souza B V, Fallavena E P, Vaillant S, and Bonino J P 2005 NiP/SiC composite coatings: the effects of particles on the electrochemical behaviour. *Corros. Sci.* **47**(3), 567-580.

Studies on Al-MWCNT MMC surface coating on Aluminium substrates through friction surfacing process

A V Sreenu[1], M Revathi [2], G Ramya Sukeerthi [3] and B Maheswari [4]

[1,2,3&4]Metallurgical & Materials Engineering Dept, RGUKT-Nuzvid, Krishna District, Andhra Pradesh-521202, India.

e-mail: *sreeiitm09@rguktn.ac.in*

Abstract. In this study, Aluminium-Multi Wall Carbon Nano Tube (Al-MWCNT) metal matrix composite surface coating was fabricated by using solid state friction surfacing technique on Al substrate. The influence of process parameters on the coating dimensions, integrity and hardness was examined. The optical microscopy studies revealed that the coating integrity with the substrate is good with Trial-2 parameters i.e RPM 2500, Axial Load 700Kg and Traverse Speed 6mm/s. Grain refining due to the dynamic recrystallization was observed in the coating microstructure. XRD studies revealed that the MW-CNTs were distributed in to the coating and no detrimental intermetallic compounds were formed between Al matrix and MW-CNTs except Al_4C_3. The Scanning Electron Microscopic Studies reveals that the Al metal was severely deformed before forming the coating which is evident from the flow lines. It is further confirmed from the SEM studies that the MW-CNTs were dispersed uniformly within the coating and the nano size was retained in the coating without significant coarsening.

1. Introduction

The outstanding mechanical properties, low thermal expansion, high thermal conductivity, good electrical properties [1-2] of carbon nanotubes make them an excellent reinforcement in Al matrix to design high-performance metal matrix composites. Several processing techniques were investigated to synthesize the Al-CNT MMCs. These techniques including cold spraying [3], spray drying [4], Plasma spraying [5], Powder metallurgical route followed by hot isostatic pressing (PM+HIP) [6], Spark Plasma Sintering followed by hot extrusion (SPS + HE) [7], Powder deformation processing [8], and friction based solid state techniques such as Friction stir processing (FSP) to form surface composites [9 & 10], The interfacial phenomena which includes wetting of the CNTs by Al matrix and chemical reactions at interface are crucial for the better functionality [11 & 12]

One of the most common problems in the processing of CNT reinforced-MMCs is the difficulty in dispersing the CNTs due to their attractive van der Waals interactions [13]. Therefore achieving an uniform dispersion of carbon nanotubes in a metal matrix is quite difficult. The other major challenges in preparing the Al-CNT composites include the following [14]
i. Fabrication of Al-CNT composites with higher vol % (>6) CNT having uniform dispersion
ii. Homogeneous dispersion while retaining the structural integrity of CNT
iii. Reduced damage of CNT during processing at higher temperatures and pressures
iv. Improving wettability of CNT with molten Al and its alloys for casting based techniques

As the Al-CNT MMCs were demonstrated for superior wear and Tensile properties [15 & 16] research was driven in to fabricate the MMC surface coatings.

Friction surfacing (FS) is a solid state surface coating technique which involves the generation of intense frictional heat between rotating consumable rod and the stable base plate. The tip of the consumable rod, after reaching a critical frictional temperature will be coated on to the substrate upon the traversing of the substrate. The process of friction surfacing is best understood with following schematic diagram in Fig. 1[17]. The coatings produced with this technique, are free from porosity, having worked structure rather than a cast structure, much finer grain size due to dynamic recrystallization effect and free from the unwanted inter-metallic compounds (IMCs) due to the higher diffusion rates.

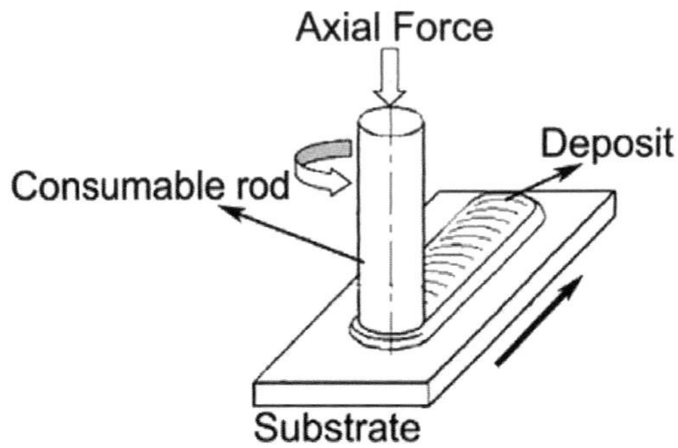

Figure 1. Friction surfacing process schematic.

In the recent studies friction surfacing is also using for fabricating the surface composites with varying volume percentage of the fiber [17 & 18]. This method of producing the surface composite coating is superior to the other conventional melting & spraying routes in the following aspects

a. Uniform distribution of the fibers due to severe plastic deformation and material flow
b. Effectively minimizes the coarsening of nano sized fibers/dispersants
c. The chemical stability of fibers and matrix are retained (or) Interfacial reactions were minimized due to lower diffusion rates at solid state processing

The present study aims for preparing in-situ, solid state Al-MWCNT composite surface coating on aluminium substrate. The MWCNTs were incorporated in to the consumable rod by drilling holes. The number and size of the holes can be varied to obtain the required volume percent (Vol%) of CNTs. In the current study it was aimed for approximately 4 Vol% of CNTs in the Al matrix. As the process is done almost in the solid state and involves severe plastic flow, the coatings are expected to have better dispersion, no interfacial reaction & nano size retention.

2. Materials and experimental methods

A friction surfacing machine (Fig. 2a) with 10kN Loading capacity and computer controlled 3 axes moment was employed to obtain the coatings. The substrate and consumable rod materials were taken in a well-polished condition and thoroughly cleaned to ensure the good bonding.

General purpose commercially pure aluminium plate with a nominal composition of Al-99.36%, Fe-0.45 and other trace elements was used as substrate. The dimensions of the substrate are Length 300mm X Width 200mm X Thickness 8mm, Al consumable rod with a nominal composition of Mg-0.55%, Si-0.75%, Mn-0.62%, Fe-0.37% and the balance is Al, in stress relieved annealing condition with Length 75mm X Diameter 20 mm were used in the present study.

IOP Conf. Series: Materials Science and Engineering 1080 (2021) 012023 doi:10.1088/1757-899X/1080/1/012023

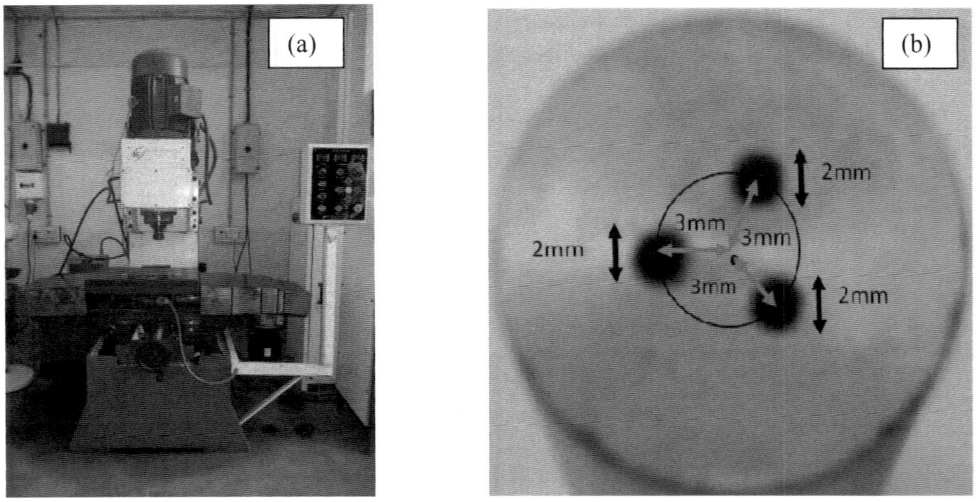

Figure 2. (a) Friction surfacing machine and (b) Hole configuration in the consumable rod.

The CNT reinforcement were packed inside 2mm diameter holes, drilled with 30mm depth from rubbing end of the consumable rod. The hole is drilled 3mm from the center of the rod. The hole configuration was shown in Fig. 2 (b).

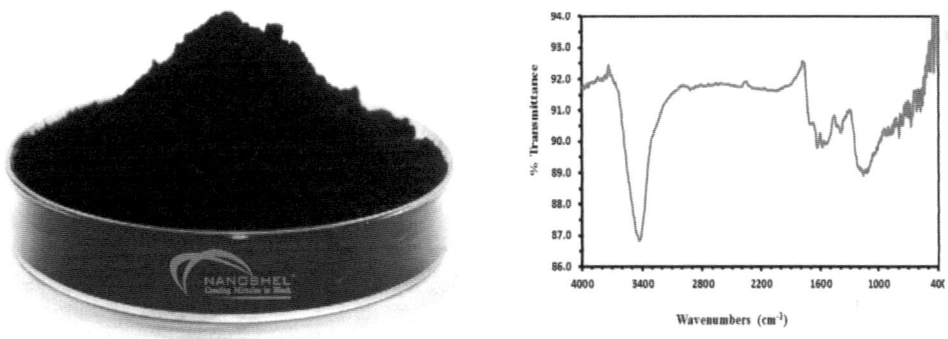

(a) (b)

Figure 3. (a) Representative image of MWCNTs. (b) FTIR spectrum of MWCNTs.

Multi wall CNTs with Diameter 10-20nm, Length 3-8μm and Purity >90% from M/s NANOSHELL India Ltd, were utilized for the study. Fig. 3 will describe the representative image and the FTIR spectrum of MWCNTs from the supplier. These CNTs were filled in to the holes drilled in to the consumable rod, carefully and sealed with a thin polymer adhesive tape while fitting in to the spindle. PLC controlled machine was employed for the purpose of friction surfacing. The Al consumable rod which was reinforced with MWCNTs being mounted in a specially made holder that fitted to the spindle of the machine. Al Substrate consisting of square plate was laid on a platform which can be traversed. Friction Surfacing was performed by lowering the spindle with Al consumable rod until a contact was made with the Al substrate. The process parameters were entered in to the user interface program with help of a computer. The following two trials were carried out mainly varying the RPM from 2000 to 2500 and traverse speed from 5 to 6 mm/s, in the trial-1 & trial-2 respectively. This is to understand the influence of RPM and traverse speed on bonding characteristics and the bead dimensions.

Table 1. The Dimensions of materials and process parameters used in Trial-1 & Trial-2.

Trial No	Rod Dimensions		Substrate Dimensions			Parameters		
	Length (mm)	Diameter (mm)	Length (mm)	Width (mm)	Thickness (mm)	RPM	Load (kg)	TS (mm/s)
1	75	20	300	200	8	2000	700	5
2	75	20	300	200	8	2500	700	6

Optical microscopy in un-etched and etched condition was carried out with Lieca metallurgical microscope. Micro hardness tester (Automatic Micro Hardness Tester, Omini Tech, India) was used for the hardness survey across the interface. Field emission scanning electron microscope (Merlin Compact, Carl Zeiss) was employed for sub microscopic images and energy dispersive spectroscopic studies. X-Ray diffraction machine (D8 advance, Bruker) with Cu Kα X-Ray radiation was used for the phase analysis. Table-3 will indicate the details of etching reagent used for the base metal and friction surfaced coating, along with its influence of revealing the micro-structure

Table 2. Etching reagents used in the study.

Material	Etching reagent	Composition	Method	Remarks
Al consumable rod	--	1gram NaOH pellets + 99ml distilled water	For Al alloys, immerse or swab the sample up to 10-30 sec	Revealed the Grain boundaries of CP Al
Al-MWCNT coating	Keller's etch	HF - 1ml HCl - 1.5ml HNO3 - 2.5ml Distilled water - 95ml	For Al alloys, immerse the sample up to 30-180 sec	Revealed the Grains and phase boundaries

Vicker's micro hardness survey was conducted for comparing the hardness of coating and the substrate. The load on the Vickers micro-hardness indenter usually ranges from a few grams to several kilograms. 200 grams load was used in the current study for both base metal and coated samples. The actual indenter used is a square based diamond pyramid with an apex angle of 136°. The hardness survey was conducted systematically starting from interface with an inter indentation distance of 100μm from both coating and substrate.

3. Results and discussion

The process parameters used in both the trials have produced the coatings with the dimensions mentioned in the Table-3. It is evident from the results that the coating thickness is inversely proportional to the travel speed.

Table 3. Dimensions of the coatings obtained in Trial-1 and Trial-2.

	Coating Dimensions		
Trial No	Thickness(mm)	Width (mm)	Length(mm)
01	9.70	25.93	26.5
02	6.52	23.1	24.4

3.1 Trial-1

A circular button type coating was deposited on the Al substrate with the experimented parameters. The features of the coating were described in Fig 4. The top view (Fig. 4a) and cross sectional view (Fig. 4b) of the coating indicates a central hollow space. There was bonding at the advancing side and the retreating side of the coating leaving an uncoated region at the central region. The central concavity of the consumable rod and its rough surface (Fig. 4c) is confirming the material transfer from rod to the coating.

Figure 4. (a) Circular coating top view (b) Cross sectional view (c) Rod tip profile at rubbed end in trail-01.

The coating can be conveniently divided in to four regions for the better understanding of the coating features. Region 1 is un-bonded region at advancing side, Region 2 is a Central uncoated region, and Region 3 is a well bonded region followed by Region 4 which is un-bonded region at retreating side. The Advancing Side (AS) is the one where rod rotation and substrate movement are in the same direction. Retreating Side (RS) is the one where the rod's rotation and substrate movement are in opposite direction. Friction surfacing process has an inherent feature of processing unbounded regions at both the AS & RS. This is due to the non uniform distribution of axial load on the consumable rod. In the present case, RS has shown less fraction of UN bonded region and more fraction of well bonded region. This is due to the more frictional heat, more plastic flow of the material and better bonding at RS [19]

Figure 5. Different regions in trail-01 coated sample.

The central unbonded region can be explained from the strengthening effect produced by the MWCNTs. The holes are nearer to the central region of the rod and the dispersion of CNTs is also more concentrated at the central region at the initial stages of the friction surfacing.

As per the studies on material transfer mechanism from rod to the coating [20] in FS process, a quasi liquid layer forms at the center of the consumable rod due to the intense frictional heat generation. This layer will be transferred from rod to the coating when the substrate is traversed. The thickness of the quasi layer also increases with increasing RPM of the rod. This mechanism is the main reason behind the slick concavity at the center of the consumable rod at the rubbing end. However in the present study, the slick concavity is not much prominent and in contrary there is a central UN coated region. This can be attributed to the fact that, the exceptional mechanical strength and the thermal conductivity of the MWCNTs present at the central region can contribute to dissipate the frictional heat to the nearby regions and retains the mechanical strength of the rod at central region. As a result the formation of the quasi liquid layer will be delayed and will be shifted to the CNTs free regions. Therefore the central region will result no materials transfer and leads to un-coated region. At the same time the side region to the center of the rod may leads to the formation of quasi liquid layer and leads to coating formation. The observed coating features are matching with the above predictions.

3.2 Trial-02

In view of the above discussion, the following process parameters were modified in the trail-02. The RPM is increased from 2000 to 2500, to promote the quasi liquid layer formation at the center. The traverse speed also increased from 5 to 6 mm/s, to reduce the thickness of the coating which in turn leads to better distribution of the axial load on the consumable rod.

Figure 6. (a) Circular coating top view (b) Cross sectional view (c) Rod tip profile at rubbed end in trail-02.

It is evident from the above Fig. 6, that the parameters in trial-02 have resulted a reasonably good coating without gross defects as observed in the trail-01. A circular button type coating without a central defect was obtained. The coating thickness was reduced from 9.7mm to 6.5mm and the slick concavity of the consumable rod tip at the central region is more prominent. It is an indication of the material transfer from the central region. It is interesting to note that the coating width was reduced from 25.93mm to 23.1mm. This can be explained based on the concept of "real rotational contact plane". Real rotational contact plane refers to the actual or instantaneous contact plane between the rotating consumable rod and the substrate during friction surfacing. The metal transfer will occur from the consumable rod to the coating through this plane. The area of the real rotational contact plane will be usually decreases with increase in rotational speed. Hence at higher rotational speeds metal transfer is restricted to a smaller contact plane leading to narrower coatings. The above observations were made at macro level without the aid of a microscope.

3.3 Microstructural Studies

The microstructures of the consumable rod before and after the coating have been observed to find out the influence of friction surfacing process on grain size. Fig.7a & b are the optical micrographs of the consumable rod in etched condition (Keller's Etch) before and after the FS. The grain size reduction from ≈ 20μ to 1-5μ as a result of FS can be observed. This will be attributed to the dynamic recrystallization effect and the severe plastic flow of the material which are inherent features of FS. The low magnification scanning electron micrograph (Fig. 7c) of the coating reveals that the traces of severe plastic flow of material is clearly visible in the form of flow lines at retreating side of the coating. The interfacial region is showing that the rubbed material between the rod and the substrate has been expelled out of the coating region in the form of flash. That is the reason behind the outside flow lines. The material transfer happens from the slick concavity of the rod and leads to the formation of the coating. The wavy nature of the interface also ensures that there is enough penetration of the coating in to the substrate which leads to better bond strength. The coating was also subjected a hand chisel test and found that the bond strength is reasonably good.

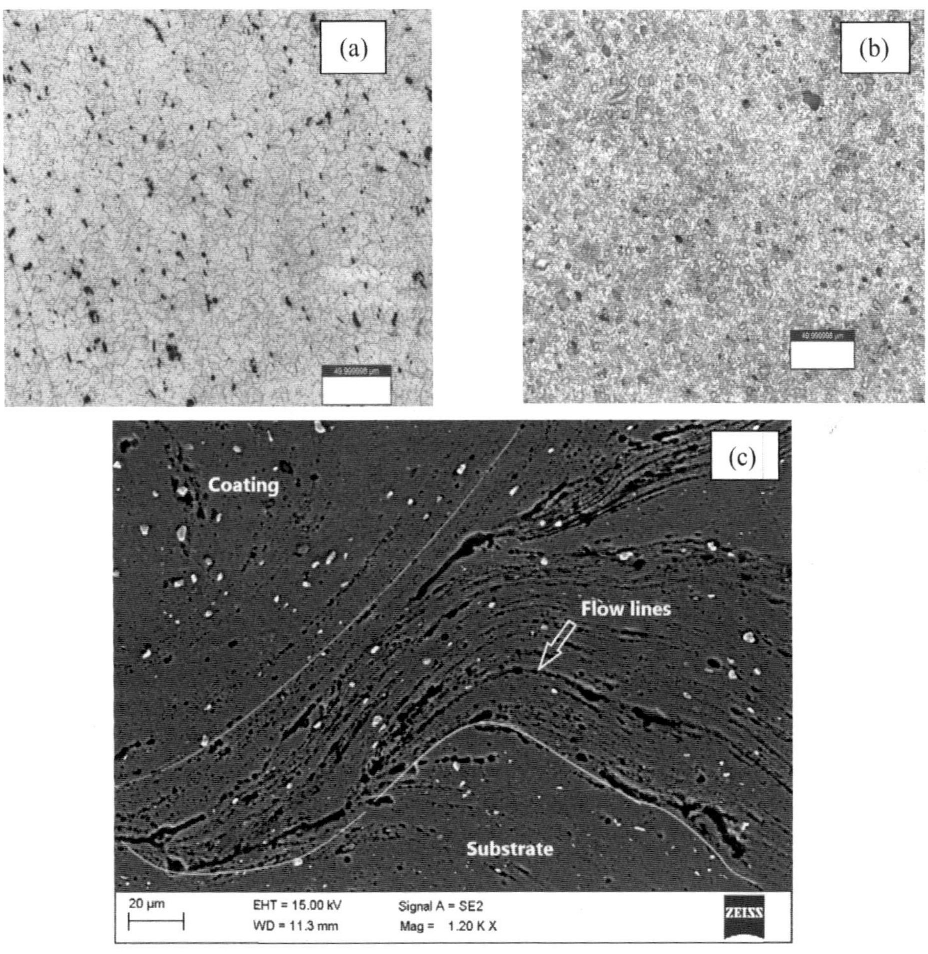

Figure 7. (a) Optical Micrographs of the consumable rod before the coating (b) FS coating (c) Low magnification SEM image of FS coating for trail 2 coating.

Other important objectives of the study are dispersing the MW-CNTs in the Al matrix uniformly and avoiding the interfacial chemical reaction between the CNTs and Al matrix. The former objective can be examined with the help of higher magnification SEM image of FS coating in Fig 8. It is evident that the CNTs distribution is mostly uniform. However some MW-CNT clusters typically of 1-2μ size were identified in the coating. Non uniform distribution of the plastic strain may be a potential reason which is resulting from the higher coating thickness. Optimization of the process parameters to produce 2-3mm coating thickness such as higher axial loads and faster traverse speeds may be useful to achieve this.

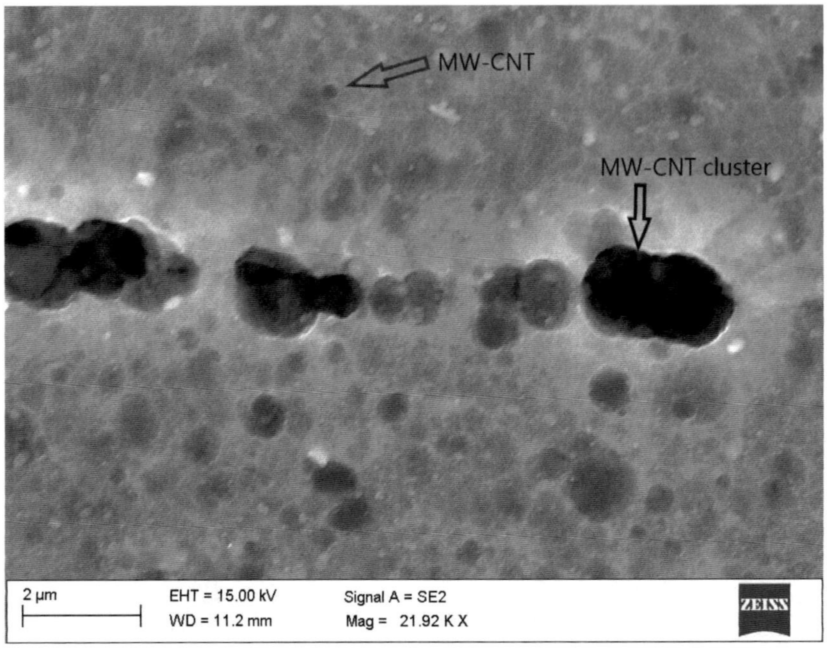

Figure 8. Higher Magnification SEM image of FS coating for trail 2 coating.

3.4 X-Ray Diffraction (XRD) Studies:

XRD studies of the Al base plate, Al consumable rod and FS coating were presented in the Fig. 9. The XRD patterns of the base plate and consumable rod shows only FCC Aluminum and the FS coating shows the Carbon and Al phases. This is a strong indication for the reinforcement of the MW-CNTs in the FS coating. The studies on the interfacial reactions between CNTs and Al [21] explains that the aluminium carbide i.e. Al_4C_3 is a very common compound forms in the Al-CNT MMCs due to the lower surface energy requirement (-12.7 kCal at 298K). This compound will usually forms on the defects such as ends of CNTs, surface of the MW-CNTs by CVD process and also on the amorphous carbon pre-curser phase. The study also concluded that the processing temperatures higher than the 450^0C will result the formation Al_4C_3 at the interfacial regions. The corresponding position of the peak for this compound in XRD is at around 36^0 of 2θ. The XRD studies of FS coating processing a peak position at 36.620 which is confirming the formation of Al_4C_3 phase. The frictional saturation temperatures in FS process, usually reaches 80-90% of the melting point of the consumable rod which is Al in the present case. Therefore it is reasonable to assume that the FS processing temperatures will easily exceeds the 450^0C and there is higher possibility for the formation of Al_4C_3 phase.

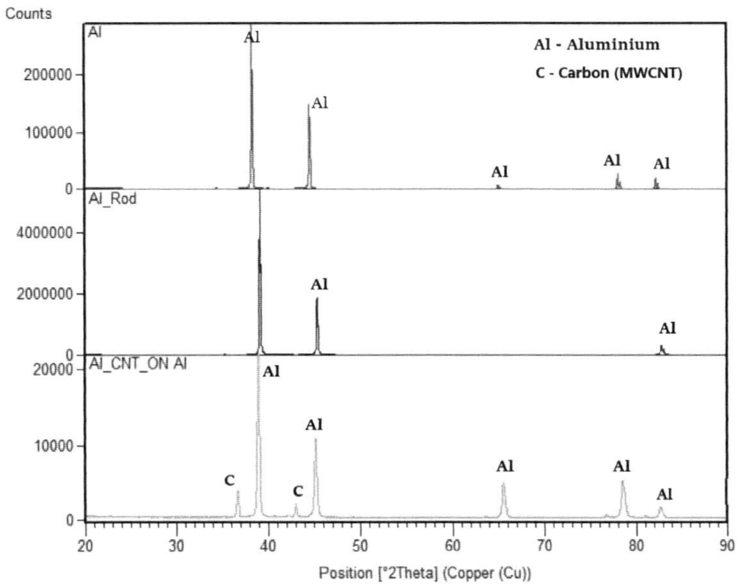

Figure 9. XRD patterns of Al base plate (Top), Al consumable rod (Middle) and FS coating with MW-CNTs for trail-02.

3.5 Micro-Hardness survey

Al-MWCNT surface coating along with the substrate was subjected to micro hardness survey (Fig. 10) across the interface. The following $HV_{0.2}$ values were measured as an average of five readings. Substrate Hardness is 38.3, Al rod Hardness is 111.2 before the FS coating, Al FS Coating Hardness is 75.33 H_V respectively. The hardness survey with square based diamond indenter with 200g load and 100µ inter indentation distance was carried from al substrate to coating across the interface. The hardness survey has revealed that the substrate hardness was increased by 96.7% and the consumable rod hardness decreased by 33% after the FS. The reasons for the decrease in hardness are uncertain. This may be due to the precipitate coarsening during FS and relaxation of dislocation density due to the dynamic recrystallization during FS. A post heat treatment after FS may require for restoring the strength and hardness. The advantage of CNT insertion in the Al matrix is not fully realized due to non uniform distribution and the cluster formation.

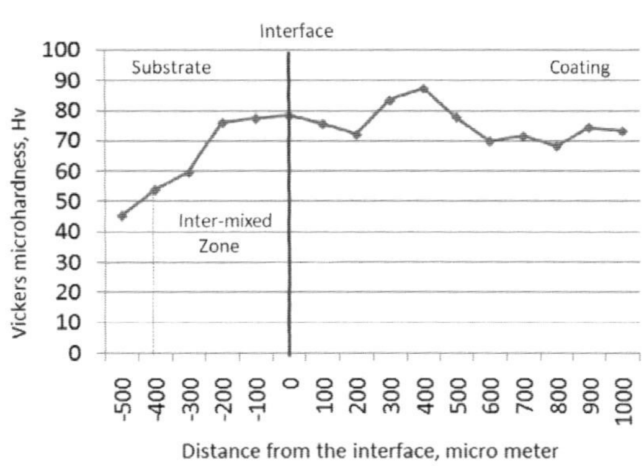

Figure 10. Micro-Hardness survey of Al-MWCNT FS coating across the interface.

It is interesting to note that there is a diffused layer or mechanically intermixed zone up to around 400μm towards the Al base metal. A gradual increase of hardness from 40 VHN to around 80 VHN can be observed in this zone at the Al substrate side. This zone is a result of consumable rod penetration in to the base metal. It can also be observed that there is no sudden raise or drop of hardness at the interfacial region, which may be an indication that there is no significant amount of hard and brittle carbides formation at the interface.

4. Conclusions

1. Al-MWCNT in-situ surface composite coating was successfully fabricated by using Friction surfacing process at the experimented trial-02 parameters i.e. 2500 rpm, 700Kg axial load and 6mm/s traverse speed
2. The coating thickness up to 6.5mm thickness can be deposited on Al substrates with the careful adjustment of the parameters
3. The Al-MWCNT MMC friction surfaced coatings are metallurgically bonded with substrates without any gross porosity or cracks
4. The distribution of the CNTs is uniform on a gross volume however some few CNT clusters of 1-2μm size range were observed may be due to the non uniform distribution of the plastic strain through the coating.
5. There is enough evidence in the XRD studies that FS also could result the formation of aluminum carbide (Al_4C_3), however further TEM studies are required to confirm the same.
6. There was an intermixed zone observed between FS coating and substrate up to 400μm due to the of high temperature strength of the coating material probably due to the MWCNTs

5. References

[1] M.D. Ganji, A. Fereidoon, M. Jahanshahi, and M.G. Ahangari, Investigation of the Mechanical Properties of Multi-walled Carbon Nanotubes Using Density Functional Theory Calculations, *J.Comput. Theor. Nanos* **9 (2012)** 980–985

[2] M.J. Oconnell, Carbon Nanotubes: Properties and Applications, CRC Press Taylor & Francis Group, Boca Raton, 2006

[3] Srinivasa R. Bakshi, Virendra Singh, Kantesh Balani, D. Graham McCartney, Sudipta Seal, Arvind Agarwal, Carbon nanotube reinforced aluminum composite coating via cold spraying, *Surface & Coatings Technology* **202 (2008)** 5162–5169

[4] Srinivasa R. Bakshi, Virendra Singh, Sudipta Seal, Arvind Agarwal, Aluminum composite reinforced with multiwalled carbon nanotubes from plasma spraying of spray dried powders, *Surface & Coatings Technology* **203 (2009)** 1544–1554

[5] Srinivasa R. Bakshi a,b, Anup K. Keshri b, Arvind Agarwal, A comparison of mechanical and wear properties of plasma sprayed carbon nanotube reinforced aluminum composites at nano and macro scale, *Materials Science and Engineering A* **528 (2011)** 3375–3384

[6] Carvalho, G. Miranda, D. Soares, F.S. Silva, CNT-reinforced aluminum composites: processing and mechanical properties, *Ciência & Tecnologia dos Materiais* **25 (2013)** 75–78

[7] H. Kurita, H. Kwon, M. Estili, A. Kawasaki, Multi-walled carbon nanotube aluminum matrix composites prepared by combination of hetero agglomeration method, spark plasma sintering and hot extrusion, *Mater. Trans.* **52 (2011)** 1960-1965

[8] J. Yuuki, H. Kwon, A. Kawasaki, A. Magario, T. Noguchi, J. Beppu, et al., Fabrication of carbon nanotube reinforced aluminum composite by powder extrusion process, *Mater. Sci. Forum* **534-536 (2007)** 889-892.

[9] Q. Liu, L. Ke, F. Liu, C. Huang, L. Xing, Microstructure and mechanical property of multi-walled carbon nanotubes reinforced aluminum matrix composites fabricated by friction stir processing, *Mater. Des.* **45 (2013)** 343-348.

[10] S. Zhang, G. Chen, J. Wei, Y. Liu, R. Xie, Q. Liu, S. Zeng, G. Zhang, Q. Shi, Effects of energy input during friction stir processing on microstructures and mechanical properties of aluminum/carbon nanotubes nanocomposites, *J. Alloy. Comp.* **798 (2019)** 523-530

[11] K. Landry, S. Kalogeropoulou, Wettability of carbon by aluminum and aluminum alloys, *Mater.Sci. Eng.A* **(1998)** 99-111.

[12] A. Abdi, K. Morsi, Long-term post-processing disintegration of aluminum carbon nanotube composites, *J. Mater. Sci.* **51 (2015)** 2049-2056.

[13] R. Pérez-Bustamante, I. Estrada-Guel, W. AntúnezFlores, M. Miki-Yoshida, P.J. Ferreira, R. MartínezSánchez, *J. Alloys Compd.* **450 (2008)**, 323–326.

[14] M. Jagannatham, Prathap Chandran, S. Sankaran, Prathap Haridoss, Niraj Nayan, Srinivasa R. Bakshi, Tensile properties of carbon nanotubes reinforced aluminum matrix composites: A review, *Carbon* **160 (2020)** 14-44

[15] Mina M.H. Bastwros, Amal M.K. Esawi, Abdalla Wifi, Friction and wear behavior of Al–CNT composites, *Wear* **307 (2013)** 164–173

[16] Srinivasa R. Bakshi, Arvind Agarwal, An analysis of the factors affecting strengthening in carbon nanotube reinforced aluminum composites, *carbon* **49 (2011)** 533 – 544

[17] A.V.Sreenu, Aluminium-boron nitride nano composite coating by friction surfacing on low carbon steel substrate – a feasibility study, *Materials Today: Proceedings* **5 (2018)** 26829–26835

[18] J. Gandra, P. Vigarinho, D. Pereira, R.M. Miranda, A. Velhinho, P. Vilaça, Wear characterization of functionally graded Al–SiC composite coatings produced by Friction Surfacing, *Materials and Design* **52 (2013)** 373–383

[19] H. Khalid rafi, krishnan balasubramaniam, G. Phanikumar and k. Prasad rao, Thermal Profiling Using Infrared Thermography in Friction Surfacing, *Metallurgical and Materials transactions A* **42A (2011)** 3425-3429

[20] X.M. Liu, Z.D. Zou, Y.H. Zhang, S.Y. Qu, X.H. Wang, Transferring mechanism of the coating rod in friction surfacing, *Surface & Coatings Technology* **202 (2008)** 1889–1894

[21] Lijie Ci, Zhenyu Ryu, Neng Yun Jin-Phillipp, Manfred Ruhle, Investigation of the interfacial reaction between multi-walled carbon nanotubes and aluminum, *Acta Materialia* **54 (2006)** 5367–5375

Unsteady development of Marangoni convection in a sidewall moving open cavity

Nirmalendu Biswas[1], Aparesh Datta[2], D. K. Mandal[3] and Nirmal K. Manna[4, a]

[1] Department of Power Engineering, Jadavpur University, Salt Lake, Kolkata–700106, India
[2] Department of Mechanical Engineering, NIT Durgapur, West Bengal–713209, India
[3] Department of Mechanical Engineering, College of Engineering and Management, Kolaghat – 721171, India
[4] Department of Mechanical Engineering, Jadavpur University, Kolkata–700032, India
[a] e-mail: *nirmalkmannaju@gmail.com*

Abstract. In this study, simultaneous effect of free, forced and Marangoni convection on fluid flow and heat transfer in a sidewall moving open cavity is investigated. The cavity is filled with low Prandtl number fluid (Pr = 0.054); such a situation can be find in the application of liquid metal and semiconductor melts. Free convection is induced due to the heating of fluid from the bottom of the cavity. Forced convection is originated due to the translation of the sidewalls either in upward or downward direction. Whereas, the Marangoni convection is generated in the upper portion of the fluid due to imposed temperature gradient along the top free surface. The transient phenomena of the combined effects are analyzed numerically in dimensionless forms utilizing an in-house CFD code. The analyses are conducted under different pertinent parameters such as Marangoni number, Biot number, Reynolds number and Richardson number. The study is conducted to explore the flow-field structures and heat transfer characteristics in the presence of three convection modes with their relative strengths.

1. Introduction

Transport phenomena over the top free surface of fluid layer are very important in numerous applications. When the top free surface of fluid interact with ambient, fluid motion induces along a free surface due to the convection induced by temperature dependent surface tension forces (contractile force), is called thermocapillary flow or Marangoni convection. It is significant in a smaller system/ negligible gravity situation. Marangoni phenomena can be found in various applications such as crystal growth, vaporizing droplet, welding process, glass manufacturing, etc. In order to improve the quality of products with increasing market demand, sufficient understanding of these physical phenomena is utmost necessary. The study of Marangoni convection was initiated many years ago, it continues to till date [1]. A comprehensive review on combined effects of buoyancy and surface tension has been reported in recent literature [2].

The convection induced by Marangoni and buoyancy under mixed convection mode has received to a great extent attention of the researchers. Marangoni mixed convection at the interface of two immiscible fluids layers have been investigated by Chamkha et al. [3]. Saleh et al. [4] analyzed surface tension and convective cooling effects on a differentially heated square enclosure. Marangoni flow coupled with mixed convection of Casson fluid subjected to thermal stratification considering viscous dissipation and Joule heating has been investigated by Hayat et al. [5]. Thereafter, Nithyadevi et al.

[6], studied buoyancy-induced flow coupled with thermocapillary flow of fluid containing in a top open cavity heated at the bottom and cooled through the sidewalls. Very recently, Marangoni natural convection of nanofluid filled cubical cavity with free top surface has been studied by Sheremet and Pop [7]. They found that intensification of Marangoni number reduces the heat transfer characteristics.

The most of the above works have considered the rectangular open cavity which is stationary. Consideration of Marangoni convection in lid-driven cavity is yet not found. Detailed review on lid-driven cavity is mentioned in a recent work [8] of us. Thus, the studied problem is framed using a bottom-heated, sides cooled square cavity with free top surface. Forced convection is imposed by the motion of cold side walls and both the side walls are translating at same constant speed either in upward or downward direction. The main aim of this work is to assess unsteady development of three different convection modes (free, forced and Marangoni convection) simultaneously to understand dynamics of flow-structure and their relative influences.

2. Problem description and solution procedure

The schematic geometry of the studied problem is depicted in Fig. 1. The geometry is a two dimensional enclosure (of height H and length L, where $L = H$), filled with liquid medium (Pr = 0.054). The bottom wall is heated isothermally (temperature of T_h). The temperature of the top and side walls are influenced by the ambient temperature ($T_c < T_h$). Side walls can translate at the same velocity (v_w) in either upward or downward direction. The sliding motions of the walls of the cavity can favor or oppose the buoyancy-driven fluid flow as well as heat transfer process. Thus, the cases of downward and upward moving sidewalls are mentioned as assisting flow and opposing flow, respectively. Top free surface can transfer heat following the Newton's law of cooling [2,4]. It is assumed that, along the top free surface the surface tension σ alters linearly with temperature $\sigma = \sigma_o[1-\gamma(T-T_c)]$, where, σ_o is the reference surface tension (at $T_o = (T_h - T_c)/2$), and γ is temperature coefficient ($\gamma = -(1/\sigma_o)(\partial\sigma/\partial T)$).This results in shear stress over the top free exterior of liquid pool ($\mu\dfrac{\partial u}{\partial y} = -\dfrac{\partial\sigma}{\partial T}\dfrac{\partial T}{\partial x}$). This stress causes a circulation, which is known as Marangoni convection and defined by Marangoni number ($\mathrm{Ma} = -\dfrac{\partial\sigma}{\partial T}\dfrac{(T_h - T_c)H}{\mu\alpha}$) [2].

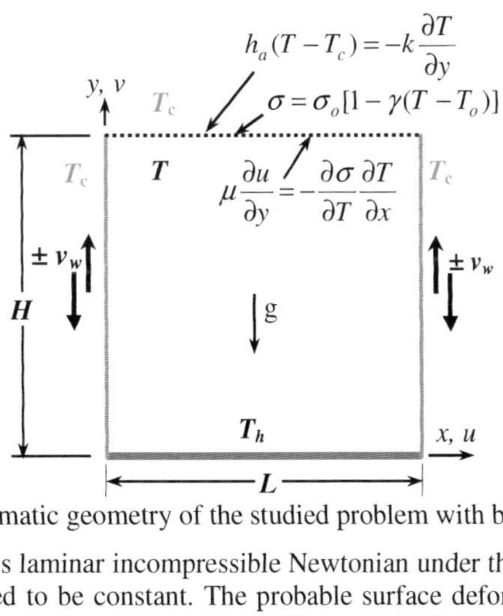

Figure 1. Schematic geometry of the studied problem with boundary conditions.

It is assumed that, flow is laminar incompressible Newtonian under the Boussinesq approximation. The fluid density is assumed to be constant. The probable surface deformation at the top fluid layer due to Marangoni convection is assumed to be insignificant [2,3–6]. Resulting transport equations are non-dimensionalized, which can expressed as

$$\frac{\partial U}{\partial X}+\frac{\partial V}{\partial Y}=0 \tag{1}$$

$$\frac{\partial U}{\partial \tau}+U\frac{\partial U}{\partial X}+V\frac{\partial U}{\partial Y}=-\frac{\partial P}{\partial X}+\frac{1}{\mathrm{Re}}\left(\frac{\partial^2 U}{\partial X^2}+\frac{\partial^2 U}{\partial Y^2}\right) \tag{2}$$

$$\frac{\partial V}{\partial \tau}+U\frac{\partial V}{\partial X}+V\frac{\partial V}{\partial Y}=-\frac{\partial P}{\partial Y}+\frac{1}{\mathrm{Re}}\left(\frac{\partial^2 V}{\partial X^2}+\frac{\partial^2 V}{\partial Y^2}\right)+\frac{\mathrm{Gr}}{\mathrm{Re}^2}\theta \tag{3}$$

$$\frac{\partial \theta}{\partial \tau}+U\frac{\partial \theta}{\partial X}+V\frac{\partial \theta}{\partial Y}=\frac{1}{\mathrm{Re}\,\mathrm{Pr}}\left(\frac{\partial^2 \theta}{\partial X^2}+\frac{\partial^2 \theta}{\partial Y^2}\right) \tag{4}$$

where,

$$(X,Y)=\frac{(x,y)}{H};(U,V)=\frac{(u,v)}{v_w};\ P=\frac{p-p_a}{\rho v_w^2};\tau=\frac{t}{H/v_w} \tag{5}$$

and $$\mathrm{Pr}=\frac{\nu}{\alpha},\ \mathrm{Re}=\frac{v_w H}{\nu},\ \mathrm{Gr}=\frac{g\beta(T_h-T_c)H^3}{\nu^2},\ \mathrm{Ri}=\frac{\mathrm{Gr}}{\mathrm{Re}^2},\mathrm{Bi}=\frac{H/k}{1/h_a} \tag{6}$$

(Pr, Re, Gr, Ri, Bi corresponds to the dimensionless Prandtl, Reynolds, Grashof numbers, Richardson and Biot number).

The applied boundary conditions for the present problem are:

at the bottom wall ($Y=0$) : $U=V=0,\ \theta=1$

at the sidewalls ($X=0,1$): $U=\theta=0,\ V=\pm1$

at the top wall ($Y=1$): $\partial U/\partial Y=(\mathrm{Ma/RePr})\partial\theta/\partial X,\ V=0,\ \partial\theta/\partial Y=-\mathrm{Bi}\theta$.

From the solved non-dimensional temperature, the heat transfer rate from the hot bottom wall calculated using average Nusselt number (Nu), as

$$\mathrm{Nu}=\int_0^1\left(-\partial\theta/\partial Y|_{Y=0}\right)dX \tag{7}$$

Coupled governing equations (1–4) with the imposed boundary conditions are solved numerically utilizing developed code based on the finite-volume method, SIMPLE algorithm [9]. The convergence of solutions is manifested when residuals and continuity mass defect < 10^{-8} and 10^{-10}, respectively. The above code is well validated with different work of published literature both with natural and mixed convective heat transfer [2,8,10]. In this study, 160×160 grid size (distributed non-uniformly) is selected for conducting extensive analysis and this grid size is selected after performing a grid independence study. For brevity same are not presented here.

3. Results and Discussions

The present work analyses three distinct convection modes (free, forced and Marangoni convection) during thermal convection from a bottom-heated cavity containing liquid metal. Heated fluid in the cavity is cooled at the side walls, which can translate in either upward or downward directions. The results reveal distinct flow patterns in the cavity for different values of relevant factors such as Biot number (Bi = 0 – 10), Marangoni number (Ma = 0 – 1000), Richardson number (Ri = 0.1 – 10) and Reynolds number (Re = 10 – 200).

3.1. Evolution of fluid-flow and temperature patterns

To get a thorough insight of the present problem, a transient study is conducted first for a set of typical values of flow parameters (Gr = 10^5, Re = 200, Bi = 1 and Ma = 1000) considering both downward and upward motion of the sidewalls. The corresponding results on the temporal evolutions of flow and thermal fields are analyzed in Figs. 2–3. Non-dimensional time instants are marked by different values of τ ranging from 0.01 to 10. First, the development of flow pattern in terms of streamlines as shown in Fig. 2 is taken up for discussion. For each time instant, the streamlines of assisting flow (with downward moving lids) are shown in Figs. 2a and 2c, whereas, the streamlines for opposing flow (with upward moving lids) are depicted in Figs. 2b and 2d. Due to the presence of heat source at the bottom wall and heat sinks at the sidewalls, there exists a symmetric flow structure with respect to the middle-vertical plane of the considered geometry, irrespective of the direction of sidewalls translation. However, the development of flow patterns under assisting and opposing flow are totally different due to the opposite translation of two sidewalls. The fluid flow in the cavity is simulated from the rest (zero velocity and zero temperature fields), as such no-streamline is noticed in the cavity at the instant of time $\tau = 0.01$. Thereafter, for $\tau = 0.1 - 0.3$, the flow circulations form and grow only near the moving sidewalls due to forced convection induced through the shear action of the sidewalls. The fluid in the cavity rotates in CCW (counter-clockwise at left wall) and CW (clockwise at right wall) for assisting flow.

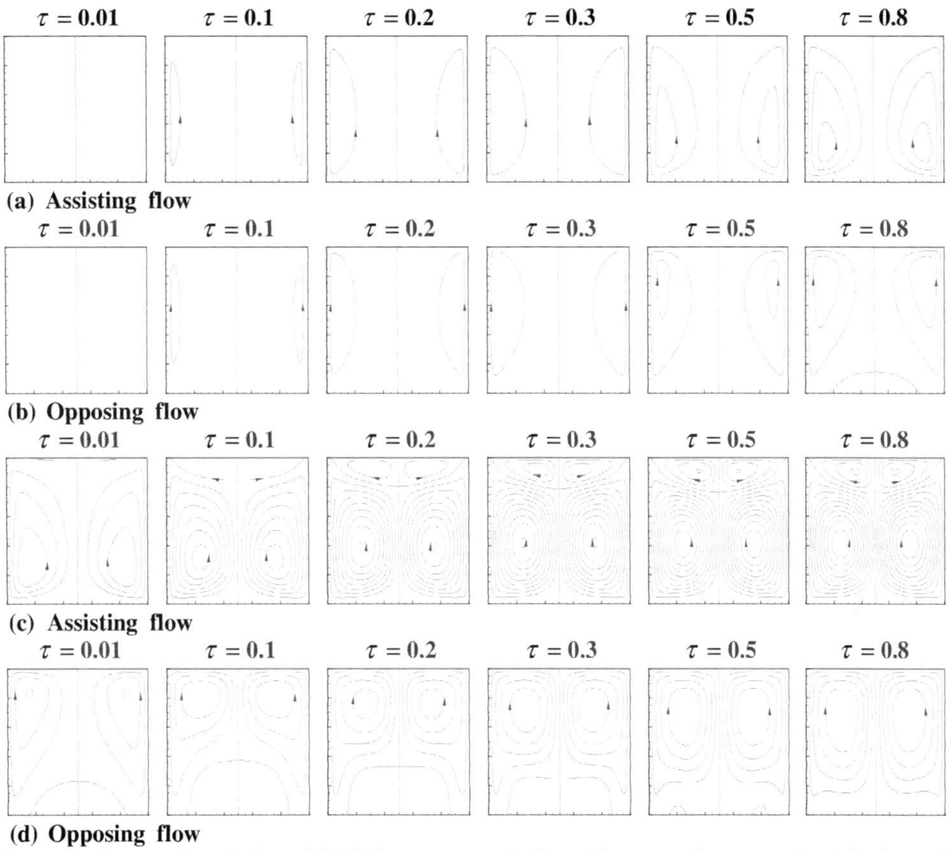

Figure 2. Temporal evolution of fluid-flow patterns (indicated by streamlines snapshots) in the cavity with downward walls motion (a, c) and with upward walls-motion (b, d) at Re = 200, Gr = 10^5, Bi = 1 and Ma = 1000. Distinct changes are noted. The growth of Marangoni convection is clearly discernible for assisting flow. However, it is indistinguishable for opposing flow for this particular data set.

For opposing flow, these directions of rotations become reversed. However, there apparent shapes, sizes and locations are found (more or less) similar, as the sidewalls are moving at the same constant speed. Afterwards, for $\tau \geq 0.5$, this similarity between assisting and opposing flow is lost. For $\tau = 0.5$ and 0.8, the fluid motion along the vertical mid-plane develops in similar way for assisting flow (rising from bottom) and opposing flow (falling from top). The shape of the streamlines changes to bulging form at bottom wall during assisting flow and at top wall during opposing flow. At $\tau = 0.8$, a pair of tiny vortices is found to form on the bottom wall of opposing flow case (Fig. 2b). It indicates the buoyancy flow now becomes sufficiently stronger and is able to counter-act the rapidly formed shear-induced primary circulation. After $\tau \geq 0.8$, rather for the period of $\tau = 0.8 - 10$, Marangoni convection at the top region of cavity appears particularly for assisting flow. However, a separate identity of Marangoni convection is totally lost for opposing flow case. It is because of the merging of co-rotating vortices (due to forced and Marangoni convection) from the very beginning of this flow development.

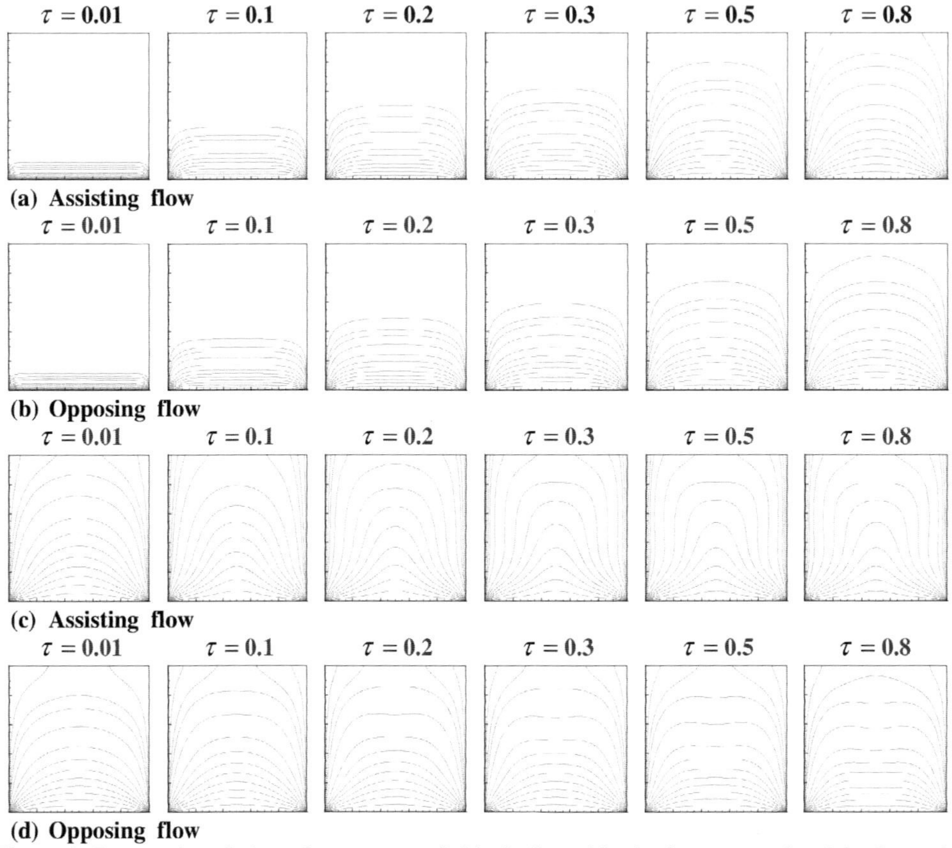

Figure 3. Temporal evolution of temperature fields (indicated by isotherms snapshots) in the cavity with downward walls motion (a, c) and with upward walls-motion (b, d) at Re = 200, Gr = 10^5, Ma = 1000, Bi = 1 and Ha = 0. Relatively, stronger changes in final temperature distribution are noted for assisting flow case.

In Fig. 2c, the growth of Marangoni convection is prominently marked at the free surface in the top region of the cavity. It gains strength and becomes larger gradually and slowly as the simultaneous effects of all three convection (free, forced and this Marangoni convection) transport the hot fluid along the top surface of the cavity. Until a temperature gradient along the free surface is developed, Marangoni convection cannot be formed. Moreover, the maturity of these convection vortices takes place after the settlement of all temperature variation, when a final state of surface tension along the free surface is reached. Thus, the formation of Marangoni convection is delayed and its adjustment in shape and magnitude takes place at the end of other convections. From Figs. 2a and 2c, it could be said that, the leader forced convection and the follower natural convection (relatively slower in onset) drive the temporal flow characteristics which is modified by Marangoni convection.

In Fig. 2d, the natural convection induced vortices in the bottom region of the cavity persist for a time duration of $\tau = 0.8 - 4$, then is almost forced to disappear due to the combined effect of forced convection and Marangoni convection. Here, both of them favor each other, and finally, jointly occupy the entire cavity. For this reason, the number of circulatory vortices reduces to two instead of four as seen for assisting flow case. It should be mentioned here that, the existence of buoyancy induced vortices strongly depend on the speed of wall motion or the strength of free convection (that is on Gr or Ri).

In Fig. 3, the temporal development of thermal fields in the cavity is demonstrated for both assisting and opposing flow. At time instant $\tau = 0.01$, a thin localized temperature fields are observed just adjacent to the heated bottom wall. With the progress of time ($\tau = 0.1 - 0.5$), it gradually spreads upward. At $\tau = 0.8$, the isotherms distribution in the cavity almost reaches the top surface. Thereafter, taking a longer period ($\tau = 1 - 10$), fine adjustments in temperature distribution are continued until the system reaches a steady-state condition. For the different types of flow (assisting and opposing), distinct changes in isotherms distribution are noted, when $\tau \geq 2$. The curving rise of isotherm contours is more prominent under downward moving system (Fig. 3c), which happens due to the formation of Marangoni convection at the upper region of the cavity. The temperature distribution for upward-moving lids is more or less horizontally stratified (Fig. 3d) at the end of the transient process ($\tau = 10$). Both assisting flow (or downward moving system) and opposing flow (or upward moving system) finally reaches the steady-state condition in all respects of streamlines, isotherms and pressure field distributions. Entire process of thermal-fluid flow development takes place symmetrically about the mid-vertical plane due to the symmetrical arrangement of boundary conditions.

3.2. Heat transfer characteristics

The progression speed and the time taken to reach the steady state solution are identified using the global parameters of maximum (and minimum) streamfunction (ψ) in the cavity, and the average Nusselt number (Nu) at the heated bottom wall, which are indicated in Figs. 4a and 4b, respectively. The figure reflects a diminishing rate of change of the process (triple convection in the cavity). From the same initial conditions (the zero state of flow-fields in the present study) with Ma=1000, Bi=1, Re = 200, and Gr = 10^5, the steady-state conditions is changes under different flow configurations (assisting and opposing flow). Here, the horizontal axis indicates the dimensionless time τ. After a certain time, ψ_{max} (and ψ_{min}) and Nu become almost invariant with time. Hence, after the initial transience, the flow structures remain invariable, stable and symmetric (as also observed from the streamline and isotherm plots in Figs. 2 and 3). Some facts are very clear from this figure. Heat transfer process at the heated bottom is quickly settled down to its steady state (or faster transience). Whereas, the flow structures takes much longer time to reach the steady state. It is because the heat transfer is calculated at on the bottom wall or in other words, for the first layer of fluid in contact the bottom surface of the cavity. The formation of thermal (and velocity) boundary layers, in absence of any external flow, quickly settles down. On the other hand, the bulk fluid-flow due to simultaneously occurring triple convection (free, forced and Marangoni convection and their mutual adjustments) takes longer time to settle down in the cavity. Furthermore, from the global ψ -plot and the inset of Nu-plot, it is clear that the opposing flow process is relatively faster. The initial very high Nu is due to very high temperature gradient at the bottom wall. During the onset of the process, maximum temperature difference exists between the wall ($\theta = 1$) and its adjacent fluid layer ($\theta = 0$) that leads to this high Nu.

The results of the selective parameters are presented here; the effect of the other parametric variation is not presented for the brevity.

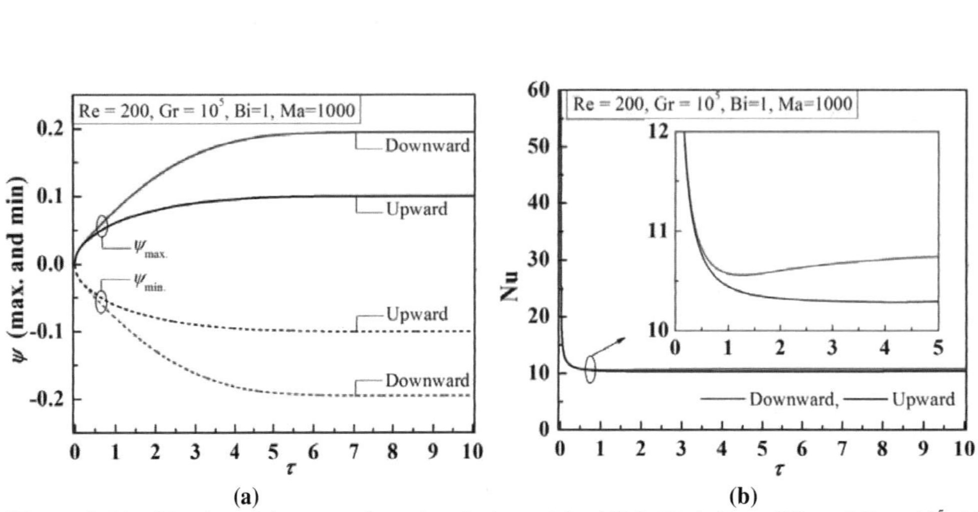

Figure 4. Identification of the state of steady solution at Ma=1000, Bi=1, Re = 200, and Gr = 10^5. (a) Temporal variations in maximum and minimum streamfunction, and (b) the average Nusselt number (Nu).

4. Conclusions

The present work explores the unsteady development of triple convective phenomena (of free, forced and Marangoni convection) of liquid metal in an open top cavity heated at the bottom and moving-sidewalls. The analysis is conducted thoroughly under wide range of parametric conditions, which alters the flow dynamics. Major observations are summarized as:

- The dynamics of flow pattern is extremely susceptible to the selection of the parametric conditions. On the other hand these parameters have minor role on the heat transfer from the hot bottom wall. Reynolds number alters the vortical structure severely, which influences on the Marangoni convection below the free the surface.
- The forced convection strongly governs the free convection or Marangoni convection.
- It is observed that, increase in the Biot and decrease in the Ma, the size of Marangoni circulating cells reduces.

References

[1] Block M J 1956 Surface tension as the cause of Bénard cells and surface deformation in a liquid film *Nature* **178** 650–651.

[2] Biswas N and Manna N K 2018 Magneto-hydrodynamic Marangoni flow in bottom-heated lid-driven cavity *J. Mol. Liq.* 251:249–266.

[3] Chamkha A J, Pop I and Takhar H S 2006 Marangoni mixed convection boundary layer flow Meccanica **41** 219–232.

[4] Saleh H, Arbin N, Roslan R and Hashim I 2012 Visualization and analysis of surface tension and cooling effects on differentially heated cavity using heatline concept *Int. J. Heat Mass Transfer* **55** 6000–6009.

[5] Hayat T, Shaheen U, Shafiq A, Alsaedi A and Asghar S 2015 Marangoni mixed convection flow with Joule heating and nonlinear radiation *AIP Advances* **5** 077140–0–10.

[6] Nithyadevi N, Begum A S and Shankar C U 2015 Buoyancy and thermocapillary driven flows in an open cavity with bottom heating and symmetrical cooling from sides *Int. J. Heat Technology* **33** 63–70.

[7] Sheremet M A and Pop I 2019 Marangoni natural convection in a cubical cavity filled with a nanofluid J. Thermal Anal. Calorim. **135** 357–369.

[8] Biswas N, Mahapatra P S and Manna N K 2016 Enhanced thermal energy transport using

adiabatic block inside lid driven cavity *Int. J. Heat Mass Transfer* **100** 407–427.
[9] Patankar SV 1980 *Numerical heat transfer and fluid flow*. New York: Hemisphere.
[10] Biswas N and Manna N K 2017 Enhanced convective heat transfer in lid–driven porous cavity with aspiration *Int. J. Heat Mass Transfer* **114** 430–452.

Nomenclature

Bi	Biot number	X,Y	dimensionless coordinates
g	acceleration due to gravity (ms^{-2})		***Greek symbols***
h	heat transfer coefficient ($Wm^{-2}K^{-1}$)	α	thermal diffusivity (m^2s^{-1})
H	height of the cavity, length scale (m)	β	volumetric thermal expansion coefficient of fluid (K^{-1})
k	thermal conductivity ($Wm^{-1}K^{-1}$)	θ	dimensionless temperature
Ma	Marangoni number	γ	temperature coefficient
Nu	average Nusselt number	μ	dynamic viscosity ($kgm^{-1}s^{-1}$)
p	pressure (Pa)	ν	kinematic viscosity (m^2s^{-1})
Pr	Prandtl number	ρ	density (kgm^{-3})
Re	Reynolds number	σ	surface tension (Nm^{-1})
Ri	Richardson number	ψ	dimensionless streamfunction
t	time (s)	τ	dimensionless time
T	temperature (K)		***Subscripts***
U,V	dimensionless velocity components	a,o	ambient, reference value
v_w	velocity of the moving wall (ms^{-1})	c,h	cold wall, hot wall

Effect of partial wall motion on MHD mixed convection heat transfer undergoing in a porous cavity filled with Cu–water nanofluid with a centrally mounted heat source

Milan K. Mondal[1], Nirmalendu Biswas[2, a], U. K. Sarkar[3], Nirmal K. Manna[4] and D. K. Mandal[5]

[1]Department of Mechanical Engineering, MSG Polytechnic, Paschim Medinipur, 721102, India
[2]Department of Power Engineering, Jadavpur University, Salt Lake, Kolkata–700106, India
[3]Department of Mechanical Engineering, Techno Main, Kolkata–700098, India
[4]Department of Mechanical Engineering, Jadavpur University, Kolkata–700032, India
[5]Department of Mechanical Engineering, College of Engineering and Management, Kolaghat – 721171, India
[a] Corresponding author's e-mail: *nirmalendubiswas@yahoo.co.in*

Abstract: The Magneto-hydrodynamics (MHD) convective heat transfer process in a partially driven cavity (PDC) containing a centrally mounted heat source is focused in the present study. The cavity is filled with Cu–water nanofluid saturated porous medium. Heated cavity fluid is cooled through the upper half of the two sidewalls. These cold sidewalls are allowed to move in upward direction at the same velocity. The horizontal walls and lower part of the sidewalls are insulated. The cavity subjected to an externally imposed uniform magnetic field. The flow structures and mixed convection heat transfer characteristics are analyzed numerically utilizing developed CFD code adopting FVM and SIMPLE algorithm. The thermo-fluid phenomena are analyzed under a range of governing parameters like Richardson number (Ri), Hartmann number (Ha), concentrations of nanoparticles (ϕ). The results reflect that the heat transfer characteristics of the PDC are greatly influenced by all of the above parameters. Heat transfer is enhanced significantly at the higher Ri and higher volume fraction.

Keywords: Partially driven cavity (PDC); Magneto-hydrodynamics (MHD); Cu-water nanofluid; Heat transfer enhancement.

1. Introduction

Because of enhanced thermal conductivity, the use of nanofluids as a working medium over the conventional working fluid is a recent technique for applying in various thermal systems. Furthermore, nanofluid flow through a porous substance can also be found in numerous applications [1,2]. Thermo-fluid flow behaviour of nanofluid flow through a porous medium becomes more complex, when it is subjected to an external magnetic field. Such a flow situation has can be found in diverse field like solar technology, material manufacturing technology, electromagnetic casting, oil and gas production process, cooling of electronic equipment, etc. When the natural convection alone is not sufficient enough for effective heat removal, mixed convection (shear induced flow due to wall translation) can efficiently improve the thermal behaviour. Many researchers have investigated the MHD mixed convective heat transfer in classical lid-driven cavity geometry. A detailed account of review on the above subject topic can be found in refs. [3,4].

Many authors have investigated the different aspects of the thermal convection of fluid saturated porous substances in a lid-driven enclosure. Rashad et al. [5] examined the MHD thermal convection of Cu-water nanofluid filled U-shaped cavity with double lids motion and heated partially from the bottom. It is observed that, the increase in the nanoparticle volume fraction leads to the increase in the heat transfer rate. MHD mixed convection in a cavity with double lids motion and heating discretely from bottom wall was examined by Hussain et al. [6]. Using Ag–MgO/water hybrid nanofluid, recently Selimefendigil and Chamkha [7] investigated the MHD thermal convection in a partitioned (in a triangular shape) cavity packed with porous layers. Gibanov et al. [8] studied the mixed convective heat transfer in a lid driven cavity filled with of ferrofluid filled and porous layers under the inclined magnetic field. Utilizing, Cu-water nanofluid flow from external source in a bottom-heated porous cavity Biswas et al. [9] have reported enhanced heat transfer (~15%). They found that impinging flow velocity and concentration of nanoparticles controls the overall thermal performance. In other class of work mixed convective heat transfer in lid-driven porous cavity has been investigated in [10,11] and others.

Above mentioned literature survey signifies the importance of the Magneto-hydrodynamic convective heat transfer in a cavity with moving lid(s). In most of the cited work, translation of the whole wall(s) of the confined space/cavity has been considered. Whereas, the overall thermal performance of the cavity under the effect of partial wall(s) translation over the whole wall(s) movement is missing. There are very few studies on the effect of partially driven walls on the thermal performance. Mondal et al. [12] recently, studied the effect of partial wall motion on the MHD thermal performance of a corner heated porous cavity. It is observed that, the partial wall(s) motion can significantly improve the heat transfer, which is further affected by the direction and speed of the partial wall(s) motion. This motivates us to undertake the present work.

In view of the above mentioned literature survey, this work is formulated to examine the effect of partial wall on the mixed convective heat transfer of Cu–water nanofluid saturated porous cavity with a centrally mounted block heated isothermally in presence of external magnetic field. The results of the analysis are visualized using streamlines, and isotherms; global thermal performance is characterized by the average Nusselt number.

2. Mathematical modelling and numerical technique

The schematic diagram of the Cu-water filled porous cavity with a centrally mounted block (having width $0.2L$ and height $0.2L$) heated isothermally (at a temperature T_h) under the influence of external magnetic field (of magnitude B_o) is depicted in Fig. 1. The domain is square cavity (2-D) of length L. Upper half portion (having dimension $0.5L$) of the sidewalls are cold (at temperature $T_c < T_h$), which are moving (with a velocity $v_w = +1$) in the upward direction at the same speed. Cu-water nanofluid flow through the porous medium is assumed to be laminar, incompressible, Newtonian, and steady within the validity of Boussinesq limit. Joule heating, induced magnetic field, viscous dissipation effects are not considered. The non-dimensional governing equations in Cartesian coordinate system are transformed [10–12] as:

$$\frac{\partial U}{\partial X}+\frac{\partial V}{\partial Y}=0 \tag{1}$$

$$\frac{1}{\varepsilon^2}\left(U\frac{\partial U}{\partial X}+V\frac{\partial U}{\partial Y}\right)=-\frac{\rho_f}{\rho_{nf}}\frac{\partial P}{\partial X}+\frac{\nu_{nf}}{\nu_f}\frac{1}{\varepsilon \mathrm{Re}}\left(\frac{\partial^2 U}{\partial X^2}+\frac{\partial^2 U}{\partial Y^2}\right)-\left(\frac{\nu_{nf}}{\nu_f}\frac{1}{\mathrm{DaRe}}+\frac{F_c\sqrt{U^2+V^2}}{\sqrt{\mathrm{Da}}\,\varepsilon^{3/2}}\right)U \tag{2}$$

$$\frac{1}{\varepsilon^2}\left(U\frac{\partial V}{\partial X}+V\frac{\partial V}{\partial Y}\right)=-\frac{\rho_f}{\rho_{nf}}\frac{\partial P}{\partial Y}+\frac{\nu_{nf}}{\nu_f}\frac{1}{\varepsilon \mathrm{Re}}\left(\frac{\partial^2 V}{\partial X^2}+\frac{\partial^2 V}{\partial Y^2}\right)-\left(\frac{\nu_{nf}}{\nu_f}\frac{1}{\mathrm{DaRe}}+\frac{F_c\sqrt{U^2+V^2}}{\sqrt{\mathrm{Da}}\,\varepsilon^{3/2}}\right)V$$
$$-\frac{\rho_f}{\rho_{nf}}\frac{\sigma_{nf}}{\sigma_f}\frac{\mathrm{Ha}^2}{\mathrm{Re}}V+\frac{(\rho\beta)_{nf}}{\rho_{nf}\beta_f}\mathrm{Ri}\theta \tag{3}$$

$$\left(U\frac{\partial\theta}{\partial X}+V\frac{\partial\theta}{\partial Y}\right)=\frac{\alpha_{nf}}{\alpha_f}\frac{1}{\mathrm{Re\,Pr}}\left(\frac{\partial^2\theta}{\partial X^2}+\frac{\partial^2\theta}{\partial Y^2}\right) \tag{4}$$

where, $\mathrm{Re}=\frac{v_w L}{\nu_f};\ \mathrm{Pr}=\frac{\nu_f}{\alpha_f};\ \mathrm{Da}=\frac{K}{L^2};\ F_c=\frac{1.75}{\sqrt{150}};\ \mathrm{Ha}=B_o L\sqrt{\sigma_f/\mu_f};\ \mathrm{Ri}=\frac{\mathrm{Gr}}{\mathrm{Re}^2};\ \mathrm{Gr}=\frac{g\beta_f(T_h-T_c)L^3}{\nu_f^2}$ (5)

are the Reynolds number, Prandtl number, Darcy number, Grashof number, Forchheimer coefficient, Hartmann number, Richardson number, and Grashof number respectively. The dimensionless equations (1)–(5) are transformed by the following factors:

$$(X,Y)=(x,y)/L;(U,V)=(u,v)L/\alpha_f;\ \theta=(T-T_c)/(T_h-T_c);P=(p-p_a)L^2/\rho_f{\alpha_f}^2 \tag{6}$$

The thermophysical properties of the base liquid (water, showed by the subscript '*f*', Pr = 6.93) and Cu nanoparticles are taken from the ref. [9,13,14]. The properties of Cu-water nanofluid (designated by the subscript '*nf*') are calculated using the volumetric concentration of nanoparticles (ϕ) suspended in the base fluid. Following relations are utilized for obtaining the nanofluid density, specific heat and thermal expansion coefficient : $\rho_{nf}=(1-\phi)\rho_f+\phi\rho_{\mathrm{Cu}}$; $(\rho C_p)_{nf}=(1-\phi)(\rho C_p)_f+\phi(\rho C_p)_{\mathrm{Cu}}$; $(\rho\beta)_{nf}=(1-\phi)(\rho\beta)_f+\phi(\rho\beta)_{\mathrm{Cu}}$

The effective dynamic viscosity, thermal diffusivity α_{nf}, thermal conductivity (k_{nf}) and effective electrical conductivity (σ_{nf}) of nanofluid [9] are given by

$$\mu_{nf}=\frac{\mu_f}{(1-\phi)^{2.5}};\ \alpha_{nf}=\frac{k_f}{(\rho C_p)_{nf}};\ k_{nf}=k_f\left[\frac{(k_s+2k_f)-2\phi(k_f-k_s)}{(k_s+2k_f)+\phi(k_f-k_s)}\right];\ \sigma_{nf}=\sigma_f\left[1+\frac{3(\sigma_s/\sigma_f-1)\phi}{(\sigma_s/\sigma_f+2)-(\sigma_s/\sigma_f-1)\phi}\right]$$

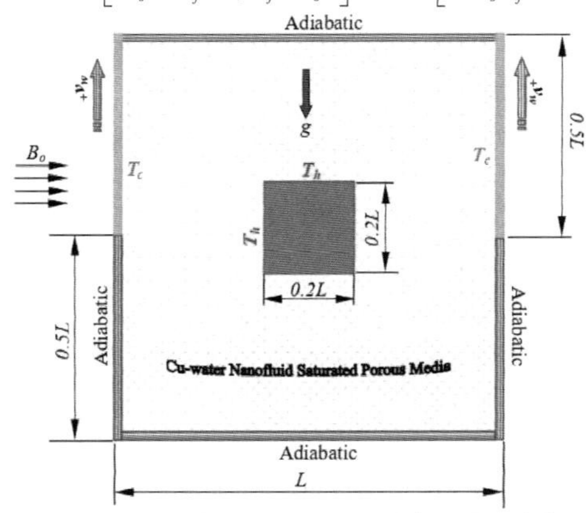

Figure 1. Schematic diagram of the computational domain of the physical problem.

The dimensionless partial differential equations (1)–(4) are solved numerically using in-house CFD code adopting the FVM through ADI sweep, TDMA solver, using the SIMPLE algorithm [15]. The converged solution is obtained when continuity mass-defect becomes less than 10^{-10}. The same code has been used in our earlier works and validated extensively under different problem of published literature [3,9,11,12,16,17]. In the present simulation, uniformly distributed 200×200 grid size is selected after conducting the gird independency study. With the grid size of 200×200, it is found the relative error in the estimation of average Nusselt number (with respect to immediate coarse grid) is less than 1%, which is within acceptable limit.

The boundary conditions in the dimensionless form are:

- at the left and right walls - lower half ($X=0,1;0\le Y<0.5$), $U=V=0,\partial\theta/\partial X=0$ and for the upper half moving cold walls ($X=0,1;0.5\le Y\le 1$), $U=0,V=+1,\theta=0$.
- for the protruded heater ($X=0.4, 0.4\le Y\le 0.6,\ X=0.6, 0.4\le Y\le 0.6$ and $Y=0.4, 0.4\le X\le 0.6, Y=0.6, 0.4\le X\le 0.6$), $U=V=0,\theta=1$.

IOP Conf. Series: Materials Science and Engineering 1080 (2021) 012025 doi:10.1088/1757-899X/1080/1/012025

- at the bottom adiabatic wall, $(Y = 0, 0 \le X \le 1)$, $U = V = 0, \partial\theta/\partial Y = 0$.
- at the top adiabatic wall, $(Y = 1, 0 \le X \le 1), U = V = 0, \partial\theta/\partial Y = 0$.

3. Results and discussion

In this the work, MHD convective heat transfer of Cu-water nanofluid saturated porous cavity containing centrally mounted heated block is examined numerically. Upper half of the cold sidewalls are moving in the upward direction (such a configuration is termed as PDC). The analysis is carried out for the range of parameters such as Richardson number (Ri = 0.1, 1, 10, 100), Hartmann number (Ha = 0, 10, 30, 50, 70, 100), nanofluid volume fraction (ϕ = 0, 0.005, 0.01, 0.02, 0.03, 0.04, 0.05). The sliding speed of the cold walls is taken as fixed value through the Reynolds number (Re = 200). The porosity (ε = 0.8) and Darcy number (Da = 10^{-3}) are also taken as fixed. The results are visualized through streamlines, isotherms, and average Nusselt number (Nu).

3.1. *Effect of Richardson number* (Ri)

On defining the operating flow regime, the choice of the Richardson number (Ri) signifies an imperative role. The impact of varying Ri values (Ri = 0.1, 1, 10 and 100) on the streamlines (top panel), and isotherms (bottom row) contours are presented in Fig. 2 for the fixed values of Ha = 30, ϕ = 0.01. As heat is released by the centrally mounted heated block surrounding fluid gets heated and moves upward (due to density difference) and then release heat through the moving cold sidewalls and then goes downward. Thus, the cavity is occupied by the two counter-rotating circulating cells. However, for the forced convection regime (Ri $\le$0.1), wall shear induced flow dominates the buoyancy effect. As a result clockwise cell (CW) appears in the left part and anti-clockwise cell appears in the right part of the cavity. At Ri = 1(mixed convection regime), in addition two circulating cells (as seen with Ri = 0.1), there also appears two more counter rotating cells (due to weaker buoyancy effect) in the lower part of the cavity. When, Ri increases to 10 and 100, buoyancy force dominates the shear force.

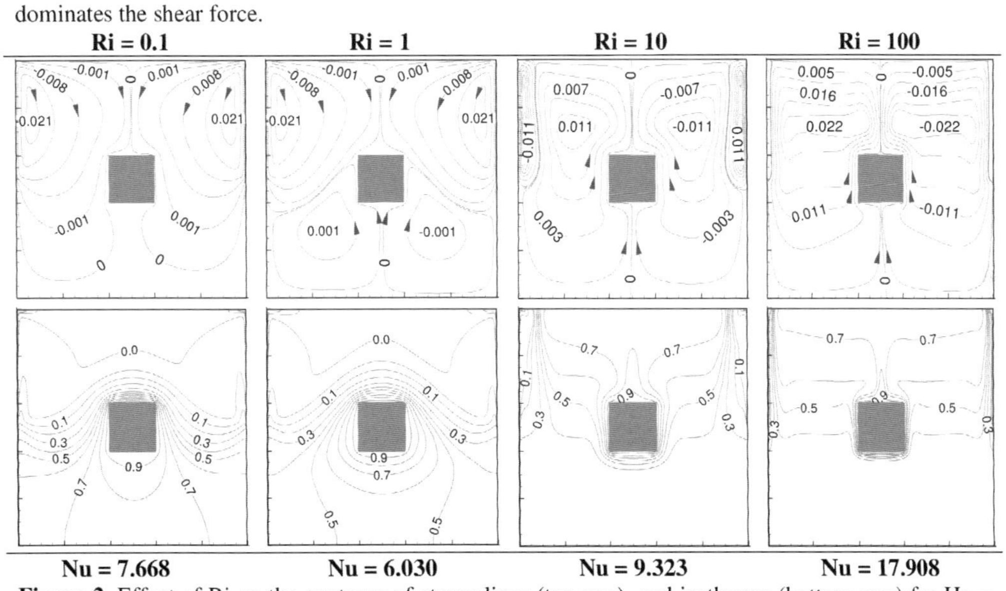

Figure 2. Effect of Ri on the contours of streamlines (top row), and isotherms (bottom row) for Ha = 30, ϕ = 0.01. Isotherms and streamline contour intervals are 0.1 and 0.001, respectively.

This results in shrinkage of shear-induced flow (towards the moving cold sidewalls) and enlargement of the buoyancy-induced flow (from lower portion to the entire cavity). With the increase in Ri value, convection becomes stronger due to increase in the strength of the fluid circulation (as denoted by the maximum streamfunction). At Ri = 100, circulating cells stretched horizontally. The effect of different Ri values on the temperature distribution can be realised from the corresponding isotherm contours. For Ri $\leq$ 1, upper region of the cavity remain colder, whereas the bottom region of the enclosure is hotter due to distribution of high temperature lines about the middle horizontal portion (due to dominating shear-driven flow). However, for Ri $\geq$10, buoyancy-driven flow governs the convection mechanism; thus high temperature lines remains in the upper part of the enclosure. Due to stronger buoyancy effect, heat transfer rate increases markedly as indicated by the average Nu value.

3.2. *Effect of Hartmann number* (Ha)

The effect of externally applied horizontal magnetic field (defined by the term Hartmann number Ha) on the contours of streamlines (top row) and isotherms (bottom row) are shown in Fig. 3 for the different value of Ha = 0, 10, 50, 100 keeping other parameters fixed at Ri = 10, ϕ = 0.01. For easy of the understanding, the case of Ha = 0 (absence of magnetic field) is analyzed first. The thermo-fluid flow structure shows a symmetric distribution with respect to the vertical mid-plane. Now, strengthening the imposed magnetic field by increasing the Ha value = 10, 50 or 100, pattern wise overall flow structure are similar with the structure at Ha = 0. However, the buoyancy-induced circulating cells become weaker (as can be seen from the maximum streamfunction value). This leads to weaker heat transfer and this reduction is maximum at higher Ha = 100. Overall, the heat transfer reduction rate is about 0.37–21.23% compared to no-magnetic field, Ha = 0.

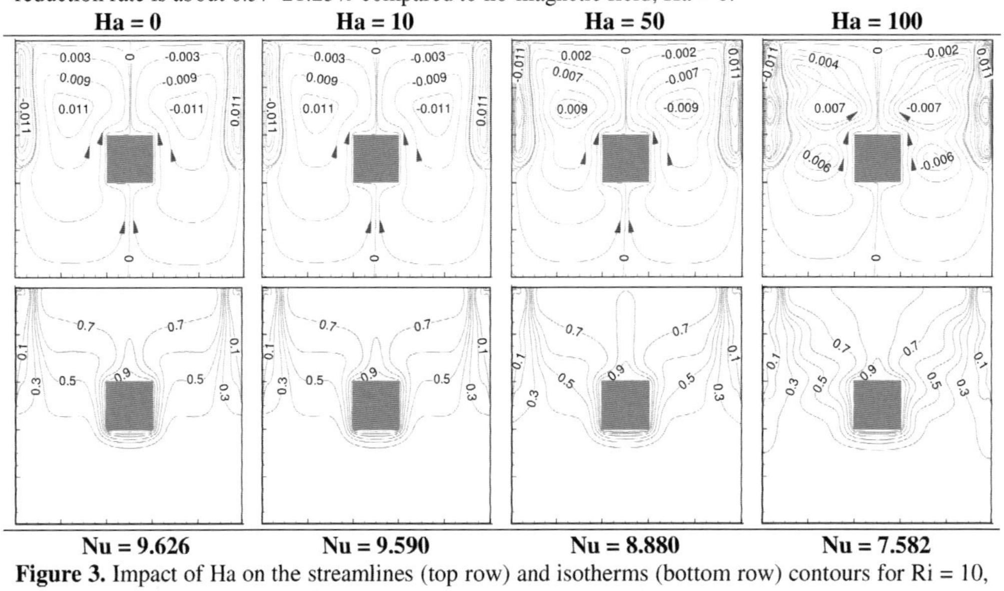

Figure 3. Impact of Ha on the streamlines (top row) and isotherms (bottom row) contours for Ri = 10, ϕ = 0.01. Isotherms and streamline contour intervals are 0.1 and 0.001, respectively.

3.3. *Effect of nanoparticle concentration* (ϕ)

The increase in Cu nanoparticle volume fraction ϕ (= 0, 0.005, 0.02, and 0.05) on the streamlines, isotherms contours and average Nu are illustrated in Fig. 4 for the fixed values of Ri = 10, Ha = 30. It is noted that, the inclusion of nanoparticles in the water leads to the increase in the strength of fluid circulation, thus improvement in the heat transfer rate. The reason behind this improvement is due to the raise in ϕ leads to the increase in the effective thermal conductivity the working fluid. Thus, improvement in the buoyancy effect, and associated heat transfer rate (as indicated by the average Nu value). In general, the heat transfer enhancement rate is about 0.46–4.46% compared to base fluid (ϕ = 0).

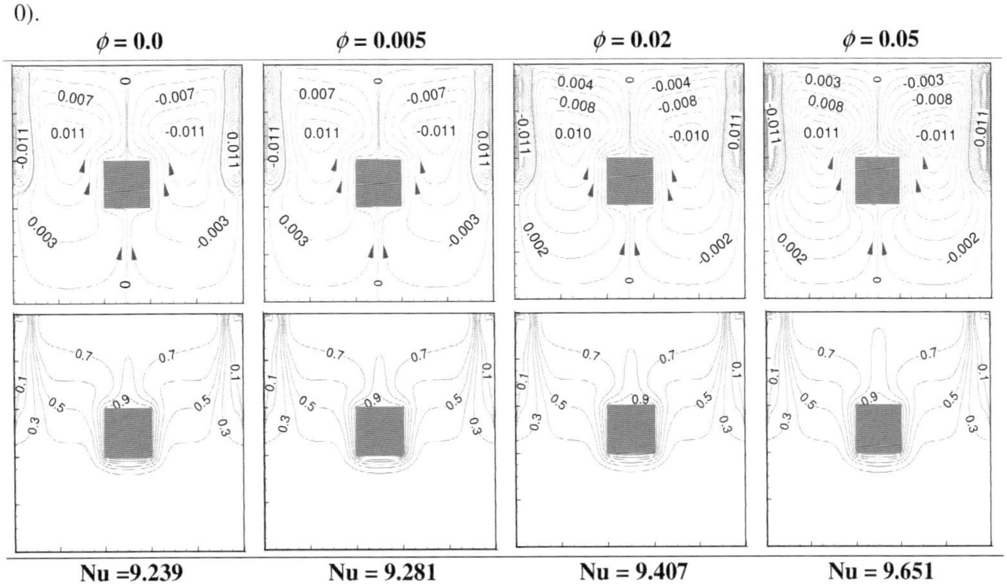

Figure 4. Effect of ϕ on the streamlines (top row) and isotherms (bottom row) contours at Ri = 10, Ha = 30. Isotherms and streamline contour intervals are 0.1 and 0.001, respectively.

3.4. Heat transfer characteristics

The average Nusselt number (Nu) is used to calculate the heat transfer characteristics at the cold sidewalls as

$$\mathrm{Nu} = \frac{k_{nf}}{k_f}\int_{0.5}^{1}\left(-\left.\frac{\partial\theta}{\partial X}\right|_{X=0}\right)dY + \frac{k_{nf}}{k_f}\int_{0.5}^{1}\left(-\left.\frac{\partial\theta}{\partial X}\right|_{X=1}\right)dY$$

The distribution of average Nu with the variation of Ri are plotted in Fig. 5 (a and b) for different Ha and ϕ keeping other parameters at fixed values. In Fig. 5a, the average Nu shows a consistently decreasing trend with the increasing Ha (for ϕ = 0.01). This happens due to the presence of negative source term with Ha in the *Y* momentum equation (3), which counteracts the positive effect of the buoyancy effect. The decreasing trend is almost similar for all Ri values except Ri = 1 (as Nu values shows a little bit increasing trend). Furthermore, the Nu curve at Ri = 0.1 remains above that of Ri = 1. The reason behind this fact may be understood by observing the flow structure at Ri = 1 (as in Fig. 2). Here the shear force is counteracted by the buoyancy force. As a result the heat transfer process is affected severely. Thus, setting Ri plays a crucial role on the thermal performance of such PDC. Furthermore, selection of the Ha value influences the heat transfer process.

The effect of volumetric concentration of Cu-water nanofluid under different Ri values are reflected in the Fig. 5b at Ha = 30. The Fig. 5b clearly reflects that, with the increase in ϕ average Nu increases for any Ri values. The heat transfer process is more influenced by the nanoparticles in the convection dominated mode (Ri = 100). Inclusion of nanoparticles leads to increase in the effective thermal conductivity of the working fluid; it leads to increase in the circulation strength. Thus more amount of heat is transferred from heat source to the heat sink. In general, the heat transfer enhancement rate varying the Cu nanoparticles volume fraction (ϕ = 0.005–0.05) is about 0.81–8.40%

(for Ri = 0.1), 1.04–11.12% (for Ri = 1), 0.61–5.94% (for Ri = 100) compared to the base fluid (ϕ = 0).

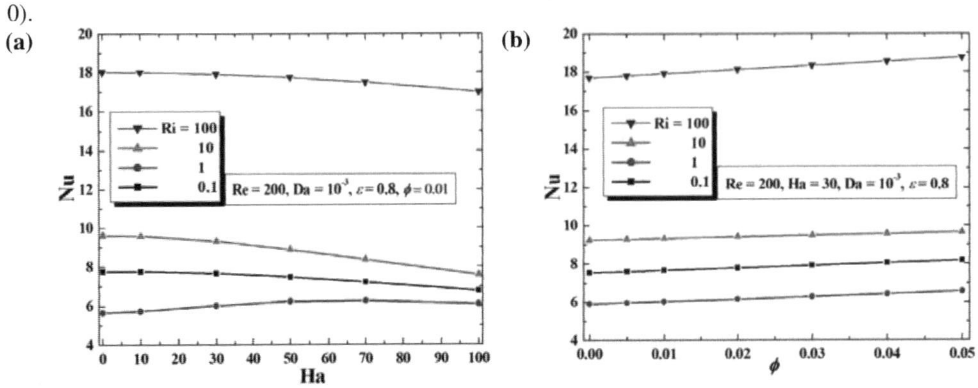

Figure 5. Heat transfer characteristics with varying Ri, (a) Ha, (b) ϕ keeping other parameters fixed.

4. Conclusions

In this study, the effects of partial wall motion on mixed convection heat transfer in Cu-water nanofluid filled porous cavity with a centrally mounted heat source under the externally applied magnetic field is investigated numerically. The thermo-fluid phenomena in a PDC are explored under different parametric influences like Ri, Ha, and ϕ. The significant conclusions are:

a) Under the partial wall motion, selection of Ri is the key factor for the convective heat transfer process. The heat transfer characteristics is increases significantly, when Ri > 1 and it is maximum at Ri = 100.

b) Increase in the magnetic field strength, buoyancy effect counteracted by the magnetic force. This results in decrement in the heat transfer rate. Such decrement is more at higher Ha = 100.

c) Increase in the Cu nanoparticle volume fraction, the heat transfer rate increases ~ 1.04–11.12% compared to base fluid (ϕ = 0).

References

[1] Nield D A, Bejan A 2006 *Convection in Porous Media*, (Springer; Berlin).

[2] Kasaeian A, Daneshazarian R, Mahian O, Kolsi L, Chamkha A J, Wongwises S, Pop I 2017 Nanofluid flow and heat transfer in porous media: a review of the latest developments *Int. J. Heat Mass Transfer* **107** 778–791.

[3] Biswas N, Manna N K 2018 Magneto-hydrodynamic Marangoni flow in bottom-heated lid-driven cavity *J. Mol. Liq.* **251** 249–266.

[4] Selimefendigil F, Öztop H F, Chamkha A J 2016 MHD mixed convection and entropy generation of nanofluid filled lid driven cavity under the influence of inclined magnetic fields imposed to its upper and lower diagonal triangular domains *J. Magn. Magn. Mater.* **406** 266–281.

[5] Rashad A M, Mansour M A, Armaghani T, Chamkha A J 2019 MHD mixed convection and entropy generation of nanofluid in a lid-driven U-shaped cavity with internal heat and partial slip *Phys. Fluids* **31** 042006-1–17.

[6] Hussain S, Mehmood K, Sagheer M 2016 MHD mixed convection and entropy generation of water–Alumina nanofluid flow in a double lid driven cavity with discrete heating *J. Magn. Magn. Mater.* **419** 140–155.

[7] Selimefendigil F, Chamkha A J 2020 MHD Mixed convection of Ag–MgO/water nanofluid in a triangular shape partitioned lid-driven square cavity involving a porous compound *J. Thermal Anal. Calorim.* doi:10.1007/s10973-020-09472-7.

[8] Gibanov N S, Sheremet M A, Oztop H F, Abu-Hamdeh N 2017 Effect of uniform inclined

magnetic field on mixed convection in a lid driven cavity having a horizontal porous layer saturated with a ferrofluid *Int. J. Heat Mass Transfer* **114** 1086–1097.

[9] Biswas N, Manna N K, Datta P, Mahapatra P S 2018 Analysis of heat transfer and pumping power for bottom-heated porous cavity saturated with Cu-water nanofluid *Powder Technol.* **326** 356–369.

[10] Khanafer K M, Chamkha A J 1999 Mixed convection flow in a lid-driven enclosure filled with a fluid saturated porous medium *Int. J. Heat Mass Transfer* **42** 2465–2481.

[11] Biswas N, Manna N K 2017 Enhanced convective heat transfer in lid–driven porous cavity with aspiration *Int. J. Heat Mass Transfer* **114** 430–452.

[12] Mondal M K, Biswas N, Manna N K 2019 MHD convection in a partially driven cavity with corner heating *SN Appl. Sci.* 1–1689.

[13] Biswas N, Manna N K, Chamkha A J 2020 Effects of half-sinusoidal nonuniform heating during MHD thermal convection in Cu–Al_2O_3/water hybrid nanofluid saturated with porous media *J Therm. Anal. Calorim.* doi.org/10.1007/s10973-020-10109-y.

[14] Biswas N, Sarkar U K, Chamkha A J, Manna N K 2020 Magneto-hydrodynamic thermal convection of Cu–Al_2O_3/water hybrid nanofluid saturated with porous media subjected to half-sinusoidal nonuniform heating *J Therm. Anal. Calorim.* doi.org/10.1007/s10973-020-10123-0.

[15] Patankar SV 1980 *Numerical Heat Transfer and Fluid Flow*, New York NY Hemisphere.

[16] Biswas N, Mahapatra P S, Manna N K 2016 Buoyancy-driven fluid and energy flow in protruded heater enclosure, *Meccanica* **51** 2159–2184.

[17] Mondal M K, Manna N K and Saha R 2013 Study of leakage flow through a spool valve under blocked-actuator port condition – simulation and experiment *Proc. IMechE Part C: J. of Mech. Engg. Sci.* **228(8)** 1405–1417.

Effect of Particle Size on fluid flow and heat transfer in a Pipe with slurry flow

D. K. Mandal[1], A. Hazra[2], and N. Biswas[3]

[1,2] Dept. of Mechanical Engg., College of Engg. & Management, Kolaghat, P.O: K.T.P.P. Township, Midnapore (E) - 721 171, West Bengal, India,
[3] Department of Power Engineering, Jadavpur University, Salt Lake, Kolkata–700098, India.Corresponding author, dipkuma@yahoo.com[1]

Abstract. A three-dimensional numerical simulation is carried out in a straight horizontal pipe to predict the impact of particle size on fluid flow and heat transfer of ash water slurry. Steady turbulent equations are solved by using commercial code Fluent. Eulerian Method is incorporated considering ash as granular material. Fly-ash particle diameter is varied with 12, 20, 28 and 34 μm suspended in water for a flow velocity 4m/s and solid concentration of 40%. The temperature of pipe wall is kept at 400K and the granular temp i.e. slurry temp is taken as 300K. Present study finds asymmetric velocity distribution at higher particle size. Pumping power increases with particle size. Granular pressure and heat transfer increases with particle size. Granular pressure is maximum at bottom of the wall.

Keywords: Slurry flow, Eulerian method, Granular pressure, Heat transfer.

Nomenclature

Roman Symbols

C_d	-	Drag Co-efficient	V	-	Mean Velocity
C_p	-	Specific heat, J/KgK	x	-	Coordinate, m
C_{vf}	-	Solid Volume concentration %			
D		Pipe diameter in m			
d_p	-	Particle diameter	Greek Symbols		
$\overline{F}_{lift}$	-	Lift Force, N	α	-	Volume fraction
$\overline{F}_{vn}$	-	Virtual mass force, N	β_T	-	Thermal expansion Coefficient, K^{-1}
f_r	-	Roughness Function	Φ	-	Dimensionless temperature
F_r	-	Froude number	ε	-	Dissipation rate for turbulent kinetic energy, m^2/s^3
g	-	Acceleration due to gravity	ϕ	-	Friction angle
g_o	-	Radial distribution function	k	-	Thermal conductivity, W/mK
h	-	Heat exchange coefficient W/m^3K	λ_s	-	Bulk viscosity of solid Kg/m-s

K	-	Momentum exchange coefficient	μ	-	Viscosity K/m-s
l_μ, l_ε	-	Length scale, m	θ_s	-	Granular temperature of solid phase, m^2/s^3
L	-	Pipe Length, m	ρ	-	Density, Kg/m^3
P	-	Pressure, Pa	$\bar{\tau}$	-	Viscous stress tensor, Pa
q	-	Heat flux, w/m^2	Subscripts		
$\vec{r}$	-	Position vector	*eff*	-	Effective
Re	-	Reynolds number	f	-	Fluid
Re_s	-	Relative Reynolds number	*in*	-	Inlet
t	-	Time, s	m	-	Mean
T	-	Temperature, K	s	-	Solid
u,v	-	Velocity, m/s	t	-	Turbulent
			w	-	Wall

1 Introduction

Slurry is a mixture of solid and liquid. There are many industries that produce slurry. Maximum flyash is produced in coal based thermal power plant. This ash is transported by mixing with water and making slurry through pipe. This transportation is reliable and safe because it has low maintenance cost rather than bulk material handling. Apart from thermal power plant, slurries transportation through pipe line is common in different industries like oil and gas, food, pharmaceutical, transportation of metal concentrate and tannings, chemicals processing, deep sea mining, mineral processing and others. The presence of ash particles in water changes its viscosity, temperature, turbulence depending on the particle size and concentration. This is essential for designing of slurry pump, heat exchanger, fluidized beds and others.

Numerical and experimental work has been carried out to study the fluid flow and heat transfer phenomenon through pipe with ash slurry flow by various researchers. Among of them, Ku et al. [1] have done experimental investigation in an 8 mm diameter pipe, mass median ranges from 4 to 78 micron with flow velocity 1m/s to study the effect of solid concentration. From the experimental investigation [2, 3, 4, 6], Horiuchi et al. [2] have shown the increase in strength of slurry with density by comparing different samples. In the study of Harada et al. [3], heat transfer coefficient is observed to be high compared to water in case of flow with glass bead particles larger than 0.35 mm diameter. Vlasak et al. [4] have observed asymmetrical velocity distribution in the upper wall of pipe for course grained with water mixture. Numerical study is carried out by various researchers [5-11], the study of Wilson et al. [5] is done to check the energy efficient transport at higher concentration. Numerical and experimental study by Kaushal et al. [6] is conducted to observe the effect of velocity concentration for sand water slurry. In the study of Ozbelge et al. [7], they have revealed that the velocity increases with solid concentrations. In a U tube pipe of Cvetkovski et al. [8], heat transfer is noticed in bend portions. The effect of particle size on flow behaviour is studied by Gopaliya et al. [9]. Kumar et al. [10] have shown that asymmetry decreases with increase in flow velocity. Pipe with 90^0 elbow pipe is taken by Yang et al. [11] to compare the pressure loss with horizontal pipe.

It is evident from the brief literature review that the works of particle size for describing the heat transfer and fluid flow analyzing in case of slurry flow is less. In this study, effect of particle size on velocity, pressure drop, granular pressure, temperature and molecular viscosity is focused for ash water slurry flow through a horizontal pipe. Particle sizes are considered as12μm, 20μm, 28μm and 34μm. Mean slurry velocity and solid volume fraction is 4 m/s and 40 % respectively.

2. Mathematical Formulation

2.1 *Problem Description*

The geometry of physical problem carried in this study is horizontal pipe. The length (L) of the pipe is taken 1.5m and diameter of the pipe is D=0. 008m [1]. The length is taken above 50D to confirm fully developed flow. Water is taken as a primary fluid and fly-ash as a secondary phase material. The temperature of pipe wall is kept at isothermal condition of 400K and the granular temp i.e. slurry temp is taken as 300K.The inlet velocity is kept constant equal to 4m/s. Volume fraction has been kept constant of 40% and the diameter of the particle of the fly-ash is varied as 12, 20, 28 and 34 μm. Properties of ash and water are given in Table.1.

Table 1. Physical Properties of fly-ash and water

Items	Test Particle	Test Liquid
Material	Fly-ash	Water
Díameter of Granular Particle (μm)	12,20,28,34	-
Density (Kg/m³)	2270	997
Specific Heat (J/KgK)	745	4,179
Thermal Conductivity (W/mK)	1.38	0.613
Viscosity (Ns/m²)	------	855×10^{-6}

The mesh is formed by using commercial software Fluent, shown in Fig. 1.

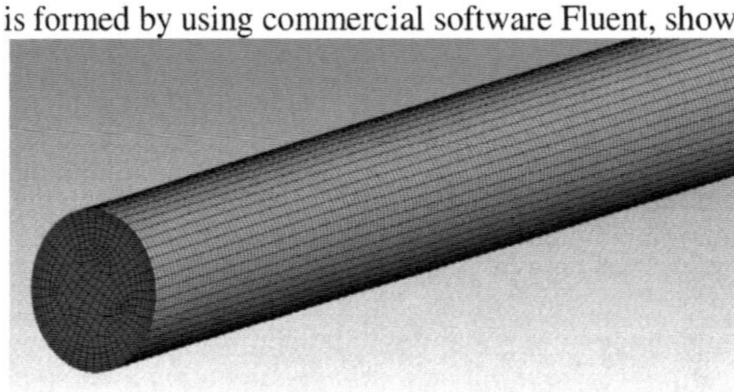

Fig.1 Mesh of computational domain

The details of mesh are given below, in Table.2. Body meshing is adopted with element size $1X10^{-4}$m.

Table 2. Mesh details

Type of mesh	Hexahedral
Transition Ratio	0.27
Maximum Layer(wall)	5
Growth Rate(Inflation)	1.2
Nodes	1386164
No of Elements	1318350
Min Aspect Ratio	1.19
Max Aspect Ratio	4.9
Skewness Factor	$2.33X10^{-2}$

2.2 Modelling and governing equations

The Eulerian model, the different phase's present in the system is treated mathematically as interpenetrating continua. Individually each phase satisfied the laws of momentum, energy and conservation of mass. Each phase contained their own physical properties, velocity, and temperature. In this model volume fraction are considered as a continuous function of space and time and their sum is equal to one. Volume fraction is nothing but the amount of volume is taken by each phase in a given control volume. In this model the solid phase property like granular viscosity, bulk viscosity, shear etc are solved by kinetic theory. Momentum equation also solved differently for solid and liquid respectively.

In this present study we have made some consideration. The mixture is homogeneous and the flow occurs at no slip condition. The governing equations [12.13] are as

Continuity Equation

For solid phase

$$\frac{\partial}{\partial t}(\alpha_s \rho_s) + \nabla(\alpha_s \rho_s \bar{v}_s) = 0$$

For solid phase

$$\frac{\partial}{\partial t}(\alpha_f \rho_f) + \nabla(\alpha_f \rho_f \bar{v}_f) = 0$$

Momentum Equations

For Solid Phase

$$\nabla(\alpha_s \rho_s \bar{v}_s \bar{v}_s) = -\alpha_s \nabla p - \nabla P_s + \nabla.\bar{\tau}_s + \alpha_s \rho_s g + \alpha_s \rho_s (\bar{F}_{lift,s} + \bar{F}_{vm,s}) + k_{fs}(\bar{v}_f - \bar{v}_s)$$

For Liquid Phase

$$\nabla(\alpha_f \rho_f \bar{v}_f \bar{v}_f) = -\alpha_f \nabla P + \nabla.\bar{\tau}_f + \alpha_f \rho_f \bar{g} + \alpha_f \rho_f (\bar{F}_{lift,f} + \bar{F}_{vm,f})$$

Turbulence Model

Turbulence model is much difficult in multiphase model than the single-phase flow. The number of terms used in momentum equation of multiphase model, make the large and complex. Here RNG $\kappa - \epsilon$ model is used to solve the flow of slurry. Fluent provides many types of turbulent model like, mixture turbulence model, dispersed turbulence model and turbulence model for each phase. Mixture model is used when phases separate, for stratified multiphase flows. Dispersed model is used in case of dilute concentration of the second phase. Primary phase influences secondary phase for getting random motion. So dispersed model is the better choice for this study.

Governing Equation

For kinetic energy

$$\nabla(\rho_m \bar{v}_m k) = \nabla(\alpha_k \mu_{m,eff} \nabla_k) + \mu_{m,t} S^2 + \beta_T g_i \frac{\mu_{m,s}}{\rho_{rt}} \frac{\partial T_m}{\partial_{xi}} - \rho_m \varepsilon$$

For dissipation rate

$$\frac{\partial}{\partial x_i}(\rho_m \bar{v}_m \varepsilon) = \nabla(\alpha_\varepsilon \mu_{m,eff} \nabla_s) + C_{1\varepsilon} \frac{\varepsilon}{k} \mu_{m,t} S^2 + C_{2\varepsilon} \rho_m \frac{\varepsilon^2}{k} - R$$

Energy equation on the basis of heat transfer

Eulerian model require an extra equation to solve enthalpy, so it has written in the followings for each phase.

For Solid Phase

$$\nabla(\alpha_s \rho_s \bar{v}_s h_s) = \bar{\tau}_s : \nabla.\bar{v}_s - \nabla.\bar{q}_s + Q_{sf}$$

For fluid phase

$$\nabla(\alpha_f \rho_f \bar{v}_f h_f) = \bar{\tau}_f : \nabla.\bar{v}_f - \nabla.\bar{q}_f + Q_{fs}$$

2.3 *Method of solution*

Present numerical study is carried out by commercial code Fluent [13]. The partial differential equations are solved control volume technique, finite volume method. Details of solution method are given in Table.3.

Table.3 Details of solution method

Solver	Pressure based, Algebraic Multigrid (AMG) solver
Multiphase model	Eulerian
Volume fractions	Implicit
Turbulence model	k-epsilon method
k-epsilon method	RNG
Turbulence multiphase model	Dispersed
Iterative algorithm	PCSIMPLE
Discretization of momentum equations	Second order upwind scheme
Discretization of kinetic energy, dissipation rate and volume fraction	First order upwind
Convergence criterion	10^{-4}
Under relaxation factors	0.5–0.7(momentum) 0.2–0.3(pressure) 0.6–0.8(turbulent kinetic energy)

3. Results and Discussion

This present study describes the effect of particle size on velocity, pressure drop, granular pressure, temperature and molecular viscosity for ash water slurry flow through a horizontal pipe. Particle sizes are considered as 12μm, 20μm, 28μm and 34μm. Mean slurry velocity and solid volume fraction is 4 m/s and 40 % respectively.

3.1 *Distribution velocity contour, pressure drop and molecular viscosity*

Figure 2 presents the velocity contour along axial direction of pipe. It is observed that velocity magnitude at middle section of pipe is more and magnitude increases with the increase in particle size. The velocity distribution of slurry flow in pipe is also important and it is noted that the velocity distribution is symmetrical in nature at 12μm and 20μm, whereas it becomes asymmetrical for the cases of 28μm and 34μm. In comparison with the velocity contour of 28μm and 34μm, contour of higher particle size shows more asymmetric in nature. The reason may be due to intensified turbulence at high particle size.

The calculation of pressure drop through slurry pipe is important for determining power required for the pump. The pressure drop in case of slurry flow is the summation of pressure drops of only water and pressure drop due to presence of solid particles. The pumping power of slurry pump is determined by multiplication of pressure drop and mass flow rate. Figure 3 shows the pressure drop in the slurry pipe with the variation of particle size. Pressure drop is observed to increase with increase in particle diameter. The possible reason may be the increase in velocity of slurry and increase in effective viscosity due to increase in particle size. At higher mean diameter of particle size, the rate of increase in pressure drop is high compared to rate of increase in pressure drop for lower size particle as noted in the Fig. 3. As the power required by pump is proportional to the pressure drop, therefore, corresponding power required for pump is more for the larger particle size.

Apart from pressure drop, molecular viscosity plays an important role for determination of pump power. The variation of molecular viscosity with particle size is shown in Fig. 4. It is revealed that the viscosity of slurry increases with particle size. This may be attributed that increased surface area with particle size enhances solid concentration and thereby it increases the viscosity.

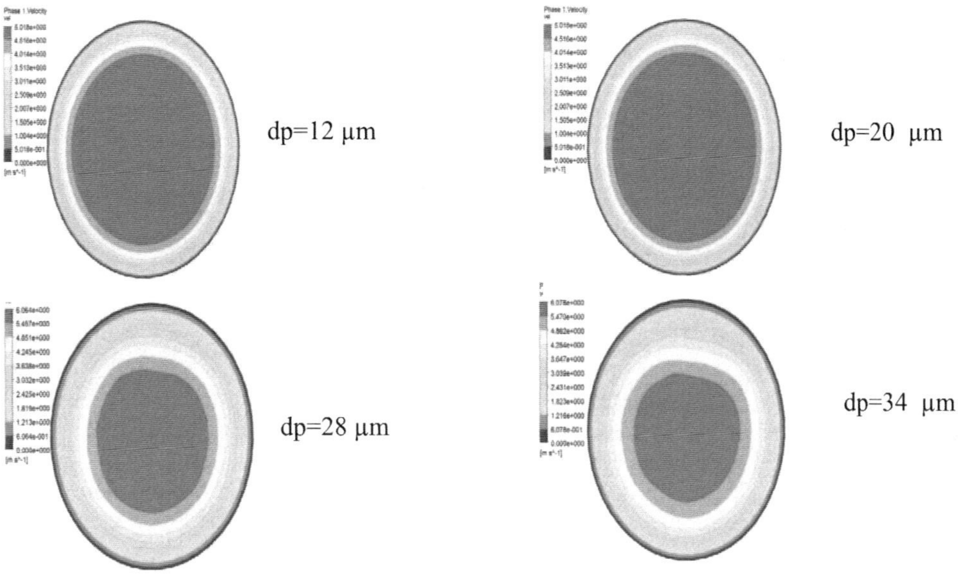

Fig. 2 Distribution of velocity in m/s at 4m/s inlet velocity for 40% volume fraction at different particle sizes

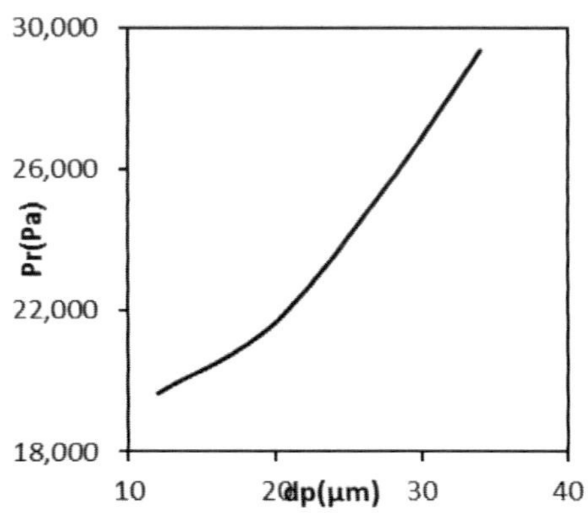

Fig. 3 Effect of particle size on pressure drop at 4m/s inlet velocity for 40% volume fraction

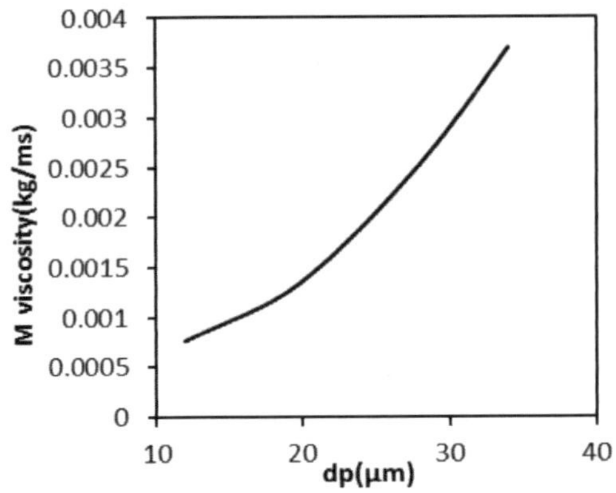

Fig. 4 Effect of particle size on molecular viscosity at 4m/s inlet velocity for 40% volume fraction

3.2 Variation of granular pressure

Since the ash is considered as granular material, the study of granular pressure may be important for discussing the erosion effect on the wall of pipe. The distribution of granular pressure for our considered particle size along the axial directions of pipe is shown in Fig.5. The granular pressure is the pressure exerted on the wall due to collision of particle. From the contour of granular pressure, it may be said that maximum granular pressure occurs towards the periphery at the bottom for all cases of particle diameter. This may be due to the fact that particles move with reduced velocity near the wall of pipe and higher gravitational effect of granular particles at bottom. The pressure is obviously becomes minimum at the core section of the pipe because of high velocity of slurry at the core. The granular pressure is revealed to be more at higher particle size as depicted in Fig.6. This figure also shows that the rate of increase of granular pressure is larger at higher particle size compared to the lower particle size.

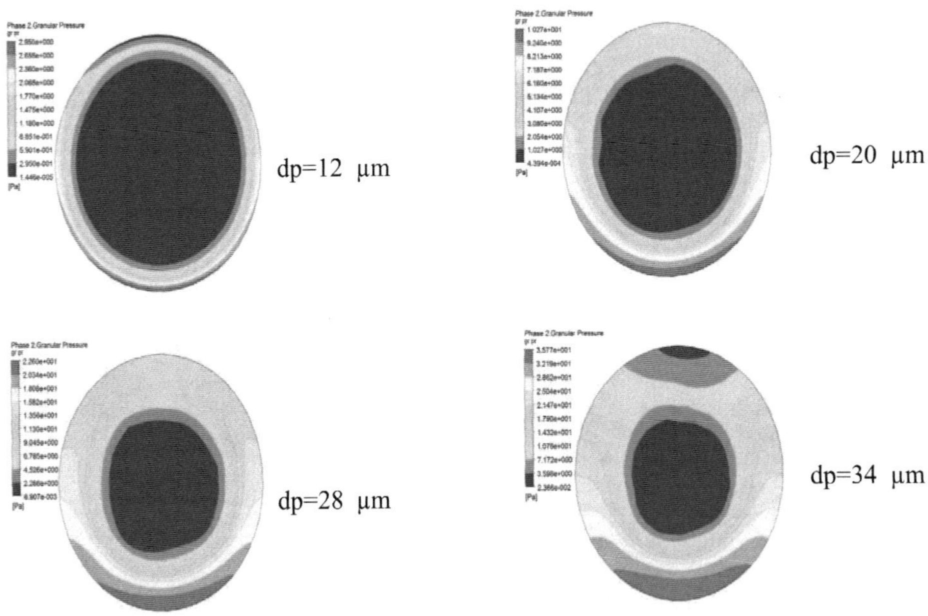

Fig.5 Distribution of granular pressure in Pa at 4m/s inlet velocity for 40% volume fraction at different particle sizes

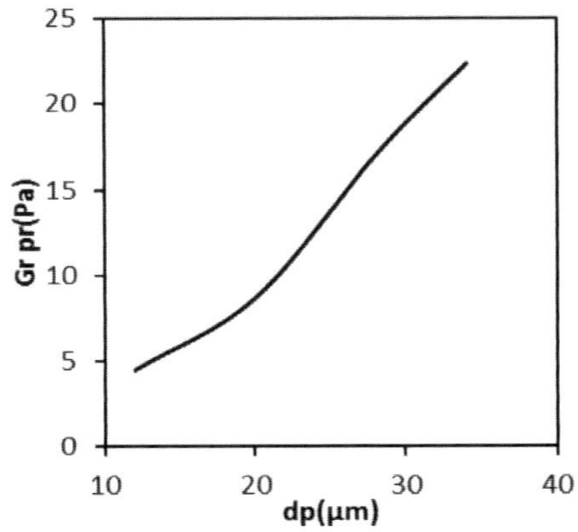

Fig. 6 Effect of particle size on granular pressure at 4m/s inlet velocity for 40% volume fraction

3.3 Variation of temperature and surface heat transfer coefficient

The temperature distribution of slurry flow at the cross section of pipe for the slurry velocity of 4 m/s and 40% volume concentrations is shown in the Fig. 7. Since the wall temperature and the fluid temperature are considered as 400 K and 300 K, therefore, heat exchange should take place from the wall to slurry due to temperature difference. It is obvious to observe maximum temperature at the vicinity to the wall. It can also be said that heat transfer should take place into the slurry mainly because of convective heat transfer. The slurry temperature difference decreases as it moves radially inward towards centre of pipe. Since the slurry velocity is more at the central zone that causes less time for heat interaction, thereby the temperature of the slurry becomes less at the central zone of pipe. The temperature distribution becomes asymmetric with the increases of particle size as observed during velocity distribution. The possible volume concentration towards downward with particle size may support this type of temperature distribution.

Surface heat transfer coefficient depends upon many factors like temperature of wall, mean slurry velocity, distribution of concentrations in the slurry pipe. The variation of surface heat transfer coefficient with particle size is shown in the Fig. 8. From the figure, it is observed that the heat transfer coefficient increases with the increase in particle size. High heat transfer coefficient results high convective heat transfer towards the central zone of pipe, depicted in the temperature contour with particle size. The asymmetric shape of temperature profile due to asymmetric shape nature of velocity, clearly noted in the temperature contour.

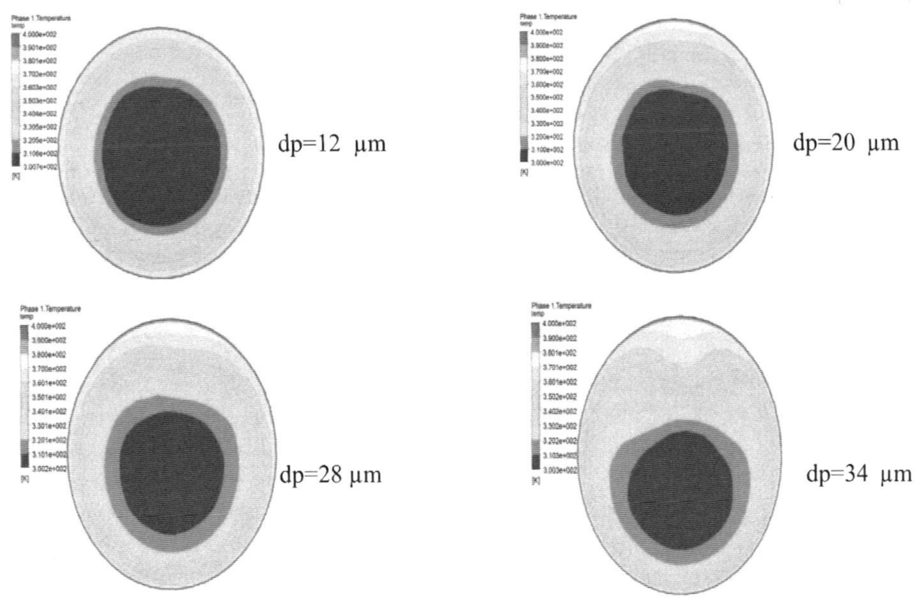

Fig. 7 Distribution of temperature in K at 4m/s inlet velocity for 40% volume fraction at different particle sizes

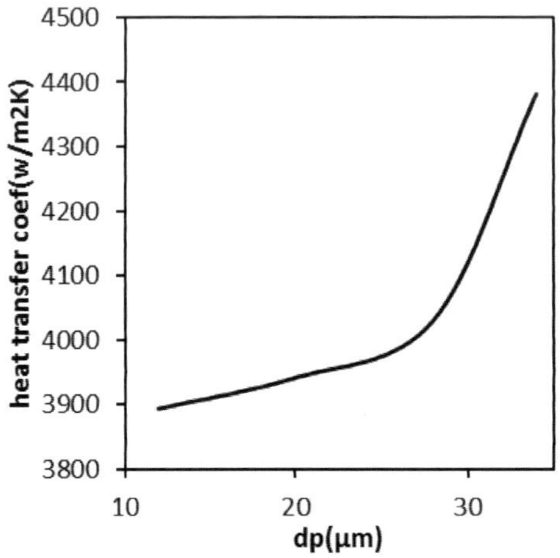

Fig. 8 Effect of particle size on surface heat transfer coefficient at 4m/s inlet velocity for 40% volume fraction

4. Conclusions

From the fluid flow and heat transfer study through a horizontal slurry pipe, the following observations are noted.

(i) Increase in particle size results in asymmetric velocity distribution.
(ii) Pressure drop through slurry pipe increases with particle size. The increment is higher at larger particle size.
(iii) Molecular viscosity increases with particle size.
(iv) More pumping power is required for higher particle size.
(v) Granular pressure is more near wall. It is maximum at bottom.
(vi) Granular pressure increases with particle size.
(vii) Heat transfer from wall increases with particle size.
(viii) Temperature distribution at core becomes asymmetric with particle size.

5. References

[1] Ku Jae-Hyun, Cho Hyun-Ho, Koo Jeong-Hwan, Yoon Suk-Goo and Lee Jae-Keuu 2000 Heat Transfer Characteristics of Lequid-Solid Suspension Flow in a Hoz. Pipe *KSME International Journal* 14(10):1159-1167.

[2] Horiuchi S, Kawaguchi M, Yasuhara K 2000 Effective use of fly ash slurry as fill material *Journal of Hazardous Materials* 301–337.

[3] Harada E, Toda M, Kuriyama M and Konno H 1985 Heat transfer between wall and solid-water suspension flow in horizontal pipes *J. Chem. Eng. Japan* 18: 33–38.

[4] Vlasak P, Kysela B and Chara Z 2012 Flow structure of coarse-grained slurry in a horizontal pipe, J. Hydrology Hydromech. 2:115–124

[5] Wilson KC , Clift R AND Sellgren A 2002 Operating points for pipelines carrying concentrated heterogeneous slurries *Powder Technology* 123:19–24.

[6] Kaushal D R, and Tomita Y 2002 Solids concentration profiles and pressure drop in pipeline flow of multisized particulate slurries *Int. J. Multiph. Flow* 28:1697–1717.
[7] Ozbelge T A and Eraslan A N 2006 A computational hydrodynamic and heat transfer study in turbulent up-flows of dilute slurries through a concentric annulus *Turkish J. Eng. Environ. Sci*. 30: 1–13.
[8] Cvetkovski C G, Reitsma S, Bolisetti T, Ting D S K 2015 Heat transfer in a U-Bend pipe: Dean number versus Reynolds number *Sust Energ Tech Assess* 11:148-158.
[9] Gopaliyal M K and Kaushal D R 2016 Modeling of sand-water slurry flow through horizontal pipe using CFD *J. Hydrol. Hydromech* 64: 261–272.
[10] Kumar N, Kaushal D R, Dwivedi V K, Singh D B, Sharma SK and Yadav J K, 2020 CFD modeling of fly-ash slurry to analyse the concentration and velocity profile *Materials Today* 21(3): 1695-1699.
[11] Yang J, Yang B, Yu M 2019 Pressure Study on Pipe Transportation Associated with Cemented Coal Gangue Fly-Ash Backfill Slurry *Applied Sciences* 9(3):512.
[12]Hiltunen K, Jäsberg A, Kallio S, Karema H, Kataja M, Koponen A, Manninen M and Taivassalo V 2009 Multiphase Flow Dynamics-Theory and Numerics, *VTT Publications, Finland.*
[13]Ansys Fluent, ver. 14.0, (2011) Theory Guide, Ansys Inc., USA.

A Numerical experimentation on fluid flow and heat transfer in a SCPP

D. K. Mandal[1], P. Goswami[2], S. Pradhan[3], R. Chakraborty[4], N. A. Khan[5] and P. Bose[6]

[1,3,4,5,6] Dept. of Mechanical Engg., College of Engg. & Management, Kolaghat, P.O: K.T.P.P. Township, Midnapore (E) - 721 171, West Bengal, India,
[2] WBPDCL, KTPP, Midnapore (E) - 721 171, West Bengal, India
Corresponding author, dipkuma@yahoo.com[1]

Abstract. Solar chimney power plant (SCPP) is a promising unit of power generation from renewable energy source in near future. Apart from climatic conditions, the performance of the unit depends on design parameters of plant. To understand the performance of the unit, fluid flow and heat transfer study is essential throughout the unit. In present work, a numerical study has been conducted to see the effect of inlet area on the performance of the unit through fluid flow and heat transfer. Numerical solutions of the steady turbulent flow are carried out by commercial software Fluent. A finite volume method has been employed to solve the governing equations. It is revealed that the considered plant having inlet height of 5cm performs better than that of having the inlet height10cm.

Keywords: SCPP, heat flux, Natural convection, velocity

Nomenclature

A_i	Inlet area, m^2
A_{ch}	Area of chimney, m^2
C_p	Specific heat of air, kJ/kgK
D_{ch}	Chimney outlet dia, m
D_{col}	Collector dia, m
H_{1col}	height of collector from ground surface at entry, m
H_{2col}	height of collector from ground surface at middle, m
H_{ch}	Chimney height, m
G	Amount of solar radiation absorbed, w/m^2
r	Radius of absorber plate, m
T_a	Ambient temperature, K
η_t	Turbine efficiency
η_{coll}	Collector efficiency
V_{ch}	Velocity through chimney, m/s

1 Introduction

Continuous increase in population and globalization causes huge demand within human beings that resulted in exploitation of natural resources, oil fields. Climate Change has become a burning question throughout the world and researchers around the globe are finding out ways to adopt alternative sources energy to meet the daily needs. Solar energy has become an alternative source to satisfy rising demands of huge power and electricity.

The SCPP model possesses three main units, a large solar collector made of greenhouse plastic, a chimney of long length built of PVC pipe, a generator run by a pressure-based wind turbine, converting potential and kinetic energy into mechanical energy. Convection mode heats the air above the bottom wall (absorber) in the ground and hot air is forced to rise through the chimney due to buoyancy. Increase in temperature and decrease in density of air is funneled towards the tower. Wind turbine present at the bottom of the tower converts energy of the air flow into mechanical energy and in turn converts into electrical energy with the aid of electric generators. Design optimization is possible by varying the diameter and height of chimney, area of absorber plate and inlet height. Solar radiation and incidence angle have a huge impact on quantity of reflected, absorbed and transmitted energy. Design and development of efficient solar chimney power project with appropriate material, shape and size in the working setup will be a great application of knowledge in the field of fluid thermal sciences

A number of studies and research on Solar Chimney Power Plant (SCPP) have been successfully completed to increase the efficiency of the chimney. Buoyancy-nature of heated air has been utilized by Saini et al. [1] to generate solar energy by discarding the use of solar panels. They have done simulation using ANSYS Fluent to calculate heat transfer during night using Roseland radiation model. The maximum chimney height for convection by neglecting negative buoyancy has been determined by Zhou et al. [2]. Desired height for highest power output has been analyzed and compared with the measurements of the model in Manzanares. Thermal analysis for laboratory type 2-D SCPP and discrete ordinance (DO) for power generation have been studied by Lal et al. [3] in a warm climate of Kota, India and compared with experimental results. Bejalwar and Belkhode [4] have constructed a SCPP on the hill top which occupies an area of near about 23 sq. m^2 and along with it 2mm thickness UV reinforced plastic, crystalline plastic and polycarbonate have been installed. Rabehi et al. [5] have constructed a Solar Chimney Power Plant (SCPP) in Algeria. Numerical study has been done in ANSYS Fluent on a two- dimensional axisymmetric model of SCPP using standard k-ε turbulence model in double precision. During simulation pressure-velocity coupling have been done by SIMPLE Arithmetic and QUICK scheme to analyze the convection terms. CFD analysis on Solar Chimney Power Plant has been done by Gholamalizadeh [6] using the k-ε turbulence and radiative transfer equations. During CFD analysis of buoyancy driven flow and heat transfer air velocity has been clearly viewed about 8.9 m/s and during turbine operation the velocity is 25% lower than no-load condition as turbine operation produces a drag against air-flow. Fasel et al. [7] have carried out numerical analysis using ANSYS Fluent and analytical steps using CFD code on a different range of tower heights in SCPP. Wide time-dependent simulations of flow have given an in-depth idea about fluid dynamics and heat transfer phenomena. A SCPP with a new idea of chimney with a hyperbolic shape was developed by Nasaroui et al. [8] and did analysis experimentally and numerically for the better performance of the hyperbolic chimney. They have shown that there is a rise in velocity in chimney inlet and decrease in the outlet due to a recirculation zone in the outlet. Xu and Zhou [9] have carried simulation in divergent chimney solar power plants (DSPPs) by varying chimney outlet- inlet area ratios (COAR) using the realizable k-ε model in 3-D model. The authors have applied the model of buoyancy- driven flow where density gap is not large and analyzed the streamline vortices, curvature and rotation.

Brief review gives parametric effects in the SCPP unit. Mostly they have varied the design parameters at different climate conditions to discuss the performance of SCPP. The physics of heat and fluid flow study is not addressed by any researchers so far. Inlet height may be one of the parameter to influence the performance of SCPP. Present work focuses of details insight of fluid flow and heat transfer study to focus on the velocity generated which in turn will drive the turbine to generate electricity. Inlet height is also varied in our simulation.

2 Numerical Modelling

2.1 *Physical Prototype*

The main components with dimensions of the solar chimney power project (SCPP) are depicted in Fig.1.

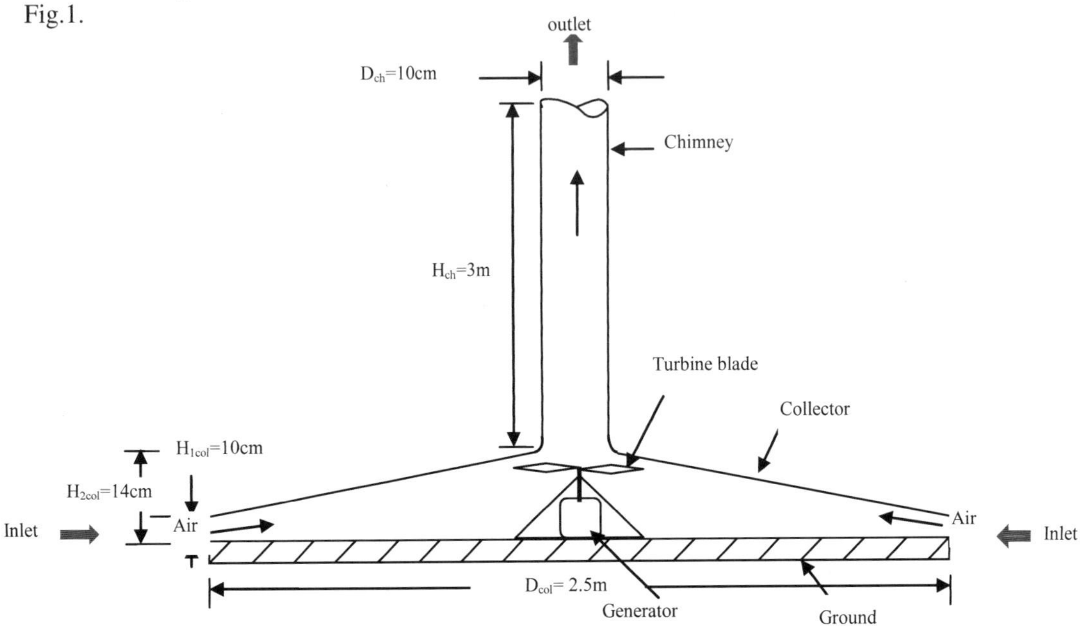

Fig.1. Prototype of solar chimney power plant

2.2 *Numerical procedure*

The mesh generation is done by Fluent commercial software, tetrahedron elements with 219803 nodes 1141776 elements is taken after grid sensitivity test. The following Fig. 2 shows the mesh of the geometry.

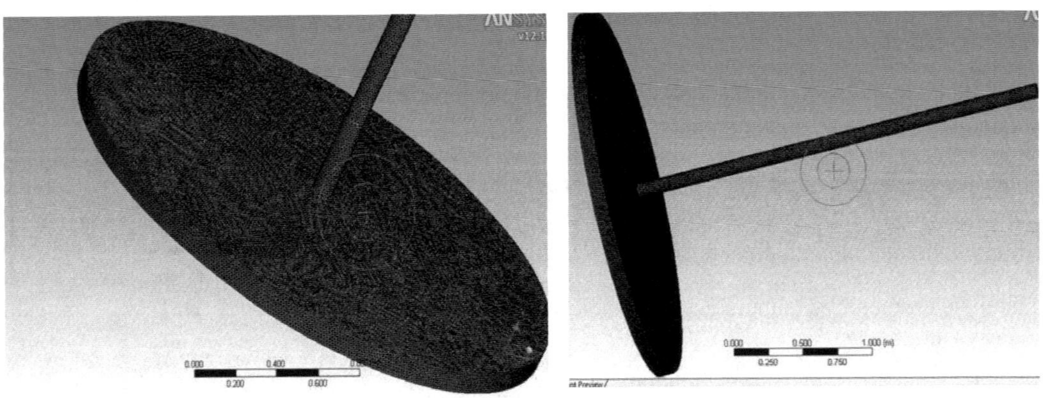

Fig.2 Mesh of computation domain

Numerical simulation is done by using Fluent with finite volume method to discretize the Navier-Stokes equations. SIMPLE algorithm is used with the pressure–velocity coupling, convective terms are discretized by 2nd order upwind scheme. All numerical calculations are done in double precision.

The iteration error for momentum is 10^{-5} for all calculations and 10^{-8} for the energy equations. Standard k-ε turbulence model is taken in this flow study. The Boussinesq model is chosen in this simulation due to less change in temperature in our study. Ambient temperature is taken as 300K and heat flux at absorber plate is taken as $600W/m^2$ respectively considering sunny conditions.

2.3 *Boundary conditions*

The boundary conditions for our numerical simulation are given below.
(1) Inlet: Pressure inlet, gauge pressure is zero, ambient temperature is 300K.
(2) Collector (transparent cover): Heat transfer coefficient is $10W/m^2K$, ambient temperature is 300K.
(3) Ground surface (absorber plate): Heat flux is taken as $600W/m^2$.
(4) Outlet: Pressure outlet boundary, gauge pressure is zero, ambient temperature is 300K.
(5) Chimney wall: Adiabatic wall.

To make the problem simple the following assumptions are considered:

i) Air flow is chosen as incompressible flow.
ii) Radiative heat transfer is negligible.
iii) Steady State is assumed in case of the mathematical model.
iv) The Boussinesq model is considered.

2.4 *Turbulence modelling*

The continuity equation, Navier-Stokes equations, energy equation, and k-ε equations are shown below:

Continuity Equation

$$\frac{1}{r}\frac{\partial(ru_r)}{\partial r}+\frac{1}{r}\frac{\partial u_\theta}{\partial\theta}+\frac{\partial u_z}{\partial z}=0$$

Momentum Equation

$$\rho\left(u_r\frac{\partial u_r}{\partial r}+\frac{u_\theta}{r}\frac{\partial u_r}{\partial\theta}+u_z\frac{\partial u_r}{\partial z}-\frac{u_\theta^{\ 2}}{r}\right)=\rho g_r-\frac{\partial p}{\partial r}+\mu\left[\frac{1}{r}\frac{\partial}{\partial r}\left(r\frac{\partial u_r}{\partial r}+\frac{1}{r^2}\frac{\partial^2 u_r}{\partial\theta^2}+\frac{\partial^2 u_r}{\partial z^2}-\frac{2}{r^2}\frac{\partial u_\theta}{\partial\theta}-\frac{u_r}{r^2}\right)\right]$$

$$\rho\left(u_r\frac{\partial u_\theta}{\partial r}+\frac{u_\theta}{r}\frac{\partial u_\theta}{\partial\theta}+u_z\frac{\partial u_\theta}{\partial z}-\frac{u_r u_\theta}{r}\right)=\rho g_\theta-\frac{1}{r}\frac{\partial p}{\partial\theta}+\mu\left[\frac{1}{r}\frac{\partial}{\partial r}\left(r\frac{\partial u_\theta}{\partial r}+\frac{1}{r^2}\frac{\partial^2 u_\theta}{\partial\theta^2}+\frac{\partial^2 u_\theta}{\partial z^2}-\frac{2}{r^2}\frac{\partial u_r}{\partial\theta}-\frac{u_\theta}{r^2}\right)\right]$$

$$\rho\left(u_r\frac{\partial u_z}{\partial r}+\frac{u_\theta}{r}\frac{\partial u_z}{\partial\theta}+u_z\frac{\partial u_z}{\partial z}\right)=\rho g_z-\frac{\partial p}{\partial z}+\mu\left[\frac{1}{r}\frac{\partial}{\partial r}\left(r\frac{\partial u_z}{\partial r}+\frac{1}{r^2}\frac{\partial^2 u_z}{\partial\theta^2}+\frac{\partial^2 u_z}{\partial z^2}\right)\right]$$

Energy Equation

$$u_r\frac{\partial T}{\partial r}+\frac{u_\theta}{r}\frac{\partial T}{\partial\theta}+u_z\frac{\partial T}{\partial z}=\frac{q_g}{c_p}+\alpha\left[\frac{1}{r}\frac{\partial}{\partial r}\left(r\frac{\partial T}{\partial r}+\frac{1}{r^2}\frac{\partial^2 T}{\partial\theta^2}+\frac{\partial^2 T}{\partial z^2}\right)\right]+\frac{\varphi}{\rho C_P}$$

Where φ=viscous dissipation rate

K-ε turbulence model based on computations of kinetic energy and dissipation rates

$$\frac{\partial}{\partial t}(\rho k)+\frac{\partial}{\partial x_i}(\rho k u_i)=\frac{\partial}{\partial x_j}\left[\left(\mu+\frac{\mu_\tau}{\sigma_k}\right)\frac{\partial k}{\partial x_j}\right]+G_k+G_b-\rho\varepsilon-Y_M$$

$$\frac{\partial}{\partial t}(\rho\varepsilon)+\frac{\partial}{\partial x_i}(\rho\varepsilon u_i)=\frac{\partial}{\partial x_j}\left[\left(\mu+\frac{\mu_\tau}{\sigma_\varepsilon}\right)\frac{\partial \varepsilon}{\partial x_j}\right]+C_{1\varepsilon}\frac{\varepsilon}{k}(G_k+G_{3\varepsilon}G_b)-C_{2\varepsilon}\rho\frac{\varepsilon^2}{k}+S_\varepsilon$$

G_k is the kinetic energy generation because of turbulence and mean velocity gradients, and G_b the kinetic energy generation due to turbulence and buoyancy. σ_T, σ_k and σ_ε are the turbulent Prandtl numbers for T, k, and ε respectively($\sigma_T = 0.9$, $\sigma_k = 1.0$, $\sigma_\varepsilon = 1.3$). C_1 and C_2 are two constants for turbulent model: $C_1 = 1.44$, $C_2 = 1.92$ and $C_\mu = 0.09$.

There are many factors which affect the electric power generation, geometrical configurations like, collector area, chimney diameter and height, climatic conditions like amount of solar radiation absorbed and ambient temperature. The expression for electric power generated by the plant is obtained by Sangi [10]

$$P_e=\frac{2}{3}\eta_{coll}\eta_t\frac{g}{C_pT_a}H_{ch}\pi r^2 G$$

Efficiency of the turbine is considered in between of 80% to 90%.
Collector efficiency can be obtained from

$$\eta_{coll}=\frac{\rho_{air}.V_{ch.}A_{ch}C_p\Delta T}{\pi r^2 G}$$

3 Results and Discussion

The phenomenon in solar chimney power plant (SCPP) may be discussed in such a way that air in contact to bottom wall (absorber) is heated up by the heat flux of absorber plate. The heat transfer to air takes place mainly due to convection force. Initially there is no flow in SCPP. At inlet and outlet, pressure is atmospheric and temperature is 300K. The temperature of air above absorber plate increases and density of air drops. Due to this, air starts to move upward the plant for natural convection and finally lighter air finds easy way to go out through chimney. Hot air in chimney develops negative pressure inside the chimney. So, air flow begins through inlet due to pressure difference and natural convection created between inlet and outlet of the plant. However, this phenomenon in SCPP behaves similar to forced convection although there is no external equipment. Since, the main physics for the performance of SCPP is heat transfer and fluid flow through the system, therefore, an attempt has been made to study the flow and heat transfer between transparent plate and absorber plate in SCPP. This may definitely help for designing and achieving the better performance of SCPP. Present study gives insight of variation of velocity and temperature in between the transparent plate and absorber plate of SCPP. Apart from variation of velocity and temperature at different sections, variation of mean temperature and mean velocity along the flow direction is also presented in this study. During numerical investigation, the parameters considered are as follows: inlet height is 10 cm and 5 cm; absorber radius is 125 cm; height of the chimney is 300 cm; radius of chimney is 5 cm; heat flux of absorber plate is 600 W/m^2.

Figure 3 shows the velocity profiles at different sections between absorber and transparent plate from inlet towards the centre of absorber plate. In the figures, notation H0 indicates inlet, H20 represents 20 cm from inlet towards the centre of absorber plate. Similarly H40, H60, H80, H100 and H120 are 40 cm, 60 cm, 80 cm, 100 cm and 120 cm distance from inlet respectively. Inlet velocity profile shows plug type turbulent flow. When the flow moves from inlet to the centre of chimney over the absorber plate, the middle zone of velocity profile is observed to become more flatten. The magnitude of velocity in the core section also increases. This increase in velocity is due to flow over heated bottom wall. Initiation of velocity begins at inlet due to negative draft created by chimney; wall

IOP Conf. Series: Materials Science and Engineering 1080 (2021) 012027 doi:10.1088/1757-899X/1080/1/012027

heat flux at absorber plate aggravates the velocity. The density of air in contact to bottom wall decreases and thereby increases specific volume. Therefore, the velocity of air rises. The asymmetric shaped profile in each case is noticed. It may be due to one side wall heat flux and density difference. Velocity profile indicates that near the centre of chimney (i. e, at H120) magnitude of velocity suddenly increases towards upper zone because of change of flow direction and phenomenon related to flow through sudden contraction. Area of flow increases from inlet to the centre over the absorber plate as the transparent plate remains at an angle. This increase in area decreases the velocity to satisfy the continuity. At the same time, velocity of air increases due to wall heat flux. The rise of velocity dominates over the decrease in velocity which finally leads to increase in velocity. The inclination in the transparent cover is allowed for smooth flow. Smooth joint between transparent cover and chimney is also provided for reduction in losses at the said joint. From the above discussion, it can be said that the average velocity should increase along the flow direction from inlet towards the centre of the chimney which has been obtained during investigation and presented in Fig. 4.

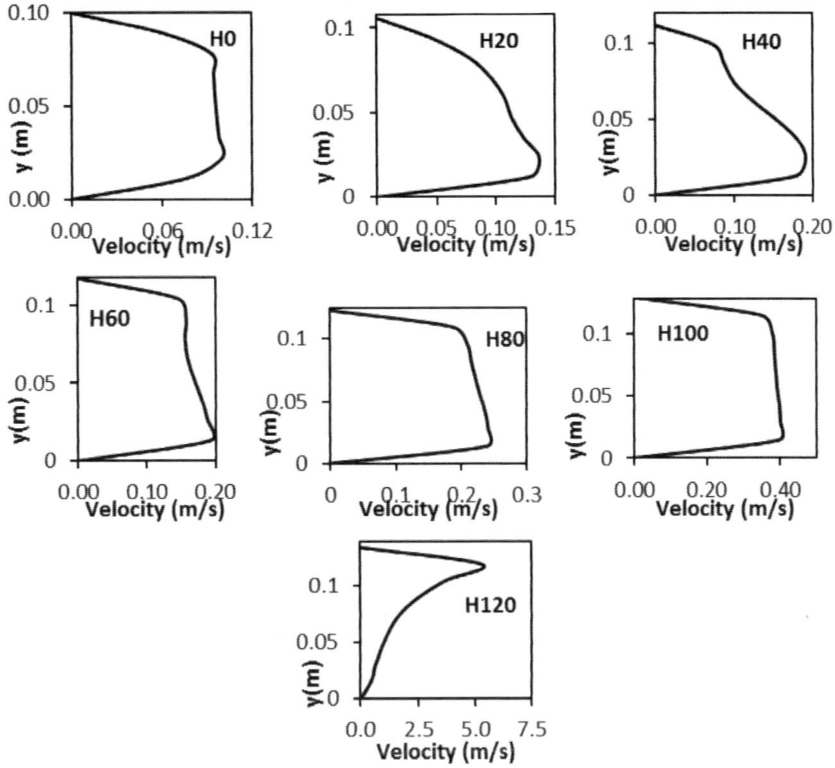

Fig. 3. Velocity profiles along the flow directions at different cross sections over absorber plate at 10cm inlet height.

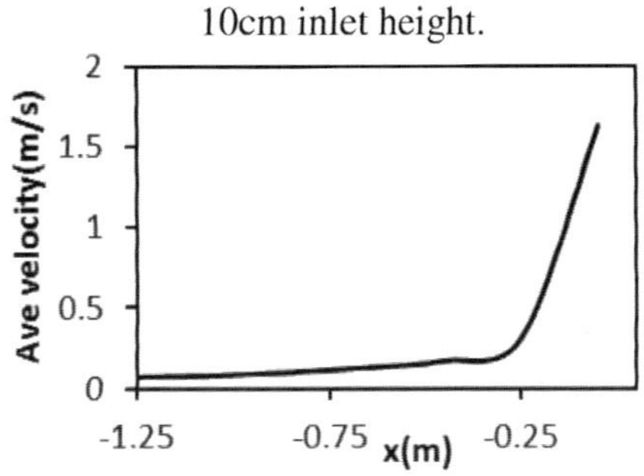

Fig. 4. Variation of average velocity along the flow directions at different cross sections over absorber plate at 10 cm inlet height.

From the temperature profiles (Fig. 5) at similar cross sections used during numerical experimentation on velocity profiles, it is evident that wall heat flux diffuses the heat transfer and increases the wall temperature from the inlet. This temperature increases along the horizontal direction of absorber plate from inlet to centre of plate. Above the thermal boundary thickness, the temperature also increases along the flow direction. This is common to attain more diffusion to the bulk fluid with the flow on absorber plate. The rate of increase in wall temperature and average temperature are shown in Fig. 6 and Fig. 7 respectively. This helps to generate velocity by buoyancy force.

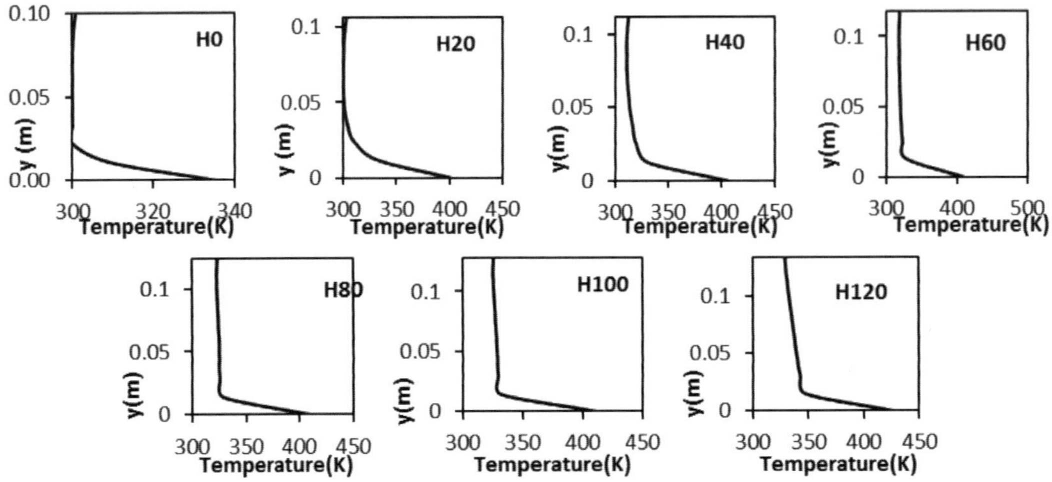

Fig. 5. Temperature profiles along the flow directions at different cross sections over absorber plate at 10 cm inlet height.

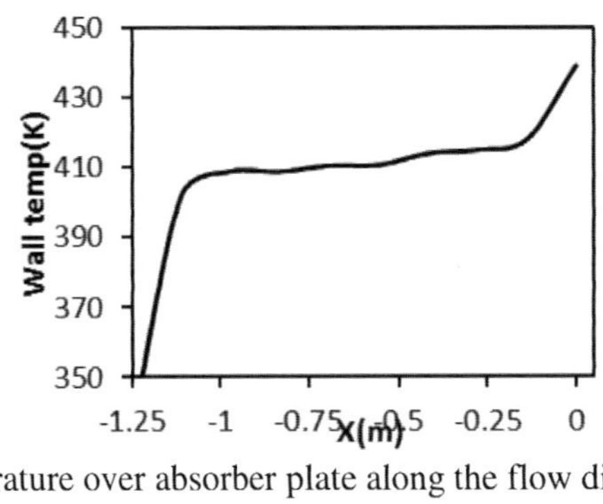

Fig. 6. Variation of temperature over absorber plate along the flow directions at 10 cm inlet height.

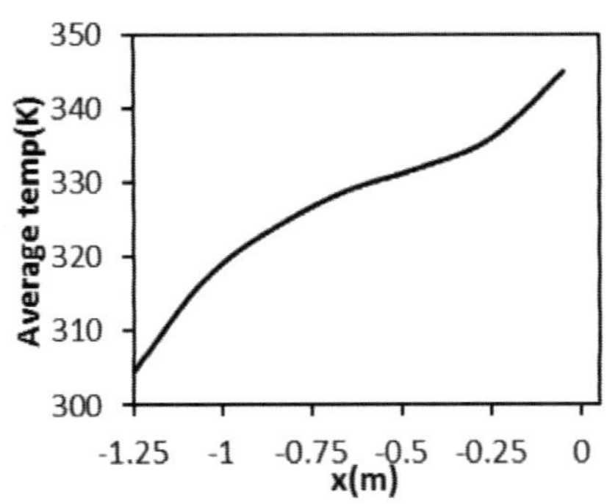

Fig. 7. Variation of average velocity over absorber plate along the flow directions at 10 cm inlet height.

From the review of literature, it is noted that all the researchers have considered diameter of chimney, height of chimney, and absorber plate area as design parameters for assessing the performance of a SCPP. But, the considered design parameters should be those which ultimately dictate fluid flow and heat transfer between transparent plate and absorber plate for the generation of air velocity through the plant. At this juncture, authors feel that effect of inlet height, as another parameter, should also have impact on the performance of the plant, which has not been investigated so far by anybody. Therefore, an attempt has also been made to study the effect of inlet height, by changing the height from 10 cm to 5 cm, on average velocity and average temperature of air along the absorber plate, keeping same inclination in the transparent plate and same wall heat flux in the absorber plate. It is clear that area of flow is more at the inlet of the plant and is less at the outlet of the plant for both the cases. For 1st case of 10 cm inlet height and 5 cm chimney radius, area at outlet becomes 100 times less than that of at inlet; in case of 2nd case of 5 cm inlet height and 5 cm chimney radius, area at outlet becomes 50 times less than that of at inlet. It is also evident that the inlet area of 2nd case becomes 0.5 times less in comparison to 1st case. After simulation for 2nd case, it has been observed that average velocity in case of 2nd case over the plate increases more than the 1st case of having 10 cm inlet height (Fig.8) and mean temperature over the absorber plate is less than that of 1st case (Fig.9).

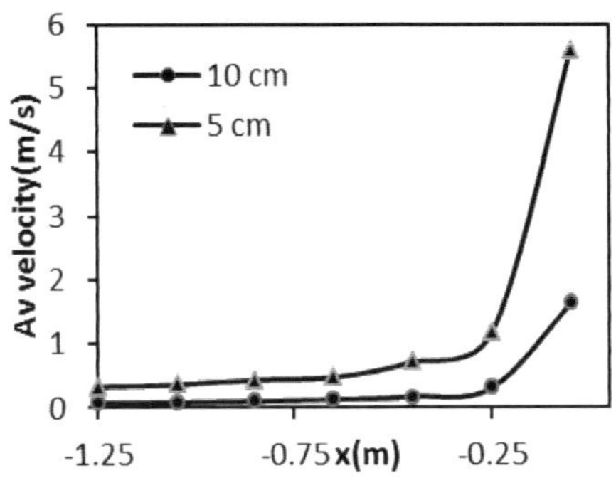

Fig.8. Variation of average velocity in between transparent cover and absorber plate.

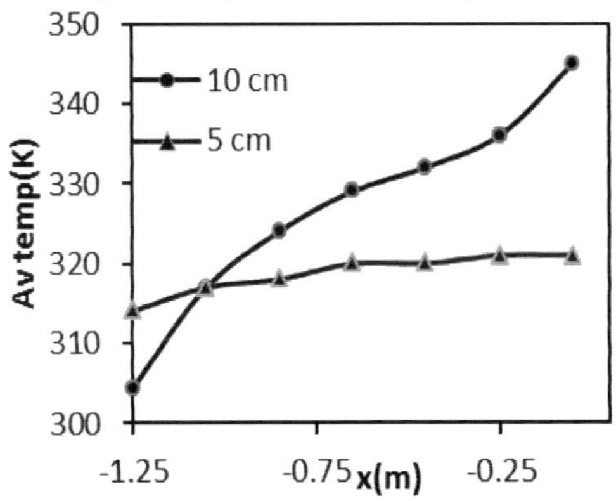

Fig. 9. Variation of average temperature in between transparent cover and absorber plate.

The simulation results of the magnitude of velocity at inlet and at outlet of the plant for two cases have been shown in table 1. Gain in velocity of air depends on velocity due to pressure drop by the chimney and due to the effect of natural convection over the absorber plate to air. In case of 2nd case having 5 cm inlet height, mean temperature is less. It may be due to insufficient space for diffusion. It reduces the chimney draft which consequently decreases the velocity of air in comparison to the 1st case with10 cm inlet height. At the same time, large temperature difference between absorber plate and air helps to enhance the velocity due to convective force. Gain in velocity also depends on inlet area. By reducing half inlet area (from case 1 to case 2) velocity at inlet increases about 5 times (Table

1). This is due to small inlet area and the dominating convective force over absorber plate. Velocity in 1st case increases approximately 100 times from inlet to outlet, whereas it increases almost 50 times in 2nd case. As the initial velocity increases 5 times for 5 cm inlet height compared to 1st case, therefore outlet velocity increases approximately 2.5 times of outlet velocity of 10 cm inlet height, noted in table.1. Therefore, it may be said that the plant of having 5 cm inlet height performs better.

Table-1. Velocity at inlet and outlet

	Inlet height	Velocity at inlet	Velocity at outlet
Case-1	10 cm	0.097 m/s	9.85 m/s
Case-2	5 cm	0.4918 m/s	24.99 m/s

4 Conclusions

Present study is focused to study the flow and heat transfer in solar chimney power plant (SCPP). Inlet height is varied from 10 cm to 5 cm. Comparison is made with mean velocity and mean temperature in each case. Lowering inlet height increases the outlet velocity thereby increases the power generation by SCPP. Further study is needed to identify the optimum height at inlet for achieving maximum power generation in a SCPP.

Acknowledgment

The authors are thankful to The Institution of Engineers (India) for their financial support under R&D grant-in-aid scheme to complete, (Ref. R.4/2/UG/2019-20/UG2020036).

5 References

[1] Saini D L, Singh A and Bhargava M 2017 Design, optimization and performance analysis of Solar chimney using CFD technique *Journal of Emerging Technologies and Innovative Research (JETIR)* 4(10):52-55.

[2] Zhou X, Yang J, Xiao B, Hou G and Xing F 2009 Analysis of chimney height for Solar chimney power plant *Applied Thermal Engineering* 29:178-185.

[3] Lal S, Kaushik S C and Hans R 2016 Experimental investigation and CFD simulation studies of a laboratory scale Solar chimney for power generation *Sustainable Energy Technologies and Assessments* 13:13-22.

[4] Bejalwar A and Belkhode P 2018 Analysis of experimental setup of a small Solar chimney power plant *Procedia Manufacturing* 20: 481-486.

[5] Rabehi R, Chaker A, Aouachria Z and Tingzhen M 2017 CFD analysis on the performance of a Solar chimney power plant system: case study in Algeria *International Journal of Green Energy* 14(12): 971-982.

[6] Gholamalizadeh E 2018 CFD simulation of a Solar chimney power plant *International Journal of Mechanical and Production Engineering* 6(4):1-5.

[7] Fasel H F, Meng F, Shams E and Gross A 2013 CFD analysis of solar chimney power plants, *Solar Energy* 98:12-22.

[8] Nasraoui H, Driss Z, Ayedi A and Kchaou H 2019 Numerical and experimental study of the aerothermal characteristics in Solar chimney power plant with hyperbolic chimney shape *Arabian Journal for Science and Engineering* 44(9):7491-7504.

[9] Xu Y and Zhou X 2018 Performance of divergent-chimney Solar power plants *Solar Energy* 170:379-387.

[10] Sangi R 2012 Performance evaluation of solar chimney power plants in Iran *Renewable and Sustainable Energy Reviews* 16 (1):704-710.

Effects of wall emissivity and fin blockage ratio on micro combustor performance

Arees Qamareen[1]Shahood S Alam[1] and Mubashshir A Ansari[1]

[1] Department of Mechanical Engineering, Aligarh Muslim University, Aligarh – 202002, India

e-mail: *aqamareen.me@amu.ac.in, areesq@gmail.com*

Abstract. Combustion system design is a complex mechanism involving several physical mechanisms such as chemical reactions, fluid dynamics and heat transfer. At micro scale, short flow residence time adds further to the complexity. Enhancement of heat transfer to the outer surface of the combustor is desirable in several power generating systems. The present work focuses on this aspect by geometrical modification in the form of wall fin and observing the change in outer wall temperature level along with heat transfer rate for various wall surface emissivities and blockage ratio (BR) of wall fin. Lower surface emissivity and higher fin blockage ratio was found to give better combustor performance in terms of outer surface temperature.

1. Introduction

An increased demand for miniaturized electronic and mechanical systems have led to widespread research activity for combustion based power supply with longer working hours, higher energy densities and shorter recharging durations as compared to conventional batteries[1], [2].The micro combustion systems are used to generate electrical power through micro gas turbines (MGT), micro thermo photovoltaic (MTPV), micro thermo electric (MTE) generators, etc[3]–[7]. The small size of combustor geometry incurs large amounts of heat and radical loss due to large surface to volume ratio. At higher velocities, the residence time of reacting species reduces remarkably and is of the same characteristic time scale of chemical reaction. This deteriorates combustion efficiency and flame stability significantly.

Various efforts in terms of geometries, materials, operating conditions have been made by a number of researchers in order to improve combustion characteristics; and recirculation zones generated because of geometrical modulation is one of them. Recirculation of reactants and products enhances mixing which helps in better heat transmission todifferent parts of micro combustors. Better heat recirculation helps in preheating of reactants coming in and more heat is transferred to outer wall of the combustor.

At high mixture velocities, stable flame sustenance becomes difficult due to reduced residence time of reactant species. Flame stabilization is then achieved by incorporating flame holders inside the combustion chamber such as backward facing step [8], [9], bluff bodies [10], [11], cavity [12], [13], porous medium [14], [15], etc.

In the present work, triangular wall fin and backward facing step configuration in a micro combustor is investigated for a range of external wall surface emissivities. The coupling effects of recirculation zones created by backward step and wall fin helps improve the combustor performance.

2. Geometrical and Computational model

Micro combustors with combined backward facing step and variable blockage ratio triangular wallfin at a distance of 2 mm from the step are investigated and the configuration is shown in Figure1(a). Blockage ratio (BR) is defined as the ratio of fin height to combustion chamber height (BR = $b/(h_s+h_i)$). Case MCW04 signifies micro combustor with blockage ratio of 0.4 and so on. Aspect ratio of the fin is kept as unity for the present case but further effects of different aspect ratios can also be investigated. Flame anchoring due to step and wall fin is expected to enhance combustion due to flow and heat recirculation. Table 1 encapsulates the geometrical considerations for simulations.

Table 1. Geometrical parameters of micro-combustors

Geometrical parameter	Parameter name	(mm)	Constant/variable
L	Combustor length	27	Constant
L_i	Inlet length	7	Constant
h	Block location	2	Constant
h_i	Inlet height	0.5	Constant
h_s	Step height	0.5	Constant
h_t	Combustor wall thickness	0.5	Constant
b	Wall fin height	0.4, 0.5, 0.6	Variable
d	Wall fin length	0.4, 0.5, 0.6	Variable
AR	Aspect ratio = d/b	1	Constant
BR	Blockage ratio = $b/(h_i+h_s)$	0.4, 0.5, 0.6	Variable

2.1. Numericalscheme

A global H_2/air reaction mechanism was implemented for combustion modelling whileignoring the surface reactions on the combustor wall. ANSYS FLUENT 16.0 was used to discretise and solve the governing conservation equations [16] of mass, momentum, energy and species along with relevant boundary conditions. Second order upwind discretization scheme and SIMPLE pressure-velocity coupling algorithm were employed. Turbulence-chemistry interaction was simulated using Finite Rate/Eddy Dissipation Model. The computational area comprises of the fluid and solid regions which makes it necessary for the heat conduction equation in the solid walls to be solved in order to account for the strong thermal interaction between solid and fluid domains in the micro combustors.
Stability and convergence issues occur as reacting flows are difficult to solve numerically due to a strong coupling between the conservation equations. Also enormous density fluctuations and accelerations in flow occur due to excessive heat release from combustion reactions.

Table 2. Boundary conditions

Boundary	Conditions applied
Fluid-solid boundaries	No-slip velocity condition Zero flux of all species Coupled heat transfer
Inlet	Uniform velocity inlet [12 m/s] Incoming H_2-air mixture temperature = 300 K

	Equivalence ratio of mixture, φ = 0.8
Outlet	Zero gauge pressure
Inlet/outlet walls	Adiabatic (Zero heat flux)
Outer wall	Mixed boundary condition (Radiation/Convection) Wall Surface emissivity, e = 0.2 – 1.0 Material (stainless steel) thermal conductivity = 5 W/m-K Heat transfer coefficient = 10 W/m^2-K Ambient temperature = 300 K

Gas mixture density was attained from ideal gas equation and mixture specific heat, viscosity and thermal conductivity from mass fraction weighted average of the species properties. Specific heat for each species is calculated by piecewise polynomial fitting of temperature[17]. Values of turbulent kinetic energy, k = 1 × 10^{-3} for and dissipation, ε = 1 × 10^{-3} for the k–ε turbulence model were defined as take by Li et al [18]. Residuals of 10^{-6} are set for convergence of all the equations. Finite Rate/ Eddy Dissipation Model is adopted for chemistry-turbulence interaction modelling.

Figure 1(a) shows the geometrical model of a simple micro combustor with a step used in validation against the experimental data of Li et al [18]. Figure 1(b) depicts the geometrical parameters used in the micro combustor. H_2/air mixture comesinto the micro combustor at a particular velocity (u = 12 m/s) and equivalence ratio (φ = 0.8) through the inlet and premixed combustion occurs in the combustion chamber causing large amounts of heat release which gets transferred through the solid wall to the ambient[18]. Exhausted gases leave the combustor from the right side located outlet.

Several assumptions are made to further simplify the problem:
(1) Dufour effects are ignored;
(2) insignificantgas radiation;
(3) nopressure and viscous stress work;
(4) steady-state combustion process.

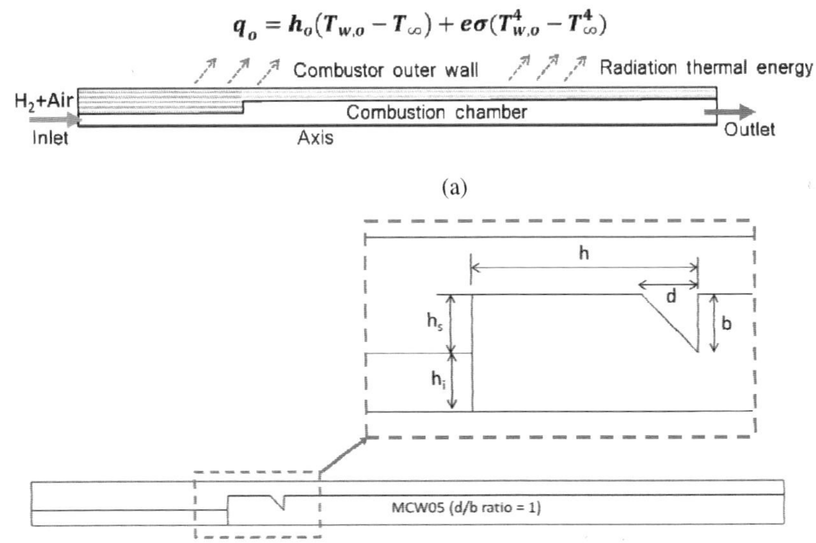

(b)

Figure 1. (a) Schematic diagram of micro combustor **(b)** Geometrical parameters of micro combustor with wall fin

2.2. *Computational grid and validation*

ANSYS Workbench was used to generate multi zone structured quadrilateral mesh (see Figure 2) with refinement near the wall fin. Three grid sizes were created to confirm mesh convergence. The outer wall temperature profiles were compared for the three mesh systems and a discreet difference for the temperature profiles was observed. Consequently, the medium sized mesh with roughly 1.1E05 number of nodes was chosen for the simulation (not shown here)to keep a balance between the computational time and precision.

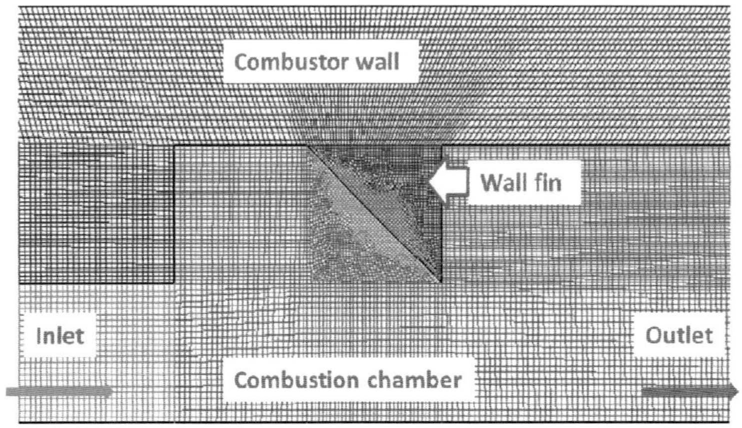

Figure 2. Computational grid system

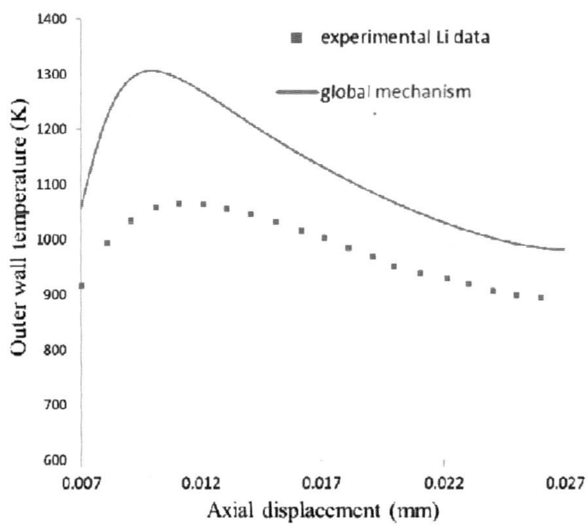

Figure 3. Validation of simple micro combustor with experimental data of Li et al. [18]

For validation of the numerical model selected for the present work, outer wall temperature of the simple micro combustor with backward facing step and without wall fin was validated against the experimental data results by Li et al. [18] for an inlet velocity of 12 m/s and equivalence ratio of 0.8.

As depicted in Figure 3, though the present simulation shows relatively larger differences compared with the experimental data, the trend of the two curves is similar. The global mechanisms over predict the temperatures but the temperature distribution trend lines are similar. Adoption of global mechanism saves large amount of computational time required in detailed reaction mechanisms while predicting better performing geometries.

3. Results and discussion

The results for qualitative as well as quantitative temperature and heat flux plots along with discussions are presented in this section.

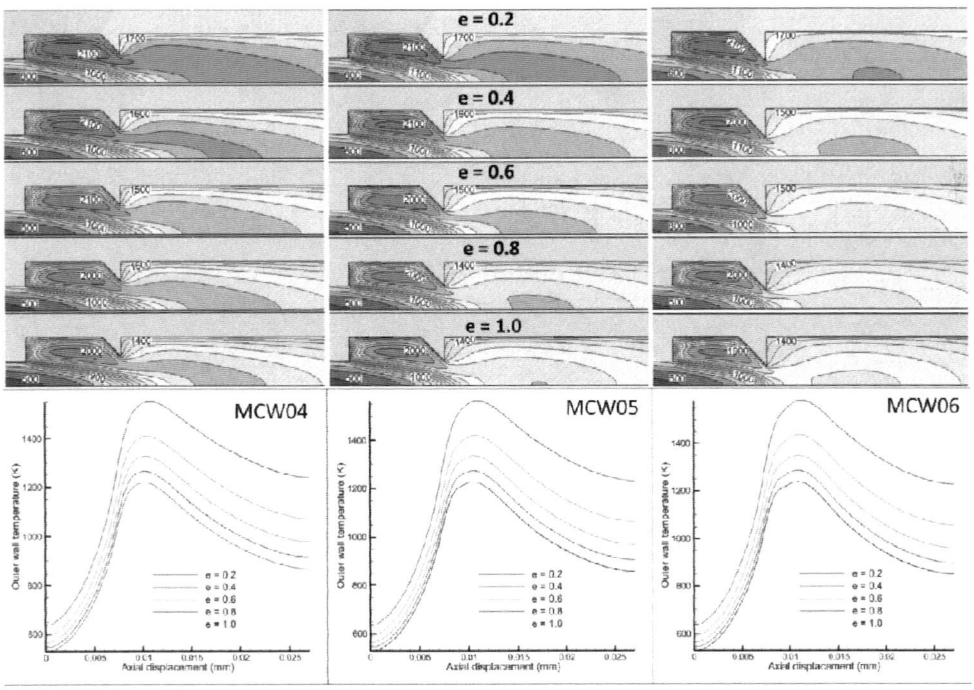

Figure 4. Temperature contours (Top) and Outer wall temperature distribution (Bottom) with respect to axial displacement for different wall emissivities (0.2 – 1)

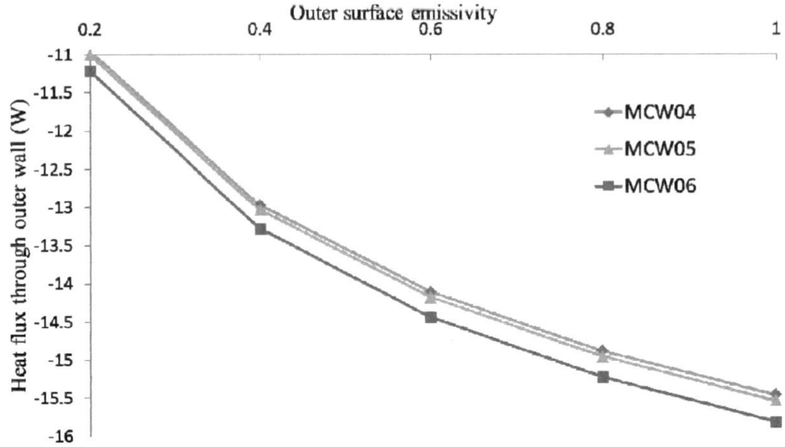

Figure 5. Heat flux through the outer surface of micro combustor at different wall emissivities (0.2 – 1) for various geometries

As can be seen in the outer wall temperature plots shown in Figure 4, the temperature levels are higher for smaller emissivities and get reduced as the emissivity value is increased. This happens because of an increased amount of heat loss due to radiation when surface emissivity is high. Since radiation heat transfer follows the fourth power law of temperature difference so for higher ε values, large amounts of heat losses can take place and this is quantitatively depicted in Figure 5. The negative values are for heat loss occurring from a system.

At the outer wall surface, the total heat loss rate (heat flux) to the ambient via natural convection and thermal radiation is given as:

$$q_o = h_o(T_{w,o} - T_\infty) + e\sigma(T_{w,o}^4 - T_\infty^4) \quad \text{--------(1)}$$

where h_o is the natural convection heat transfer coefficient (10 W/(m^2 K), $T_{w,o}$is the outer wall temperature, T_∞ is the ambient temperature (300 K), e is the emissivity of the solid surface and σ is the Stephan Boltzmann constant, 5.67×10^{-8} W/(m^2 K^4).

It is observed that more amount of heat loss is occurring from MCW06 design which may be due to more amount of heat transferred to the solid wall from the combustion zone as explained below.

In Figure 6, the recirculation zones formed in micro combustors with different blockage ratios at an inlet mixture velocity of 12 m/s, equivalence ratio of 0.8 and external surface emissivity of 0.6 are shown with the help of flow path lines. Two recirculation zones are formed in the micro combustor – one near the step wall and another behind the wall fin. The recirculation zone helps in better mixing of fluids causing enhanced heat transfer from one region to another. As the blockage ratio of fin is increased, the recirculation zone is seen to grow larger while no significant changes are observed in the zone created near the step wall. The enlargement of this zone is the main reason for increased outer wall temperatures for a particular emissivity.

In Figure 7, it is observed that for a particular value of emissivity, MCW06 has highest peak temperature for the whole range of ε. This is due to the enlarged recirculation zone formed as explained previously which causes better heat recirculation along with the radicals. Thus larger sized wall fin is an effective means for enhancing outer wall temperature.

Two important aspects need to be discussed at this point. One is that for small ε, radiation heat loss is smaller. Secondly, when heat loss is less, the combustion process intensifies and more heat gets generated. And more heat is available to be conducted upstream through the solid wall which is evident from the temperature contours and plots shown.

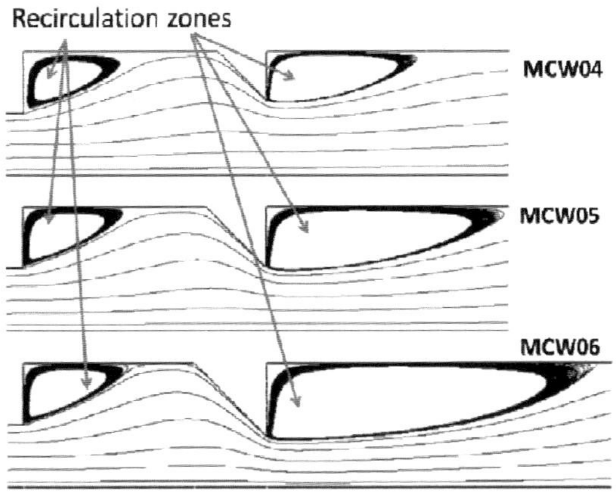

Figure 6. Streamlines and recirculation zone depiction in different micro combustor geometries at ε = 0.6, u = 12 m/s

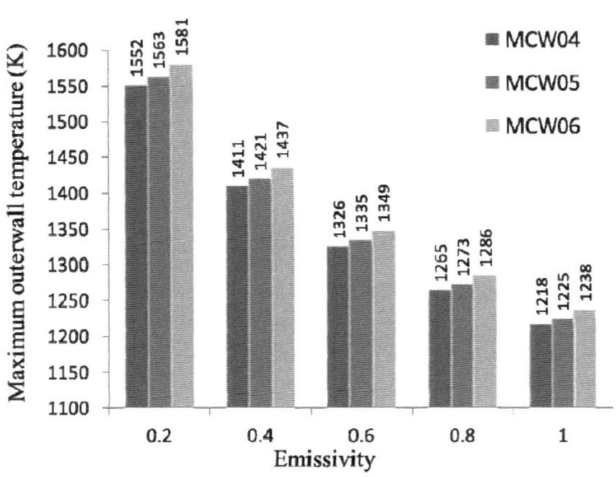

Figure 7. Comparison of maximum outer wall temperatures at different wall emissivities for various geometries

Conclusions

Micro combustors having a backward facing step and triangular wall fin were studied numerically for different blockage ratios (0.4, 0.5 and 0.6) with premixed H_2–air combustion for a range of outer wall surface emissivities (0.2 – 1).

- The outer wall temperature was observed to be higher for lower emissivities due to lesser amount of radiation heat loss through the outer surface of combustor.

- For a particular outer wall surface emissivity, MCW06 displayed highest peak temperatures for the whole range of emissivities due to enlargement of recirculation zone behind the wall fin when blockage ratio is increased.
- From the results, it can be concluded that lower emissivity and higher blockage ratiocontributeto greater outer wall temperatures and would provide better MTPV performance.

References

[1] Block M J 1956 Surface tension as the cause of Bénard cells and surface deformation in a liquid film *Nature***178** 650–651.

[1] Tang A, Pan J, Yang W, Xu Y, and Hou Z 2015Numerical study of premixed hydrogen/air combustion in a micro planar combustor with parallel separating plates *Int. J. Hydrogen Energy***40** 2396–2403,.
[2] YangW, FanA, WanJ, and LiuW, 2015 Effect of external surface emissivity on fl ame-splitting limit in a micro *Appl. Therm. Eng.* **83** 8–15.
[3] MansouriZ 2019Combustion in wavy micro-channels for thermo-photovoltaic applications – Part I: Effects of wavy wall geometry, wall temperature profile and reaction mechanism *Energy Convers. Manag.* **198** 0–1.
[4] PanJ, ZhuJ, LiuQ, ZhuY, TangA, and LuQ 2017 Effect of micro-pin-fin arrays on the heat transfer and combustion characteristics in the micro-combustor *Int. J. Hydrogen Energy* **42** 23207–23217.
[5] ChanniwalaS A 2011 Hydrogen Fuelled Micro Gas Turbine Combustion Chamber," *Hong Kong Inst. Eng. Trans.***18** 19–25.
[6] AravindB, SainiD K and Kumar S 2018 Experimental investigations on the role of various heat sinks in developing an efficient combustion based micro power generator *Appl. Therm. Eng.*
[7] AravindB, RaghuramG K S, KishoreV R and KumarS 2018 Compact design of planar stepped micro combustor for portable thermoelectric power generation *Energy Convers. Manag.***156**224–234.
[8] AkhtarS, KhanM N, KurniaJ C, and Shamim T 2017 Investigation of energy conversion and flame stability in a curved micro-combustor for thermo-photovoltaic (TPV) applications *Appl. Energy***192** 134–145.
[9] Yang W M, Chou SK, Shu C, Li Z W, Xue H 2002 Combustion in micro-cylindrical combustors with and without a backward facing step *Appl. Therm. Eng.***22** 1777–1787.
[10] BagheriG, HosseiniS E, and WahidM A 2014 Effects of bluff body shape on the flame stability in premixed micro-combustion of hydrogen-air mixture *Appl. Therm. Eng.***67** 266–272.
[11] ZhangZ, WuK, Yuen R, YaoWand Wang J 2010 Numerical investigation on the performance of bluff body augmented micro cavity-combustor *Int. J. Hydrogen Energy***45** 4932–4945.
[12] Hai C and Wei-qiang L I U 2019 Numerical study of effect of front cavity on hydrogen / air premixed combustion in a micro-combustion chamber *J. Cent South Univ* 0–1.
[13] WanJ, FanA, YaoH, and LiuW 2015 Flame-anchoring mechanisms of a micro cavity-combustor for premixed H 2 / air flame*Chem. Eng. J.* **275** 17–26.
[14] MarbachT L 2006 Heat-Recirculating Combustor Using Porous Inert Media for Mesoscale Applications*J. Propuls. Power* **22**145–150.
[15] PengQ, JiaqiangE, ChenJ, ZuoW, andZhao X 2018 Investigation on the e ff ects of wall thickness and porous media on the thermal performance of a non-premixed hydrogen fueled cylindrical micro combustor *Energy Convers. Manag..* **155** 276–286.
[16] Bazooyar B and DarabkhaniH G 2019 Analysis of flame stabilization to a thermo-photovoltaic micro-combustor step in turbulent premixed hydrogen flame*Fuel***257** 115989.

[17] Akhtar S, KurniaJ C, and Shamim T 2015 A three-dimensional computational model of H_2 – air premixed combustion in non-circular micro-channels for a thermo-photovoltaic (TPV) application *Appl. Energy***152** 47–57.
[18] LiJ, ChouS K, YangW M, and Li Z W 2009 Experimental and numerical study of the wall temperature of cylindrical micro combustors*J. Micromechanics Microengineering*. **19**.

Nomenclature

q_o convection and radiation heat transfer through the outer wall
h_o convective heat transfer coefficient
eouter wall surface emissivity
σ Stephan- Boltzmann constant = $5{:}67 \times 10^{-8}$ W/(m^{2}K^4)
$T_{w,o}$ Outer wall surface temperature
T_∞ Ambient temperature
k turbulent kinetic energy
ε energy dissipation

Thermal convection in an inclined cavity under the influence of partial magnetic field

Shreyasi Maitra[1], Nirmal K. Manna[2], Dipak Kumar Mandal[3], Nirmalendu Biswas[1, a]

[1]Department of Power Engineering, Jadavpur University, Salt Lake, Kolkata–700106, India
[2]Department of Mechanical Engineering, Jadavpur University, Kolkata–700032, India
[3]Department of Mechanical Engineering, College of Engineering and Management, Kolaghat – 721171, India
[a] corresponding author's e-mail: *nirmalendubiswas@yahoo.co.in*

Abstract: This work is focused on the implication of a partial magnetic field on a square cavity heated isothermally at the bottom wall and cooled at the sidewalls. The top wall is insulated. The enclosure is filled with electrically conducting fluid. The coupled transport equations are solved numerically using a written computing code adopting FVM and SIMPLE algorithm applying specific boundary conditions. The analysis is carried out for different involved parameters like Rayleigh number (Ra), Hartmann number (Ha), and angle of inclination (γ). The results are illustrated using contours of streamline, isotherms, and heat transfer characteristics. It is found that the flow structure, the temperature distribution is severely affected by the above-mentioned parameters. The rate of heat transfer is significantly reduced with the increasing Ha.

Keywords: Natural convection; Partial magnetic field; Inclined cavity; Heat transfer.

1. Introduction

Natural convection has high relevance in natural as well as engineering applications. Mantle convection within the earth's mantle is the driving force of the tectonic plates, sea breeze formation and climatic advances also depend on the rate of convection. Besides its implementation in nature, it consumes a variety of applications in thermal engineering, nuclear reactors, heating and cooling systems, solar power, and many more. It provides immense scope for the researches to conduct wide ranges of analysis and experiments. A detailed review of free convection inside an enclosure can be found in Refs. [1,2].

The application of the magnetic field has numerous aspects in the field of heat transfer and fluid flow. Magnetic fields are highly used in biomedical, thermal designs to control the dissipation of heat in microprocessors, electronics, aviation, and other engineering processes. In this context, fundamentals about the subject topic magnetic convection can be found in [3]. The consequence of the imposed magnetic field is associated with the Lorentz force and it is known as magnetohydrodynamics (MHD). Several researchers have investigated the thermo-magnetic convection. For example, Rudraiah et al. [4] studied the influence of the imposed magnetic field during natural convective heat transfer from a rectangular enclosure. One of the results was as the Hartmann number increased Nusselt number decreased. Tagawa et al. [5] examined the influence of the direction of a magnetic field on free convective heat transfer from a square cavity. Magnetic convection in a heated cubical cavity filled with paramagnetic fluid is observed by Bednarz et al. [6]. Ganguly et al. [7] investigated the magneto-hydrodynamic convective heat transfer process in an enclosure adopting variable magnetic fields utilizing line dipole. The application of a partial magnetic field on thermal convection has received considerable attention for controlling various engineering processes. The thermo-magnetic convection in a differentially heated square cavity subjected to partially active magnetic fields has been examined by Jalil et al. [8]. It is observed that the thermo-fluid flow structure is severely affected by the imposed magnetic fields. The impact of non-uniform magnetic fields during the thermal convection from a cavity has been studied by Szabo and Früh [9]. In this study, it is found that buoyancy-induced thermal convection dominates the magnetic force and associated rate of heat transfer. Very recently, Geridonmez and Oztop [10] have investigated the partial magnetic fields effect on thermal convection in a differentially heated air filled porous cavity. It is observed that the strength, location of the partial magnetic fields can alter the convective heat transfer process. A study related to Magnetohydrodynamic (MHD) influence on natural convection inside a closed cavity was conducted by Matt et al. [11].

Taking into account an extensive literature survey, it is observed that there are no works about the thermal convection in an inclined chamber subject to a partial magnetic field. Thus, this study aims to explore the application of a partial magnetic field on thermal convection in an inclined cavity heated isothermally at the bottom and cooled at the sidewalls. The investigation is conducted numerically and the results are presented and described with the help of streamlines, isotherms, and average Nusselt numbers. The inclination of the cavity along with the magnetic field has a finite amount of research which creates a new horizon for more exploration. In due course, the present work could be helpful is thermal engineering and medicine.

2. Problem description and numerical procedure

The geometry of the problem is comprised of a square enclosure ($L \times H$) with isothermally heated bottom wall (at temperature T_h). The sidewalls are kept cold (at temperature T_c, where $T_c < T_h$) and top wall is insulated as described in Figure1. The fluid within the cavity is taken as air (Pr = 0.71) and it is assumed that the fluid is steady, laminar, Newtonian, and incompressible within the validity of Boussinesq approximation. An external magnetic field acts (perpendicular to the sidewalls) partially over the effective length 0.5H (about the middle point of the sidewalls). It is important to note that, for obtaining the magnetic field along the x direction, electric field or voltage is applied perpendicular to the x-y plane. The study involves the inclination of the cavity (γ) with respect to the horizontal direction. As the cavity inclination angle is changing with an angle γ with respect to the horizontal direction, magnetic field remains perpendicular to the sidewalls. However, the gravity force acts into sin and cos components of the buoyancy force as given in the momentum equations.

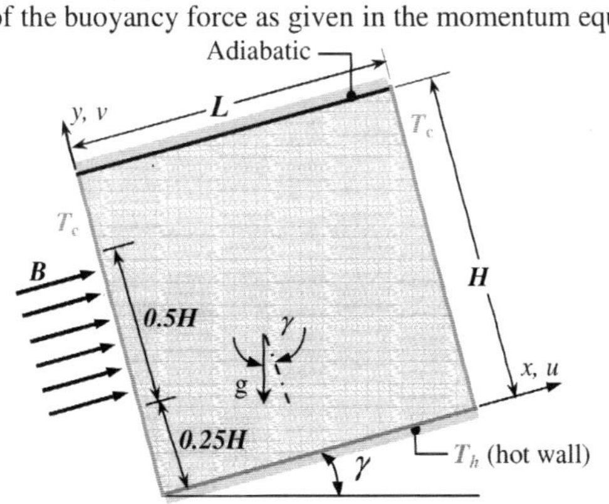

Figure 1. Schematic diagram of the computational domain of the physical problem.

Defining the following non-dimensional parameter, the dimensional transport equations are transformed into the dimensionless form:

$$(X,Y)=(x,y)/H;(U,V)=(u,v)H/\alpha;\theta=(T-T_c)/(T_h-T_c);P=((p+\rho gy)-p_a)H^2/\rho\alpha^2$$

Where (u, v) are dimensional velocity components, (U, V) are dimensionless velocities. The dimensional and non-dimensional axes are (x, y) and (X, Y) respectively. The non-dimensional pressure P constitute of ambient pressure (pa) and pressure change along the vertical (ρgy), respectively. The non-dimensional governing equations in the Cartesian coordinate system are mentioned below:

$$\frac{\partial U}{\partial X}+\frac{\partial V}{\partial Y}=0 \tag{1}$$

$$\left(U\frac{\partial U}{\partial X}+V\frac{\partial U}{\partial Y}\right)=-\frac{\partial P}{\partial X}+\Pr\left(\frac{\partial^2 U}{\partial X^2}+\frac{\partial^2 U}{\partial Y^2}\right)+\Pr\mathrm{Ra}\theta\sin\gamma \tag{2}$$

$$\left(U\frac{\partial V}{\partial X}+V\frac{\partial V}{\partial Y}\right)=-\frac{\partial P}{\partial Y}+\Pr\left(\frac{\partial^2 V}{\partial X^2}+\frac{\partial^2 V}{\partial Y^2}\right)-\lambda_B\mathrm{Ha}^2\Pr V+\Pr\mathrm{Ra}\theta\cos\gamma \tag{3}$$

$$\left(U\frac{\partial \theta}{\partial X}+V\frac{\partial \theta}{\partial Y}\right)=\left(\frac{\partial^2 \theta}{\partial X^2}+\frac{\partial^2 \theta}{\partial Y^2}\right) \tag{4}$$

where,

$$\Pr=\frac{\nu}{\alpha};\ \mathrm{Ra}=\frac{g\beta(T_h-T_c)H^3}{\nu^2}\Pr;\ \mathrm{Ha}=BH\sqrt{\sigma/\mu}$$

The fluid properties (such as thermal diffusivity α, kinematic viscosity ν, and volumetric expansion coefficient) are assumed to be constant except density (ρ) variation, which is taken care by the Boussinesq approximation. The partial magnetic fields adopted through the term λ_B by setting 1 and 0 for the active and inactive zones, respectively.

The thermal condition of the top wall is adiabatic ($\partial\theta/\partial Y=0$), sidewalls are isothermally cold ($\theta=0$) and the bottom wall is isothermally heated ($\theta=1$). All the walls are set at zero cavity velocity (U = V = 0). Along with the mentioned boundary conditions, an in-housed computing code adopting the Finite Volume Method (FVM), and SIMPLE algorithm [12], has been employed iteratively to evaluate the governing equations. The above-mentioned code has been validated extensively under natural and mixed convective heat transfer [13–19]. After conducting a grid independence study, the grid size for the present numerical study is chosen as 200 × 200. The grids are distributed uniformly to capture the partial magnetic field effect correctly. The maximum residual and continuity mass defect is chosen as 10^{-8} for the convergence of the result.

The equations of streamfunction have been evaluated and availed to present the flow field pattern in terms of streamlines. The stream function (ψ) is expressed as

$$-\frac{\partial\psi}{\partial X}=V;\ \frac{\partial\psi}{\partial Y}=U$$

In order to analyze the heat transfer characteristics for the bottom heated inclined square enclosure, the average Nusselt number is calculated on the hot wall as

$$\mathrm{Nu}=\int_0^1\left(-\left.\frac{\partial\theta}{\partial Y}\right|_{Y=0}\right)dX$$

3. Results and discussion

In the present study, thermomagnetic convection of a square inclined enclosure with heated bottom, an adiabatic top wall, and cold sidewalls is examined under the impact of a partial magnetic field. The analysis is carried out for a wide range of parameters like Rayleigh number (Ra = 10^3, 10^4, 10^5, 10^6), Hartmann number (Ha = 0, 30, 50, 70, 100), and cavity inclination angle (γ = 0°, 30°, 60°, 90°, 150°,

180°). The results of the study are illustrated by the visualization of streamlines, isotherms, and average Nusselt number.

3.1. *Effect of Rayleigh number* (Ra)

Figure 2 illustrates the streamlines, isotherms and average Nu for Ra = 10^3, 10^4, 10^6 keeping Ha = 50 and γ = 0° fixed. For the different magnitudes of evolved parameters, two numbers convective circulation cells are formed inside the enclosure, namely, clockwise (CW) at the right half and anticlockwise (CCW) at the left half of the cavity. These circulation cells are formed due to the heating at the bottom and cooling at the sidewalls. Due to the bottom heating, heated fluid rises upwards and then obstructed by the top adiabatic wall and then directed towards the side walls, where it cools and goes downward. As magnetic fields act partially over the length 0.5H (about the middle point of the sidewalls), fluid flow faces resistance (due to the Lorentz force) over this area of imposed magnetic fields. As a result, flow structure as well as static temperature distribution distorted in this zone. At the lower value of Ra (= 10^3), conduction mode dictates the heat transfer mechanism. Thus, fluid circulation is poor. Two strong counter-rotating circulations cells appear in the lower part and two weaker counter-rotating circulations cells appear in the upper part of the enclosure. The contours of isotherm are concentrated near the heated bottom wall in a certain symmetrical pattern. As the convection mechanism increases with the increasing Ra and becomes maximum at Ra = 10^6, the circulation strength increases significantly. The buoyancy effect dominates the Lorentz force at higher Ra. Thus, upper counter-rotating circulations cells become weaker. The isotherm contours remains clustered near the active walls. There is a significant rise in the heat transfer by the fluid as indicated by the increasing average Nusselt number as Rayleigh number is increased.

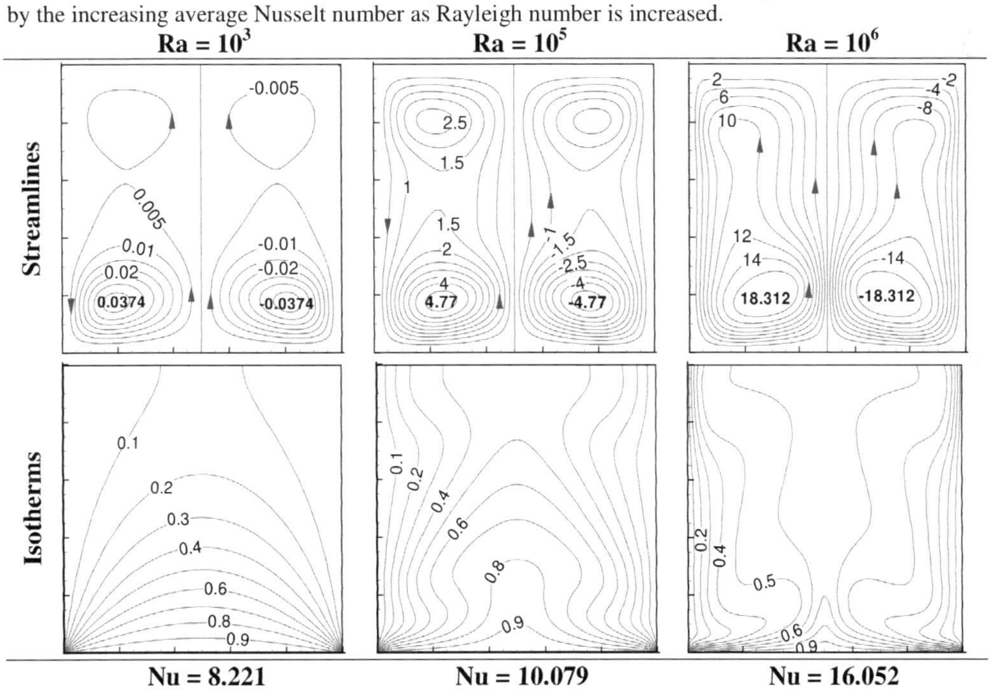

Figure 2. Effect of Ra on the contours of streamlines (top row), and isotherms (bottom row) for Ha = 50, $\gamma = 0$.

3.2. *Effect of Hartmann number* (Ha)

The effect of partially imposed magnetic field intensity is described by the Hartman number (Ha). The contours of streamlines, isotherms, and average Nu are shown in Figure 3 for Ha = 0, 30, 70 keeping Ra = 10^5 and $\gamma = 0°$ fixed. In order to understand the effect of the imposed magnetic field, the case of without magnetic field (Ha = 0) is introduced first. Eventually, the magnetic field strength is increased by increasing the value of Ha = 30 and 70. In the absence of a magnetic field, the fluid flow shows two symmetrically cells (without any distortion) and isotherms distributed symmetrically about the mid vertical plane. When the magnetic field is imposed with intensity Ha = 30, two circulation cells stretched in a vertical direction (with a reduction in the circulation strength) and divided into two pairs of counter-rotating circulation cells (one pair at the upper and other pair at the lower part of the cavity). The isotherm lines are distorted about the mid-horizontal zone of the cavity. The associated heat transfer rate is decreased due to a reduction in fluid circulation. Although the heat transfer is symmetrical in the cavity but declines as the partial magnetic strength is further increased to Ha = 70. Flow structure affected significantly with a reduction in the circulation strength. The average Nusselt number decreases substantially with the increasing magnetic field strength. Therefore, a partial magnetic field, when used wisely, is able to control the heat propagation in an enclosure.

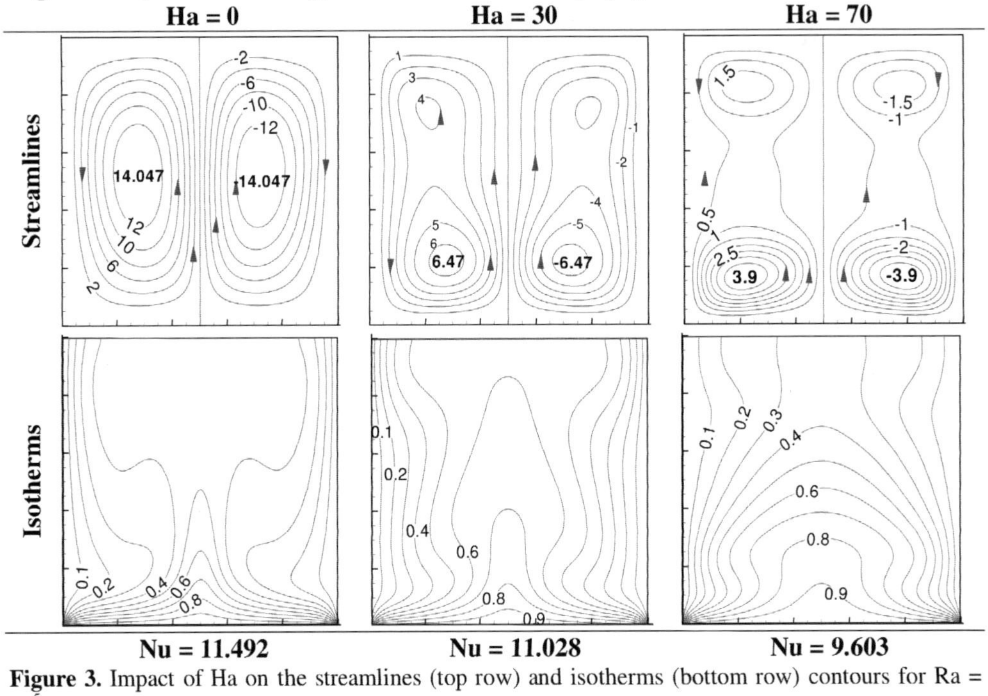

Figure 3. Impact of Ha on the streamlines (top row) and isotherms (bottom row) contours for Ra = 10^5, $\gamma = 0$.

3.3. *Effect of cavity inclination* (γ)

The influence of the cavity inclination angle on the thermo-fluid phenomena is investigated by varying the inclination γ = 30°, 60°, 90°, 150° keeping the Ha = 30 and Ra = 10^5 fixed as shown in Figure 4. Now, with the change in the inclination angle (γ) from 0 to 30° flow-structure alters significantly compared $\gamma = 0°$ (as in Fig. 3, second column). Left CCW circulation becomes larger and merged into a single cell; whereas right circulation cells split into two weaker cells. Corresponding isotherm lines change and show an asymmetric pattern. Further increase in γ from 30° to 60° all the circulating cells merged into single-cell rotating in CCW direction. The contours of isotherms have accumulated towards the right half of the cavity due to the inclination of the cavity in the counter-clockwise direction. Now, an increase in γ to 90°, the heated wall of the cavity is oriented like a right vertical wall, whereas cold walls are positioned as horizontal top and bottom walls. As a result, isotherms of high temperature are clustered near the heated wall and low-temperature lines distributed throughout the cavity. Here single circulating cells occupy the entire cavity. With the continual increase in γ heat transfer rate decreases gradually. This happens due to the change in the flow structure as well as temperature distribution. Further increase in γ to 150°, the heated wall of the cavity is positioned towards the top horizontal direction, whereas the cold walls are positioned as vertical walls. As a result, a single circulating cell shifts towards the right portion of the cavity and breaks into multiple cells. Heat transfer is decreased significantly.

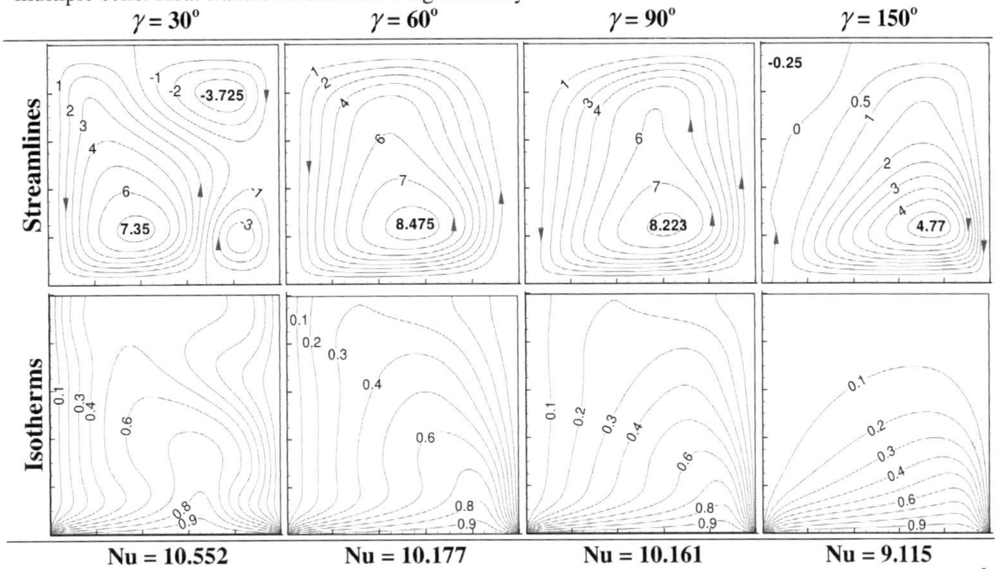

Figure 4. Effect of γ on the streamlines (top row) and isotherms (bottom row) contours at Ra = 10^5, Ha = 30.

3.4. Heat transfer characteristics

Now, the global thermal performance of the cavity is evaluated using the average Nu with the variation of Ra (= 10^5, 10^6) and is depicted in Figure 5 (a and b) for different Ha and γ keeping other parameters at fixed values. In Figure 5a, the average Nu shows a consistently decreasing trend with the increasing Ha for both Ra = 10^5, 10^6. This happens due to the presence of negative source terms associated with Ha as in the Y momentum equation (3). Thus, the magnetic force weakens the buoyancy effect. Furthermore, the Nu curve at Ra = 10^6 remains above that of Ra = 10^5. The reason behind this fact may be understood by observing the flow structure for Ra (as in Fig. 2). Higher Ra, results in stronger buoyancy force. As a result, heat transfer is higher. Furthermore, the selection of the Ha value influences the heat transfer process. In Figure 5b, the average Nu deceases continually as the inclination angle (γ) gradually increases. When the angle rises, fluid circulation velocity within the cavity reduces leading to lower heat transfer. Thus, the decreasing trend for Ra= 10^5 and 10^6 is depicted in the figure. The Nu curve at Ra = 10^6 remains well above Ra = 10^5 with a steeper slope. It is interesting to note that, as the as the inclination angle (γ) increases till 180°, the heated wall of the cavity is oriented like a top horizontal wall, whereas cold walls are positioned as vertical sidewalls. It results in reduced heat transfer. Thus, average Nu decreases continually.

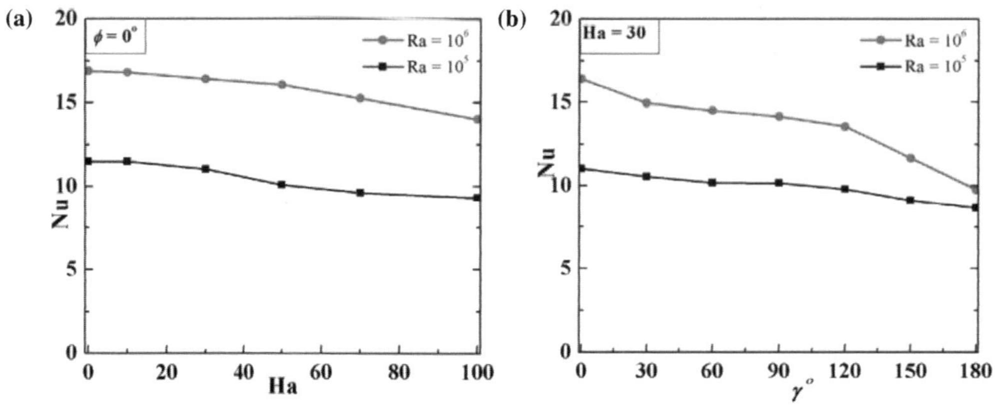

Figure 5. Heat transfer characteristics with varying Ra, (a) Ha, (b) γ keeping other parameters fixed.

4. Conclusions

The present study investigates the effect of a partial magnetic field on the thermal convection in an inclined cavity heated from the bottom and cooled from the sidewalls. The thermomagnetic convection is explored under different parameters like Ra, Ha and γ. The thermal propagation and fluid flow were analyzed thoroughly using streamlines, isotherms, and average Nusselt number. Mentioned below are some of the major observations:

a) An increase in the partial magnetic field strength significantly alters the convective motion and it results in a decrement in heat transfer in the cavity. Hence, the usage of appropriate magnetic field strength can effectively control the thermal convection within the enclosure.
b) The gradual increase in the inclination angle significantly alters the thermo-fluid structure and associated heat transfer characteristics. The fluid flow becomes asymmetrical and the heat propagation inside the cavity reduces significantly.
c) Along with a change in Ha, the heat transfer rate can also be altered by changing the Ra.

References

[1] Narasimham GSVL (2010) Natural convection from discrete heat sources in enclosures: an overview *Vivechan Int J Res* **1** 63–78.
[2] Biswas N, Mahapatra PS, Manna NK (2016) Buoyancy-driven fluid and energy flow in protruded heater enclosure *Meccanica* **51** 2159–2184.
[3] Ozoe H (2005) *Magnetic Convection* Imperial College Press, Singapore.
[4] Rudraiah N, Barron RM, Venkatachalappa M, Subbaraya CK (1995) Effect of a magnetic field on free convection in a rectangular enclosure *Int J Eng Sci* **30** 1075–1084 .
[5] Tagawa T, Shigemitsu R, Ozoe H (2002) Magnetizing force modeled and numerically solved for natural convection of air in a cubic enclosure: effect of the direction of the magnetic field *Int J Heat Mass Transfer* **45** 267–277.
[6] Bednarz T, Fornalik E, Tagawa T, Ozoe H, Szmyd J S (2005) Experimental and numerical analyses of magnetic convection of paramagnetic fluid in a cube heated and cooled from opposing verticals walls *Int J Therm Sci* **44** 933–943.
[7] Ganguly R, Sen S, Puri IK (2004) Thermomagnetic convection in a square enclosure using a line dipole *Phys. Fluids* **16**(7) 2228–2236.
[8] Jalil JM, Al-Tae'y KA, Ismail SJ (2013) Natural convection in an enclosure with a partially active magnetic field *Numer Heat Transfer A* **64**(1) 72–91.

[9] Szabo PSB, Früh W-G (2018) The transition from natural convection to thermomagnetic convection of a magnetic fluid in a non-uniform magnetic field *J Magn Magn Mater* **447** 116–123.
[10] Geridonmez BP, Oztop HF (2019) Natural convection in a cavity filled with porous medium under the effect of a partial magnetic field *Int J Mech Sci* **161–162** 105077.
[11] Matt CFT, Quaresma JNN, Cotta RM (2017) Analysis of magnetohydrodynamic natural convection in closed cavities through integral transforms *Int J Heat Mass Transfer* **113** 502–513.
[12] Patankar SV (1980) *Numerical heat transfer and fluid flow* Taylor and Francis, London.
[13] Biswas N, Manna N K (2018) Magneto-hydrodynamic Marangoni flow in bottom-heated lid-driven cavity *J Mol Liq* **251** 249–266.
[14] Biswas N, Manna NK, Chamkha AJ (2020) Effects of half-sinusoidal nonuniform heating during MHD thermal convection in Cu–Al_2O_3/water hybrid nanofluid saturated with porous media *J Therm Anal Calorim* doi.org/10.1007/s10973-020-10109-y.
[15] Biswas N, Sarkar UK, Chamkha AJ, Manna NK (2020) Magneto-hydrodynamic thermal convection of Cu–Al_2O_3/water hybrid nanofluid saturated with porous media subjected to half-sinusoidal nonuniform heating *J Therm Anal Calorim* doi.org/10.1007/s10973-020-10123-0.
[16] Biswas N, Mahapatra PS, Manna NK (2016) Buoyancy-driven fluid and energy flow in protruded heater enclosure *Meccanica* **51** 2159–2184.
[17] Manna NK, Mahapatra PS, Biswas N, (2019) Convective heat transfer enhancement: effect of multi-frequency heating *Int J Numer Meth Heat Fluid Flow* **29**(10) 3822–3856.
[18] Biswas N, Manna NK, Datta A, Mandal DK, Benim AC (2020) Role of aspiration to enhance MHD convection in protruded heater cavity *Prog Comp Fluid Dyn* **X**(Y) xxx–xxx
[19] Biswas N, Mahapatra PS, Manna NK (2017) Enhanced convective heat transfer in lid-driven porous cavity with aspiration *Int J Heat Mass Transfer* **114** 430–452.

Nomenclature

B	magnetic fields (Tesla, N A^{-1} m^{-2})		***Greek symbols***
g	acceleration due to gravity (m s^{-2})	α	thermal diffusivity (m^2 s^{-1})
H	height of the cavity/length scale (m)	β	thermal expansion coefficient (K^{-1})
Ha	Hartmann number	θ	dimensionless temperature
L	length of the cavity (m)	γ	inclination angle of the cavity
Nu	average Nusselt number	λ_B	magnetic field parameter
p	pressure (Pa)	μ	dynamic viscosity
P	dimensionless pressure	ν	kinematic viscosity (m^2 s^{-1})
Pr	Prandtl number	ρ	density (kg m^{-3})
Ra	Rayleigh number	σ	electrical conductivity (μ Scm^{-1})
T	temperature (K)	ψ	dimensionless stream function
u, v	velocity components (m s^{-1})		***Subscripts***
U, V	dimensionless velocity components	a	ambient
x, y	Cartesian coordinates (m)	c	cold
X, Y	dimensionless coordinates	h	hot

Effect of partial magnetic field on thermo gravitational convection in an inclined cavity

Deep Chatterjee[1], Nirmal K. Manna[1], Dipak Kumar Mandal[2], Nirmalendu Biswas[3, a]

[1]Department of Mechanical Engineering, Jadavpur University, Kolkata–700032, India
[2]Department of Mechanical Engineering, College of Engineering and Management, Kolaghat – 721171, India
[3]Department of Power Engineering, Jadavpur University, Salt Lake, Kolkata–700106, India
[a] corresponding author's e-mail: *nirmalendubiswas@yahoo.co.in*

Abstract: The study presented gives a succinct overview of the effect of partially applied magnetic field on natural convective heat transfer in a differentially heated inclined cavity. The left and right sidewalls are differentially heated such that their temperatures are maintained (at T_h and T_c respectively), while the top and bottom walls have been assumed to be adiabatic, such that heat transfer through those walls can be neglected. The results are supposed to elucidate the effect of changing Rayleigh number (Ra), Hartmann number (Ha), and cavity inclination angle (ϕ) on the thermo-fluid phenomena. The analysis is carried out numerically utilizing finite volume based computing code. The results have been demonstrated in the form of streamlines, isotherms, and average Nusselt number. From the analysis, it is found that the flow structure and associated heat transfer characteristics are severely influenced by the studied range of governing parameters.

Keywords: Partial magnetic field; Inclined cavity; Natural convection; Heat transfer.

1. Introduction

The application of the magnetic field alters the thermo-fluid phenomena and this can be a very useful technique taking into an opportunity for improving the thermal performance in many engineering processes. The magnetic field induces the Lorentz force in a fluid in motion which is further affected by the thermal gradient and other flow conditions. Furthermore, fluid flow parameters and the type of fluid used, introduction of porous media, or externally driven flow along with the application of magnetic field, and various boundary layer conditions can greatly affect the thermal performance and hence the convective heat transfer efficacy, which leads us again to the practical importance of this field of study. Fundamentals on the above subject topics can be found in Refs. [1,2].

The effect of the magnetic field on buoyancy-driven flow has been studied before in different forms by multiple authors. Detailed research in this area has been conducted by Giwa et al.[3] where the effect of several variables and changing flow parameters have been reviewed. The inclination of the magnetic field, the introduction of porous media, nanofluids, thermal and concentration boundary conditions, Brownian motion as well as some other variable conditions were briefly studied and their effects were discussed. Thermo-magnetic convective heat transfer in a confined cavity has been studied by several researchers considering uniform magnetic fields [2], line dipole [4], and inclined magnetic fields [5]. In other class of work, the magnetohydrodynamics (MHD) convection has been investigated considering partial magnetic fields [6–8] instead of whole domain magnetic fields.

Buoyancy-induced convective phenomena in a cavity subjected to a partial magnetic field have been analyzed by Jalil et al. [6]. The influence of the non-uniform magnetic field using permanent magnets of varying strength, and hence the transition from natural convection to thermomagnetic convection was studied by Szabo et al. [7]. Very recently, Geridonmez and Oztop [9] studied natural convection in a differentially heated porous cavity where a partial magnetic field was applied horizontally. They concluded that depending on the impact area of the partial magnetic field, the fluid flow is affected. A further conclusion was that at higher values of Rayleigh number, the influence of the magnetic field is more pronounced (due to an increase in Lorentz force experienced). Magneto-hydrodynamic convection in a corner heated cavity has been studied by Mondal et al. [10].

In the aforementioned literature survey, some of the researchers have focused on the technique of using partial magnetic fields, while most of the researchers have used the whole domain magnetic field to control the thermal performance during buoyancy-induced convection. With this introduction on the importance of the study of magnetohydrodynamics (MHD) the present work is formulated considering a differentially heated inclined cavity under the impact of a partially applied magnetic field. The purpose of this study is to highlight and contrast the effect of the magnetic field versus simple buoyancy-driven convective flow and to further analyze the conditions which favor a faster convective heat transfer rate.

2. Mathematical modeling and numerical technique

A schematic figure of the model is shown in Figure 1. The geometry is a square cavity (of side H) containing air as the working fluid under the effect of a partially applied uniform magnetic field. The cavity is inclined to the horizontal direction and the inclination angle is represented by ϕ. The left sidewall is the hot wall (at a temperature T_h) which is isothermally heated, while the right sidewall is cold (at a temperature T_c). The upper and lower walls of the cavity are kept adiabatic. The magnetic field is perpendicular to the sidewalls over a length of $0.5H$ (positioned at the center of the sidewall) and the intensity of the field has been varied (which is reflected by changing the Hartmann number). Here, the magnetic field along the x direction is generated due to the influence of electric field or voltage applied perpendicular to the x-y plane. In this study, the external magnetic field is acting perpendicular to the sidewalls and partially. As the cavity inclination angle is changing with an angle ϕ with respect to the horizontal direction, magnetic field remains perpendicular to the sidewalls. However, the gravity force acts into sin and cos components of the buoyancy force (which are appeared in the momentum equations).

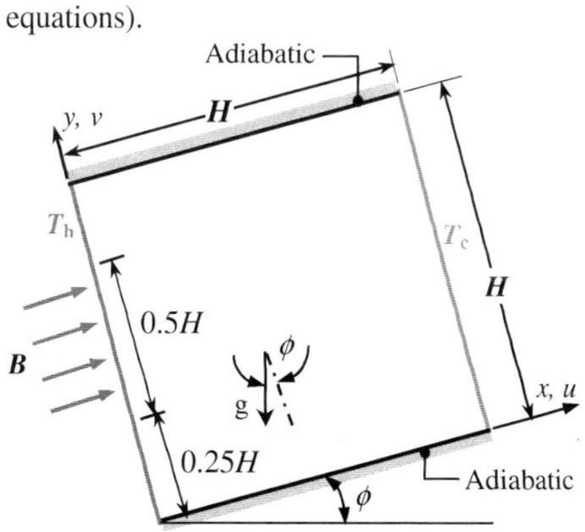

Figure 1. Schematic diagram of the computational domain.

The flow field is considered as two-dimensional (the perpendicular direction is considered infinitely long). The flow is assumed to be laminar, steady, and incompressible while Boussinesq approximations are taken to hold good. The working fluid (of Pr = 0.71) is assumed to be electrically conducting. The dimensionless governing equations for the problem are as follows:

$$\frac{\partial U}{\partial X}+\frac{\partial V}{\partial Y}=0 \tag{1}$$

$$U\frac{\partial U}{\partial X}+V\frac{\partial U}{\partial Y}=-\frac{\partial P}{\partial X}+\Pr\left(\frac{\partial^2 U}{\partial X^2}+\frac{\partial^2 U}{\partial Y^2}\right)+\Pr\,\mathrm{Ra}\,\theta\sin\phi \tag{2}$$

$$U\frac{\partial V}{\partial X}+V\frac{\partial V}{\partial Y}=-\frac{\partial P}{\partial Y}+\Pr\left(\frac{\partial^2 V}{\partial X^2}+\frac{\partial^2 V}{\partial Y^2}\right)-\lambda_B \mathrm{Ha}^2 \Pr V+\Pr\,\mathrm{Ra}\,\theta\cos\phi \tag{3}$$

$$U\frac{\partial \theta}{\partial X}+V\frac{\partial \theta}{\partial Y}=\frac{\partial^2 \theta}{\partial X^2}+\frac{\partial^2 \theta}{\partial Y^2} \tag{4}$$

where,

$$\Pr=\frac{\nu}{\alpha};\ \mathrm{Ra}=\frac{g\beta(T_h-T_c)H^3}{\nu^2}\Pr;\ \mathrm{Ha}=BH\sqrt{\sigma/\mu}$$

The non-dimensional governing equations have been constructed from the following scale factors:

$$(X,Y)=(x,y)/H;(U,V)=(u,v)H/\alpha;\ \theta=(T-T_c)/(T_h-T_c);\ P=(p-p_a)H^2/\rho\alpha^2 \tag{5}$$

The partial magnetic fields were adopted through the term λ_B by setting 1 and 0 for the active and inactive zones, respectively. The dimensionless partial differential equations (1)–(4) are solved numerically using an in-house computing code adopting the FVM through ADI sweep, TDMA solver, using the SIMPLE algorithm [11]. The converged solution is obtained when continuity mass-defect becomes less than 10^{-10}. The same code has been used in our earlier works and validated extensively under the different problem of published literature [2,12–18]. In the present simulation, a uniformly distributed 200×200 grid size is selected after conducting the gird independency study.

The boundary conditions are implemented by setting $\theta = 1$ and $\theta = 0$ respectively at the left and right walls, $\partial\theta/\partial Y = 0$ at the top and bottom walls, and zero-velocity ($U = V = 0$) for all the walls.

3. Results and discussion

The present work investigated the thermo-magnetic natural convective heat transfer in an inclined cavity subjected to a magnetic field applied partially. The effects of various changing parameters are depicted here. The Rayleigh number (Ra = 10^3, 10^4, 10^5, 10^6), Hartmann number (Ha = 0, 30, 50, 70), and cavity inclination angle (ϕ = 0°, 30°, 60°, 90°, 150°, 180°) have been changed individually and the effect is observed for each variation while the other two parameters remained constant. The results are visualized through streamlines, isotherms, and average Nusselt number (Nu) values.

3.1. *Effect of Rayleigh number* (Ra)

Figure 2 shows the variation of streamlines (first row) and isotherms (second row) for changing the Rayleigh number (Ra = 10^3, 10^4, 10^6) while the cavity remains horizontal (ϕ = 0) and the Hartmann number constant at Ha = 50. Due to differential heating at the cavity, fluid circulation establishes in the clockwise direction (CW). The variation of streamlines distribution with Rayleigh number shows that at a low value of Ra = 10^3 (which is indicative of low buoyancy forces due to conduction dominance), the flow structure is almost symmetric, with the reduced flow strength. There are two smaller vortices placed vertically away from the mid-horizontal plane of the cavity. This happens due to the imposed magnetic field partially about the middle zone of the cavity (about 0.5*H* length). Also, the streamlines are more widely spaced in the middle zone for low Ra values, which is indicative of low fluid velocity in the region. Thus, fluid flows faster away from the middle of the cavity. As Ra increases, the two rotating cells towards the top and bottom wall shift and align themselves such that they move farther away from the center and at Ra = 10^6, the lower cell disappears. The flow strength obviously increases with an increasing Rayleigh number, which is reflected by the increasing magnitudes of the maximum value of streamfunction. The flow field is pushed towards the periphery as the value of Ra increases, moving away from the zone of influence of the magnetic field, with overall flow velocity increasing due to an increase in buoyant forces as the Rayleigh number increases. Also, the streamlines are denser in the zone of the magnetic field for Ra = 10^6 compared to the previous cases showing that the buoyant forces are causing an increased flow velocity while the flow is receding away from the central zone of the cavity under the effect of increased Lorentz forces that is restricting fluid flow. Thus, it can be concluded that with increasing Rayleigh number, the magnitude of flow velocity increases and leads to an increase in heat transfer.

The isotherm lines at Ra = 10^3 are nearly all vertical with very slight changes in orientation throughout, which can be assumed to be uniform. The isotherms vary with increasing values of Ra and gradually become horizontal (at Ra = 10^6). At lower Ra, the heat transfer mechanism is governed by the conduction mode, thus lesser heat transfer as indicated by average Nu. The temperature distribution is more affected by the increased Ra of the flow as compared to the increased magnitude of Lorentz forces accompanying it. At higher Ra (> 10^4), the convection mode of heat transfer dominates the conduction mode. Leading to the higher fluid velocity and associated heat transfer. Thus, the average Nu value increases markedly with the increasing Ra.

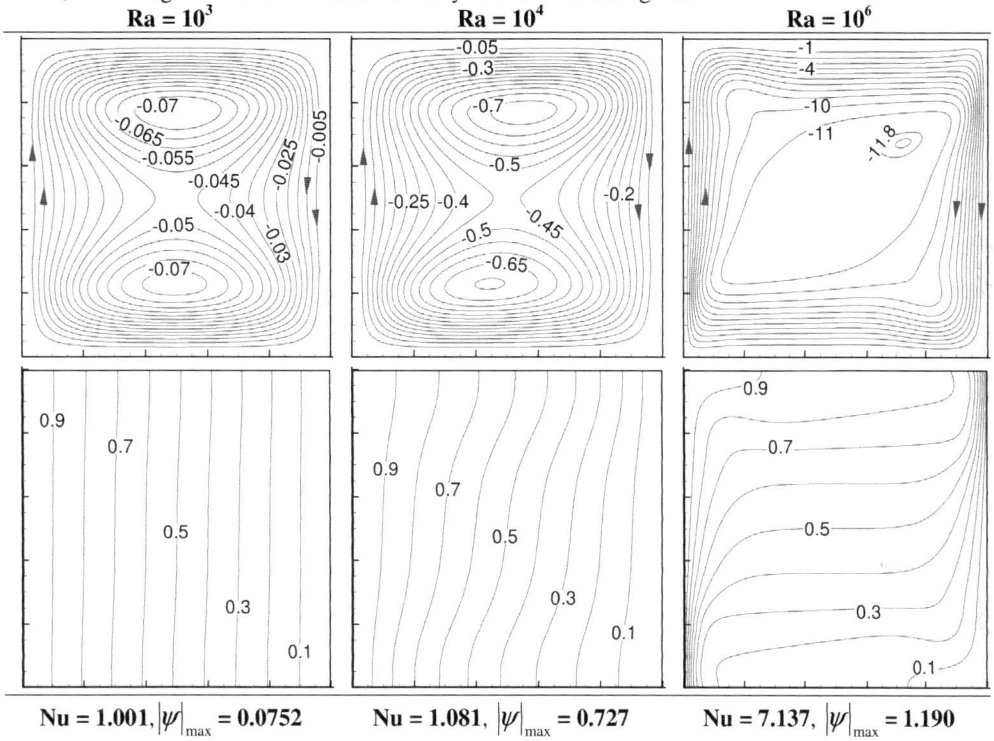

Figure 2. Effect of Ra on the contours of streamlines (top row), and isotherms (bottom row) for Ha = 50, $\phi = 0^o$.

3.2. *Effect of Hartmann number* (Ha)

Figure 3 shows the variation of streamlines and isotherms with changing the Hartmann number (Ha = 0, 30, 70), keeping the Ra = 10^5) and $\phi = 0^o$ fixed. The case of no magnetic field (Ha = 0) resents natural convection under no influence of the magnetic field. Clockwise streamlines are formed, with two small cells aligned diagonally in the middle of the square cavity. As Ha value increases, the strength of the flow decreases due to increasing opposing Lorentz force that affects the flow field, as is evident from the decreasing value of the absolute magnitude of streamfunction. The alignment of the fluid flow also changes with changing the intensity of the magnetic field. At Ha = 0, the two central vortices are aligned along the 'top-left bottom-right' diagonal. But, as the Ha increases, the fluid flow (perpendicular to the magnetic field direction) in the middle zone of the cavity is arrested-evident by the fact that the streamlines are more diffused in the middle zone, which indicates a lower fluid velocity. Hence, the fluid velocity is cut down in the active zone of the imposed magnetic field and is increased towards the periphery (denoted by denser streamlines) with increasing Ha. Also, the two rotating cells in the middle that was inclined along the 'top-left bottom-right' diagonal for the case of Ha = 0, become inclined along the opposite diagonal when the Hartmann number reaches an appreciable value (Ha = 70).

The isotherms, also show marked change with increasing Ha, such that, for the case of Ha = 0, the isotherms that are almost horizontal become almost vertical for when the Ha = 70. The isotherms, therefore, orient themselves perpendicular to the direction of magnetic field lines in the middle zone where the magnetic field acts as the field becomes more pronounced. The temperature gradient, therefore, becomes more horizontal as the isotherms become vertical. The heat transfer rate is somewhat compromised with increased Ha values, which is also indicated in the figure below.

Ha = 0 **Ha = 30** **Ha = 70**

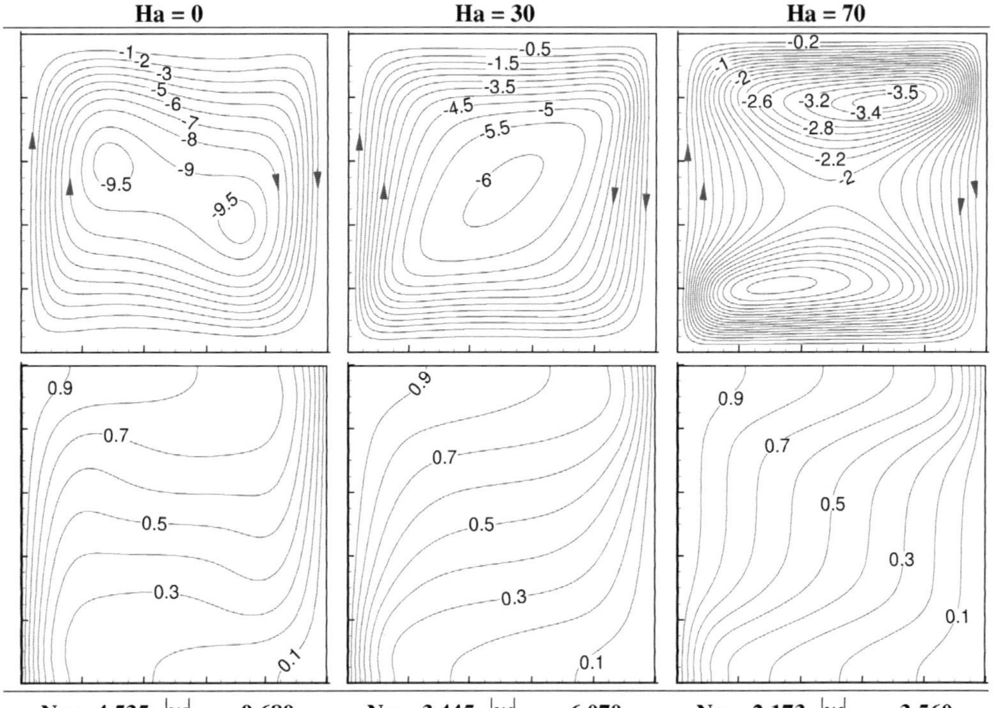

Nu = 4.535 , $|\psi|_{\max}$ = 9.680 **Nu = 3.445, $|\psi|_{\max}$ = 6.070** **Nu = 2.173, $|\psi|_{\max}$ = 3.560**

Figure 3. Effect of Ha on the contours of streamlines (top row), and isotherms (bottom row) for Ra = 10^5, $\phi = 0^{\circ}$.

3.3. *Effect of cavity inclination* (ϕ)

Figure 4 illustrates the effect of cavity inclination angle (ϕ = 30°, 60°, 90°, 150°) with the horizontal direction, keeping Ra = 10^5, Ha = 50 fixed. The flow field change markedly due to the influence of changing orientation of the hot and cold walls positions. As the cavity inclination increases from $\phi = 30^{\circ}$ to 60°, we see that the flow field shifts towards the left slightly; and towards the middle, the streamlines come closer and become more concave. Also, it is observed that the flow velocity shows a slight increase, which essentially indicates a better heat transfer rate through convection. At $\phi = 90^{\circ}$, the streamlines change markedly, such that clockwise and anticlockwise rotating cells are separately formed and coexist. At the position (ϕ = 90°), the hot wall is at the bottom and the cold wall is at the top and the unique streamline pattern comes as a consequence of the arrangement. The flow velocity remains almost the same ϕ = 90° as in the previous case (ϕ = 60°). The flow regime shows a distinct shift to the left as the angle of inclination becomes obtuse with the two core rotating cells shifting to the left side. The flow circulation also changes to an anticlockwise direction at ϕ = 150°. The flow regime at ϕ = 150° is a perfect mirror image of the flow pattern at ϕ = 30° with a similar magnitude of the absolute maximum values of the streamfunction.

The isotherms show little variation from ϕ = 30° to ϕ = 60°, the isotherms near the cold wall become slightly more concave. At ϕ = 60°, however, they show a marked change and assume a more symmetrical structure with a high concavity of the lines towards the middle of the cold wall. The isotherms at ϕ = 150° is essentially a mirror image of the isotherms at ϕ = 30°. The effective heat transfer rate for both cases is also likely to be the same.

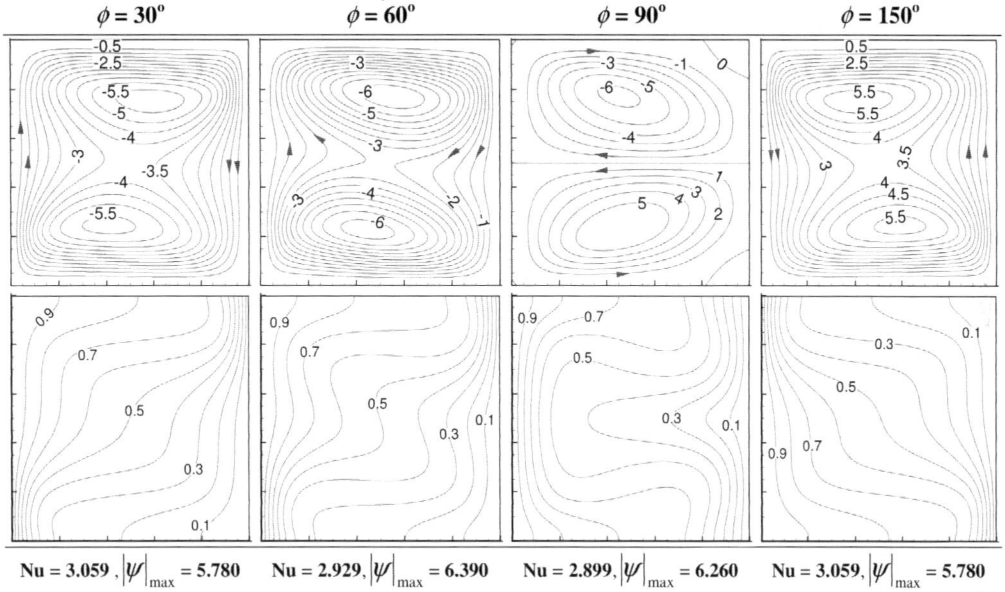

Figure 3. Effect of ϕ on the streamlines (top row) and isotherms (bottom row) contours at Ra = 10^5, Ha = 50.

3.4. Heat transfer characteristics

The average Nusselt number (Nu) is utilized to calculate the heat transfer characteristics at the heated wall as

$$\mathrm{Nu} = \int_0^1 \left(-\left.\frac{\partial \theta}{\partial X}\right|_{X=0}\right) dY$$

Figure 5a depicts the variation of global heat transfer characteristics (average Nu) as one changes the Hartmann number values (that is the magnetic field strength) for Ra = 10^5 and 10^6 respectively. From the figure, it is observed that the average Nu for all the values of Ha in the case of Ra = 10^6 is greater than for the case of Ra = 10^5. This indicates that convective heat transfer is largely increased with an increase in Rayleigh number value. In other words, buoyant forces aid convection. The general trend for both the plots is that with increasing values of Hartmann number, the average Nusselt number gradually decreases. Thus, convective heat transfer becomes less predominant as the magnetic field strength is increased. This can be attributed to the Lorentz forces which reduce the flow velocity.

The curve for Ra = 10^6 has a steeper downward slope than the one for Ra = 10^5. This indicates that the effect of Lorentz force is greater in a flow condition with a higher Rayleigh number value.

Figure 5b represents the variation of average Nusselt number values with changing the angle of inclination of the square cavity w.r.t the horizontal. The Rayleigh number has been assumed fixed at 10^5. From Figure 5b, it is observed that the average Nusselt number values for Hartmann number 50 are always less than the corresponding values for Hartmann number 0. Thus, increased magnetic field strength decreases the convective heat transfer rate. For Hartmann number 0, the average Nu value increases very little as the inclination angle increases from horizontal to 30°. It reduces from 30° till 90°, after which it shoots up as the inclination reaches 150°. For Hartmann number 50, the figure shows a distinctly different plot. The Nusselt number increases markedly from the angle of inclination 0° to 30°. The Nusselt number value remains fairly constant untill the inclination value reaches 150°. Unlike the plot at Hartmann number 0, even when the angle increases from 90° to 150°, the Nusselt number only shows a negligible increase-meaning that under the presence of the magnetic field, Nusselt number and hence heat transfer by convection is affected with increasing angle of inclination but after a certain inclination value, the convection heat transfer rate becomes somewhat stabilized and shows lesser dependency on any further inclination changes when compared to the case of Ha = 0.

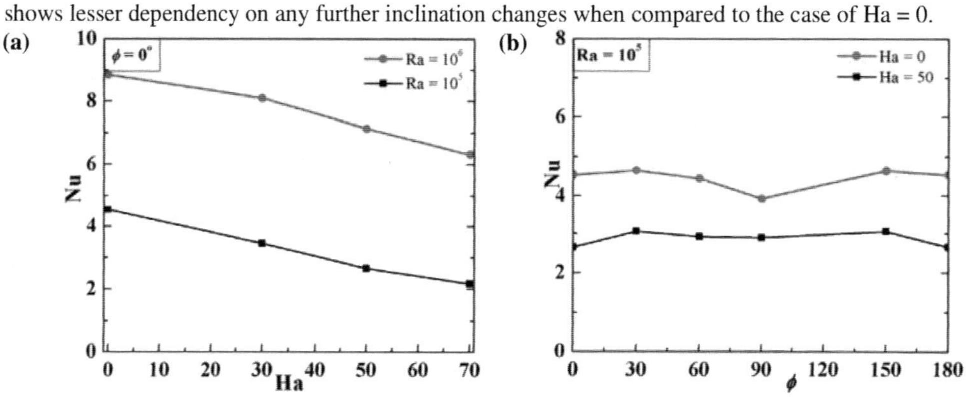

Figure 4. Heat transfer characteristics with varying Ra, (a) Ha, (b) ϕ keeping other parameters fixed.

4. Conclusions

In this study, the impact of partial magnetic fields on the thermal convective heat transfer in an inclined differentially heated cavity is investigated numerically. The thermo-fluid phenomena are explored under different parametric influences like Ra, Ha, and ϕ. The following conclusions may be drawn:

a) As the Rayleigh number increases, convection mode governs the heat transfer mechanism with increased flow velocity and a higher rate of heat transfer. Flow structure as well as static temperature distribution affected severely in the zone of the imposed partially magnetic field.

b) With increasing Hartmann number, the Lorentz force counteracts the buoyancy force. The fluid velocity is markedly higher away from the zone of magnetic field compared to the fluid flow under the influence of magnetic forces, for a high value of Ha. The isotherms also align themselves in a way that the temperature gradient variation occurs in the direction of the magnetic field for such cases. Higher is the Ha, leads to the decreasing flow velocity and thus decreasing rate of heat transfer.

c) With the change in the inclination angle, fluid and temperature distribution changes significantly. At smaller angles of cavity inclination (below 30°) for an appreciable value of Ha, there is a significant effect on the average Nu (and hence the convective heat transfer rate). Further increase in the angle, the heat transfer rate decreases. The change in average Nu is minimal for changes in angle of inclination for Ha = 50 compared with Ha = 0. In the absence of a magnetic field, the Nu value shows larger variations against changing values of angle of inclination (for inclination values greater than 30°).

References

[1] Davidson PA (2001) *An introduction to magnetohydrodynamics* Cambridge Texts in Applied Mathematics, Cambridge, UK: Cambridge University Press.
[2] Biswas N, Manna N K 2018 Magneto-hydrodynamic Marangoni flow in bottom-heated lid-driven cavity *J. Mol. Liq.* **251** 249–266.
[3] Giwa SO, Sharifpur M, Ahmadi MH, Meyer JP (2020) A review of magnetic field influence on natural convection heat transfer performance of nanofluids in square cavities *J. Thermal Anal. Calorim.* doi.org/10.1007/s10973-020-09832-3.
[4] Ganguly R, Sen S, Puri IK (2004) Thermomagnetic convection in a square enclosure using a line dipole *Phys. Fluids* **16**(7) 2228–2236.
[5] Al-Balushi LM, Rahman MM (2019) Convective heat transfer utilizing magnetic nanoparticles in the presence of a sloping magnetic field inside a square enclosure *J. Thermal Sci. Eng. Appl.* **11** 041013-1-19.
[6] Jalil JM, Al-Tae'y KA, Ismail SJ (2013) Natural convection in an enclosure with a partially active magnetic field *Numer Heat Transfer A* **64**(1) 72–91.
[7] Szabo PSB, Früh W-G (2018) The transition from natural convection to thermomagnetic convection of a magnetic fluid in a non-uniform magnetic field *J Magn Magn Mater* **447** 116–123.
[8] Al-Rashed A A AA, Kolsi L, Oztop HF, Aydi A, Malekshah EH, Abu-Hamdeh N, Borjini M N (2018) 3D magneto-convective heat transfer in CNT-nanofluid filled cavity under partially active magnetic field *Physica E* **99** 294–303.
[9] Geridonmez BP, Oztop HF (2019) Natural convection in a cavity filled with porous medium under the effect of a partial magnetic field *Int J Mech Sci* **161–162** 105077.
[10] Mondal M K, Biswas N, Manna N K 2019 MHD convection in a partially driven cavity with corner heating *SN Appl. Sci.* 1-1689.
[11] Patankar SV 1980 *Numerical Heat Transfer and Fluid Flow*, New York NY Hemisphere.
[12] Biswas N, Mahapatra PS, Manna NK (2016) Buoyancy-driven fluid and energy flow in protruded heater enclosure *Meccanica* **51** 2159–2184.
[13] Biswas N, Mahapatra PS, Manna NK (2017) Enhanced convective heat transfer in lid-driven porous cavity with aspiration *Int J Heat Mass Transfer* **114** 430–452.
[14] Biswas N, Mahapatra PS, Manna NK (2016) Enhanced thermal energy transport using adiabatic block inside lid driven cavity *Int. J. Heat Mass Transfer* **100** 407–427.
[15] Biswas N, Manna NK, Chamkha AJ (2020) Effects of half-sinusoidal nonuniform heating during MHD thermal convection in Cu–Al_2O_3/water hybrid nanofluid saturated with porous media *J Therm Anal Calorim* doi.org/10.1007/s10973-020-10109-y.
[16] Biswas N, Sarkar UK, Chamkha AJ, Manna NK (2020) Magneto-hydrodynamic thermal convection of Cu–Al_2O_3/water hybrid nanofluid saturated with porous media subjected to half-sinusoidal nonuniform heating *J Therm Anal Calorim* doi.org/10.1007/s10973-020-10123-0.
[17] Biswas N, Manna NK, Datta A, Mandal DK, Benim AC (2020) Role of aspiration to enhance MHD convection in protruded heater cavity *Prog Comp Fluid Dyn* **X**(Y) xxx–xxx.
[18] Manna NK, Mahapatra PS, Biswas N, (2019) Convective heat transfer enhancement: effect of multi-frequency heating *Int J Numer Meth Heat Fluid Flow* **29**(10) 3822–3856.

Nomenclature

B	magnetic fields (Tesla, N A^{-1} m^{-2})		***Greek symbols***
g	acceleration due to gravity (m s^{-2})	α	thermal diffusivity (m^2 s^{-1})
H	height of the cavity/length scale (m)	β	thermal expansion coefficient (K^{-1})

IOP Conf. Series: Materials Science and Engineering 1080 (2021) 012030 doi:10.1088/1757-899X/1080/1/012030

Ha	Hartmann number	θ	dimensionless temperature
L	length of the cavity (m)	ϕ	inclination angle of the cavity
Nu	average Nusselt number	λ_B	magnetic field parameter
p	pressure (Pa)	μ	dynamic viscosity
P	dimensionless pressure	ν	kinematic viscosity ($m^2 s^{-1}$)
Pr	Prandtl number	ρ	density (kg m^{-3})
Ra	Rayleigh number	σ	electrical conductivity (μ Scm^{-1})
T	temperature (K)	ψ	dimensionless stream function
u, v	velocity components (m s^{-1})		***Subscripts***
U, V	dimensionless velocity components	a	ambient
x, y	Cartesian coordinates (m)	c	cold
X, Y	dimensionless coordinates	h	hot

IOP Conf. Series: Materials Science and Engineering 1080 (2021) 012031 doi:10.1088/1757-899X/1080/1/012031

Computational analysis of heat transfer due to turbulent annular jet impingement

Prasun Dutta[1] and Himadri Chattopadhyay[2]

[1, 2] Department of Mechanical Engineering, Jadavpur University, Kolkata–700032, India

e-mail: *pd.iiest@gmail.com*

Abstract. Present paper deals with the numerical study of fluid flow and heat transfer due to an annular turbulent jet impingement on a flat heated plate. The geometrical configurations of annular jet was selected such a way that the mass and momentums efflux at the annular jet exit will be same as a circular jet having same inner diameter at a particular Reynolds number. Three dimensional turbulent flow field was resolved using Transition SST Model by solving the mass, momentum and energy equation with SIMPLE algorithm. A highly refined mesh was used for numerical computation after validation of present model with established results of same geometrical configuration. The flow structure and heat transfer characteristic of the annular jet was compared with the conventional circular jet and also with an established two dimensional study of same configurations and reported in this paper. It was observed that that two dimensional studies on annular jet under predicts heat transfer characteristics.

1. Introduction

It is well known that, high heat transfer rate can easily obtained by implementation of the jet impingement technique on a hot surface. For that reason, impingement of jet is extensively used in numerous industry. The applications contain moderating of glass, paper drying, chemical vapor removal, cooling of small electronic component to large turbine blade. In the past few decades enormous studies have been done to scrutinize the prime factors of impinging jet analogous to flow structure and heat transfer enhancement [1–3].

In a number of fluid-thermal application, consistency of heat transfer with an annular jet over a broad area are beneficial as an alternative of localised heat transfer near the impingement area with the circular jet. Jet impingement technique using an annular jet probably a prospective method of succeeding more consistency and enrichment in terms of heat transfer than the circular jet impingement method. The flow and geometric parameters for annular jet impingement is mainly categorized by; (a) "blockage ratio $BR = D_i/D_o$", where D_i and D_o are the inner and outer diameters of the annular jet, (b) "non-dimensional jet exit to impingement plate distance, $H = h/D_i$", where h is the dimensional distance from jet exit to impingement plate, and (c) "Reynolds number, *Re*". In earlier effort of Maki and Yabe [4], it was revealed that a possibility of occurrence of the "reverse stagnation point" may develop in the impingement surface subject to the jet height from the surface. They also observed that the Reynolds number does not play a major role on heat transfer at this particular reverse stagnation point where the characteristic length of the jet is defined by the jet width. Meanwhile Ramos [5] reported one dimensional analytical and numerical studies on liquid annular jet. The effect of nozzle arrangement was of subject of interest in some of the previous works of Shuja et al. [6] where they observed higher heat transfer for the conical nozzles as compared to the annular nozzles.

The prime objective of the present study is to estimate consistently the performance of the turbulent annular jet through numerical simulations with three dimensional model and also compare with the circular jet as well as established two dimensional model keeping same mass and momentum efflux at the nozzle exit. Recently, some similar studies on the annular jet was performed by Chattopadhyay [7,8] considering two dimensional axisymmetric case. They suggested to consider the inner diameter of the nozzle as characteristics length to define the Reynolds number (*Re*) and to choose the outer diameter of the nozzle in a systematic manner so that the same level of mass and momentum efflux can be achieved at the nozzle exit for both annular and circular jet same inner diameter. They also suggested that it would be more meaningful to define *Re* with the inner diameter of the nozzle as characteristics length and choose the outer diameter in such a way so that the mass and momentum efflux at the nozzle exit is at the same value as in the case of a circular jet. Recently, Dutta and Chattopadhyay [9] performed numerical simulations on annular jet impingent over a hot moving surface. They observed that, the flow structure over the impinging surface was strongly influenced by the surface velocity which leads to the enhancement of heat transfer.

In the present study, inner and outer the diameter of the annular jet are considered as D_i and $\sqrt{2}D_i$ respectively, so that the area under the annular portion is same as the area of a circle with diameter D_i. Utilising that concept, a three dimensional model is created for present study. The schematic diagram of present model is shown in figure 1. It was recognized by the literature that present computational domain is adequate to exclude the end effects [10]. The comparison of performance of the annular jet of inner diameter D_i was done with a circular jet of same inner diameter $D = D_i$ and the impingement height $h = 2D_i$.

2. Problem description and solution procedure

The flow is assumed to be incompressible with constant properties. Air is considered for present study with Prandelt number (*Pr*) of 0.71. The flow is considered in turbulent regime with Reynolds number (Re) of 20,000 which is defined on the basis of velocity and characteristic length of the jet. To measure the heat transfer characteristic of the jet, local temperature gradient is used to calculate the local Nusselt number $\mathrm{Nu} = -\partial T/\partial Z$. The area averaged Nusselt number is calculated by $\mathrm{Nu}_{avg} = \mathrm{Nu}(r) \times \partial r / L$ (L denotes the length of the domain in r direction).

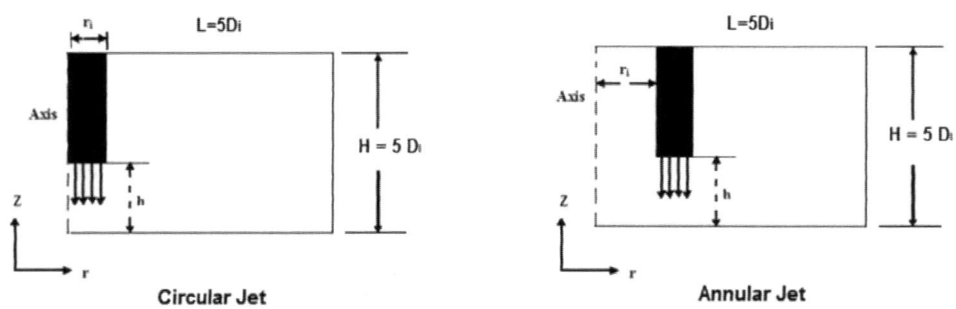

(a)

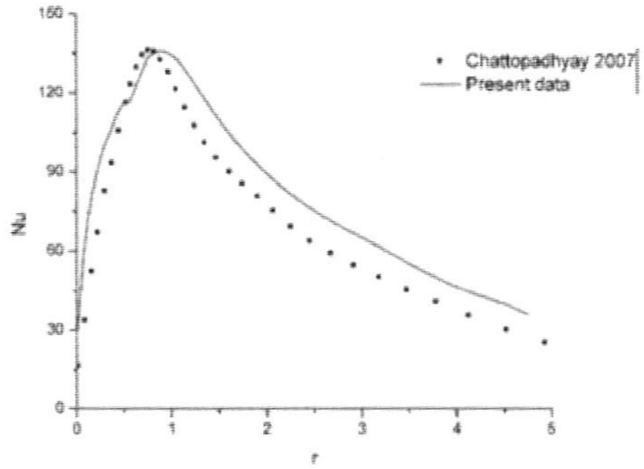

(b)

Figure 1: (a) Axisymmetric schematic diagram of circular and annular jet, (b) validation of present model with published data

The Transition SST model [11] is used in the present study to solve the governing equations; i. e, mass, momentum and energy {Reynolds-averaged Navier-Stokes (RANS) equations} with SIMPLE formulation [12] using the commercial CFD code ANSYS Fluent [13]. The detailed information about RANS equations and Transition SST model with it's additional equations for turbulence closure are accessible in ANSYS Fluent user manual [13] and not reported here.

Non-uniform structured mesh is used for simulations. Dense grids are placed near the surface of jet impinging region as well as near the jet exit. For selecting appropriate mesh, a grid independence study has been carried out with surface averaged Nusselt number for different grid size. After getting final mesh which gives grid independent results, present model & simulation setup ware validated against published data [8]. The result from validation study shows a good agreement with published data.

3. Results and Discussions

For getting a clear picture about flow structure, streamlines on symmetry mid plane passing the centre of nozzle at H=2D is shown in figure.2 for both circular and annular jet flow. In case of annular jet, first circulation was found just after the jet exit i.e, in the initial merging zone; a secondary circulation was also observed near the flat plate. On the other hand no such recirculation zone was found. This circulation sustained near the plate upto 1D in the radial direction from the center of annular jet which can be clearly seen by the radial velocity distribution plot in figure 3.

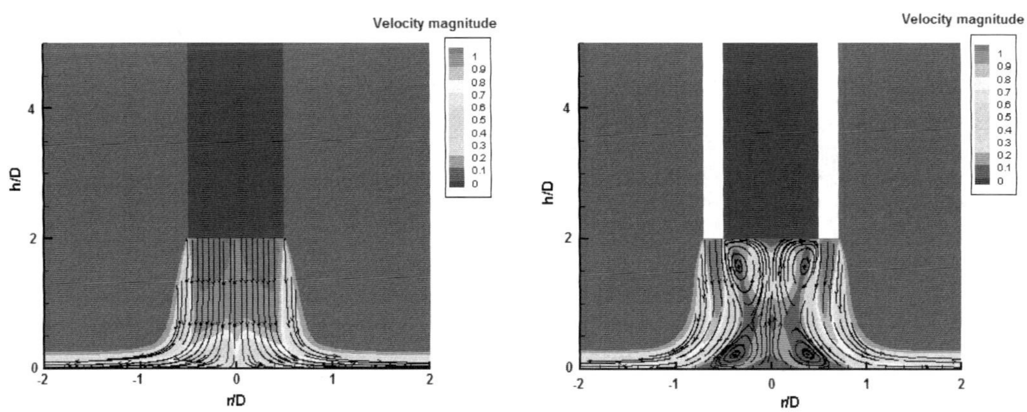

Figure 2: Streamlines of jet at symmetry mid plane, (left: circular jet; right: annular jet).

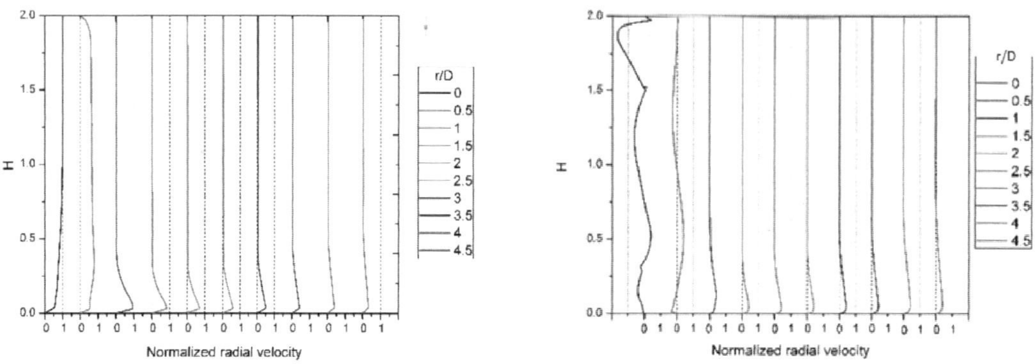

Figure 3: Compression of velocity profiles at symmetry mid plane, (left: circular jet; right: annular jet).

In previous studies [7,8] (2 dimensional cases) it was found that, the averaged Nusselt number of the annular jet is near 20% lesser than circular jet for same mass & momentum efflux condition. But three dimensionally, a clear picture of the distribution of Nusselt number on the flat impingement surface is needed for a concluding remark. For that purpose distributions of local Nusselt number on the surface of jet impingement from present numerical simulations are presented in figure 4. The peak of Nusselt number is found near the center of the plate for circular jet and it shifted away from the center of plate for annular jet. The values of area-averaged Nu at the jet height of h = 2.0, is presented in table 1 along with the previous two dimensional study results [8]. It is observed that similar to the two dimensional study, averaged Nusselt number of annular jet is lower compare to the circular jet. It is also observed that the two dimensional studies on annular jet under predicts heat transfer characteristics.

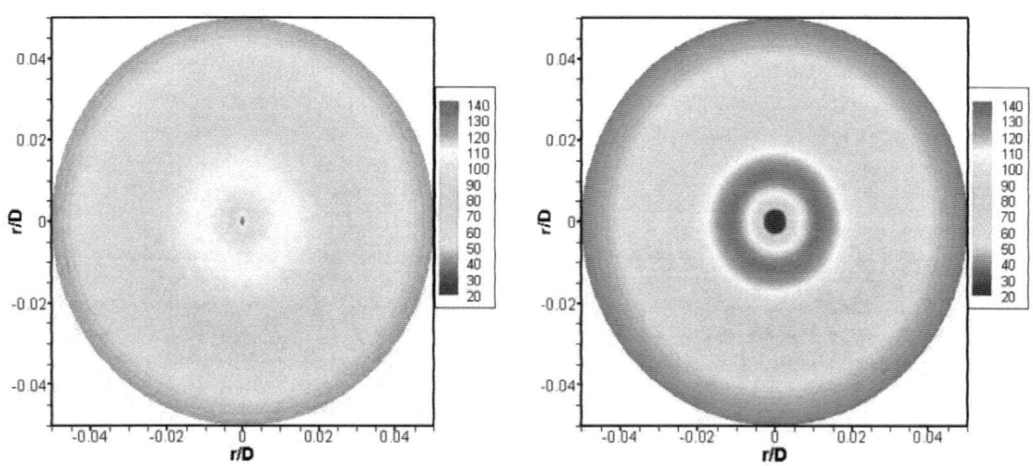

Figure 4: Compression of Nu_{avg} distribution on flat pale, (left: circular jet; right: annular jet).

Table 1: Heat transfer results

	Re	Nu_{avg} (circular jet)	Nu_{avg} (annular jet)	Difference(%)
(Chattopadhyay 2007) [8] (two dimensional)	20,000	59.61	54.85	7.98
Present study (three dimensional)	20,000	67.26	60.34	10.29

4. Conclusions

Heat transfer from both circular and annular jet keeping same mass and momentums efflux at a particular Reynolds number in the turbulent regime has been simulated using the Transition SST Model. Compressions between Three dimensional flow fields of circular jet with annular jet in terms of streamline with radial velocity profile plot shows the presences of circulation zones only for annular jet. From the heat transfer results of present study it was found that two dimensional studies on annular jet under predicts the heat transfer characteristic.

References

[1] Jambunathan K, Lai E, Moss M A and Button B L 1992 A review of heat transfer data for single circular jet impingement *Int. J. heat fluid flow* **13** 106–15
[2] Weigand B and Spring S 2011 Multiple jet impingement– a review *Heat Transf. Res.* **42**
[3] Carlomagno G M and Ianiro A 2014 Thermo-fluid-dynamics of submerged jets impinging at short nozzle-to-plate distance: a review *Exp. Therm. fluid Sci.* **58** 15–35
[4] Maki H and Yabe A 1989 Heat transfer by the annular impinging jet *Exp. Heat Transf.* **2** 1–12
[5] Ramos J I 1996 One-dimensional models of steady, inviscid, annular liquid jets *Appl. Math. Model.* **20** 593–607
[6] Shuja S Z, Yilbas B S and Budair M O 2001 Local entropy generation in an impinging jet: minimum entropy concept evaluating various turbulence models *Comput. Methods Appl. Mech. Eng.* **190** 3623–44
[7] Chattopadhyay H 2004 Numerical investigations of heat transfer from impinging annular jet *Int. J. Heat Mass Transf.* **47** 3197–201
[8] Chattopadhyay H 2007 Impinging Heat Transfer Due to a Turbulent Annular Jet. *Int. J. Transp. Phenom.* **9**
[9] Dutta P and Chattopadhyay H 2019 NUMERICAL SIMULATION OF HEAT TRANSFER DUE TO ARRAY OF LAMINAR ANNULAR JETS IMPINGING ON HOT MOVING SURFACE *PROCEEDINGS BOOK OF FIFTH INTERNATIONAL CONFERENCE ON ADVANCES IN MECHANICAL ENGINEERING: ICAME 2019* (Istanbul) pp 162–70
[10] Laschefski H, Cziesla T and Mitra N K 1997 Evolution of flow structure in impinging three-dimensional axial and radial jets *Int. J. Numer. methods fluids* **25** 1083–103
[11] Abraham J P, Sparrow E M and Tong J C K 2009 Heat transfer in all pipe flow regimes: laminar, transitional/intermittent, and turbulent *Int. J. Heat Mass Transf.* **52** 557–63
[12] Patankar S 2018 *Numerical heat transfer and fluid flow* (CRC press)
[13] Fluent A 2016 Fluent 15 users guide *Lebanon, USA*

THERMODYNAMIC ANALYSIS OF DOWNDRAFT BIOMASS GASIFIER TO STUDY THE EFFECT OF TEMPERATURE USING DIFFERENT FEEDSTOCK

Sampad Kumar Das[1], P. C. Roy[2]

[1] Department of Mechanical Engineering, Jadavpur University, Kolkata–700032, India
[2] Department of Mechanical Engineering, Jadavpur University, Kolkata–700032, India
e-mail: er.sampad1@gmail.com, prokash.roy@gmail.com

Abstract. In the present paper, effect of gasification temperature on gasification process in a downdraft gasifier using different biomass feedstock has been examined. An equilibrium model of the downdraft biomass gasifier has been developed to carry out the study. The model has been validated with experimental results. The producer gas yield has been analysed by using stoichiometric equilibrium model. The lower heating value (LHV) calculation has been done for different biomass with respect to temperature. It has been observed that the hydrogen production rate and *CO* production rate increases with respect to increase in temperature. On the other hand the Methane production decreases with increase in temperature. The lower heating value (LHV) for the studied sample has been observed to decrease with respect to the increase in gasification temperature.

Keywords: Downdraft biomass gasifier, equilibrium model, effect of temperature

1. Introduction

In the present context of world environment and need of energy, it is important to develop sustainable and renewable energy option. For such criteria of energy demand the biomass gasification is very important. Therefore, in the present study a downdraft biomass gasifier model has been considered and the gasification process has been analysed according to the schematic given in the block diagram of figure 1.

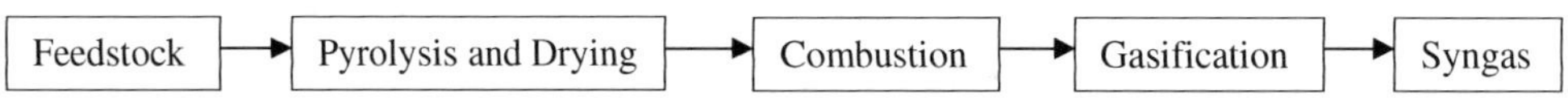

Figure 1: Block diagram of downdraft biomass gasifier for the present model of biomass gasification.

In gasification process the equivalence ratio of oxygen and gasification temperature have been found with great influence. The gasification process has been modelled in different ways in different studies [1-3]. However, for thermodynamic analysis the stoichiometric equilibrium model [4-5] and minimization of Gibbs free energy has been found convenient and reliable with respect to the producer gas yield. Therefore, in the present study the modified stoichiometric equilibrium model has been used to check the different gas yield and lower heating value (LHV) of produced gas.

2. Model description and methodology

In the present study stoichiometric equilibrium model has been applied with modified reaction constants, which vary with respect to gasification temperature. The developed model has been validated with experimental literature and then the model has been utilized to predict the composition of produced gas mainly hydrogen, carbon mono oxide and methane for different biomass feed stock. For biomass ultimate analysis Douglas fir, Paper, Maple wood and Redwood feedstock has been utilized. The ultimate analysis of this biomass has been given in the table 1.

Table 1: Ultimate analysis of biomass on dry basis (wt %)

Biomass	C	H	O	N	Ash	HHV(MJ/Kg)
Maple wood	50.6	6	41.7	0.3	1.4	19.958
Douglas fir	52.3	6.3	40.5	0.1	0.8	21.051
Paper	43.4	5.8	44.3	0.3	6.2	17.613
redwood	53.5	5.9	40.3	0.1	0.2	21.028

To run the calculation Carbon, Hydrogen, Oxygen and Nitrogen ratio along with the moisture content of the biomass has been considered as important input. The moisture content of the different biomass has been given in table 2. After biomass consideration model has been developed in section 2.1;

Table 2: moisture content of different biomass (wt % of dry biomass)

Biomass	Maple wood	Douglas fir	Paper	Redwood
MC	35	27	45	25

2.1. *Assumptions and model formulation*

The analysis has been done with the thermodynamic equilibrium model i.e. stoichiometric equilibrium model (calculation has been done on the basis of equilibrium constant and stoichimetric ratios). To develop such a model following assumptions has been made. The reaction time has been considered very high so that all reactions are in steady state and all carbon molecules converts into gas. Tar content has been neglected and ash has been considered as inert in every gasification reactions. Along with this all the produced gas has been considered as ideal gas. To analyse the process the main gasification equation has been considered in a very simplified form and it has been given in eq. (1);

$$CH_xO_yN_z+w_kH_2O+m(O_2+3.76N_2)\rightarrow \dot{N}_{H_2}H_2+\dot{N}_{CO}CO+\dot{N}_{CO_2}CO_2+\dot{N}_{H_2O}H_2O+\dot{N}_{CH_4}CH_4+\dot{N}_{N_2}N_2 \quad (1)$$

Here the unknown parameters are $\dot{N}_{H_2}$, $\dot{N}_{CO}$, $\dot{N}_{CO_2}$, $\dot{N}_{H_2O}$, $\dot{N}_{CH_4}$, $\dot{N}_{N_2}$. These values of unknown parameter will give the composition of producer gas yield. To get these unknown, the mass balance of carbon, hydrogen, oxygen and nitrogen has been used along with methanation reaction and water-gas shift reaction.

$$\dot{N}_{CO}+\dot{N}_{CO_2}+\dot{N}_{CH_4}=1 \quad (2)$$

$$2\dot{N}_{H_2}+2\dot{N}_{H_2O}+4\dot{N}_{CH_4}=x+2w_k \quad (3)$$

$$\dot{N}_{CO}+2\dot{N}_{CO_2}+\dot{N}_{H_2O}=y+w_k+2m \quad (4)$$

$$2\dot{N}_{N_2}=z+7.52m \quad (5)$$

$$w_k=\frac{M_B\times MC}{(1-MC)\times 18} \quad (6)$$

$$m=ER(1+0.25x-0.5y) \quad (7)$$

The value of different parameter of eq. (6 and 7) has been calculated from the ultimate analysis of different biomass feedstock.

$$C+CO_2\rightarrow 2\,CO \quad (8)$$

$$C+H_2O\rightarrow CO+H_2 \quad (9)$$

The first two reactions shown above eq. (8 and 9) can be combined into eq. (10) known as water-gas shift reaction:

$$CO+H_2O\rightarrow CO_2+H_2 \quad (10)$$

$$C+2H_2\rightarrow CH_4 \quad (11)$$

The main reactions in gasification process have been considered as water-gas shift reaction eq. (10) and methanation reaction eq. (11). The reaction rate constants have been given in eq. (12-17);

$$K_{mn} = \frac{\dot{N}_{CH_4}}{(\dot{N}_{H_2})^2} \quad (12)$$

$$K_{wgs} = \frac{\dot{N}_{H_2} \times \dot{N}_{CO_2}}{\dot{N}_{CO} \times \dot{N}_{H_2O}} \quad (13)$$

It has been found that the equilibrium constants (K_{mn} and K_{wgs}) are functions of reaction temperature T and Gibbs free energy ($\Delta g^0_{mn}, \Delta g^0_{wgs}$) of the corresponding reactions. On the other hand the Gibbs free energy has been found as a function of reaction temperature. Therefore the reaction constants have been calculated from the gasification temperature by using the simplified reaction rate constant equation of Zainal *et al.* [5].

$$K_{mn} = exp\left(-\frac{\Delta g^0_{mn}}{RT_G}\right) \quad (14)$$

$$K_{wgs} = exp\left(-\frac{\Delta g^0_{wgs}}{RT_G}\right) \quad (15)$$

$$K_{mn} = exp\left(\frac{5878}{T} + 1.86 \ln T - 0.27 \times 10^{-3} T - \frac{58200}{T^2} - 18\right) \quad (16)$$

$$K_{mn} = exp\left(\frac{7082.842}{T} - 6.567 \ln T + \frac{7.467 \times 10^{-3}}{2} T - \frac{2.167 \times 10^{-6}}{6} T^2 + \frac{0.702 \times 10^{-5}}{2T^2} + 32.541\right) \quad (17)$$

The six eq. (2, 3, 4, 5 and 12, 13) has been solved by the standard Newton-Raphson method. After calculation of six unknown $\dot{N}_{H_2}, \dot{N}_{CO}, \dot{N}_{CO_2}, \dot{N}_{H_2O}, \dot{N}_{CH_4}, \dot{N}_{N_2}$ the lower heating value (LHV) of the producer gas has been calculated from eq. (18)

$$LHV = 10.78 \times \% H_2 + 12.63 \times \% CO + 35.88 \times \% CH_4 \ \ [MJ/m^3] \quad (18)$$

3. Model validation

The produced hydrogen gas percentage from model calculation has been compared with experimental results of Sittisun et al. [6]. and the root mean square error (RMSE) has been calculated to check the deviation. The root mean square error has been found as 5.92 for present study. To compensate the non-equilibrium conditions present in gasification, the reaction constants K_{mn} and K_{wgs} have been multiplied by non-equilibrium coefficients 20 and 0.91 for minimization of errors. In several literatures such method has been applied to develop their models and resulted in better prediction [6]. The result of present study as compared with the experimental result has been given in figure 2.

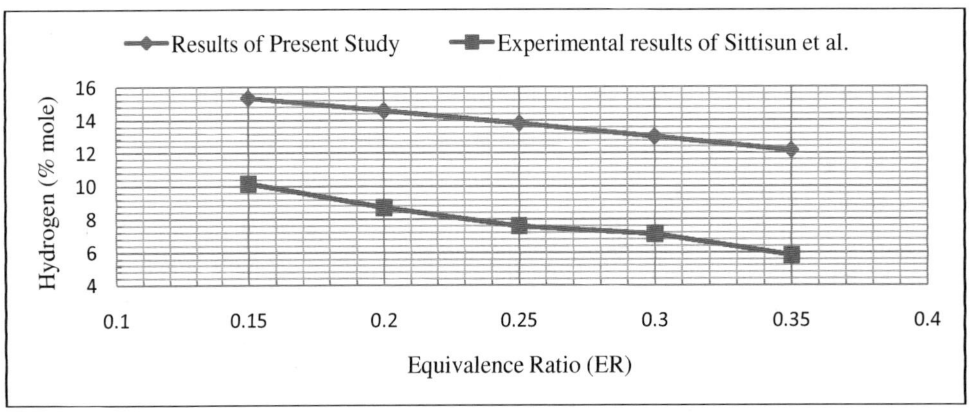

Figure 2: Hydrogen production with respect to the equivalence ratio.

4. Results and Discussions

The percentage of H_2, CO and CH_4 has been calculated using the developed model based on the thermodynamic equilibrium model and stoichiometric equilibrium model. Figure 3 represents the gas production percentage and LHV with respect to the temperature for different biomass.

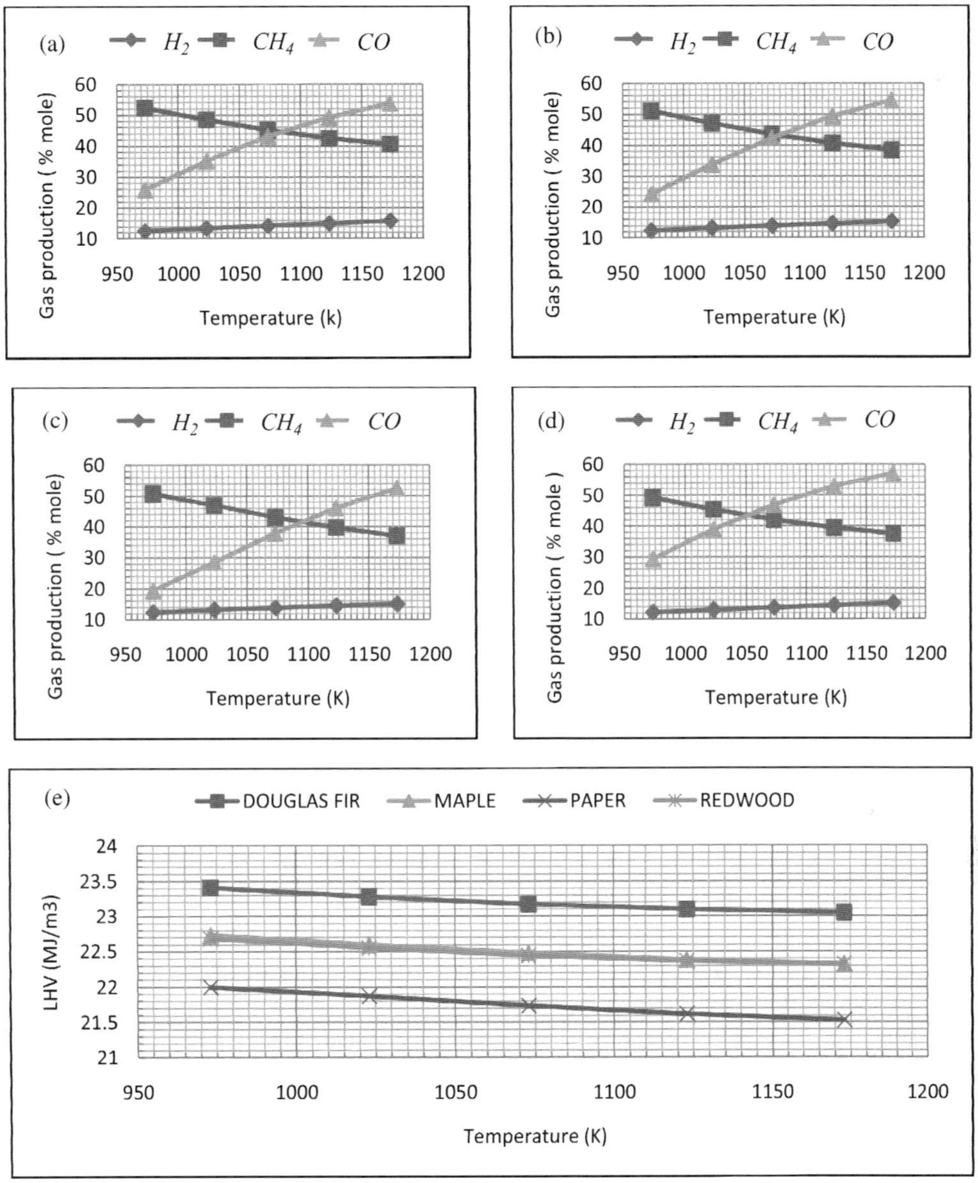

Figure 3: Effect of gasification temperature on gas production for (a) Douglas fir (b) Maple (c) Paper (d) Redwood (e) Effect of gasification temperature on LHV of produced gas.

Composition analysis of different gas with respect to temperature has been done. The equivalence ratio has kept fixed as 0.25 and the temperature have varied from 973K to 1173K.The LHV calculation has also been made for different biomass with respect to temperature. It has been observed that the hydrogen production rate and *CO* production rate increases with respect to increase in temperature. On the other hand the Methane production decreases with increase in temperature. The LHV for the studied sample have found to decrease with respect to the increase in temperature.

5. Conclusions

From the present work it can be concluded that the LHV for different biomass is decreasing with temperature,

- Because there is less production of CH_4 at higher temperature and CH_4 contributes a large factor nearly 40 times of its percentage in the produced gas for the calculation of LHV. This is due to the fact that with the increase in temperature the methanation reaction rate constant has decreased sharply from 33.45 to 16.33, as a result less reaction took place producing less CH_4.
- On the other hand the H_2 and CO production rate increases because with the increase in temperature the oxidation rate of carbon to CO_2 has increased rapidly due to less requirement of reaction activation energy along with more char and moisture conversion to CO_2 and H_2O.
- But the increase in H_2 and CO is not sufficient with respect to the decrease in CH_4 production. More precisely their contribution to LHV is nearly 10-12 times of their percentage in produced gas, which is comparatively very less than that of CH_4 contribution (nearly 40 times of its percentage in total gas). Hence overall LHV is decreasing with increase in temperature.

Acknowledgement: Authors acknowledge the funding agency **SERB (DST, GOI)** for financial support to carry out this research.

References

[1] Jarungthammachote S and Dutta A 2007 Thermodynamic equilibrium model and second law analysis of a downdraft waste gasifier *Energy* **32(9)** 1660-1669
[2] Barman N S Ghosh S and De S 2012 Gasification of biomass in a fixed bed downdraft gasifier a realistic model including tar *Bioresource technology* **107** 505-511
[3] Janajreh I and Al Shrah M 2013 Numerical and experimental investigation of downdraft gasification of wood chips *Energy Conversion and Management* **65** 783-792
[4] Mendiburu A Z Carvalho Jr J A Zanzi R Coronado C R and Silveira J L 2014 Thermochemical equilibrium modeling of a biomass downdraft gasifier constrained and unconstrained non-stoichiometric models *Energy* **71** 624-637
[5] Zainal Z A Ali R Lean C H and Seetharamu K N 2001 Prediction of performance of a downdraft gasifier using equilibrium modeling for different biomass materials *Energy conversion and management* **42(12)** 1499-1515
[6] Sittisun P Tippayawong N and Pang S 2019 Biomass gasification in a fixed bed downdraft reactor with oxygen enriched air a modified equilibrium modeling study *Energy Procedia* **160** 317-323

Nomenclature

x	Hydrogen to carbon mole ratio in dry biomass
y	Oxygen to carbon mole ratio in dry biomass
z	Nitrogen to carbon mole ratio in dry biomass
w_k	Number of water molecule in dry biomass
m	Number of oxygen molecule
M_B	Mass of dry biomass
MC	Moisture content

EXPERIMENTAL INVESTIGATION ON A DOWNDRAFT BIOMASS GASIFIER USING TREE WASTE AND SAWDUST PELLET

A. Dutta[1], Sampad Kumar Das[2], P. C. Roy[3*]

[1]Department of Mechanical Engineering, Jadavpur University, Kolkata–700032, India
[2]Department of Mechanical Engineering, Jadavpur University, Kolkata–700032, India
[3]Department of Mechanical Engineering, Jadavpur University, Kolkata–700032, India
e-mail: er.sampad1@gmail.com, prokash.roy@gmail.com*

Abstract. In this article, investigation and analysis of an experiment for downdraft biomass gasification has been conducted using two different feed stocks namely tree waste and sawdust pellet at different equivalence ratio. The different gas yield due to gasification mainly hydrogen, carbon monoxide, methane, carbon dioxide and nitrogen have been observed by the use of online gas analyser. The equivalence ratio of air has been varied from 0.3 to 0.4 for both of the biomass feedstock. On the basis of producer gas yield, the LHV of producer gas have also been calculated for different equivalence ratio and for both of the biomass. After that the LHV has been compared between the two biomass feedstock for equivalence ratios of 0.3, 0.33 and 0.4. The trend of LHV on the basis of different equivalence ratio has been examined and plotted.

Keywords: Downdraft biomass gasifier, Experimental Investigation, Effect of equivalence ratio

1. Introduction

The growing shortage of fossil fuels accompanied with increasing concerns about the environmental impacts resulting from combustion of such fuels has resulted in the adoption of renewable technologies throughout the world. Biomass gasification is one of the most promising technologies to provide sustainable energy solutions. Biomass can be defined as all the biologically produced matter. Biomass energy is harnessed from all the plant sources like wood from natural forest reserves and agricultural outputs. The energy generated utilizing biomass is renewable in nature and does not add to the carbon dioxide level in the atmosphere. Gasification is a process of conversion of solid bio-fuels into producer gas with a heating value. During the process of gasification the temperatures are maintained at a higher level so as to optimize the gas production. The producer gas comprises of carbon monoxide, hydrogen, methane, carbon dioxide and nitrogen.[1,2,3]Apart from the fact that it is carbon neutral, the advantages of the utilization of biomass is that it is readily available everywhere and is very suitable for distributed generation purpose using locally available feedstock. Biomass gasification when applied to power generation in distributed generation mode helps to achieve environmental impact mitigation, gainful utilization of domestic un-utilized resources, savings in economy, reduced energy dependency on grid, boost to agriculture, creation of local employment, replacement of fossil fuels and improving the energy security in areas located far away from the national grid. Among all the gasifiers available for the thermo-chemical conversion of biomass, the downdraft gasifier [4] is found to be the most suitable option for small scale power generation due to its simple construction, proven technology, easier operation and ease of maintenance. The carbon conversion is also high combined with higher char and tar conversion rates. Downdraft gasifier has good fuel flexibility along with coupling up options with producer gas engines, solid oxide fuel cell and gas turbines. Due to high residence time of solids, the overall carbon conversion is higher. However to generate producer gas with optimum quality for turbines and internal combustion engines, it is pertinent to carry out proper selection of various parameters like gasifier type (updraft, downdraft), gasifier design, gasifying agent (air, steam), temperature range and equivalence ratio (ER). With an increased thrust placed on generation of sustainable energy means and reduction of carbon footprint, it is pertinent to focus our attention on further research and development of biomass gasification techniques and up gradation options so as to ensure a better future for the generations to come. Biomass energy has the potential to play a pivotal role in addressing the environmental concerns and energy security issues [5,6]. Many researchers throughout the world have carried out characterization studies on downdraft gasifier on the pilot as well as laboratory installations. Dong et al.[7]studied the variance of gas composition and tar generation by the varying equivalence ratio (ER) and biomass (corn stalk) feed rate in a three air staged fixed bed downdraft reactor. With biomass feed rate of 7.5 kg/h and ER at 0.25-0.27, the LHV (lower heating value) of the producer gas was 5400 kJ/Nm3 and cold gas efficiency was 65%. At increased ER of 0.32, the tar yield was 0.52 g/Nm3.With ER at 0.27, an increase in biomass feed rate from 5.8 kg/h to 9.3 kg/h, the LHV decreased from 5455.5 kJ/Nm3 to 5253.2 kJ/Nm3 and tar yield increased from 0.82 g/Nm3to 2.78g/Nm3. Ram et al. [8] designed and developed a 75 kWe downdraft gasifier integrated Internal Combustion (IC) engine with the key target to optimize overall system performance through proper selection of ER, increasing the biomass to gas conversion efficiency by minimizing the charcoal removal rate by incorporating a vibrating ash removal grate. Waste heat recovery system was also adopted comprising of heat exchanger and blower in order to supply pre-heated air into the gasifier so as to minimize the heat loss of the system and improve overall thermal performance. The power output varied from 65 kWe and 73 kWe. At ER of 0.35, the heating value of the syngas was 5.7 MJ/Nm3.Generally it has been observed from literature, gasifiers are designed for a specific biomass to be utilized in the gasification process. But in the present work, experimental investigation has been carried out in a downdraft gasifier using different biomass wastes. Effect of equivalence ratio on producer gas composition and its heating value have been reported and discussed.

2. Experimental Set up and Methodology

A complete experimental set up consist of gasifier unit along with its cleaning and drying system and associates accessories shown in Fig.1 (a&b).

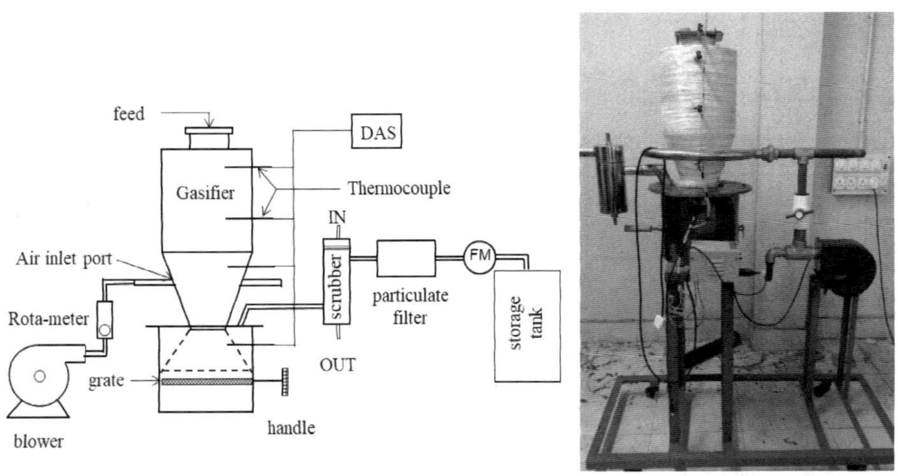

Fig. 1. (a) Schematic of experimental setup and (b) Photographic view of the experimental setup.

Two types of biomass feedstock have been utilized in the present work: one is tree waste and another one wood waste (sawdust) shown in Fig. 2. Tree waste has been collected from Jadavpur University campus as twig and branches fallen from trees. Twig and branches are cut into proper size suitable for gasifier unit shown in Fig. 2 (a). Sawdust has been collected from nearby sawmills. But saw dust received from sawmills are not suitable for downdraft gasifier application due to its lower size. Saw dust pellet has been prepared by biomass pellet machine to provide required shape and size shown in Fig. 2(b). Proximate and ultimate analyses of the sample have been done. Proximate and ultimate analyses of the sample are presented in the table 1.

Fig. 2. (a) Tree waste

Fig.2.(b) Sawdust Pellet

Biomass Feedstock	Proximate Analysis,% Dry basis			Ultimate Analysis, % by weight			
	Volatile Mater	Fixed Carbon	Ash	C	H	N	O
Tree Waste	80.2	15.3	4.5	45.1	6	1.6	41.6
Sawdust Pellet	81.4	15.1	3.5	44.5	5.5	1.8	40.4

Table 1. Proximate and ultimate analysis of biomass feedstock.

3. Results and Discussions

Experiments have been carried out at different equivalence ratio by measuring the air supply using Rota meter. After stabilizing the gasifier run composition of the producer gas has been measured using an online producer gas analyser after successive intervals. Composition of the producer gas has been shown in Table 2.

Biomass Feedstock	Equivalence Ratio	Producer gas compositions, %				
		CO	H2	CO2	CH4	N2
Tree Waste	0.3	17.5	16.5	11.5	0.85	53
	0.33	18.1	15.6	10.6	1.2	52
	0.4	18.8	15.2	10.3	1.3	51
Sawdust Pellet	0.3	16.5	15.5	12.5	1.1	54
	0.33	17.1	15.1	11.6	1.2	53
	0.4	17.8	14.3	10.7	1.2	52

Table 2. Producer gas compositions at different equivalence ratio during trial run

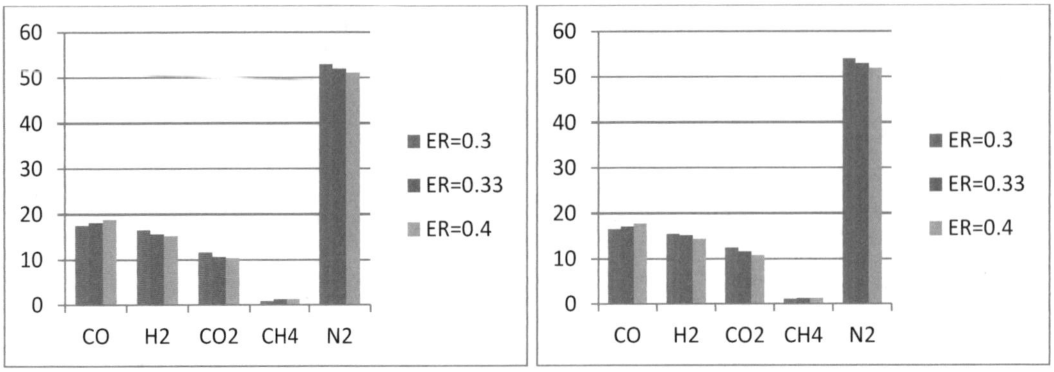

(a) Tree waste (b) Saw dust Pellet

Figure 3: Effect of ER on Producer gas composition for tree waste and sawdust pellet.

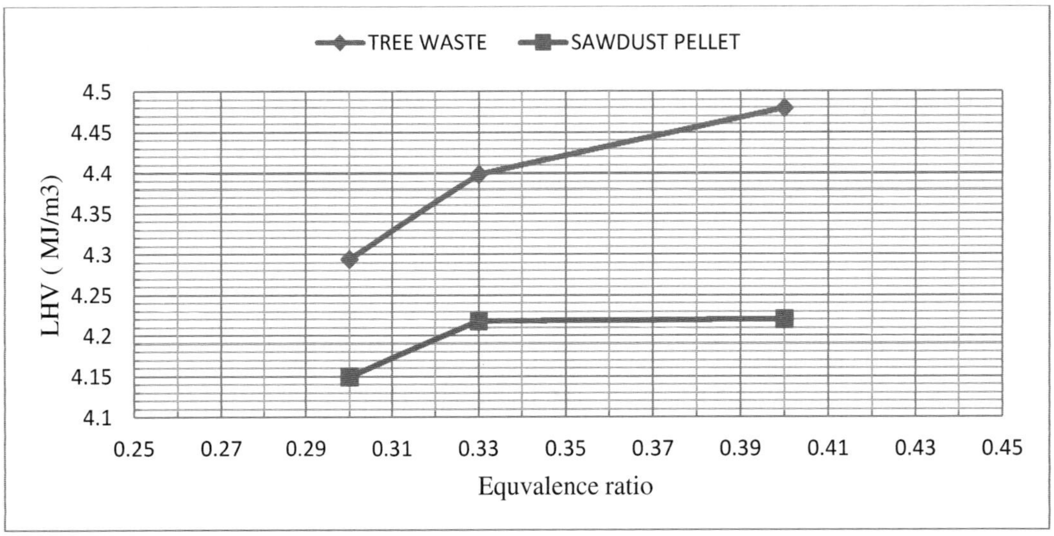

Figure 4: Effect of ER on LHV for tree waste and sawdust pellet.

The average of LHV for producer gas in the present study has been found 4.19 MJ/m3 on the other hand in the study of Sittisun et al. [10] the LHV of producer gas using same sawdust has been found 4.06 MJ/m3. Hence it can be concluded that the present study has found standard results as compared with the established literature. Through the experimental results it has been observed that the CO yield is more with the increase of ER from 0.3 to 0.4 for both of the feedstock. In case of tree waste it is ranging from 17.5 to 18.8 percent of the producer gas and in case of saw dust pellet it is 16.5 to 17.8 percent of the producer gas. On the other hand the production of H2 has decreased with respect to the increase in ER. For tree waste the percentage decreases from 16.5 to 15.2 and for saw dust pellet it is from 15.5 to 14.3 percent. For CH4 the production rate very nominal, it is increasing but the magnitude of percentage and increase is not that much prominent as it has been observed for the CO. The CH4 percentage in case of tree waste is ranging from 0.85 to 1.3 and for saw dust it is 1.1 to 1.2. Therefore an increasing effect has been observed in the trend of LHV for both of the gas. Here the percentage of CH4 production has played a very important role in the calculation of LHV. By the presented result the dominance of CO percentage in the gas composition is observed as compared to the H2 and CH4 production.

4. Conclusions

In this experimental study following observation has been made along with their conclusions.

- It is observed that the LHV of producer gas is increasing for tree waste sample with respect to the equivalence ratio. This kind of observation may be due to the fact of the increasing percentage of CO production as compared with the H2 and CH4 in the produced gas from tree waste sample, which has contributed nearly 12.63 times of its percentage in producer gas to LHV.
- On the other hand LHV for sawdust pellet has increased for the equivalence ratio 0.3 to 0.33 and remains constant form 0.33 to 0.4; This may be due to the same amount of CH4 nearly 1.2 percent of producer gas and decrease in H2 percentage, though there is an increase in CO percentage.
- Therefore on the basis of producer gas yield it is clear the percentage of CH4 and CO yield is the important contributor to the calculation of LHV. It is also concluded that the LHV is in the very near range of other experimental results of different literature.

Acknowledgement: Authors acknowledge the funding agency SERB (DST, GOI) for financial support to carry out this research.

References

[1] Ramadhas A S Jayaraj S and Muraleedharan C 2006 Power generation using coir-pith and wood derived producer gas in diesel engines *Fuel Processing Technology* **87(10)** 849-853
[2] Balat M Balat M Kırtay E and Balat H 2009 Main routes for the thermo-conversion of biomass into fuels and chemicals part 2 Gasification systems *Energy conversion and management* **50(12)** 3158-3168
[3] Balat M 2009 New biofuel production technologies *Energy Educ Sci Technol Part A* **22(2)** 147-61
[4] Buragohain B Mahanta P and Moholkar V S 2010 Thermodynamic optimization of biomass gasification for decentralized power generation and Fischer Tropsch synthesis *Energy* **35(6)** 2557-2579
[5] Xu R Ferrante L Briens C and Berruti F 2009 Flash pyrolysis of grape residues into biofuel in a bubbling fluid bed *Journal of Analytical and Applied Pyrolysis* **86(1)** 58-65
[6] Kobayashi N and Fan L S 2011 Biomass direct chemical looping process a perspective *Biomass and Bioenergy* **35(3)** 1252-1262
[7] Guo F Dong Y Dong L and Guo C 2014 Effect of design and operating parameters on the gasification process of biomass in a downdraft fixed bed an experimental study *International journal of hydrogen energy* **39(11)** 5625-5633
[8] Raman P and Ram N K 2013 Design improvements and performance testing of a biomass gasifier based electric power generation system *Biomass and bioenergy* **56** 555-571
[9] Gai C and Dong Y 2012 Experimental study on non-woody biomass gasification in a downdraft gasifier *International Journal of hydrogen energy* **37(6)** 4935-4944
[10] Sittisun P Tippayawong N and Pang S 2019 Biomass gasification in a fixed bed downdraft reactor with oxygen enriched air a modified equilibrium modeling study *Energy Procedia* **160** 317-323

IOP Conf. Series: Materials Science and Engineering 1080 (2021) 012034 doi:10.1088/1757-899X/1080/1/012034

Waste Cooking Oil Biodiesel: Its Testing, Performance and Emission in an Unaltered Diesel Engine

Abhishek Samanta[1], Anindya Adhikary[2]and Prokash C Roy[2*]

[1] Department of Automobile Engineering, Ranaghat Govt. Polytechnic, West Bengal.
[2] Department of Mechanical Engineering, Jadavpur University, Kolkata–700032, India
e-mail: *prokash.roy@gmail.com**

Abstract. Rapid climate changes with global warming are the primary dispute of the recent era cause of diesel engine emission. Among the probable fuels, biodiesel is the most promising one as the alternative of petroleum diesel. Biodiesel can be formed from residue of cooking oil by transesterification. The reaction rate can be increased with alkali Catalyst at a 55℃ temperature. The actions of waste cooking oil biodiesel (WCOB) compared with petroleum diesel were performed to observe the nature of an unaltered stationary diesel engine. In this study the specific fuels are: B10 biodiesel-diesel blend contained 10% WCOB and petroleum diesel (B0).The performance of an unaltered diesel engine fuelled with WCOB blend is examined in this study. In order to find out the performance results constant engine speed (1500 rpm) is being maintained during the experiments under 0%, 25%, 50% and 100% load conditions. It has been observed that when the engine setup was fuelled with WCOB blend B10, the performance depleted at mid to top load conditions but no drastic reduction in emission.

1. Introduction

The growing modernisation and industrialisation in the world led to the rise in demand and consumption of petroleum based fuel more and more. This causes air pollution and consequently accumulation of green house gases into the atmosphere. Two kinds of internal combustion engine e.g. petrol engine and diesel engine contribute for growing the air quality in the urban area. There are 22% of green house gas emits from the transportation sector for increasing demand of vehicles. IEA renowned as the international Energy Agency forecasts that the emission of carbon dioxide from transportation sector will increase by 92% between 1990 to 2020.It is also reported that average global temperature is hiked by 20 centigrade during this period [1].Again the petroleum based fuels have limited reserve potent in specific sector of the globe. Many research experts says that the supply of petroleum based commodity can be drained within 40 to 60 years and also their increasing price make alternative fuel more and more attractive. The form of renewable energy is hydropower, wind power, solar energy, and biomass and bio fuels. The contribution of all these resources is important because of the economic and environmental reasons, and biodiesel could be one of the solutions. In an article, it indicates that India can produce about 30-40% and rest 70% petroleum is imported, which will cost around Rs.10,00000 million per year. On the other hand 5% biodiesel blend can save Rs.40,000 million per year [2].Typically in the production of biodiesel edible vegetable oils are used [3, 4]. In modern era waste cooking oil (WCO) was targeted for its low cost and disposal problems. The primary obstacle for neat WCO is viscosity when applied directly in unaltered engines [5, 7], also blockage in the delivery valve of fuel line, poor atomization and incomplete combustion are the associated drawbacks [8]. To overcome this problem transesterification process is used with catalyst to yield biodiesel from neat WCO. The neat WCO reacts with methanol and converts into fatty acid methyl esters having more influence and realized as 'Green' to the environment compared to fossil fuels. Not merely, fuel recuperated from waste oil could balance the energy demand but also it could provide a better waste management solution. Addition to that WCO transformation into biodiesel is renewable and eco-friendly, where throwing it away without taking any measurement is a serious hazard to the environment [9].WCO can be reprocessed into profitable commodity as glycerol, soap, biodiesel etc. A huge amount of WCO can be possessed from various food chain industry, restaurants and hotels. In India a single unit of renowned food chain centre can produce 15 liters of WCO per day [10].In this study WCO was collected from local guest house kitchen at Jadavpur University and converted into biodiesel (WCOB) by transesterification process. The produced biodiesel yielding was optimized by varying catalyst concentration, reaction temperature and molar ratio of the reactant. The fuel properties were also tested (Table 3) for the requirement as engine fuel. Outcome of fuel blend B10 WCOB at the zero load conditions and higher load conditions has been analyzed with reference to the performance and emission.

2. Problem description and solution procedure

In this study biodiesel was produced by 'Lab scale batch reactor' method, the WCO was collected from local guest house kitchen at Jadavpur University. With the reagents as waste cooking oil, methanol, base catalyst (KOH) and distilled water, biodiesel was prepared by following the procedure of treatment of raw materials, alcohol-catalyst mixing, chemical reaction, separation of the reaction products, purification of the reaction products and removal of moisture. Depending on the optimal results by varying catalyst concentration, reaction temperature and molar ratio the yield percentage were observed at catalyst concentration (0.5-1.5 %), reaction temperature (50-65℃) and molar ratio (1:8 to 1:15). Single cylinder VCR (variable compression ratio), 4- stroke DI (direct injection) diesel engine is used to find out combustion fluctuation and nature of performance with Diesel and B10 WCOB biodiesel blends. An eddy current dynamometer is used (Technomech, 10 BHP) for loading the test engine through a controller (230V, AX155, Apex). Combustion pressure, mass burn fraction is detected by a pressure sensor (Piezotronics, S111A22, up to 5000psi). A data retrieval system (NI USB-6210, 16-bit, 250kS/s.) collects spontaneous temperature, mass burn fraction, pressure, air flow, fuel flow & volume information from the engine cylinder. All the engine cylinder information is monitored using a real time software interface. During experiments test engine is being operated at 1500 rpm & 18:1 compression ratio but at particular loads (0 %, 25%, 50%, 75% & 100%). A gas analyzer is used to measure the emission of test fuel (Manatec make Eco Gas 100).

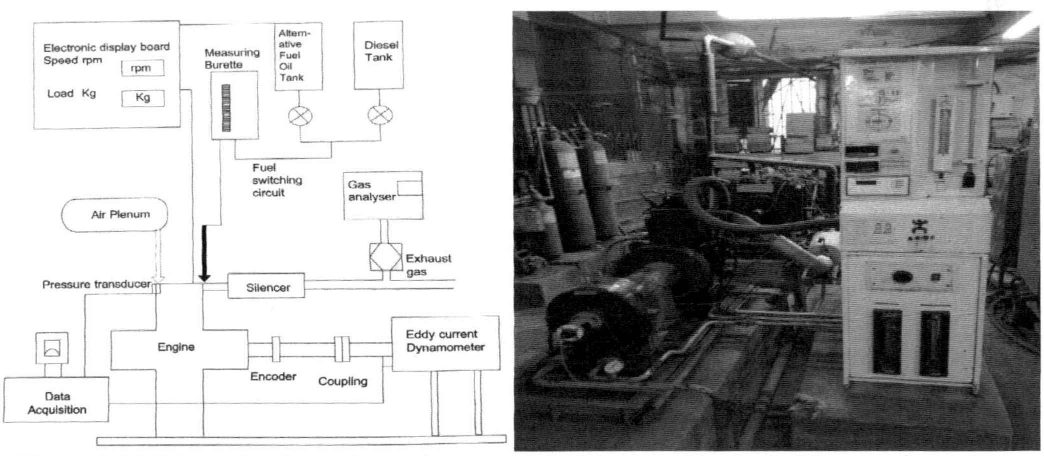

Figure 1.(a) Schematic of experimental setup and **(b)** Photographic view of the experimental setup.

Table 1: Experimental setup specifications

Diesel Engine specifications	Make and model	Make Kirloskar, Model TV1, 1cylinder, 4 stroke.
	Bore X Stroke	87.5 mm X 110 mm
	Compression ratio	18
	Max power,	3.6 kW
	Operating speed	1500 rpm
	Injection type & timing	Direct Injection at 23°CA BTDC
Gas Analyzer specifications	Make and model	Make Manatec, Model Eco Gas 100.
	Technology & Operating temperature	Non dispersive Infrared (NDIR) technology, + 5℃ to + 45℃
Measurement Parameters	CO (Carbon Monoxide)	Range: 0 - 15% , Resolution: 0.01%
	HC (Hydro Carbon)	Range: 0 - 20000 PPM, Resolution: 1 PPM
	NOx (Oxides of Nitrogen)	Range: 0 - 5000 PPM, Resolution: 1 PPM

3. Results and Discussions

Optimum reaction parameters for biodiesel production from waste cooking oil on a batch scale reaction method have been shown in Table 2.Yielding of biodiesel is specified as the ratio of actual pure biodiesel produced from the amount of oil used interaction. The yield percentage is dependent on many factors such as methanol against oil molar ratio, reaction temperature, amount of catalyst and speed used in experiment. The optimal yielding was obtained at 1% Concentration, 55℃ Reaction Temperature and 1:8 Molar ratio of the reactant. The performance characteristics of B10 WCOB & B0 (Diesel) fuels blends have been tested in a single-cylinder (661.45cc), four-stroke, direct-injection, compression-ignition engine (3.50 kW @ 1500) experimental setup coupled with a dynamometer (TMEC10).All the performance and emission results were collected at 18 compression ratio. Relevant engine performance criterion like Brake power, BTE (brake thermal efficiency) and BSFC (brake specific fuel consumption) have been examined at various load percentage at 1500 rpm.

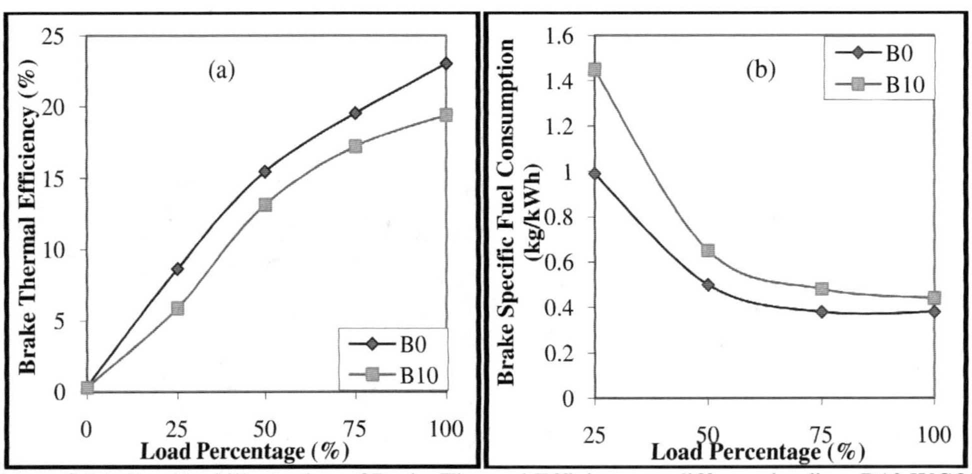

Figure 2.(a) Graphical illustration of Brake Thermal Efficiency at different loading, B10 WCOB biodiesel blend & B0 Diesel, **(b)** Graphical illustration of brake specific fuel consumption at different loading, B10 WCOB biodiesel blend & B0 Diesel at compression ratio:18.

The deviation in BTE for the test fuels are illustrated in figure 2.(a). High BTE for petroleum diesel at mid to high engine loads conditions has been observed as 15.46 % for B0 and 13.14% for B10, also 23.05% for B0 and 19.39% for B10 respectively. It is natural as the petroleum diesel has better spray atomization characteristics which lead to enhanced combustion compared to B10 blend. Though at low engine load conditions BTE for B10 blend has similar trends compared to petroleum diesel, this may be the reason of additional lubricity and presence of oxygen in biodiesel.

Table 2: Yield of biodiesel from WCO with change in reaction parameters

Catalyst Concentration (%)	Reaction Temperature (℃)	Reaction Time (hr)	Molar Ratio	Speed (rpm)	Yield (%)
1	50	1	1:6	1000	87.5
	55	1	1:6	1000	89.4
	60	1	1:6	1000	92
	65	1	1:6	1000	82
	55	1.5	1:8	1000	93.2
	55	1.5	1:10	1000	90
	55	1.5	1:12	1000	88.5
	55	1.5	1:15	1000	88

In figure 2.(b), it was noticed that as the engine load raised BSFC for test fuels fall off. Marginal differences were observed in BSFC for all the test fuels at mid to high engine loads conditions as 0.65 kg/kWh for B10 and 0.5 kg/kWh for B0, also, 0.44 kg/kWh for B10 and 0.38 kg/kWh for B0 respectively. This may be the reason of combined effort of greater density, low calorific value and viscosity of WCOB.

In figure 3.(a), it is noticed that as the engine load was raised, the carbon monoxide (CO) emission was decreased then lastly increased. At low engine load condition the cylinder temperature is low which leads to inadequate combustion and comparatively high CO emission. During the gradual increase in engine load conditions, the cylinder temperature also increases which provides better combustion and leads to lower CO emission as the oxidation rate of CO is enhanced at relatively higher temperature. At peak engine load condition the amount of fuel entering the cylinder is very high, thus leads to drastic increase in CO emission for all the test fuels. It is observed that at peak engine load condition CO emission for biodiesel is marginally high; this may be the reason of poor atomization, irregular air-fuel mixing.

Table 3: properties of produced waste cooking oil biodiesel at optimal yielding conditions

Property	Unit	Experimentally determined value
Density at 200C	gm/cc	0.964
Acid Value	mg KOH/gm	0.898
Free Fatty Acid	Wt %	0.451
Saponification Value	mg KOH/gm	185.13
Molecular weight	gm/mol	913.52
Heating value	MJ/Kg	38

In figure 3.(b), it is noticed that as the engine load was raised Hydrocarbon (HC) emission was increased with load as the fuel quantity was increased. Though HC emission was marginally low compared to petroleum diesel; this may be the reason of oxygen presence in fuel.

Oxides of nitrogen (NOx) emission depend on combustion temperature and duration, along with oxygen presence. The deviations in NOx emission at various engine load conditions are presented in figure 4 for the test fuels. Initially NOx emission was high for B10 blend due to high combustion duration and pressure which leads to high combustion temperature. However, at peak load condition

NOx emission for B10 blend was marginally low compared to B0 petroleum diesel, this may be the reason of low calorific value of WCOB which leads to low combustion temperature.

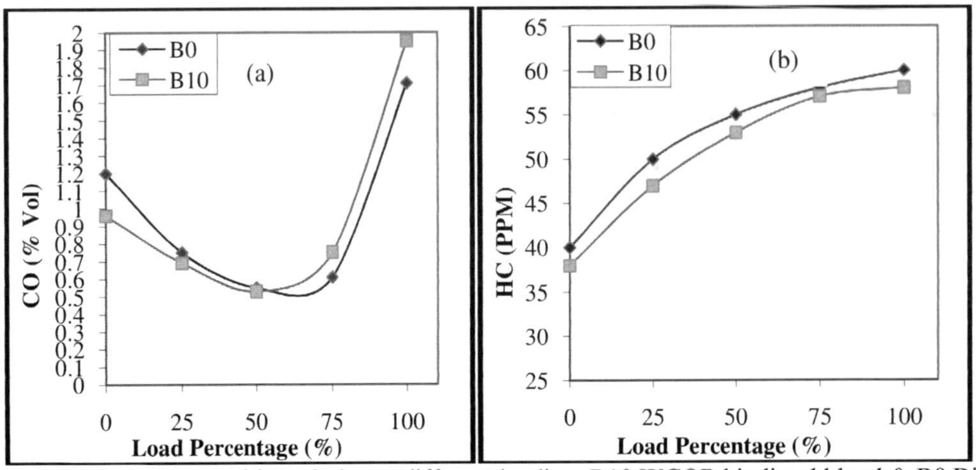

Figure 3.(a) Carbon monoxide emission at different loading, B10 WCOB biodiesel blend & B0 Diesel **(b)**Unburnt hydrocarbon emission at different loading, B10 WCOB biodiesel blend & B0 Diesel

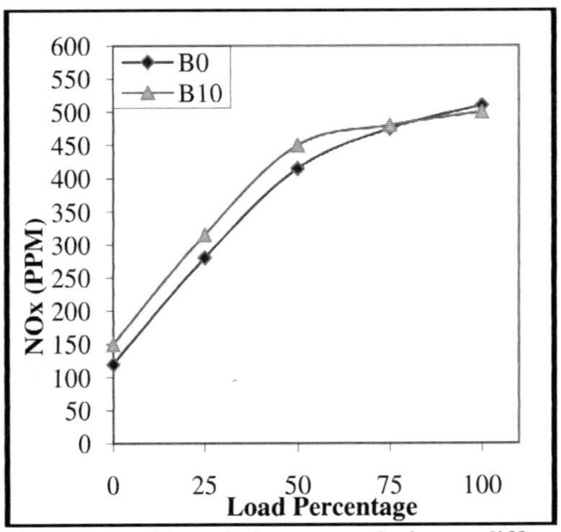

Figure 4. Graphical illustration of oxides of nitrogen emission at different loading, B10 WCOB biodiesel blend & B0 Diesel

4. Conclusions

In this study, it has been identified that waste cooking oil could be a sustainable source of waste cooking oil biodiesel. Optimization has been done to increase the production and maintain the quality of biodiesel. The performance study indicates WCOB biodiesel blend B10 can be used to reinstate petroleum diesel at low engine load conditions. Although there are concerns regarding poor performance criterion at mid to high engine loads. The viscosity variation, poor atomization and low heating value of fuel in WCOB blends affects the performance quality for B10 blends compared to Diesel (B0). The main conclusions are summarized as follows:

- Optimum yielding of biodiesel is about 93.2% for WCOB when the average heating value is about 38 MJ/Kg.
- Results reveal that quality of performance the sink around 15% in mid to high engine loads conditions compared to petroleum diesel fuel in terms of BTE (brake thermal efficiency).

- Marginal differences were observed in BSFC for B10 blend compared to petroleum diesel B0 at mid to high engine loads conditions.
- It is observed that at peak engine load condition CO emission for B10 biodiesel blend is marginally high about 14% compared to diesel.
- HC emission was marginally low about 6% compared to petroleum diesel.
- At peak load condition NOx emission for B10 blend was marginally low about 2% compared to B0 petroleum diesel.
- Though no drastic reduction in emission was observed for B10 WCOB biodiesel blend.

Acknowledgement:
Authors wish to acknowledge financial support received from RUSA 2.0 (GOI) to carry out the research.

References
[1] Bhuyia MMK, Mofijur M, Rasul MG and Hyde J 2015 Role of biodiesel on IC Engine emission reduction *Energy procedia* **75** 886-892.
[2] Gupta R, Gupta P and Bhalla J 2016 Performance Analysis of a Diesel Engine using the Soybean Oil based Biodiesel *Indian Journal of Science and Technology* Vol **9** (36),DOI: 10.17485/ijst/2016/v9i36/102153.
[3] Ramadhas A, Jayaraj S and Muraleedharan C 2004 Use of vegetable oils as I.C. enginefuels—a review, *Renewable Energy* **29** 727-742.
[4] Nurun Nab Md, MustafizurRahman Md and ShamimAkhter Md 2009 Biodiesel fromcotton seed oil and its effect on engine performance and exhaust emissions *Appl. Therm. Eng.* **29** 2265–2270.
[5] Shahir SA, Masjuki HH, Kalam MA, Imran A, Rizwanul Fattah IM and Sanjid A 2014 Feasibility of diesel–biodiesel–ethanol/bioethanol blend as existing CI enginefuel: an assessment of properties, material compatibility, safety andcombustion *Renew. Sustain. Energy Rev.* **32** 379–395.
[6] Zheng M, Mulenga MC, Reader GT, Wang M, Ting DS and Tjong J 2008 Biodieselengine performance and emissions in low temperature combustion, *Fuel* **87** 714–722.
[7] Mirzajanzadeh M,Tabatabaei M, Ardjmand M and Rashidi A 2015 A novel soluble nano-catalysts in diesel–biodiesel fuel blends to improve diesel engines performance and reduceexhaust emissions *Fuel* **139** 374–382.
[8] Gumus S, Ozcan H, Ozbey M and Topaloglu B 2016 Aluminum oxide and copper oxidenano diesel fuel properties and usage in a compression ignition engine *Fuel***163** 80–87.
[9] Rajesh A, Gopal K, Melvin Victor DP, Rajesh Kumar B, Sathiyagnanam AP and Damodharan D 2020 Effect of anisole addition to waste cooking oil methyl ester on combustion, emission and performance characteristics of a DI diesel engine without any modifications.*Fuel***278** 118-315.
[10] Avinash A, Sasikumar P and Murugesan A 2018 Understanding the interaction among the barriers of biodiesel production from waste cooking oil in India- An interpretive structural modeling approach *Renewable Energy***127** 678-684.

Nomenclature

WCOB	Waste cooking oil biodiesel	WCO	Waste cooking oil
B0	Petroleum Diesel	KOH	Potassium hydroxide
B10	Blend contained 10% WCOB	VCR	Variable compression ratio
rpm	Revolutions per minute	NOx	Oxides of nitrogen
BTE	Brake thermal efficiency	BSFC	Brake specific fuel consumption
HC	Unburnt Hydrocarbon	CO	Carbon monoxide

Experimental analysis of small scale water cooler equipped with TTHC heat exchanger

Rajesh Kumar[1], Kumar Gaurav[2] and Prakash Chandra[2]

[1]Mechanical Engineering Department, Sandip University Sijoul, Madhubani-847235, India,
[2]Mechanical Engineering Department, National Institute of Technology Patna-80006, India
E-mail: *rajeshknitp@gmail.com*

Abstract. In the present work, experiments were conducted to recover waste heat from water cooler condenser. A Tube in Tube Helical Coil (TTHC) heat exchanger was made from cu pipe of 4.47 mm outer diameter of inner tube and 8.30 mm inner diameter of outer tube replaced the water cooler condenser to recover the waste heat. A new empirical correlation was suggested to predict the annulus Nusselt number for laminar flow of water with Reynolds number ranging from 1500 to 3900 and Prandtl number approx 1.6. The results reveal that the annulus Nusselt number varies with volume flow rate through annulus of helical tube significantly in laminar flow region. The small scale water cooler equipped with TTHC heat exchanger recovered maximum 371.5 W heat by cooling water.
Key Words: Equivalent diameter, Reynolds number, TTHC heat exchanger, Water cooler,

1. Introduction

Helical coiled tube heat exchanger is one of the passive methods to enhance the heat transfer which do not require external power for enhancement of heat transfer. A passive method commonly uses geometrical modification to the flow channel for enhancement and compact in size [1-4]. Helical coiled heat exchanger is used to transfer heat between two or more fluid for numerous heating and cooling in industrial and engineering applications, such as cooling in electronic industries, refrigerator & air-conditioning units, thermal power plant, heat recovery system, domestic water heater, space vehicles, automobiles, etc. The current work mainly focus on heat recovery by cooling water from small scale water cooler equipped with TTHC heat exchanger.

Many researchers Prabhanjan et al. [5], Mao et al. [6], Dravid et al. [7] have reported that the secondary flow occurs in helical tube due to centrifugal force which enhances the heat transfer.

Elsayed et al. [8] studied small scale cooling system with helical coil condenser and evaporator. A mathematical model was developed for this system and result found that the COP increases as the coil diameter decreases. Seban and Mclaughlin [9] Studied on heat transfer in helical coiled tube with laminar flow of medium heavy freezene oil and turbulent flow of water experimentally and developed a correlation for two coil diameter to tube diameter ratios are 17 and 104. Similar work was done by Rogers and Mayhew 1964 for three different coil diameters to tube diameter ratio such as 10.8, 13.3 and 20.12 and compared with the result of Kalb and Seader [10]. Xin and Ebadian [11] discussed the effect of Prandtl numbers on convective heat transfer coefficients in helical tubes, for three different fluids air, water and ethylene glycol and developed a empirical expression from experimental data and some data from previous work.

The above literature shows that the numerous studies carried out on helical coil coiled tube as well as tube-in tube helical coil heat exchanger. Although there are number of correlations were developed for inner tube side helical coil tube but very few correlations are available on annulus region of tube-in tube helical coil heat exchanger. In the present work, Nusselt number was analyzed through annulus of Tube-in Tube Helical Coil (TTHC) heat exchanger by using experimental data from suggested correlation.

2. Material and methods

The test section used in the present work consists of soft copper tubing as shown in figure 1. Initially the equal lengths (L) of both inner as well as outer straight tube of 9 m were taken. The geometrical parameters of TTHC heat exchanger is given in table 1.

Figure 1.TTHC heat exchanger

The inner tube of outer diameter 4.47 mm is inserted into outer tube of inner diameter 8.3 mm and fine sand particles filled in inner tube as well as annulus space in order to maintain the smoothness of the inner surface. The TTHC heat exchanger of 282 mm coil diameter is formed with the help of wooden pattern and flushes the sand particles after coiling.

Table 1 Physical dimension of TTHC heat exchanger used in the experiment

Dimensions	L (m)	$d_{1,i}$ (m)	$d_{1,o}$ (m)	$d_{2,i}$ (m)	$d_{2,o}$ (m)	D (m)	p (m)
TTHC heat exchanger	9	0.00442	0.00447	0.0083	0.00952	0.282	0.034

3. Experimental setup and test section

The flow diagram of the experimental setup is shown in figure 2. The test facility was made-up of a primary open loop and a secondary closed loop. The primary open loop consists of cooling water tank, pipe fittings, measuring instruments, 12 volt DC brushless water pump, flow control valve, rotameter with accuracy within 1.5 % of full scale and an outer helical tube. The secondary closed loop consists of a complete set of water cooler but the present work considering only condenser part that is TTHC heat exchanger. The test section is a TTHC heat exchanger consisting of an inner helical tube in which refrigerant flow at a constant rate for maintaining its outer surface at 42±0.2°C and an outer helical tube (annulus passage) through cooling water as working fluid flow at different rate. The outer surface of TTHC heat exchanger is insulated to minimize the heat loss to the surrounding. In this experiment, there is five k type thermocouples were used out of these; one is fixed in cooling water tank and remaining four were inserted into drilled hole at each inlet and outlet of inner and outer helical test section. The epoxy is used at the drilled hole to prevent the leakage of fluid. Two four channel temperature indicator were used to measure the temperature at each point.

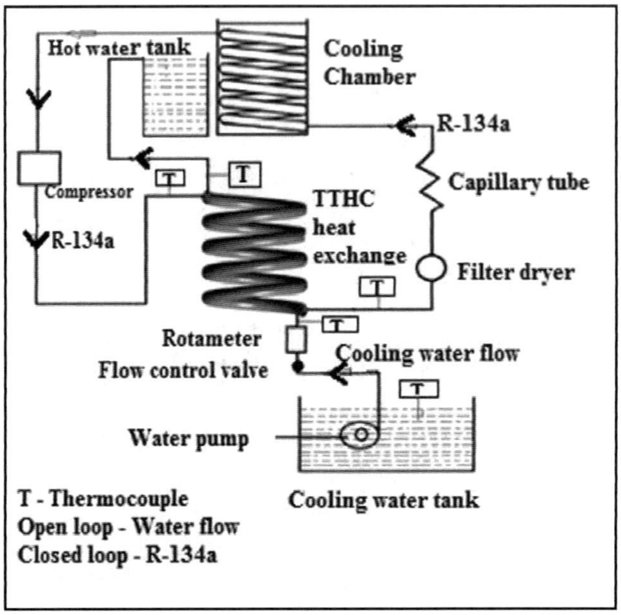

Figure 2. Flow diagram of experimental set-up

The experiments were performed under steady state condition. Switch on the compressor of water cooler, refrigerant flow through inner tube of TTHC heat exchanger. Tap water is pumped from water bucket at 26±1 °C to the annulus passage of helical tube in counter clockwise direction of refrigerant at six different flow rates 15 l/h, 20 l/h, 25 l/h, 30 l/h, 35 l/h and 40 l/h by flow regulating valve. The temperatures are recorded after steady state achieved by 4 channel temperature indicator with 0.1 °C precision.

4. Analysis of experiment

In the present work, thermo-physical properties of water are obtained as a function of mean film temperature Kays et al. [12].

Heat recovered by cooling water from refrigerant R-134a

$$Q_c = \rho_w V c_{p,w} \left(T_{w,o} - T_{w,i}\right) \quad (1)$$

Reynolds number of water flow in annulus of helical tube based on equivalent diameter is calculated from Eq. (2)

$$\mathrm{Re} = \frac{\rho_w V D_{eq.}}{\mu_w A} \quad (2)$$

Cross sectional area of annulus helical tube and annulus equivalent diameter as given by Kumar et al. [13].

$$A = \frac{\pi}{4}\left(d^2_{2,i} - d^2_{1,o}\right) \quad (3)$$

$$D_{eq.} = \frac{4 \times A}{wettedperimeter} = \frac{4\pi\left(d^2_{2,i} - d^2_{1,o}\right)}{4\pi d_{1,o}} = \frac{\left(d^2_{2,i} - d^2_{1,o}\right)}{d_{1,o}} \quad (4)$$

where $d_{1,i}, d_{1,o}, d_{2,i}, d_{2,o}$ are the inner diameter of inner tube, outer diameter of inner tube, inner diameter of outer tube, outer diameter of outer tube of TTHC heat exchanger respectively.

Dean number of water flow in annulus of helical tube based on equivalent diameter is defined by Eq. (5)

$$De = \mathrm{Re}\left(\frac{D_{eq.}}{D}\right)^{1/2} \quad (5)$$

5. Results and discussion

Figure 3. illustrates the variation of Nusselt number with Reynold number in annulus of TTHC heat exchanger. It can be clear that annulus Nusselt number increases with an increase in Reynold number. Figure 4 also shows the comparison of present calculated annulus Nusselt number with the data of Seban & McLaughlin, Xin & Ebadian and Garimella et al. The calculated Nusselt number in the present work is higher as compared to Seban & McLaughlin and Garimella et al. but lower than that of Xin & Ebadian. This may be due to curvature ratio of the helical tube.

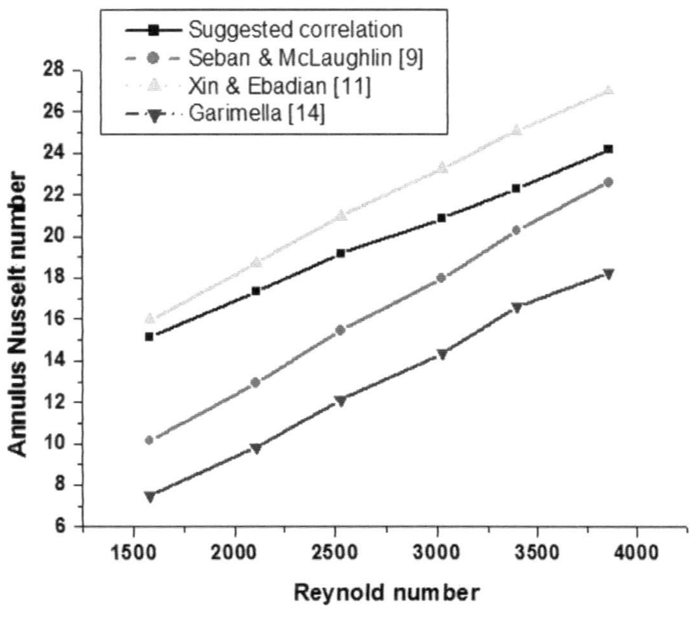

Figure 3. Effect of annulus Nusselt number versus Reynold number

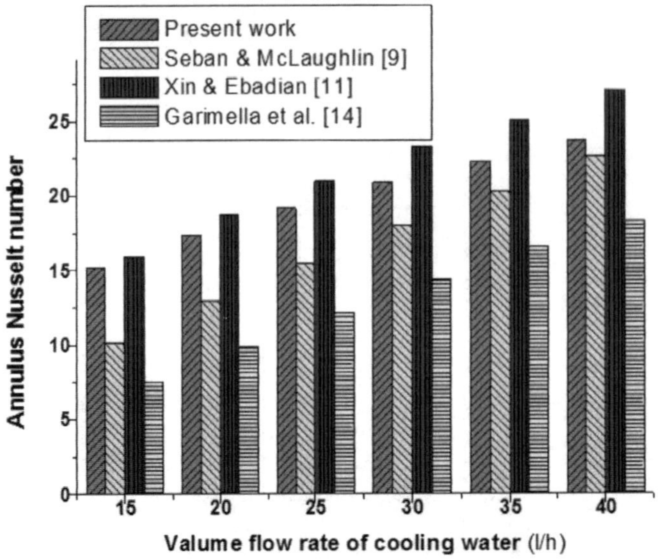

Figure 4. Variation of annulus Nusselt number versus volume flow rate of cooling water

Figure 4 shows the variation of annulus Nusselt number at different volume flow rate of cooling water. It can be seen that annulus Nusselt number increased with the volume flow rate of cooling water. The present results compared with the previous works and found appropriate results.

A new correlation Eq. (7) was recommended for annulus Nusselt number of TTHC heat exchanger by modifying the existing correlation Eq. (6) developed by Kalb and Seader [10].

$$Nu = 0.913\mathrm{Re}^{0.476}\,\mathrm{Pr}^{0.2}\left(\frac{d_i}{D}\right)^{0.238} \tag{6}$$

Most of the previous works used hydraulic diameter in place of inner tube diameter for annulus Nusselt number; however in the present suggested correlation d_i should be replaced by $D_{eq.}$ and Eq. (7) becomes a new set of correlation for annulus Nusselt number calculation.

$$Nu = 0.913\,\mathrm{Re}^{0.476}\,\mathrm{Pr}^{0.2}\left(\frac{D_{eq.}}{D}\right)^{0.238} \tag{7}$$

for $1500 < \mathrm{Re} < 3900$ and $\mathrm{Pr} = 1.6$

The experiments were carried out to obtain annulus Nusselt number from Eq. (7) using experimental data. The annulus Nusselt number of TTHC heat exchanger were compared with the previous developed correlations by Seban & McLaughlin [9], Xin & Ebadian [11] and Garimella et al. [14] correspondingly.

$$Nu = 0.023\mathrm{Re}^{0.85}\,\mathrm{Pr}^{0.4}\left(\frac{d_i}{D}\right)^{0.1} \tag{8}$$

for $12 \le \mathrm{Re} \le 5600$ laminar flow, $6000 \le \mathrm{Re} \le 65000$ turbulent flow and D/d = 17 and 104

$$Nu = \left(2.153 + 0.318De^{0.643}\right)\mathrm{Pr}^{0.177} \tag{9}$$

for $20 \le \mathrm{De} \le 2000$, $0.7 \le \mathrm{Pr} \le 175$ and $0.0267 \le \delta \le 0.0884$

$$Nu = 0.027De^{0.94}\,\mathrm{Pr}^{0.69}\left(\frac{d_{2,i} - d_{1,o}}{D}\right)^{0.01} \tag{10}$$

for $300 \le \mathrm{De} \le 4002$

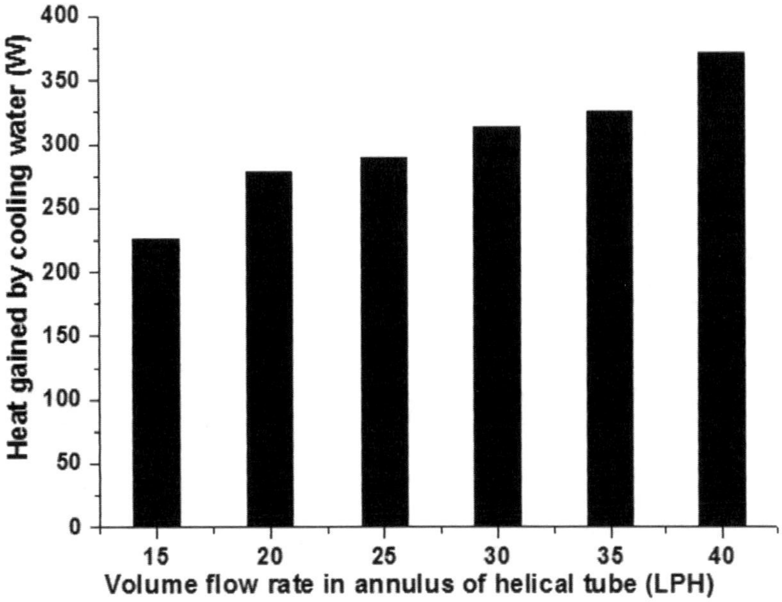

Figure 5. Variation of heat recovery by cooling water from R-134a versus volume flow rate of cooling water

Figure 5 shows the effect volume flow rate in annulus of helical tube on heat recovery from R-134a. Heat recovery from refrigerant increased as the water flow through annulus of helical tube increases.

6. Conclusion

In this study, experiments were conducted on TTHC heat exchanger. The maximum 371.5 W heat recovered by cooling water at 40 l/h from small scale water cooler equipped with TTHC heat exchanger. Based on equivalent diameter, a new set of correlation was suggested for calculating the annulus Nusselt number for laminar flow of cooling water. The present suggested correlation results were found to be better agreement with the previous suggested correlations within $1500 < Re < 3900$, $Pr = 1.6$ and $D_{eq}/d = 25.77$.

References

[1] Dewan A, Mahanta P, Raju K S and Kumar P S 2004 Review of passive heat transfer augmentation techniques *Proceedings of the Institution of Mechanical Engineers, Part A: J. of Power and Energy* **218**(7) pp 509-527.

[2] Yildiz C, Biçer Y, Pehlivan D 1995 Heat transfers and pressure drops in rotating helical pipes *Applied Energy* **50**(1) pp 85-94.

[3] Kumar R, Chandra P and Raj G 2020 Natural Convection Heat Transfer from Circular Finned Helical Coil Heat Exchanger in Air. *Trends in Manufacturing Processes (Singapore)* pp 93-100.

[4] Kumar R and Chandra P 2020 Innovative method for heat transfer enhancement through shell and coil side fluid flow in shell and helical coil heat exchanger *Arch. of Thermodyn.* **41**(2) pp 239-256.

[5] Prabhanjan D G, Raghavan G S V and Rennie T J 2002 Comparison of heat transfer rates between a straight tube heat exchanger and a helically coiled heat exchanger *Int. Commun. in Heat and Mass Transfer* **29**(2) pp 185-191.
[6] Mao Y, Guo L, Bai B and Zhang X 2010 Convective heat transfer in helical coils for constant-property and variable-property flows with high Reynolds numbers *Frontiers of Energy and Power Eng. in China* **4**(4) pp 546-552.
[7] Dravid A N, Smith K A, Merrill E W and Brian P L T 1971 Effect of secondary fluid motion on laminar flow heat transfer in helically coiled tubes *AICHE J.* **17**(5) pp 1114-1122.
[8] Elsayed A, Al-Dadah R.K, Mahmoud S and Rezk A 2012 Experimental and theoretical investigation of small-scale cooling system equipped with helically coiled evaporator and condenser *Proceedings of the Institution of Mech. Engineers, Part C: J. of Mech. Eng. Science* **226**(3) pp 724-737.
[9] Seban R A and McLaughlin E F 1963 Heat transfer in tube coils with laminar and turbulent flow *Int. j. of heat and mass transfer* **6**(5) pp 387-395.
[10] Kalb C E and Seader J D 1972 Heat and mass transfer phenomena for viscous flow in curved circular tubes *Int. J. of Heat and Mass Transfer* **15**(4) pp 801-817.
[11] Xin R C and Ebadian M A 1997 The effects of Prandtl numbers on local and average convective heat transfer characteristics in helical pipes *J. of Heat Transfer* **119**(3) pp 467-473.
[12] Kays W, Crawford M and Weigand B 2005 Convective Heat and Mass Transfer *4th Ed. McGraw Hill (Singapore).*
[13] Kumar V, Saini S, Sharma M and Nigam K D P 2006 Pressure drop and heat transfer study in tube-in-tube helical heat exchanger *Chemical Eng. Science* **61**(13) pp 4403-4416.
[14] Garimella S, Richards D E and Christensen R N 1988 Experimental investigation of heat transfer in coiled annular ducts *J. of heat transfer* **110**(2) pp 329-336.

Nomenclature

A Cross sectional area of annulus helical tube, m^2
D Helical coil diameter, m
d Tube diameter, m
De Dean number
$D_{eq.}$ Annulus equivalent diameter, m
k Thermal conductivity, W/mK
L Length of tube, m
Nu Nusselt number
p Pitch of helical coil, m
Pr Prandtl number
Q Heat transfer rate, W
Re Reynold number
T Temperature, °C
V Volume flow rate of water, m^3/sec

Greek Symbols

ρ Density, kg/m^3
C_p Specific heat, J/kgK
θ Bulk mean temperature,°C
μ Viscosity, Ns/m^2

Subscripts

c Cooling
i Inlet/Inner
o Outlet/Outside
w Water

Feasibility study of a liquid desiccant based marine air conditioning system using waste heat of diesel alternators

P K Iyer[1] and A. Ganguly[2]

[1]CGRA, Nanganallur, Chennai, India.
[2]Department of Mechanical Engineering , Indian Institute of Engineering Science and Technology, Shibpur, PO -B. Garden, Howrah, West Bengal, India
e-mail: *aritra78@gmail.com*

Abstract. The present article proposes an alternative form of an air conditioning system for ships using a packed bed liquid desiccant based dehumidification system, followed with regenerative evaporative cooling of air. Abundantly available waste heat from the hot water of Diesel Alternators (DA) is proposed for the regeneration of the desiccant, and seawater for cooling the hot desiccant. A thermal model of the system has been developed and analyzed. The study revealed that the proposed packed bed liquid desiccant system can dehumidify the outdoor air to well below the design humidity ratio (11g kg_{da}^{-1}). Also, the regeneration of the desiccant through hot water from the DA system is feasible, with a temperature range of 73°C to 90°C. The target enthalpy (43.84 kJ kg^{-1}) can be achieved with a branching ratio of 0.25 to 0.4 at the regenerative evaporative cooler.

1. Introduction

Ventilation and air conditioning systems have a significant influence on factors such as fatigue, seasickness, fire hazard, safety, and protection of sensitive navigation, communication, and control equipment. In recent times, there has been a greater push to look into green environment friendly systems across the globe. This is very relevant for ship-based air conditioning as well. The 3rd greenhouse gas emission study by the International Maritime Organization [1] indicated an average annual greenhouse gas emission of around 15 million tons due to the leakage of refrigerants alone. The high frequency of refrigerant leakage at sea is attributed to the continuous vibration/movement of the systems due to the sea waves in high seas and the cumbersome process of identifying the leakage source at sea [2]. One of the major decisions taken in recent years was also to include the HFCs to the list of gases to be controlled to prevent the ozone depletion, in the Kigali amendment of Montreal Protocol on October 15, 2016 [3].

Accordingly, various research works are being undertaken on the alternative systems, which can help to reduce the global warming/ozone depletion potential and the power consumption of the air conditioning systems used in ships. A nascent technology for shipboard air-conditioning is the use of dedicated outdoor air system (DOAS). This system aims at utilizing parallel systems to remove the latent heat loads, and the sensible loads. DOAS was proposed for ships using solid desiccant wheels and a parallel system using seawater cooling along with VCS -Vapour Compression System [4, 5]. These papers highlighted the feasibility of regeneration of the desiccant through the heat produced from the exhaust of main engines. The models used a hybrid DOAS and VCS model, wherein the latent heat was catered to by the solid desiccant wheels and the sensible heat was handled by the VCS. The present paper proposes a novel liquid desiccant based dehumidification system for handling the latent load, followed with a regenerative (dew point) evaporative cooling system for handling the sensible load in ships. The liquid desiccant system uses waste heat emanating from the hot water of diesel alternators to regenerate the weak desiccant solution. The schematic of the proposed system is shown in Fig. 1. The air circuit passes through three stages.

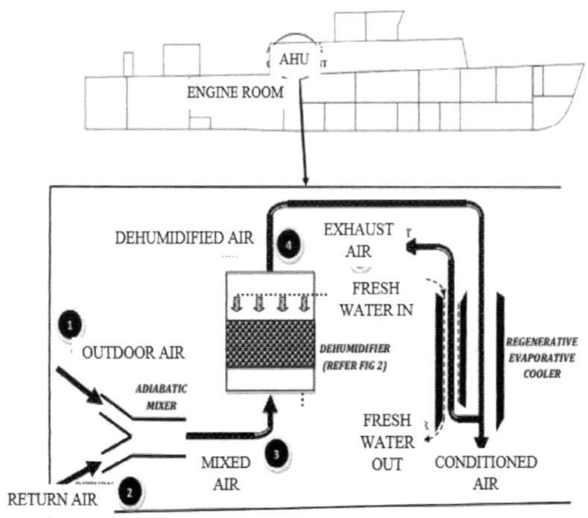

Figure 1. Schematic of proposed system.

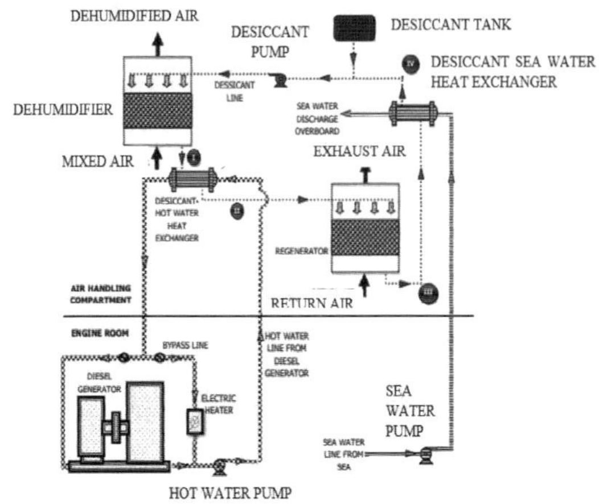

Figure2. Schematic of dehumidification system.

In the first stage, the outdoor air is adiabatically mixed with the return air from the ship spaces (State 1, 2 to State 3), resulting in a mixed air with reduced DBT and humidity ratio of the outdoor air, theoretically by a fraction equal to the ratio of outdoor air and return air. In the second stage, the mixed air passes through a packed bed liquid desiccant dehumidifier (State 3 to State 4). In the third stage, the dehumidified air is passed through a regenerative evaporative cooler (State 4 to State 5). It is a form of an indirect evaporative cooler, wherein a part of the air is branched into the wet passage. The conditioned air with lower DBT and enthalpy is then finally transported to the ship spaces through the ducts. In addition to the air circuit, there also exists a desiccant circuit, as shown in Fig. 2. The desiccant solution, diluted by water from moist air in the dehumidifier, passes through a shell and tube heat exchanger (Fig. 2 – State I to State II), where it is heated. The heating is provided through the use of waste heat of hot water used in the cooling system of the diesel alternators (DA). This causes an increase in solution water vapour pressure [6]. The heated desiccant is then re-concentrated in the packed bed regenerator (State II to State III) due to its increased vapour pressure (being more than the partial pressure of water vapour content of the incoming air). The incoming air used for the regeneration of the desiccant solution is part of the return air from the ship spaces, which is then discharged overboard. The concentrated desiccant solution is then cooled by seawater in another shell and tube heat exchanger (State III to State IV) to its original state. The cooled desiccant solution is pumped back into the dehumidifier (State IV to State I).

2. Development of thermal model

This section covers the procedure adopted to estimate the thermal load conditions of the ship, and equations used to undertake the thermal model development. The input conditions of the ambient air have been assumed from Indian Standards for air conditioning on board ships - IS 14808:2000. The target parameters to be achieved in the ship spaces and the parameters of return air from the ship spaces have been assumed from an air conditioning manual of a ship [7]. These parameters have been illustrated in Table 1. Following assumptions are made while developing the thermal model:

(a) All heat exchange processes have been assumed to be adiabatic, with no heat transfer to surroundings.

(b) Branched air has evenly wetted in the wet passage of regenerative evaporative cooler.

(c) Concentration of regenerated desiccant solution is assumed to remain constant at 35%.

The representation of the various state points of air are done in the psychrometric chart shown in Fig.3.

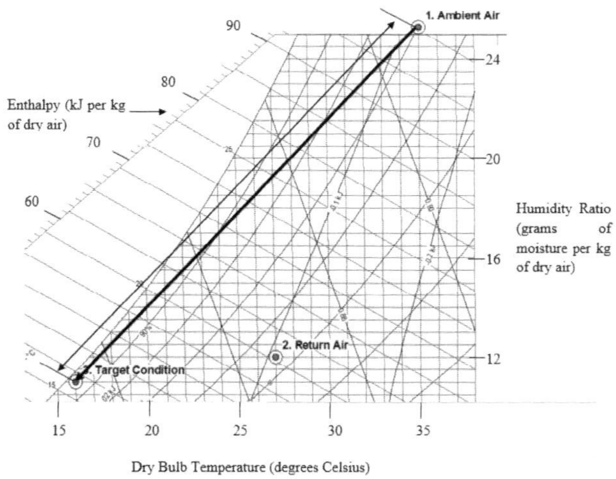

Figure 3. Schematic of the Process in Psychrometric Chart.

Table 1. Input and target conditions of air.

Sl	Parameter	Design Value	Unit
(a)	Ambient air DBT	35	°C
(b)	Ambient Air Humidity Ratio	0.025	$kg_{water}\ kg_{dryair}^{-1}$
(c)	Return Air Humidity Ratio	0.012	$kg_{water}\ kg_{dryair}^{-1}$
(d)	Target (Conditioned Air) DBT	16	°C
(e)	Target (Conditioned Air) Humidity Ratio	0.011	$kg_{water}\ kg_{dryair}^{-1}$
(f)	Target (Conditioned Air) Enthalpy	43.84	$kJ\ kg^{-1}$

This paper proposes to analyze whether it would be feasible to account for the entire reduction of ambient air humidity ratio from 25 g $kg_{da\text{-}1}$ to 11 g $kg_{da\text{-}1}$ using the liquid desiccant based dehumidification system, and reduction of the remaining enthalpy to 43.84 kJ $kg_{\text{-}1}$ through regenerative evaporative cooling. The major governing equations are given in this section. The effectiveness of the dehumidifier was obtained from [8]:

$$\varepsilon_{deh} = \frac{1-\{\frac{0.024\cdot\left(m_3/m_{d,in}\right)\cdot\exp\left[1.057\cdot\left(T_3/T_{d,in}\right)\right]}{(A_t\cdot Z)^{-0.185}\cdot\pi^{0.638}}\}}{1-\{\frac{0.192\cdot\exp\left[0.615\cdot\left(T_3/T_{d,in}\right)\right]}{\pi^{-21.498}}\}} \tag{1}$$

The output humidity ratio of the dehumidified air can be found as [9]:

$$W4 = W3 - \varepsilon_{deh}\cdot(W3 - W_{eq}) \tag{2}$$

Regenerator equations were also calculated in a similar manner. Target (conditioned air) temperature is calculated from a modified heat and transfer mass transfer equation [10]:

$$T5 = T4 - (C_{min}/\ C_h)\cdot\varepsilon_{evp}\cdot(T4 - T_{s,iwb}) \tag{3}$$

where, $T_{s,iwb} = T5_{wb}$ since the same air is branched in the secondary channel. The effectiveness of the regenerative evaporative cooler ε_{evp} was also obtained from a modified form of heat exchanger equation which caters to the mass transfer in the wet secondary channel.

3. Results and Discussions

A computer code using MATLAB software version R2014a has been developed based on the heat load estimations and equations described in the previous sections. Figure 4 shows the variation of dehumidifier effectiveness with D/A Ratio (Ratio of desiccant mass flow rate to mixed air mass flow rate), which has been validated against a similar study available in the literature [11]. It is observed that the dehumidifier effectiveness increases with the D/A ratio for both the models. The possible reason for the trend is that with the increase in the D/A ratio, more desiccant is allowed to be sprayed over a given amount of air, thereby increasing the potential of moisture transfer from the air to the desiccant. Figure 5 shows the different mass flow rates of return air needed to regenerate the desiccant solution to a concentration of 0.35 for the various hot water temperature. It is seen from Fig. 5 that the mass flow rate of return air needed for regeneration keeps on reducing with an increase in the hot water temperature. It may be also noted that the mass flow rate of return air for regeneration must not be increased beyond 1.3 kg s^{-1} (shown as dotted line in Fig. 5), since the remaining part of the return air is also required for adiabatic mixing with the outdoor air. This necessitates that the hot water temperature must not be lowered below 73°C for the proposed model.

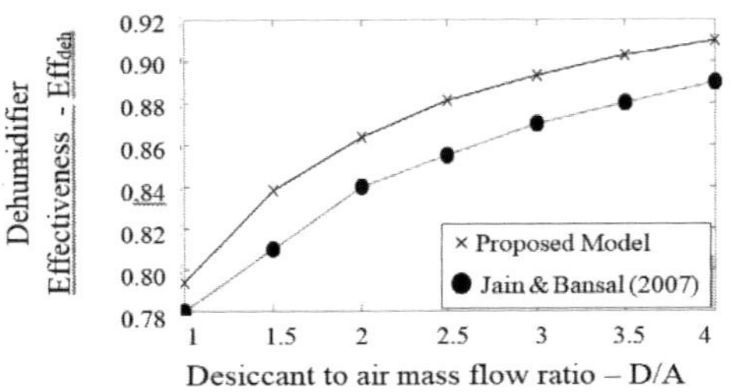

Figure 4. Dehumidifier Effectiveness vs. Desiccant to Air Mass Flow Ratio.

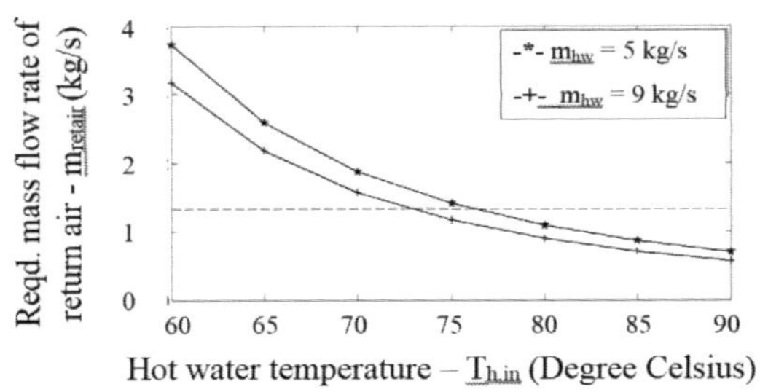

Figure 5. Required Mass Flow Rate of Return Air vs. Hot Water Temperature.

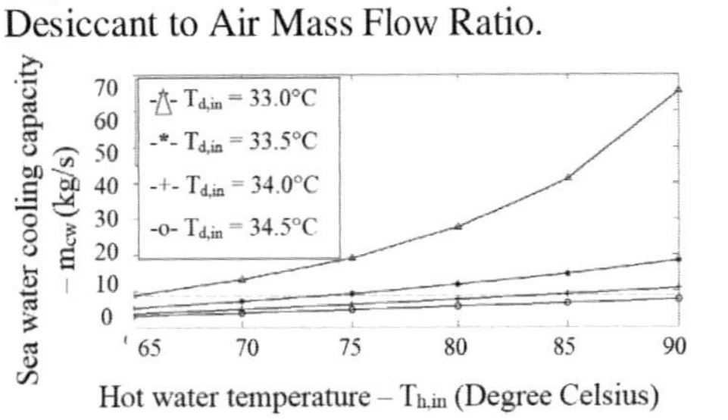

Figure 6. Seawater Cooling Capacity vs. Hot Water Temperature.

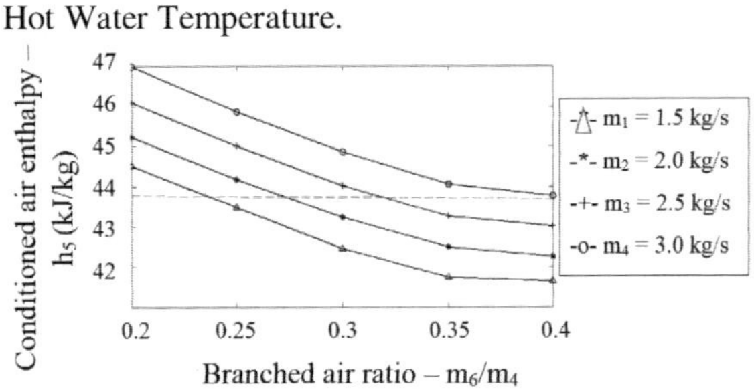

Figure 7. Conditioned Air Enthalpy vs. Branch Air Ratio.

Figure 6 shows the capacity of the sea water pump needed to cool the desiccant at various temperature of hot water. It is seen from Fig. 6 that increasing the hot water temperature increases the capacity of the seawater cooling pump needed to cool the desiccant, especially to lower temperature. It is practically recommended to have seawater cooling pump capacity to be less than 9 kg s^{-1} for lesser power consumption, as shown by the dotted line in Fig. 6. Accordingly, hot water temperature needs to be lesser than 90°C, and $T_{d,in}$ greater than 33.5°C for the proposed model. Figure 7 shows the variation of conditioned air enthalpy with branched air ratio m_6/m_4 for different values of outdoor air m_1. The figure shows that the conditioned air enthalpy decreases with increased branching of dehumidified air for a given amount of outdoor air. It is deduced that if outdoor air m_1 is operated between 1.5 kg s^{-1} and 3 kg s^{-1}, and if the branched air ratio is maintained between 0.25 and 0.4, then it is feasible to implement the proposed model for the given class of ships.

4. Conclusion

A novel liquid desiccant based alternative air conditioning system for ships has been proposed in this paper, which includes regeneration of desiccant solution by using waste heat of hot water from the Diesel Alternator system, while the sensible cooling of air is carried out by regenerative evaporative cooling. Following points have been revealed from the present study:

(a) A branching of 25-40% of dehumidified air at regenerative evaporative cooler is observed to be suitable for the proposed model to achieve target enthalpy of 43.84 kJ kg^{-1} for ship conditions.

(b) Hot water temperature needs to be maintained in the range from 73°C to 90°C for the present model to attain a 35% concentration of desiccant solution post- regeneration, which can be easily met with the DA system held on board the ships.

(c) The study reinforces the need and the viability of the proposed desiccant based air-conditioning system for maintaining a suitable micro-climate in the ships.

References

[1] Smith T W P, Jalkanen J P and Anderson B A 2015 Third IMO GHG Study. International Maritime Organization.

[2] Hafner A, Gabrielii C and Widell K 2019 Refrigeration Units in Marine Vessels - Alternatives To HCFCS And High GWP HFCS. Nordic Council of Ministers.

[3] Polonara F, Kuijpers L J M and Peixoto R A 2017 Potential impacts of the Montreal Protocol Kigali Amendment to the choice of refrigerant alternatives *Int. J. of Heat and Technology* **35** 1–8.

[4] Guojie Z, Chaoyu Z, Guanghai Y and Wu C 2012 Development of a New Marine Rotary Desiccant Air- conditioning System and its Energy Consumption Analysis *Energy Procedia* **16** 1095–1101.

[5] Digiovanni J A, Colahan J D, Knauss D T 2005 Dynamic Desiccation Cooling System for Ships, United States Patent US **6** 854,279 B1.

[6] Ertas A, Anderson, E E and Kiris I 1992 Properties of a new Liquid Desiccant Solution - Lithium Choride and Calcium Chloride Mixture *Solar Energy* **49** 205–212.

[7] Heinen and Hopman, 2015 *Heat Load Calculation for A/C System.* (C. S. Ltd, Ed.). Heinen and Hopman Engineering (I) Pvt. Ltd.

[8] Chung T W and Luo C M 1999 Vapor pressures of the aqueous desiccants *J. of Chem and Engg. Data* **44** 1024–1027.

[9] Kim M H, Park J S and Jeong J W 2013 Energy saving potential of liquid desiccant in evaporative-cooling-assisted 100% outdoor air system *Energy* **59** 726–736.

[10] Woods J and Kozubal E 2013 A desiccant-enhanced evaporative air conditioner: Numerical model and experiments *Energy Conversion and Management* **65** 208–220.

[11] Jain S and Bansal P K 2007 Performance analysis of liquid desiccant dehumidification systems *Int. J. of Refrigeration* **30** 861–872.

Nomenclature

W	Humidity Ratio (kg $kg_{dry\ air}^{-1}$)	Z	Packing Height (m)
m	Mass flow rate (kg s^{-1})	ε	Effectiveness
t	Temperature (°C)	π	Vapor pressure depression
A_t	Dehumidifier / Regenerator Packing Size (m^{-1})	des	Desiccant

Flakner-Skan equation flow of Newtonian fluid past a moving wedge with suction/injection

Abeetath Ghosh[1] and Nitesh Mondal[2]

[1] School of Nuclear Studies and Application, Jadavpur University, Salt Lake, Kolkata–700106, India
[2] Department of Mechanical Engineering, Techno India University, Salt Lake, Kolkata –700091, India
e-mail: *niteshju@yahoo.com*

Abstract. In this study, numerically analysed the Falkner-Skan equation in the two-dimensional laminar flow past a moveable wedge for a Newtonian fluid with suction/injection is investigated through Matlab environment. The boundary conditions are scrutinized and applied in this equation for different conditions. The simultaneous effect of the parameter which used to measure the pressure gradient along x-axis (β), velocity ratio between the wedge and mainstream (λ), suction and injection are inspected on the momentum boundary layer thickness or dimensionless velocity profile.

1. Introduction

Boundary layers are the tinny fluid layers nearby the surface of a body in which healthy viscous effects exist. The nature of such a flow field is known from information gathered from large numbers of experiments. Generally, a dotted line that originates at the front stagnation point and moves downstream near the top and bottom surfaces of the body is used to show the boundary [1]. The viscous effect inside the boundary is very strong where the outside is almost free from viscous effects [2]. Many researchers give their time on it in different ways to study the effects of it in the field of fluid mechanics and thermodynamics. In 1930 Falkner and Skan [3] exemplified the application of Prandtl's boundary layer theory on steady and laminar flow passing a stationary wedge. Ishik et al. [4] worked on moving wedge and the conditions applied of the flow are steady, laminar, two-dimensional boundary layer flow and the fluid is viscous as well as irrational. Using the finite difference approach numerically solved the transformation of the boundary layer for the moveable wedge with suction. Afzal [5] analytically worked on the Falkner–Skan equation when flow passes through a stretching surface and studied the effect of suction or blowing considering the laminar boundary layer flow of an incompressible and viscous fluid. Hussaini et al. [6] investigated a boundary layer problem where the flat plate moves with a uniform velocity in the opposite direction of the mainstream. In this investigation, an important outcome has come that the solution of the boundary layer problem depends on the parameter which is the ratio of the velocity of plate and mainstream and the solution occurs within the certain value of the parameter. Watanabe [7] theoretically explored the properties of the incompressible laminar boundary layer for a special case of force flow over a wedge assuming unvarying suction or injection. The clarifications of the resulting equation are presented in integral equations and represented effects of the suction parameter on velocity and temperature distribution, Skin friction coefficient and obviously on heat transfer. Kudenatti and Awati [8] worked on the stretching plate considering the suction effects to find out the solution of pressure gradient and deal with two parameters for the governing problem Falkner - Skan equation with β, the pressure gradient along streamwise and γ velocity at suction. The effect of β and γ on the velocity profile and found out an excellent match of the result with the solution which comes from numerical. Heeg at al. [9] also worked on the Falkner-Skan equation numerous inflection points with respect to some stable properties and find out the effects of inflection points on the Reynolds number. Mansutti et al. [10] worked on a steady flow of different types of non-linear fluids pass over a porous plate considering constant injection or suction. To solve the problem on it numerically modelled the non-Newtonian fluid, momentum and mass balance equation with proper boundary conditions and put some idea to solve the resulting equation by finite difference method or by collocation method with a special function B-spline; defined the manner through which the material parameters disturb the boundary layer structure. Garg et al. [11] established a pseudo-similarity solution of a laminar incompressible fluid past over wage and studied the effect of normal stress modulus of fluid and wage angle on fluid flow, wall shear stress. A new method known as the optimal homotopy asymptotic method independent of small parameters is developed by Marinca eta al. [12] to solve the Falkner-Skan nonlinear differential equation. A great agreement has established between the result obtained by this method and numerical solution. Khan and pop [13] concentrated on the problem where the stretching wedge has a uniform velocity not along the free stream. Using the mass, momentum and energy equation in the ODE form along with the proper boundary condition and solve these equations by finite difference method. Ullah et al. [14] worked on the flow through a porous media having heat interaction and viscous dissipation properties over a moving wedge and established the relation between fluid velocity and Eckert number, Prandtl number with heat transfer rate. Recently Xu [15] worked on the accelerating fluid past on a wedge to study the change of behavior of vortex dynamic and material moving rate close to the wedge tip. Based on the available literature, this work presents the development of the boundary layer or dimensionless velocity profile for different conditions flow past over a wedge. The problem formulation, the mathematical model of the problem, interesting occurrence with explanations and the outcome of the work is presented.

2. Problem description and solution procedure

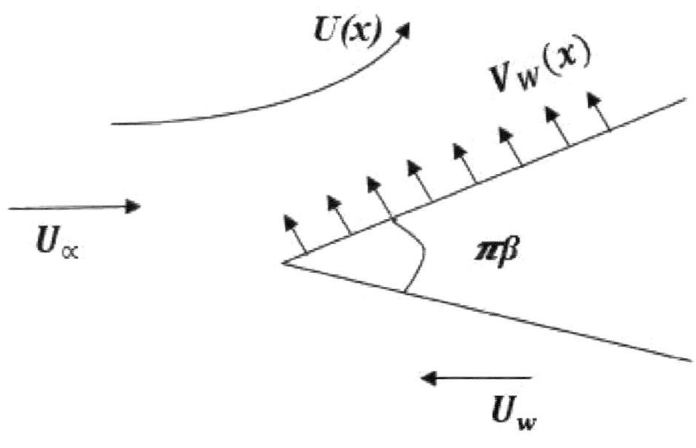

Figure 1. Physical architect of the model and co-ordinate consideration

Figure 1 represents the physical architect of the model and the coordinate system of the model. Cartesian co-ordinate is considered for this problem, x- axis is considered along the wedge and the normal to the wedge represent y-axis. There are some assumptions to form the governing equations-

1) The flow is steady.
2) The flow is two-dimensional and laminar.

3) The working fluid is viscous and incompressible.

The wedge has a uniform velocity U_w opposite to the free stream velocity U_α, the lateral mass flux from the moving wedge with a non-uniform velocity $V_w(x)$ and the exterior flow velocity is $U(x)$. Operating equations for continuity and momentum, under boundary layer estimation presented by [16]

$$\frac{\partial u}{dx}+\frac{\partial v}{dy}=0 \ , \tag{1}$$

$$\rho\left(u\frac{\partial u}{\partial x}+v\frac{\partial v}{\partial y}\right)=U\frac{du}{dx}+\mu\left(\frac{\partial^2 u}{\partial x^2}\right) \ , \tag{2}$$

where u and v represent the velocity components of the fluid along x and y axis respectively. μ is the dynamic viscosity of the fluid.

Now consider the boundary conditions for these equations are

for $x>0$

$$u(x,0)=u(x)=-\mathrm{U_w}\ (x/L)^{\mathrm{m}} \ ,$$

$$v(x,0)=V_w(x),$$

$$u\to U(x)=U_\alpha(x/L)^m \text{ as } \mathrm{y}\to\infty \ , \tag{3}$$

Applying the boundary conditions (3) to solve the equations (1) and (2) and similarity variables are chosen as following-

$$\psi=x^{\frac{1+m}{2}}\sqrt{\frac{2vU_\alpha}{(1+m)L^m}}f(\eta) \ , \tag{4}$$

$$\text{where } \eta=\sqrt{\frac{(1+\mathrm{m})U_\alpha}{2vL^m}}\left(\frac{y}{x^{(1-m)/2}}\right) . \tag{4a}$$

ψ is defined as the stream function which means

$$u=\frac{d\psi}{dy} \text{ and } v=-\frac{d\psi}{dx} . \tag{4b}$$

Now adopting the correlation among u,v and η from [4] and use this correlation and substitute it on Eqn 2 to establish a ordinary differential equation as

$$f'''+ff''+\beta[1-(f')^2]=0. \tag{5}$$

This equation is a very well-known equation in the field of boundary layer theory; known as the Falkner-Skan boundary-layer equation with boundary conditions

$$f(0)=f_0 \ , f'(0)=-\lambda \text{ and } f'(\alpha)=1, \tag{5a}$$

$$\text{where } \beta=\frac{2m}{m+1} \text{ and } \lambda=\frac{U_w}{U_\alpha} , \tag{5b}$$

β is the parameter which used to measure the pressure gradient along x-axis and λ denotes relation between velocity of the surface and the mainstream.

3. Results and Discussions

In this part, the numerical results are placed from Fig. 2 to Fig.6. The effects of positive and negative β, positive and negative λ for positive β and the suction or injection effect on the velocity with respect to a similarity variable (η) has been described in a set of boundary conditions. The variation scale of the wedge angle is very large. In this study the variation is considered in between -0.66 to 0.33.

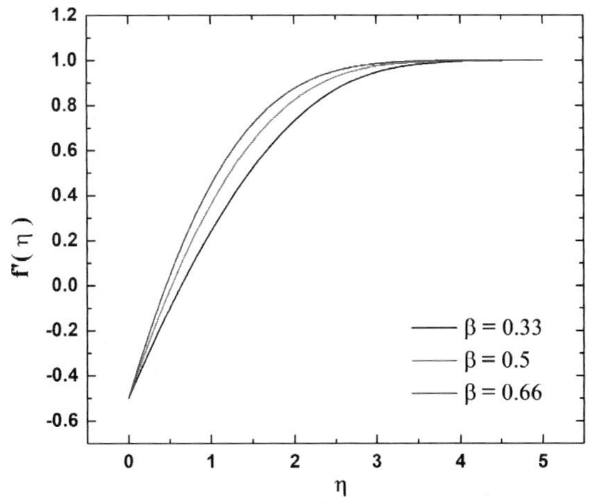

Figure 2. Dimensionless velocity profiles for different positive values of β

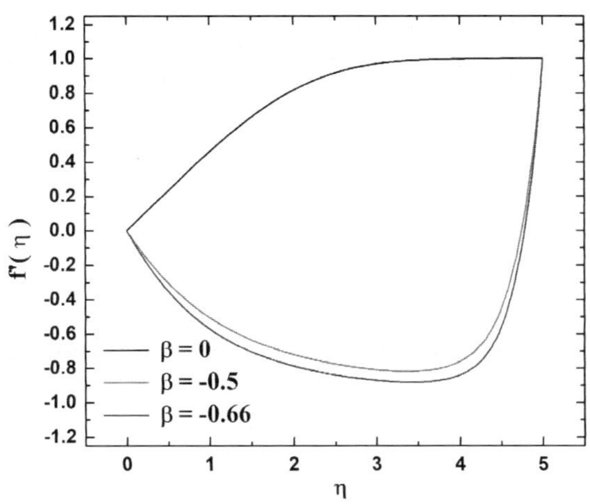

Figure 3. Dimensionless velocity profiles for zero and different negative values of β

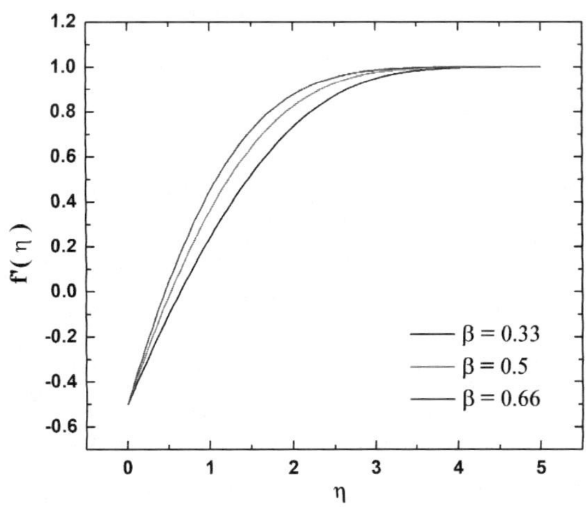

Figure 4. Dimensionless velocity profiles for positive velocity ratio for different β

Figure 2 describes the effects of positive wedge angle without injection or suction and zero velocity ratio between surface and mainstream. It is observed from the Fig.2 that the as wedge angle increases the boundary layer thickness is reduced. Applying the same conditions the reverse phenomenon is noticeable as evident Fig.3. $\beta < 0$ indicates the flow is under as adverse pressure gradient. Due to this unfavourable or adverse pressure gradient effect the flow is reversed near the wall of a stationary wedge. In case of zero wedge angle the wedge acts as a flat plate and the nature of the boundary layer is same as flow past over flat plate. Figure 4 and 5 indicates the variation of velocity profile for positive velocity ratio (+1) and a negative velocity ratio (-1) between wedge and mainstream without injections or suction.

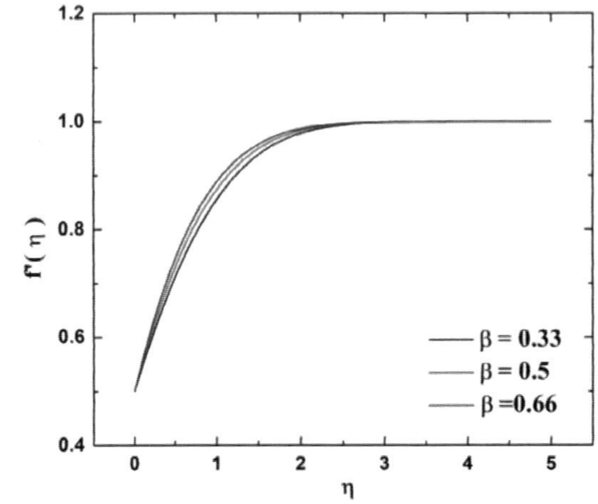

Figure 5. Dimensionless velocity profiles for negative velocity ratio for different β

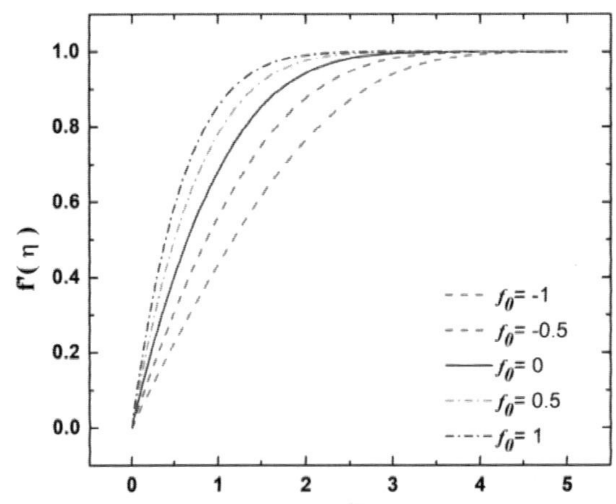

Figure 6. Effect of suction and injection on dimensionless velocity profiles

The positive wedge angle is considered because the positive wedge angle always gives a favourable pressure gradient. In this case, positive λ means the wedge moves opposite to the mainstream flow. The same pattern is reflected as previous that with the increasing wedge angle the boundary layer gets thinner. When the mainstream and wedge moves in the same direction the nature of the boundary layer is same as positive velocity ratio, but it is noticeable that the effects of the wedge angle on boundary layer thickness is less prominent in this case as evidenced Fig.5. Figure 6 represents the effects of suction and injection on the flow past wedge. The negative f_0 indicates the suction effects and positive f_0 indicates injection. The suction and injection on the wedge has an important role in the development of the boundary layer. In case of injection, it may happen that extra momentum is added to the system which intern increases the velocity of the fluid particle inside the boundary layer, so it achieves the free stream velocity much faster than without injection, the effect is reversed for suction. So, the boundary layer thickness is increasing with increasing of suction effects and opposite phenomena is taking place in the case of injection.

4. Conclusions

The present work explores the feature of the dimensionless velocity profile of a Newtonian fluid past over a moving wedge considering the effects of suction and injection. The importance of pressure gradient along with the flow and how it depends on the wedge angle is recognized. The relation between the dimensionless velocity with the velocity ratio is captured for the positive β, so the importance of wedge velocity with respect to mainstream velocity on the velocity profile is established. Lastly, the effect of suction and injection of the wedge may ensure stress development. The mathematical analysis of the Falkner-Skan boundary-layer equation is conducted thoroughly under a wide range of parametric conditions that explore the parametric effects on the dimensionless velocity which explore the effect of various parameters on momentum boundary layer thickness.

References

[1] Massey B S and Ward Smith J 1998 Mechanics of fluids (Vol. 1) Eight Edition *Taylor & Francis London and New York* 299-302.
[2] White F M 1979 Fluid mechanics Fourth Edition Tata McGraw-Hill Education 475-477.
[3] Falkner M V and Skan W S 1931 Some approximate solutions of the boundary-layer equations *Phiols Mag.* **12** 865-896.
[4] Ishak A, Nazar R and Pop I 2007 Falkner-Skan equation for flow past a moving wedge with suction or injection *J. of Applied Mathematics and Computing* **25(1-2)** 67-83.
[5] Afzal N 2010 Falkner–Skan equation for flow past a stretching surface with suction or blowing analytical solutions *Applied mathematics and computation* **217(6)** 2724-2736.
[6] Hussaini M Y, Lakin W D and Nachman A. 1987 On similarity solutions of a boundary layer problem with an upstream moving wall. *SIAM J. on Applied Mathematics* **47(4)** 699-709.
[7] Watanabe T. 1990 Thermal boundary layers over a wedge with uniform suction or injection in forced flow *Acta Mechanica* **83(3-4)** 119-126.
[8] Kudenatti R B and Awati V B. 2009 Solution of pressure gradient stretching plate with suction *Applied mathematics and computation* **210(1)** 151-157.
[9] Heeg R S, Dijkstra D and Zandbergen P J 1999 The stability of Falkner-Skan flows with several inflection points Zeitschrift für angewandte *Mathematik und Physik ZAMP* **50(1)** 82-93.
[10] Mansutti D, Pontrelli G, and Rajagopal K R 1993 Steady flows of non-Newtonian fluids past a porous plate with suction or injection *Int. J. for Numerical Methods in Fluids* **17.11** 927-941.
[11] Garg V K and Rajagopal K R 1991 Flow of a non-Newtonian fluid past a wedge *Acta Mechanica* **88(1-2)** 113-123.
[12] Marinca V, Ene R D and Marinca B 2014 Analytic approximate solution for Falkner-Skan equation *The Scientific World Journal* AID **617453** 1-22.
[13] Khan W A and Pop I 2013 Boundary layer flow past a wedge moving in a nanofluid *Mathematical Problems in Engineering*, AID **637285** 1-4.
[14] Ullah I, Khan I, and Shafie S 2016 Hydromagnetic Falkner-Skan flow of Casson fluid past a moving wedge with heat transfer *Alexandria Engineering Journal* **55(3)** 2139-2148.
[15] Xu L. 2020 Accelerating viscous flow past a wedge *Physics of Fluids* **32(1)** 013604.
[16] Kuo B L, 2003 Application of the differential transformation method to the solutions of Falkner- Skan wedge flow *Acta Mechanica* **164** 161-174.

Nomenclature

L	Characteristic length, m^2
m	Flakner-Skan power law parameter
u,v	Velocity component along x and y axis, m/s
U_w, U_α	Wedge velocity and main stream velocity respectively m/s
V_w	Suction or injection velocity, m/s
μ	Dynamic viscosity, Pa-s
β	Parameter for the pressure gradient along x-axis, Pa/m
λ	Velocity ratio between wedge velocity to mainstream velocity

Design of Solar Powered Desiccant based Air Handling Unit

Vyom Chaturvedi, Devendra Kumar Vishwakarma, Suvanjan Bhattacharyya*

Department of Mechanical Engineering, Birla Institute of Technology and Science, Pilani, Pilani Campus, Pilani 333031, Rajasthan, India.
*e-mail: suvanjan.bhattacharyya@pilani.bits-pilani.ac.in

Abstract. Using an air handling unit (AHU) is a suitable method to meet the ventilation requirements in a conditioned space. This paper proposes two novel design configurations of AHUs, which ensure that the excess load on the AHU cooling coil is minimized using air to air heat recovery units, a desiccant based rotary dehumidifier, and an evaporative cooler. The proposed design is also capable of harvesting surplus water in the process. The impact assessment of the proposed designs is illustrated taking the example of a typical high-humidity design case. Simplified mathematical models were used to analyze the power-saving and moisture harvesting potential of the two configurations. Calculations show that both the configurations reduce the excess load on the cooling coil by a significant amount, while also co-producing surplus water for domestic use.

1. Introduction:

Ventilation requirements are one of the major concerns while designing an HVAC system [1]. Indoor air quality and proper ventilation have become even more relevant with the advent of deadly pandemics like COVID-19. Generally, depending on the occupancy of the building, the requirement of outdoor air (fresh air) is specified. Using an air handling unit (AHU) is a suitable method to meet the air quality and flow requirements in a conditioned space. Due to the difference in temperature and humidity content between the conditioned space and ambient, excess cooling capacity has to be installed over and above the sensible and latent heat gains of the building. To meet the fresh air requirements with the minimum amount of energy, the outdoor air is to be brought closer in temperature and humidity to the conditioned space. This can be accomplished using heat recovery wheels and multiple other innovative technologies. One such technology is the use of desiccant dehumidification, utilizing solar energy. Here, the supply air is dehumidified in the desiccant wheel without cooling the air below its dew point. The process requires low-grade energy in the form of heat, which can be generated from renewable and waste heat sources. Multiple simulation-based studies are available to illustrate the potential of this technology in energy-efficient space cooling applications [2]. In this paper, two innovative designs that couple desiccant dehumidification and air to air heat recovery are studied.

2. Configurations:

The following two configurations are proposed, varying the arrangement of components and air streams. The impact of both these configurations is then compared. The excess moisture in the outlet stream of regenerative air through the desiccant dehumidifier in both the configurations is condensed to produce water, a part of which is used in the evaporative cooler. The regenerative air stream is preheated in the heat recovery unit and further heated using a solar thermal collector. Water in heating the coil has been assumed to be around 60-70°C. Other means to heat the air can also be used subject to the local availability of the resources.

2.1. Configuration 1:

As illustrated in Figure 1, the outdoor air passes through a rotary desiccant dehumidifier where it loses moisture and gains heat. Thereafter, it exchanges sensible heat successively with ambient air and cold air from the evaporative cooler, before passing through the cooling coil. The desiccant dehumidifier is regenerated by the hot air stream coming from heat recovery wheel 1, which is additionally heated using the solar thermal heating coil.

2.2. Configuration 2:

Configuration 2, as shown in Figure 2 is similar to Configuration 1, different primarily in only two aspects. The inlet air streams for the evaporative cooler and regenerative stream for the desiccant dehumidifier have been interchanged. Also, the supply air only goes through a single heat recovery wheel (in contrary to two heat recovery wheels in Configuration 1).

3. Design Conditions:

To illustrate the usefulness of these configurations, a typical high-humidity design condition is assumed. All calculations and results are based on the following conditions.

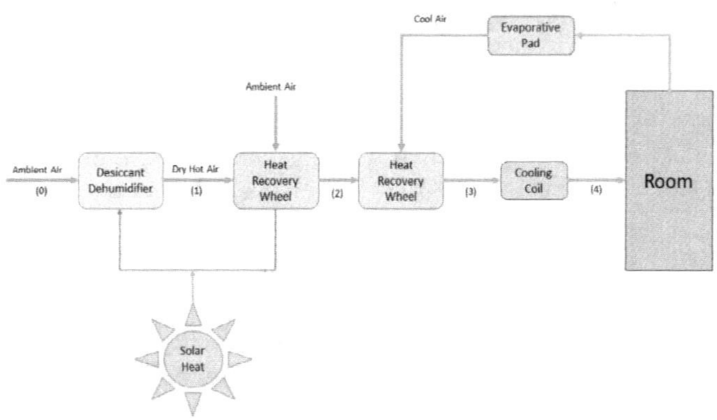

Figure 1. Schematic of configuration 1.

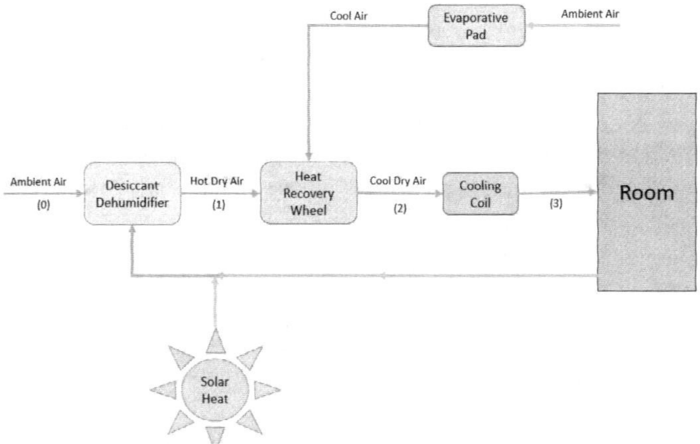

Figure 2 Schematic of configuration 2

Indoor Temperature: 24°C
Indoor RH: 50%

Outdoor Temperature: 35°C
Outdoor RH: 50%

Sensible cooling load (Q_h): 400kW
Latent cooling load (Q_l): 100kW

Number of occupants: 90
Fresh air requirement: 30 CFM per person (From ASHRAE Ventilation Standard)
Inlet mass flow required: 1.5 kg/s

4. Mathematical Models:

For the sake of simplistic analysis, the following mathematical models are used to make the calculations. The assumptions are carefully taken referring to similar studies and other research articles.

4.1. Rotary Desiccant Dehumidifier:

A lot of studies are being done on desiccant dehumidifiers with very complex mathematical models. This paper uses the following simplified equations [3] which were experimentally validated with good accuracy by the authors. The enthalpy (h) and relative humidity (RH) of the outlet stream (represented as *out* in subscript) are calculated from the properties of the inlet (represented as *in* in subscript) and regenerative (represented as *reg* in subscript) air streams. Other parameters were calculated using a psychrometric chart.

$$h_{out} = (0.1312h_{reg} + 0.8688h_{in})$$

$$RH_{out} = (0.9428RH_{reg} + 0.0572RH_{in})$$

4.2. *Evaporative Cooler:*

Evaporative cooling was modeled as an isenthalpic humidification and cooling process. The outlet RH was taken to be 90% which is a safe assumption and has been used in multiple research articles. The process is illustrated for both the configurations in Figure 3 and Figure 4.

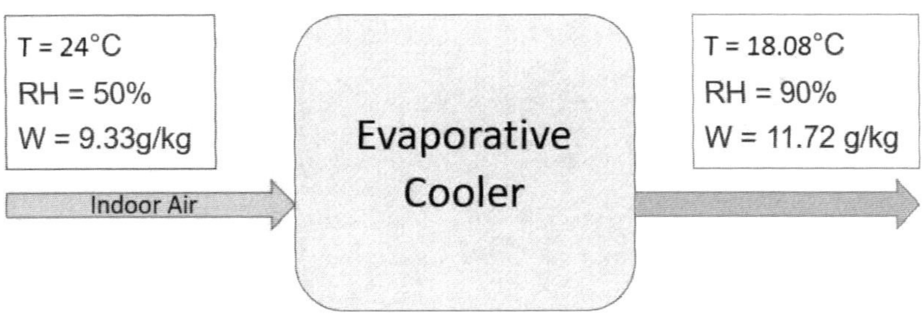

Figure 3 . Cooling Process for configuration 1.

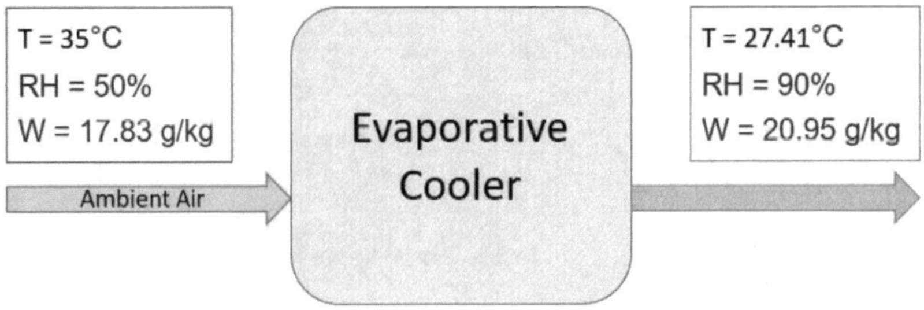

Figure 4 Cooling process for configuration 2.

4.3. Heat Recovery Wheel:
The heat recovery wheel was modeled using sensible heat exchange with no moisture transfer assumption and the efficiency of the process was assumed to be 0.6.

$$Q = \varepsilon . \min\,(C_A m_A, C_B m_b)(T_A - T_B)$$

5. Results:
Applying the aforementioned mathematical models to the proposed configurations yields the following results. The final state of supply air before passing through the cooling coil:

- Configuration 1:
  Temperature (T) = 27.55ºC; Moisture (W) = 12.8 g/kg-da; Enthalpy (h) = 60.48 kJ/kg
- Configuration 2:
  Temperature (T) = 32.72ºC; Moisture (W) = 9.98 g/kg-da; Enthalpy (h) = 58.46 kJ/kg

The results show that both these configurations significantly reduce the excess load on the cooling coil and have great energy-saving potential. Moreover, the water consumed by the evaporative cooler can be harvested from the moisture in the regenerative stream, making it a completely sustainable system in all aspects as shown in Figure 5.

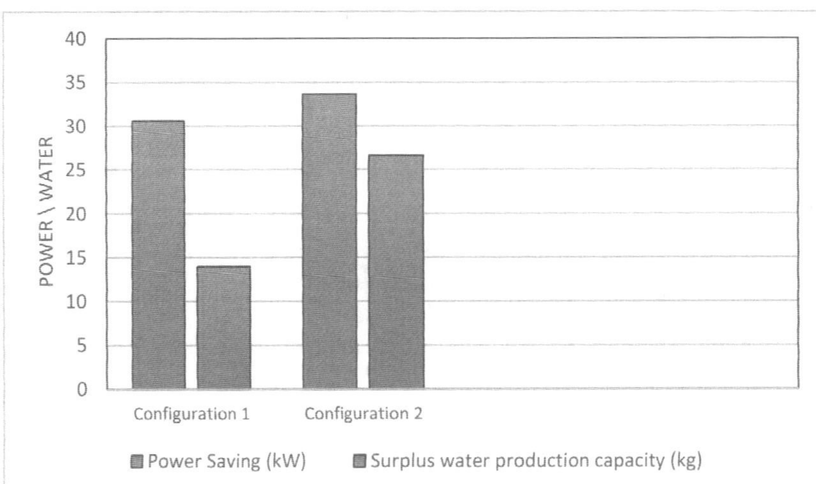

Figure 5. Power and water saving over conventional systems.

The power saving capacity and surplus water consumption of the two configurations are indicated by blue and red bars respectively. Power saving represents the **excess** power that would have been used by a conventional air conditioning system using only a cooling coil.

- Implementing the proposed AHU designs, more than **30kW** power can be saved in both the configurations over a conventional refrigeration cycle–based system.

Surplus water production capacity represents the total moisture that can be harvested from the desiccant dehumidifier minus the water consumed by the evaporative cooler in a particular configuration.

- The graph indicates that both configurations are capable of generating **excess** water. Hence, water scarcity will not be an issue for the evaporative cooler in either configuration.

6. Conclusion and Scope for Future Work

Research studies predict that ventilation is going to be a major requirement in almost every closed space. Getting fresh outdoor air into the conditioned space comes at the expense of a great deal of energy. Heat recovery and desiccant dehumidification combined with evaporative cooling has a great potential for energy saving. Moreover, these technologies can be operated with low-grade energy like solar thermal heat and are hence sustainable. There can be multiple configurations in which these technologies can be arranged and used according to the local ambient conditions.
Furthering this work, a generalized study will be done to predict the ideal configuration for a given set of ambient conditions, using more accurate mathematical models.

References

[1] Shady Attia, Chapter 5 - Occupants Well-Being and Indoor Environmental Quality, Net Zero Energy Buildings (NZEB), Butterworth-Heinemann, 2018
[2] F. Calise, M. Dentice d'Accadia, C. Roselli, M. Sasso, F. Tariello, Desiccant-based AHU interacting with a CPVT collector: Simulation of energy and environmental performance, Solar Energy, Volume 103, 2014
[3] Beccali, M., Butera, F., Guanella, R. and Adhikari, R. (2003), Simplified models for the performance evaluation of desiccant wheel dehumidification. Int. J. Energy Res., 27: 17-29. doi:10.1002/er.856
[4] Amirreza Heidari, Ramin Roshandel, Vahid Vakiloroaya, An innovative solar assisted desiccant-based evaporative cooling system for co-production of water and cooling in hot and humid climates, Energy Conversion and Management, Volume 185, 2019.

Nomenclature

HVAC	Heating Ventilation & Air Conditioning
AHU	Air Handling Unit
RH	Relative Humidity
Q_h	Sensible cooling load
Q_l	Latent cooling load
CFM	Cubic Feet per Minute
ASHRAE	American Society of Heating, Refrigerating and Air-Conditioning Engineers
h	Enthalpy kJ/kg
W	Moisture in g/kg-da
T	Temperature in °C

Air Quality Index Effect on Particles Deposition in Human Respiratory Tract under Natural Breath

Digamber Singh[1]

[1,] Department of Applied Mechanics, MNNIT, Prayagraj, Uttar Pradesh–211004, India
e-mail: *digamber35@gmail.com*

Abstract. The huge demands of material goods decayed the breathing air quality. Now, it is reflecting on the clinical respiratory test of humans. Therefore, it is a serious issue that needs to be taken care of. The article throws-up light on important factor such as a deposition of inspired toxic particles under natural breath. The inspired particles are having an aerodynamic diameter 2.5 µm and 10 µm. A three-dimensional CT-scan based Human Respiratory Tract is successfully re-constructed nasal cavity to 7th generation bronchi. The flow is developed during inhalation laminar to turbulent from external nostrils to the tracheobronchial airway due to anatomical effect. It is affected the particles deposition on the various regions due to inertial impaction or curved shape surface of the airway model. The appropriate turbulence model quantifies the internal flow features so that an accurate trajectory of particles transport pattern can be presented. The discrete phase model (DPM) is used for particles tracing on Language frame, which gives the combined effect of anatomical structure and airflow impact on the deposition of particles. It could be useful for the diagnosis of Human airways illness.
Keywords: Air quality index (AQI), Computer tomography (CT), Human Respiratory Tract (HRT) , Computational fluid dynamics (CFD), Discrete phase model (DPM).

1. Introduction

The inhaled micro-size particles deposited into the Human airways. In Asian countries, many states peoples are breathing air below the safety standards. According to the national air quality index in India. The breathing air quality index should below for $PM_{2.5}$ (41 µg/m^3) and PM_{10} (154 µg/m^3). In the particle deposition analysis, nasal cavity played a vital role. Nasal cavity is the prime defence mechanism which filters the inspired air before distributed into the Human airways. However, in that mechanism, the anatomical structure of airways has put great influence on particle deposition. The minor changes in the airway structure are putting a significant effect on particle deposition and characteristic flow (laminar, transitional and turbulent) on different segments [1]. The sudden expansion and contraction in airway passage are increasing the deposition efficiency due to inertial impaction and also increases the computational efforts [2]. With the Weibel model it is very difficult to predict the realistic particle deposition efficiency and airflow characteristics [3]. However with the different breathing conditions, breath hold of 2 s, cannot accomplish particle deposition efficiency in the airway. But it definitely increases deposition efficiency in the bifurcation junctions [4, 5]. The particles deposition efficiency is connected with the inertial force of particles. It is a key factor for the small sizes particles deposition [6]. The particles deposition efficiency in diseased airway relatively with high Reynolds number flow comparing with the asthmatic and healthy airway. Also it is decreasing the lungs function mechanism [7]. The micro-sized particle deposition efficiency is higher in comparison of nano-size particles in the airway [8]. The maximum deposition efficiency was occurs at a high airflow rate during inhalation [9]. In past studies, the particles deposition in-vitro and in-vivo model including with the different size of particles listed in the literature. Although, steady-state numerical findings are reported. The particle deposition with the natural breath in-vivo model of nasal cavity was found very less. Also, in this regards many health risk studies were performed without considering the AQI and natural breathing. The article concern the importance of particle deposition in the HRT model at moderate and very unhealthy AQI. There is no literature available on such a topic. The numerical results are presented the regional particle transport and deposition pattern during peak inspiratory airflow.

2. Human Respiratory Tract Model and Grid Generation

The total 1024 number of CT-scan slices arranged together. The field view of slices 31.60 cm and a pixel size of 0.623 mm. The three-dimensional mask of selected images are created by changing the Hounsfield Unit (HU) (-1024, -200) of tissues. The airways model is ready for computer graphical user interface. In figure 1 the three-dimensional HRT model is shown with descriptive validation.

To discretize the flow domain the unstructured grid is constructing using "ICEM" software. The grid indecency test are performed for the grid elements size of 4.5 mm, 4 mm, 3.5 mm and 3 mm at the airflow rate of 30 L/min. The numerical grid is independent in 4 mm element size. The final grid is having 68, 90,000 elements.

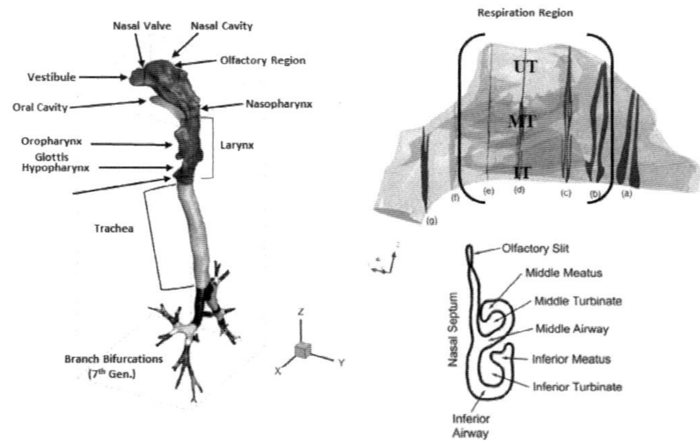

Figure 1. CT-scan based three-dimensional Human Respiratory Tract.

3. Numerical Methodology

The Natural breathing airflow (10 L/min) is simulating shown in figure 2. The following assumptions are considered, (1) Unsteady state incompressible flow, (2) Flow domain wall considered as rigid hydraulically. The conservation of mass (continuity equation) and momentum equation can be written as:

$$\frac{\partial \rho}{\partial t}+\frac{\partial(\rho u_i)}{\partial x_j}=0 \text{ .... (1)}$$

Momentum equation:

$$\frac{\partial u_i}{\partial t}+\frac{\partial(u_i u_j)}{\partial x_i}=-\frac{1}{\rho}\frac{\partial p}{\partial x_i}+\frac{\mu}{\rho}\frac{\partial^2 u_i}{\partial x_i \partial x_j} \text{ .... (2)}$$

Where μ is the dynamic coefficient. The ui, uj, (i, j = 1, 2, 3) are velocity components in the cartesian coordinate in x,y,z directions. The p, ρ, ν stand for the pressure, air mass density (ρ =1.25 kg/m^3) and kinematic viscous coefficient of air.

3.1 Assessments of Turbulence Model

The (Realizable) k–ε model investigates the detailed internal flow information of airways. The similar turbulence model was used [10] in past research. The turbulence model transport equations for k and ε in the realizable are:

$$\frac{\partial}{\partial t}(\rho k)+\frac{\partial}{\partial x_j}(\rho k u_j)=\frac{\partial}{\partial x_j}\left[(\mu+\frac{\mu_t}{\sigma_k})\right]+G_k+G_b-\rho\varepsilon-Y_M+S_k \text{ .... (3)}$$

$$\frac{\partial}{\partial t}(\rho\varepsilon)+\frac{\partial}{\partial x_j}(\rho\varepsilon u_j)=\frac{\partial}{\partial x_j}\left[(\mu+\frac{\mu_t}{\sigma_\varepsilon})\right]+\rho C_1 S_\varepsilon-\rho C_2\frac{\varepsilon^2}{k+\sqrt{v\varepsilon}}+C_{1\varepsilon}\frac{\varepsilon}{k}C_{3\varepsilon}G_b+S_\varepsilon \text{ .... (4)}$$

Where constant:

$C_1=\max\left[0.43,\frac{n}{n+5}\right], n=S\frac{k}{\varepsilon}, S=\sqrt{2S_{ij}S_{ij}}$ .... (5), $C_2=1.9$ and $C_\mu=0.09$, where, S is the modulus of the mean rate of strain tensor, defined as $S=\sqrt{2S_{ij}S_{ij}}$ .....(6), μ is the dynamic viscosity of the fluid. Eddy viscosity of the fluid is given as $\mu_t=\rho C_\mu\frac{k^2}{\varepsilon}$ ......(7), the k–ε turbulence model appears to be capable of reproducing the limiting laminar and fully turbulent flow.

3.2 Discrete Phase Model

To solve the trajectory of particles are considering the effect of drag force and gravity at integrating force balance equation, therefore numerical equation is written on a Lagrangian frame.

$$\frac{du_p}{dt} = F_D(u-u_p) + \left[\frac{g(\rho_p-\rho)}{\rho_p}\right] \quad \ldots\ldots (8)$$

Here, U is the fluid velocity. Up is the particle velocity. μ is the molecular viscosity of the fluid. ρ_p is the density of the particle. dp is the particle diameter. Where F_D (U- Up) is the drag force per unit injected particle mass in the airway model, written as

$$F_D = \frac{18\mu}{\rho_p d_p^p}\frac{C_D \mathrm{Re}}{24} \quad \ldots\ldots\ldots (9)$$

R_e is the relative Reynolds number, written as

$$\mathrm{Re} \equiv \frac{\rho d_p |u_p - u|}{\mu} \quad \ldots\ldots\ldots\ldots (10)$$

The density of particles is considered as equivalent to water (1000 kg/m^3). The mass of PM $_{2.5}$ and PM_{10} are injecting according to moderate and very unhealthy AQI.

3.3 Numerical Schemes

The individual velocity inlet UDF (10 L/min) is defined at both nostrils with hydraulic diameter 0.0141 m and 0.0126 m left and right nostril respectively. The trapped boundary conditions are imposing on the flow wall. At the inlet and all outlets defined as reflecting boundary conditions. Zero-Gauge pressure has established at the outlets. The internal fluid dynamics is simulated by the finite-volume based solver. The pressure-velocity coupling SIMPLE-C was adopted. The QUICK transport equations were discretized up to second-order accurate in space. The adaptive time method is used to capture the local features of the airway.

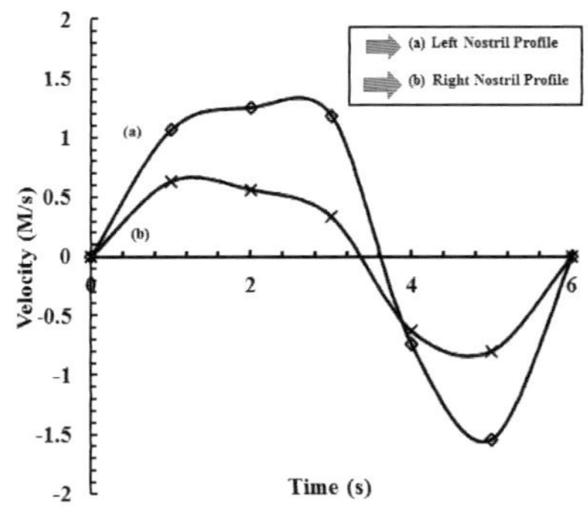

Figure 2. Natural breathing profile for normal volumetric flow (10 L/min).

4. Results and Discussions

4.1 Particle Transport and Used Turbulence Model Comparison

The transport and deposition efficiency of particles in the nasal cavity is compared with the experimental results of Kelly et al. [11] shown in figure 3 (A). The CT-scan based nasal cavity is used for the validation. It is necessary to know the quantified value of particle mass deposited inside the airway. The regional deposition efficiency is a good parameter. The deposition efficiency is defined as the ratio of particle mass deposited in the trap region/mass of particles injected. In HURT model, the sudden anatomical changes occurred. If the flow magnitude is increasing due cross-sectional area the particle deposition efficiency increases due to inertial impaction (IP). The overall CFD results are showing reasonable agreement with experimental data. Also during inhalation, the higher resistances occurred at the nasal valve because of narrow cross-section. Which means pressure drop occurs at the nasal valve in [12]. The flow validation is carried out with the computational results of Wang et al. [10]. The average pressure drop is calculated at the nasal valve with various inspiratory airflow rate. The flow characteristics show a good agreement with the simulated results are shown in figure 3 (B).

4.2 Inspired Particles Transport and Deposition Pattern

In figure 4 show the airflow streamlines and transport pattern of the inspired particles at moderate and very unhealthy AQI for $PM_{2.5}$ and PM_{10} during peak inspiratory flow at t = 2.1 s. The particles flow inertia, geometrical shape and gravitational force are affected the transport pattern of particles within the airways.

During normal breathing, airflow is laminar [13] in the nasal cavity. As the airflow magnitude increases in distinct segments of airways.

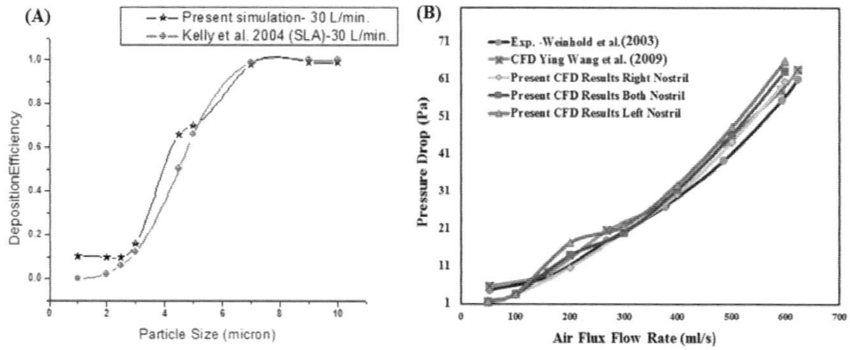

Figure 3. (A) The deposition efficiency in the nasal cavity due to the inertial mechanism, (B) The avg. pressure drop across the nasal valve during inhalation.

The particles deposition efficiency is varying due to asymmetric velocity inlet. It is observed that initially, the left and right nostril particles transport pattern are not similar. Although inspired air is the carrier for the injected mass of inspired particles. As the airflow streams are moving the particles also moving and deposited on the flow wall as come in to contact due to mucous. The airflow streamlines interact with the transport pattern in the middle turbinate and olfactory region. As increasing the AQI (201-300) the transport pattern of particles is changed in the nostrils and olfactory region as well. The airflow is progressed towards the nasopharynx region flow streams are merging in that region. Which is coming from both the nostrils cavities. Due to the sudden proximal bend of 90^0 particles start accelerating because of high flow magnitude. The anatomical structure of pharynx has influenced the transport and deposition patterns. Due to high flow velocity magnitude, the particle is concentrating middle of the glottis because of the nozzle effect it is moving towards the anterior side of the pharynx. The deposition of particles must be increased in pharynx larynx. The trachea has a straight pipe. But in the real model, it has the cartilage rings which is affecting the airflow streamlines locally.

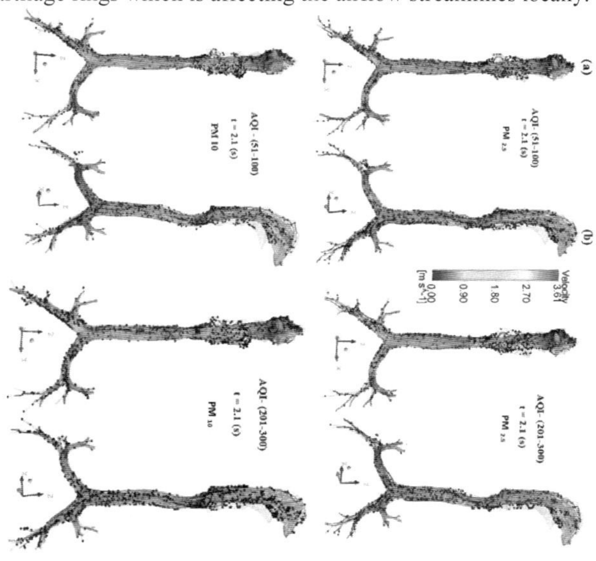

Figure 4. The inspired particles transport pattern and velocity streamlines during moderate and very unhealthy AQI for PM $_{2.5}$ and PM_{10} at [t = 2.1 s] respectively: (a) Front view. (b) Isometric view.

Therefore at the entrance of trachea, the particles influenced by glottis nozzle effect. In the other side the development of inspiratory flow in trachea prominently particles concentration at the posterior side. Also, the particles transport pattern shifted towards the right side instead of left side bronchus and low flow velocity zone is formed in the right side before the first bifurcation seen in figure 4. The inspired particles are non-uniformly strike to main bronchus carina and partitioned left and right main lungs asymmetrically. However, the inspired flow inertia is the primary mechanism to penetrated particles in subsequent branches of lungs. Therefore more mass of inspired particles transported and deposited in the left lungs that's why total deposition efficiency is higher in left lungs bifurcation. The PM_{10} particles trends are different as compared with $PM_{2.5}$. It shows a higher deposition in the upper airways. The particles deposition efficiency is compared for 2.5 μm and 10 μm particles. The left lunges have higher deposition efficiency 1.2% as compared to right lungs during AQI (51-100) similarly as increase the AQI (201-300) it is increasing up to 1.38% for $PM_{2.5}$. Also for PM_{10}, it will become 2.24% higher in left lungs during AQI (51-100) and 2.48% during AQI (51-100) and AQI (201-300) for PM_{10}. At a particular instant significant differences occurred in transport and deposition pattern of particles in airway after changing AQI moderate to very unhealthy.

Conclusions

The current study shows the following conclusions.

1. The anatomical structure of HURT model distinct the transport pattern of injected mass in left and right side nostrils in moderate and very unhealthy AQI.
2. The development of airflow in the trachea, the inspired mass of particles shifted towards the right side because of low flow velocity zone in trachea. The particles are deposited in left lungs.
3. The higher deposition efficiency is found in daughter branches due direct impact of mass on cranial. The prominent difference occurred in deposition of particles after changing the AQI moderate to very unhealthy.

Reference

[1]. Jeffry D. Schroeter, Guilherme J.M. Garcia, Julia S. Kimbell, (2011). Effect of surface smoothness on inertial particle deposition in human nasal cavity, *J.Aero.Sci.*, 42, 52-93.

[2]. Doorly, D. J., Taylor, D. J., Gambaruto, A. M., Schroter, R. C., Tolley, N. (2008a). Nasal architecture. Philosophical Transactions, *Royal Society A*, 366, 3225–3246.

[3]. Zhe Zhang, Clement Klinstreuer, Chong S. Kim, (2009). Comparison of analytical and CFD models with regard to micron particle deposition in a human 16-generation tracheobronchial airway model, *J. Aero. Sci.*, 40, 16-28.

[4]. Kio Inthavong, Lok-Tin Choi, Jiyuan Tu, Songlin Ding, Francis Thien, (2010). Micron particle deposition in the tracheobronchial airway model under different breathing conditions, J. *MED. Engg. & Phys,*32, 1198-1204..

[5]. Vivek Kr. Srivastava, A.R. Paul, Anuj Jain, (2012). Computational fluid dynamics study of airflow and particle transport in third to sixth generation human respiratory tract, Int. *J. of Emg. Multi fluid Sci*, 3, 227-234.

[6]. Huang Jian Hua, Zhang Lian Zhong, (2011). Micro-particle deposition and lobar distribution of mass flow in the human upper respiratory tract model, Chinese Sci Bull, 56, 380-385.

[7].Wei-Hsin Chen, Kuo-Hao Lee, Justus Kavita Mutuku, Chii- Jong Hwang, (2018). Flow dynamics and PM 2.5 deposition in healthy and asthmatic airway at different inhalation statues, Aerosol and air quality research, 18, 886-883.

[8]. Y. D. Shang, K. Inthavong, J.Y.Tu, (2015). Detailed micro-particle deposition patterns in the human nasal cavity influenced by breathing zone, Computers and fluid, 114, 141-150.

[9]. M. Pirhadi, B. Sajadi, G. Ahmadi, D. Malekin, (2018). Phase change and deposition of inhaled droplets in the human nasal cavity under cyclic inspiratory airflow, Journal of aerosol science, 118, 64-81.

[10]. Ying Wang, Yingxi Liu, Xiuzhen Sun, (2009). Numerical analysis of respiratory flow patterns within human upper airway, Acta Mech Sin., 25, 737-746.

[11]. James T. Kelly, Bahman Asgharian, Julia S. Kimbell, and Brain A. Wong, (2004). Particle deposition in human nasal airway replicas manufactured by different methods. Part 1, Aerosol science and technology, 38, 1063-1073.

[12]. Ivo Weinhold, Gunter Mlynski, (2003). Numerical simulation of airflow in the human nose, Eur Arch Otorhinolaryngol, 261, 452-455.

[13]. Debo Li, Qisheng Xu, Yaming Liu, Yin Libao, Jin Jun, (2014). Numerical simulation of particles deposition in a human upper airway, Advances in mechanical engineering volume 14.

[14]. Calmet, H., Gambaruto, A. M., Bates, A. J., Vázquez, M., Houzeaux, G., & Doorly, D. J. (2016). Large-scale CFD simulations of the transitional and turbulent regime for the large human airways during rapid inhalation. Computers in Biology and Medicine, 69, 166–180.

IOP Conf. Series: Materials Science and Engineering 1080 (2021) 012040 doi:10.1088/1757-899X/1080/1/012040

Mathematical Model Development of a Solar Assisted Liquid Desiccant Air Conditioning System with Fin Coil Dehumidifier

S Paul[1], S Chowdhury[2], T Saha[3] and A Ganguly[1]

[1, 2,3]Department of Mechanical Engineering,
Indian Institute of Engineering Science and Technology, Shibpur, Howrah 711103, India
e-mail: paulsubhechchha@gmail.com

Abstract: A novel scheme of a solar assisted, liquid desiccant supported comfort air conditioning system has been proposed in this paper. The primary intent of the system is to supply air at 25°C and 52% relative humidity, even during the hot and humid summer months prevailing in the plains of sub-tropical countries like India. A mathematical model has been put forward to predict the output air temperature, humidity ratio and the effectiveness of dehumidifier as well as the regenerator. The predicted air humidity ratio has been validated against data available in the literature. The model has been used to predict the optimum mass flow rates required to achieve the desired output conditions for the selected location (Kolkata). The study thus strengthens the need and feasibility of solar assisted desiccant based air conditioning systems for the sub-tropical climates as a replacement of the traditional VCR (vapor compression refrigeration) systems.

1. Introduction

Economic development and rising standards of living have made heating, ventilation and air conditioning (HVAC) a necessity, rather than a luxury. Air conditioning is an integral part of HVAC sector and studies have shown that about 50% of the total energy expended in buildings, is consumed by air conditioning systems alone [1]. This figure is even higher for countries witnessing tropical or sub tropical climates. The conventional VCR systems have exhibited poor control capacity, when it comes to variations in latent and sensible heat loads [2]. Damped spaces are often formed within such systems, which enhance the likelihood of mould and bacterial infestations. Moreover, refrigerants used by vapor compression systems have been known to cause ozone layer depletion and are responsible for making detrimental changes to the environment. As energy and environmental issues are ubiquitous, developing environmentally benign and sustainable energy technologies have become the need of the hour. Against this backdrop, liquid desiccant air conditioning (LDAC) system can be a promising technology to address this problem. Unlike conventional vapor compression systems, LDAC can handle both the latent and sensible loads independently, thereby reducing energy consumption up to 50% [3]. Although initial cost of LDAC systems have been found to be high, yet studies show that they have the potential to lower the overall cost by 54%[4]. Regeneration of liquid desiccant solution at temperatures lower than 80^{0}C has provided such systems an edge over the solid desiccant air conditioning systems [5]. At the same time such systems can be powered using the solar energy, which is available in plenty in the plains of the Gangetic Bengal in the Indian sub-continent. With 250-300 sunny days per year and an average solar radiation of 200 MW/m^2, India has a huge demand for air conditioning systems and the potential to power such systems using the solar energy [6].

In the past studies several novel designs have been proposed by researchers, but very few research works are available that discuss the aspect of integration of the fin-coil type of dehumidifier in a LDAC system that too for the hot and humid climate prevailing in the plains of India. The purpose of this current study is to propose a mathematical model of a LDAC system using fin-coil type of dehumidifier to provide conditioned air at a temperature of 25^{0}C and 50-55% relative humidity which falls well within the standards of human comfort set forth by ASHRAE standard 55-2010. Calcium chloride solution with specific heat capacity of 2.5 kJ/kg°C and, equilibrium enthalpy and equilibrium specific humidity of 31.84 kJ/kg and 11.4 g/kg has been used as the desiccant for the present study.

2. System Description

The layout of the proposed solar driven LDAC system has been shown in Fig. 1. The system essentially comprises a dehumidifier, a regenerator, a cooling tower and heat exchangers. A fin-coil type internally cooled dehumidifier (shown in Fig. 3) as developed by Liu et al.[7] has been incorporated in the proposed system. A forced flow flat plate solar collector has been used as the regenerator for the system (shown in Fig. 4). The collector prototype similar to what has been proposed by Alizadeh and Saman [8] has been adopted for the present study. The water from a cooling tower which supplies chilled water at 12^{0}C [9] has been used in the heat exchangers to cool desiccant and air.

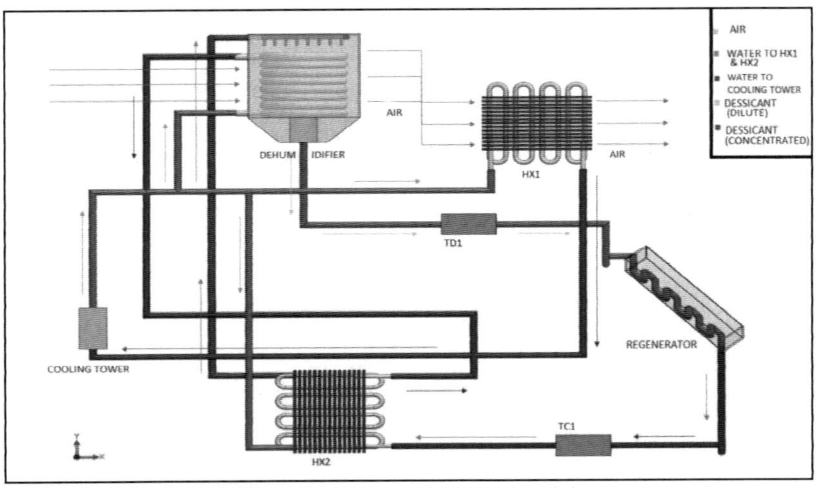

Figure 1. Proposed LDAC model

The schematic consists of two air streams, one of which carries process air (i.e. air to be dehumidified) to the dehumidifier, while the other supplies ambient air to the regenerator. In the dehumidifier, the process air has a higher vapor pressure than the solution of desiccant. As a result of this vapor pressure difference, the moisture flows from the air to the desiccant solution and the air gets relatively dehumidified. The cooling water prevents the desiccant solution from getting heated up, thereby ensuring that with the progress of the dehumidification process, the vapor pressure difference is maintained. The air coming out of the dehumidifier is dry, but still at a higher temperature than that desired to reduce the temperature in the conditioned space. Thus, the same air is passed through heat exchanger HX_1 to supply cool and dry air to the conditioned space. The desiccant solution exiting the dehumidifier is dilute and is stored in a dilute solution tank TD_1. In the regenerator, the dilute desiccant solution has a higher vapor pressure than the incoming ambient air and thus the moisture transfer occurs from the desiccant to the air. The exhaust air is released into the atmosphere and the concentrated desiccant solution is stored in tank TC_1. The concentrated desiccant solution from TC_1 is passed on to heat exchanger HX_2, where the desiccant solution is cooled using chilled cooling water coming from the cooling tower. Thereafter, the relatively cool and concentrated desiccant is pumped into the dehumidifier and the cycle continues.

3. Mathematical Model

Assumptions made while developing the mathematical model are as follows [9][8]:

i. Desiccant is well spread between the tubes, and there is no direct heat transfer between the cooling water and the incoming humid air.
ii. The regenerator unit is considered to be in steady state, and the temperature gradient in the thin desiccant film in the regenerator is not considered.
iii. The heat of vaporization for the water from the weak solution and the specific heat capacities of air and desiccant are constant for the operating range of temperatures.
iv. The rear loss coefficient of the collector is neglected since the bottom end is sufficiently coated.

Based on the assumptions made, three types of heat and mass transfer processes take place inside an internally cooled dehumidifier. For heat and mass transfer processes taking place in the dehumidifier, the equations for energy conservation are given as follows [9]:

$$\dot{m}_a dh_a + d\left(\dot{m}_s h_s\right) - c_{p,w}\dot{m}_w dT_w = 0 \quad (1)$$

Where ṁ, h, c_p and T represent mass flow rate, enthalpy, specific heat capacity and temperature respectively. Throughout the paper, the subscripts a, s and w have been used to denote air, desiccant and water respectively. Whereas subscripts i and o has been used to denote inlet and outlet respectively.
The heat transfer between cooling water and desiccant can be expressed as follows [9]:

$$c_{p,w}\dot{m}_w \frac{dT_w}{dx} = \frac{K_h F_h}{H}(T_w - T_s) \quad (2)$$

K_h and F_h denote heat transfer coefficient and heat transfer area between cooling water and desiccant. H is the height of the dehumidifier and x represents distance along the height of dehumidifier.
Along the height of the dehumidifier, the heat transfer and moisture removal can be denoted in terms of change in humidity ratio and enthalpy of air as follows [9]:

$$\dot{m}_a \frac{\partial h_a}{\partial x} = \frac{K_m F_m}{H}(h_a - h_e) \quad (3)$$

$$\dot{m}_a \frac{\partial \omega_a}{\partial x} = \frac{K_m F_m}{H}(\omega_a - \omega_e) \quad (4)$$

Where Km and Fm denote mass transfer coefficient and mass transfer area between air and desiccant. h_e and ω_e represent enthalpy and specific humidity of air in equilibrium with the desiccant.
In the regenerator, the mass balance equation can be represented using Eq. (5), where $\dot{m}_r$ represents the rate of desorption.

$$\dot{m}_r = \dot{m}_a(\omega_{ao} - \omega_{ai}) \quad (5)$$

Equation (6) shows the relations between the concentrations of desiccant (X_s) with the rate of desorption, mass flow rate of desiccant and the initial desiccant concentration [8].

$$\frac{1}{X_s} = \frac{1}{X_{si}}\left(1 - \frac{\dot{m}_r}{\dot{m}_s}\right) \quad (6)$$

Energy balance can be represented by Eq. (7), where I_L denotes the mean solar power collected by the collector.

$$I_L = \dot{m}_r h_{fg} + \dot{m}_s c_{ps}\Delta T_s + \dot{m}_a c_{pa}\Delta T_a \quad (7)$$

The effectiveness of dehumidifier (η_{dh}) and effectiveness of regenerator (M_{ev}) has been computed as follows:

$$\eta_{dh} = \frac{\omega_{ai} - \omega_{ao}}{\omega_{ai} - \omega_{e}} \text{ and } M_{ev} = m_s\left(1 - \frac{X_{in}}{X_{out}}\right) \tag{8}$$

4. Results and Discussions

A Computer code in MATLAB 2019 R3 has been developed based on the model discussed in the previous section. The program takes ambient air temperature and humidity ratio as input and predicts output temperature and humidity ratio of air keeping mass flow rates of air, desiccant and water as well as concentration of desiccant as constant. The air humidity ratio at the outlet has been validated (shown in Fig 2) against a data available in the literature [9]. The predicted results deviate only by 3.47% and hence show good agreement with data available in literature. Figures 3(a) and 3(b) represent the hourly variation of temperature and humidity inside the conditioned space with the proposed LDAC system respectively for the mean day of the months during May, June and July in Kolkata. These months correspond to the summer and the monsoon season for the place when both the surrounding relative humidity and temperature levels are high. It is observed that the variation in the temperature and relative humidity closely follow the daylong pattern of the variation in the ambient temperature and humidity, first increases, reaching a maxima around the noon and then again diminishes. However, it is found that with the proposed system, both the temperature and the humidity can be restricted within the target value (25^{0}C with a relative humidity of 51.19% that is equivalent to around 10.3 g/kg of specific humidity) for major time of the day for the months under consideration. Fig. 3(c) and 3(d) depicts the hourly variation of dehumidifier efficiency and evaporation rate in regenerator, respectively, for a representative day in May, June and July keeping all mass flow rates and desiccant concentration as constant. The system could achieve dehumidifier efficiency of about 64%, significantly higher than 49% as presented in[7]. This increase in efficiency can be attributed to higher concentration of desiccant (45% $CaCl_2$ solution) and higher mass flow rates of water and desiccant (0.28 kg/s and 0.1 kg/s).The evaporation rates achieved by the model for the set parameters is higher than that achieved by [8]. The cooling capacity has been found to be approximately 3 kW.

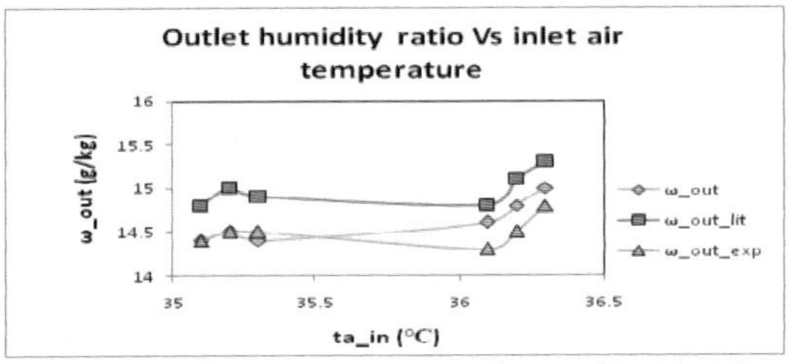

Figure 2. Outlet air humidity ratio Vs inlet air temperature for predicted and published data

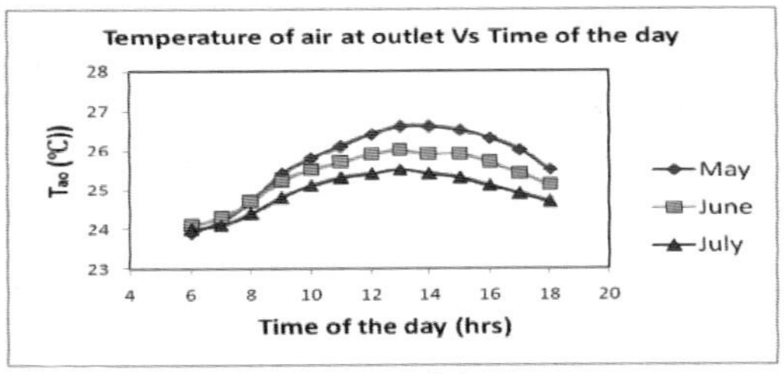

(a)

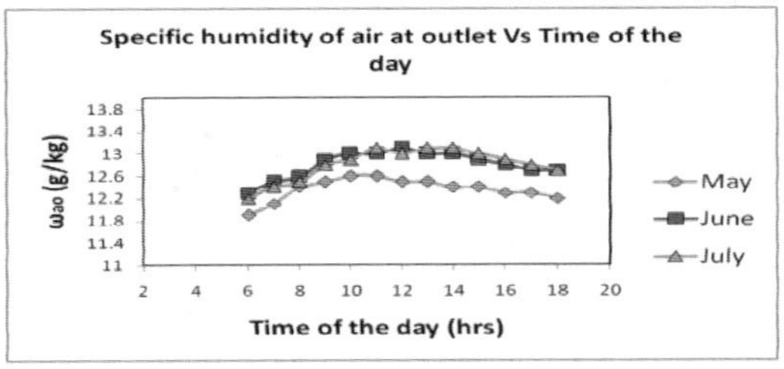

(b)

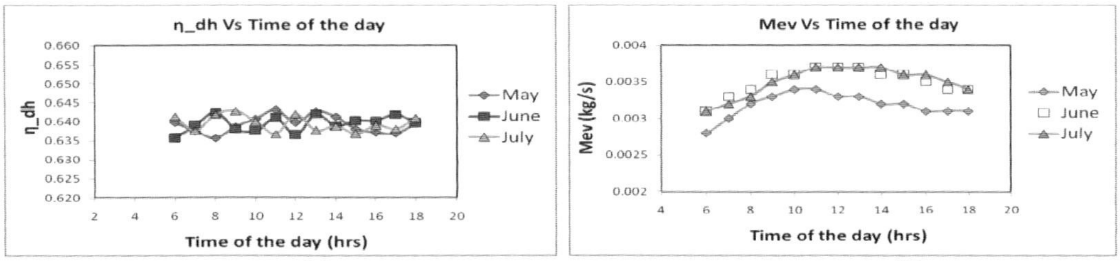

(c) (d)

Figure 3. Hourly variation of (a) output air temperature, (b) output air humidity, (c) dehumidifier efficiency and (d) evaporation rate of regenerator, for the mean day of the month for May, June and July in Kolkata.

Figure 4 shows the variation of outlet air temperature and relative humidity versus air mass flow rate keeping desiccant and water mass flow rates as constant. The desiccant concentration is also kept constant. It is observed that with increase in the mass flow rate of air, the relative humidity of output air increases and output temperature of air decreases. While decrease in temperature is desirable, increase in humidity levels is not. Hence an optimum value of mass flow rate needs to be chosen in order to balance the humidity and temperature effects. It can be observed that for an air mass flow rate of 0.06 kg/s, optimum conditions with respect to the target temperature and humidity levels can be achieved.

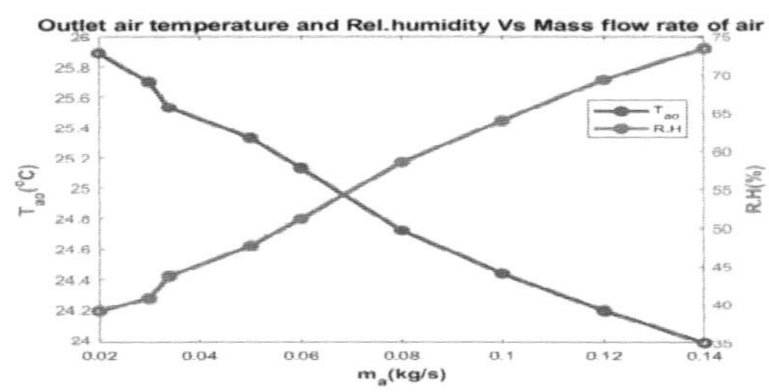

Figure 4. Variation of output air humidity and output air temperature with mass flow rate of air

5. Conclusion

The study reveals the following:

1. With the proposed system, the target temperature and humidity levels (25℃, 50-55%) can be maintained for major part of the day even for the months of May, June and July when the climatic conditions are much more severe.
2. The proposed system could achieve dehumidifier efficiency of about 64%, significantly higher than 49% with that available in the literature.
3. Optimum value of air mass flow rate has been obtained to be 0.06 kg/s.
4. The study buttresses the viability and need of solar assisted LDAC system with fin-coil dehumidifier to maintain the target comfort conditions for the plains of country like India.

References

[1] Pérez-Lombard L, Ortiz J and Pout C 2008 A review on buildings energy consumption information *Energy Build.* **40** 394–398.

[2] Giampieri A, Ma Z, Smallbone A and Roskilly A P 2018 Thermodynamics and economics of liquid desiccants for heating, ventilation and air-conditioning – An overview *Appl. Energy* **220** 455–479.

[3] Yutong L and Hongxing Y 2008 Investigation on solar desiccant dehumidification process for

energy conservation of central air-conditioning systems *Appl. Therm. Eng.* **28** 1118–1126.
[4] Henning H M, Erpenbeck T, Hindenburg C and Santamaria I 2001 The potential of solar energy use in desiccant cooling cycles *Int. J. Refrig.* **24** 220–229.
[5] Buker M S and Riffat S B 2015 Recent developments in solar assisted liquid desiccant evaporative cooling technology—A review *Energy Build.* **96** 95–108.
[6] Das R S and Jain S 2017 Experimental investigations on a solar assisted liquid desiccant cooling system with indirect contact dehumidifier *Sol. Energy* **153** 289–300.
[7] Liu J, Zhang T, Liu X and Jiang J 2015 Experimental analysis of an internally-cooled/heated liquid desiccant dehumidifier/regenerator made of thermally conductive plastic *Energy Build.* **99** 75–86.
[8] Alizadeh S and Saman W Y 2002 Modeling and performance of a forced flow solar collector/regenerator using liquid desiccant *Sol. Energy* **72** 143–154.
[9] Liu J, Liu X and Zhang T 2016 Performance comparison of three typical types of internally-cooled liquid desiccant dehumidifiers *Build. Environ.* **103** 134–145.

IOP Conf. Series: Materials Science and Engineering 1080 (2021) 012041 doi:10.1088/1757-899X/1080/1/012041

Multiphase dynamics in a three dimensional branching network

Agnivo Ghosh, Daipayan Sen, Nirmal Kumar Manna and Sandip Sarkar*

Department of Mechanical Engineering, Jadavpur University, Kolkata-700032
*e-mail: thesandipsarkar3@gmail.com

Abstract - The article aims at understanding the complex unsteady fluid dynamics in six generations of a human bronchial tree, which comprises of 63 straight sections and 31 bifurcation modules in a complete breathing cycle. The effects of 3-D arrangement on oscillatory flow is analyzed. Unsteady effects are maximum during the shift of expiration to inspiration and vice versa. It is also observed that the symmetry of the flow in cased of fully developed flow is destroyed due to unsteady effects. The flow division at the bifurcations causes a large asymmetry in the flow field during the inspiration than during the expiration process at the same cross-sections, thus displaying irreversibility of fluid dynamics. The second part of the study is aimed at analyzing the multiphase fluid dynamics inside a six generation bronchial tree for the case of an active smoker. Both the velocity magnitude contours and volume fractions of air and smoke is analyzed and it is observed that the general symmetry of the flow in the main two halves of the branching network is lost due to the interaction of smoke and air in the branches during inhalation and exhalation. It is also observed that higher mass of smoke is accumulated within the daughter branches which are at a lesser angular deviation with the trachea, whereas the branches at the extreme ends have little to no smoke deposit. This phenomenon can be attributed to the high smoke particulate density compared to that of air.

1. Introduction

The process of breathing is an unsteady flow phenomenon involving periodic reversal of the predominant flow direction. The total breathing cycle comprises of essentially two distinct phases viz. inhalation /aspiratory phase during which air is sucked inside the respiratory system and an exhalation/expiratory phase during which air is released out. The present study is aimed at analyzing the fluid dynamics at four specific positions namely peak inhalation, end of inhalation, peak exhalation and end of exhalation. It is important to realize that apparent symmetry in the geometric structure doesn't automatically imply symmetry in flow field.The combined effects of flow path curvature, flow division at a bifurcation and inertia of flow result in skewed velocity profiles which are not symmetric. It should also be noted that viscous effects present try to establish a symmetric paraboloidal profile but the length of the branch is not sufficient for complete removal of asymmetry. A thorough explanation and discussion on this matter is done in [1]. Before indulging into the complexities of oscillatory flow in branching networks it is important to acknowledge the literature available for oscillating flow in a pipe. Womersely [2-3] showed that the interaction between inertial and viscous effects in oscillatory pipe flow would alter the velocity profile such that it would not remain parabolic.However it has been shown here that the flow field is developing and non- axisymmetric and is a function of all three space coordinates and time. Since the geometric complexity of branching networks renders an analytical solution practically impossible, so in such a case numerical and experimental results are the way forward. Fresconi and Prasad [4] experimentally investigated the unsteady flow field in 4 generation branching network. Nowak, Kakade and Annapragada[9] put forward computational fluid dynamics simulations for air flow and particle deposition in a human trachea upto 4th generation. Soni and Thompson[8] identified the crucial aspects of particle transport in the bronchial tubes due to unsteady flow. Martonen ,Quan, Zhang and Musante[7] performed flow simulations in the upper respiratory tract. Elcner, Lizal and Jicha[10] performed a numerical simulation of air flow inside an idealized geometry of tracheobronchial airways and verified it with experiments. Gorji and Pourmehran[12] studied the airflow behaviour and particle transport and deposition in different breathing conditions. Lieber and Zhao[13] experimentally studied the oscillatory flow inside a bifurcation model. Pradhan and Guha[14] studied the effects on oscillatory flow in a three dimensional branching network. Most authors in the above mentioned literatures have adhered to using the Weibel structure and have taken the branches to be rigid. So to keep the focus of the discussion the present article has also used the same. The present computational study has adopted tight spatial and temporal accuracy (with double precision Arithmetic) and convergence at each time step with 450000 nodes and time stepsize of 0.01s.

2. Numerical Model Development

A particular generation in the network is referred to here as 'Gn' where *n* takes values from 0 to 5. The six generation network considered here starts from G0 to G5.The dimensions of the branches are given in the Table 1 below which is in order accordingto Weibel [11]. The bifurcation angle between two daughter branches is set at 70°. All the branches are denoted by four characters 'GnBk' where 'Gn' represents the generation and 'Bk' denotes the branch number at a particular generation. The cross-sectional plane at the end of each straight section of a branch is denoted by "GnPk". Fig 1 shows a three dimensional view of the symmetric model of the human bronchial network.

Table 1: Geometrical parameter details of the model

Generation number	Diameter (mm)	Length (mm)
G0	18	120
G1	12.20	47.60
G2	8.30	19.00
G3	5.60	7.60
G4	4.50	12.70
G5	3.50	10.70

The governing equations for the multiphase flows are used in Eulerian-Eulerian platform and are detailed as follows:

Continuity and momentum equations of gas phase:

Continuity: $$\frac{\partial(\alpha_g \rho_g)}{\partial t} + \nabla \cdot (\alpha_g \rho_g \vec{u_g}) = 0$$

Momentum:

$$\frac{\partial(\alpha_g \rho_g \vec{u_g})}{\partial t} + \nabla \cdot (\alpha_g \rho_g \vec{u_g} \vec{u_g}) = -\alpha_g \nabla p + \nabla \cdot \left[\alpha_g \mu_{eff,g} \left(\nabla \vec{u_g} + \nabla \vec{u_g}^T \right) \right] + \vec{F_{G,g}} + \alpha_g \rho_g g$$

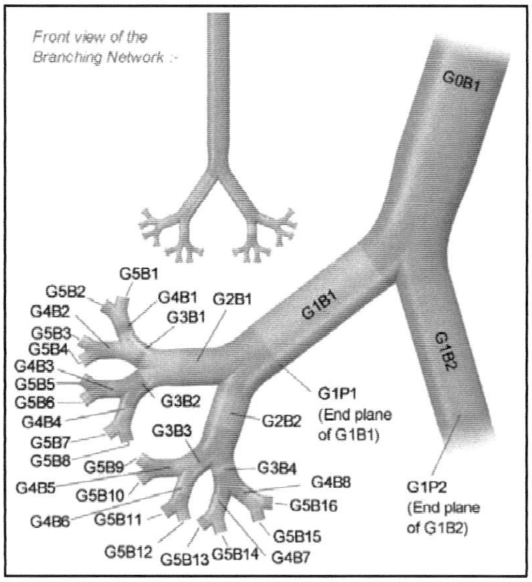

Figure 1: Six Generation Branching Network

Here α_g represent volume fraction of gas phase, $\mu_{eff,g}$ is the effective viscosity of gas, $\vec{F}_{G,g}$ is the interfacial source terms between the two phases.

The 3-D model of the branching network is built in SpaceClaim and Design Modeller in-built within ANSYS Workbench 18. All the meshing and simulations are performed on ANSYS Mesh Modeller and Fluent respectively on computer with i5 processor and 16GB RAM. A Viscous Laminar model is adopted. The scheme used for momentum is First order upwind and gradient being Least Square Cell Based.

A time varying velocity field given by $V = V_{max} \sin(2\pi ft)$ is prescribed at the cross sectional boundary of the first branch where V_{max}= 1.235m/s, f = frequency of breathing cycle = 12/minute.
For the multiphase simulation, following conditions are adopted: Puff Volume Rate, Q = 100sinθ ml/s (From Ref [5]), Area of inlet = A = 0.000254 m^2.
For active smoker the volume fraction of smoke in mixture is taken to be 0.1 i.e. 10%. Volume flux of air = 1.235sinθ m/s, phase velocity of air = 1.235sinθ/0.9 = 1.372sinθ m/s, volume flux of smoke = Q/A = 0.393sinθ m/s, phase velocity of smoke = 0.393sinθ/0.1 = 3.93sinθ m/s.

In the present case, the Womersley number turns out to be 2.64 which is greater than 1 thus depicting the dominance of unsteady effects in branching networks. The maximum Reynolds number attained is about 1550, which is safely below the critical value for the onset of turbulence, thus justifying the assumption of viscous laminar model in this article.

IOP Conf. Series: Materials Science and Engineering 1080 (2021) 012041 doi:10.1088/1757-899X/1080/1/012041

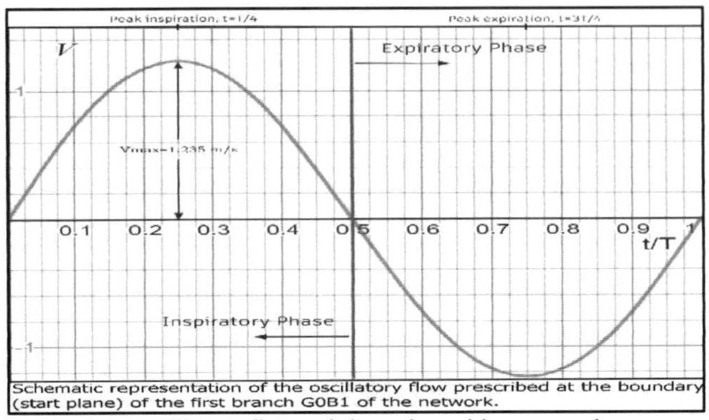

Figure 2: Oscillatory flow at inlet to branching network

3. Results & Discussion
3.1 Normal Breathing Cycle

At T/4

Fig 3 shows the contour of velocity magnitude at specific locations like G0P1, G1P1, G2P1 and so on. The primary velocity distribution at G0P1 is somewhat similar to pipe flow with maximum velocity occurring at the centre. The velocity distribution forG1P1 is skewed with maximum velocity occurring at the inner edge of the bifurcation. Due to asymmetry the contours of G2P1 and G2P2 are different. Similar kinds of results are obtained for remaining cross-sections.

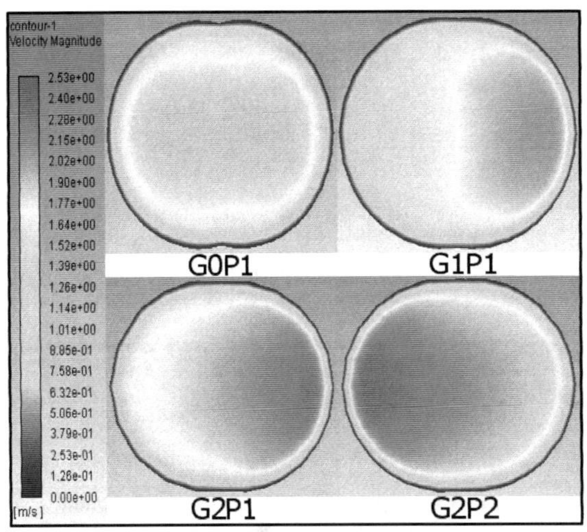

Figure 3: Velocity contours at different locations.

At 3T/4

The contours are shown in Fig. 4.Two striking differences between flow structures at this instant is that it is characterized by more symmetrical flow as compared to the instant at t=T/4. This may be attributed to the difference in geometry of flow path during inhalation and exhalation. During inhalation, the flow at the centre of the flow path of the parent branch, is divided on reaching the bifurcation such that the maximum cross-sectional velocity is shifted toward the inner edge of the bifurcation module in the daughter branches, resulting in a skewed velocity distribution in the daughter branches which, in turn, leads to asymmetric mass-flow distribution among the downstream branches of the daughters.Whereas, during exhalation, as the flow reaches a bifurcation joining two daughter branches to a parent branch, the curvature of the flow path pushes the maximum velocity towards the centre of the cross section in the parent branch, leading to more symmetric velocity distribution in the parent branch andhence a reduction in the asymmetry in the mass flow distribution among the downstream branches.

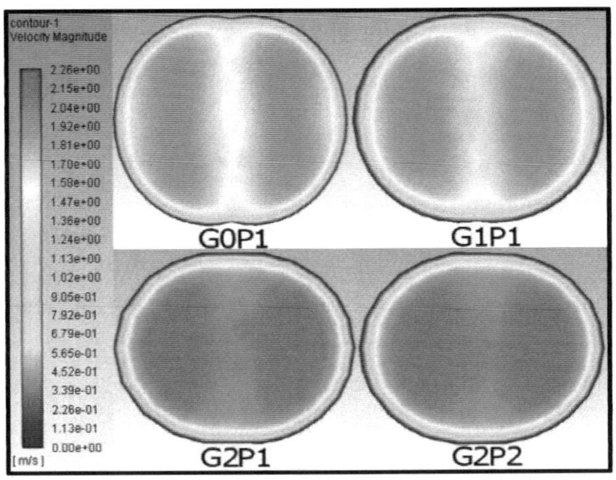

Figure 4: Velocity contours at different locations.

3.2 Active Smoker

At T/4

Figure 5 show the mid plane section of the model at t=T/4, i.e. during peak inhalation. For an active smoker, the volume fraction of smoke is taken to be 0.1, or 10% of total volume of respiratory fluid.Figure 5 also depict the air velocity and smoke velocity contours which depict the maximum velocity around 6m/s and 4m/s, occurring at the daughter branches respectively.It can be seen that these two flows are complementary to each other. The characteristics observed here may be attributed to the physical properties of smoke and air.Smoke, having a high particulate density, acts as a heavier fluid, forcing most of the air towards the walls of the branches.The remaining mass of air in mixture with smoke flows in unison along the inner edges of the bifurcation modules. It is also noted that the volume fraction and thereby mass percentage of smoke decreases as the angular deviation from the root (trachea) increases.This effect can also be attributed to the density of smoke which facilitates flow in the downstream region and impedes flow as the branches become increasingly steep. The volume fraction contours has also been in Fig 5.

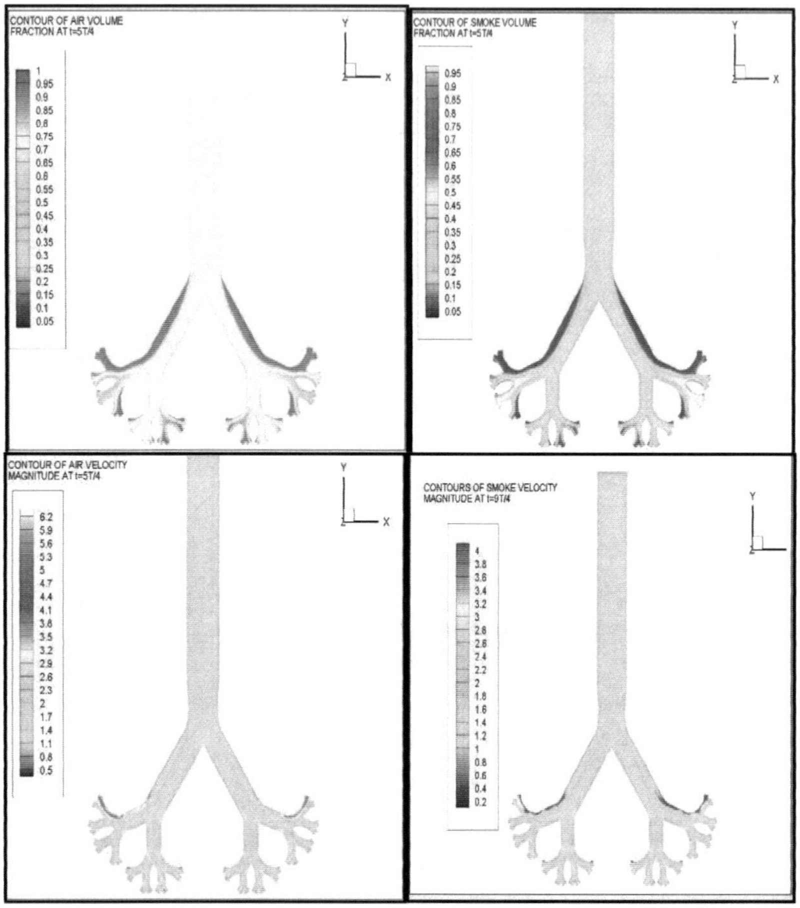

Figure 5: Velocity (Top) and Gas volume fractions (bottom) contours at the middle plane cross section.

At 3T/4

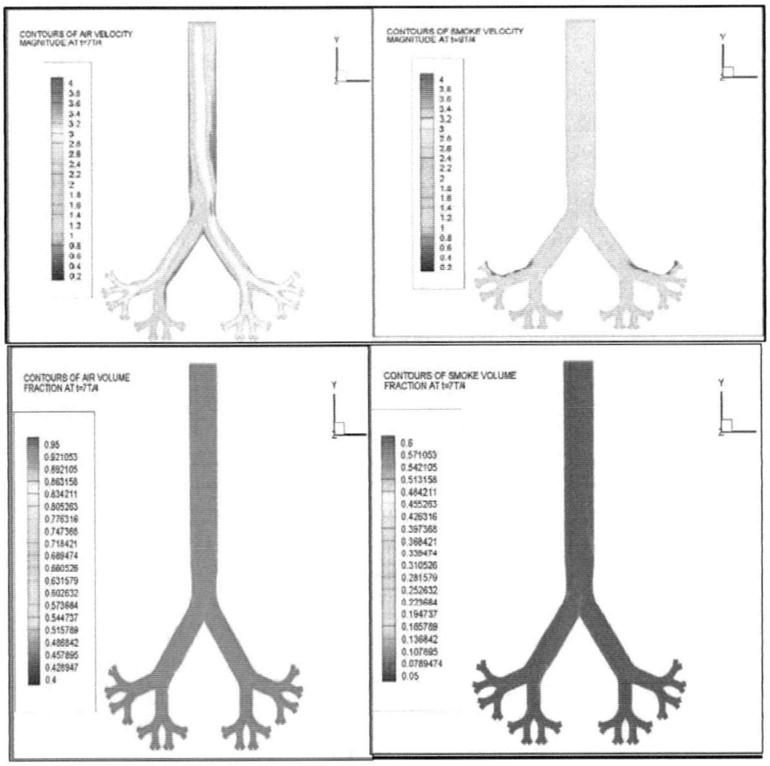

Figure 6: Velocity (Top) and Gas volume fractions (bottom) contours at the middle plane cross section.

The results of peak expiration have been shown above in Fig. 6. The initial cycle has been simulated and left out so as to remove any initial effects. It is observed that the air velocity remains nearly about 3m/s along the core of all the branches whereas it is about 2m/s as we approach the walls. Any cross-section in the branching network is therefore characterized by a centrally located peak thereby giving rise to a plateau like distribution.In case of smoke, velocity reaches a maximum of about 2.5m/s nearer to the walls which are at a less angular deviation from the trachea, and the velocity drops to 1-1.2m/s as we further move away from the center line.The volume fraction nearly remains constant in the entire expiration cycle (about 0.9 of air and 0.1 of smoke) at each cross-section. While majority of the smoke is exhaled out into the atmosphere, a certain amount of smoke remains as residue along the walls of the branches.This residue may be explained by the viscous and wall friction due to which smoke, in contact with the inner edges of the bifurcations, come to rest.The force exerted by the expiratory air is not sufficiently high enough to push out the smoke along the walls owing to its higher density and hence a residual layer of smoke remains in the branches.

4. Validation of Data

The results obtained in the present article have been validated through comparative study with the results obtained by Pradhan and Guha[14]. The contours of cross section of the branches at similar positions have been taken and compared. The results obtained in this study agree closely with those of the previous works. The contours show reasonably close profiles at all the critical flow times in the breathing cycle. The deviation in data is about 4%-10%. As an example, the maximum velocity magnitude obtained by Pradhan and Guha during exhalation (without smoke) is 2.1m/s, whereas it comes to 2.26m/s in our current study.This minimal difference can be attributed to the difference in mesh structure and complexity and time step size. The study with smoke is based on the same model, with the addition of multiple phases. As mentioned in Section 2, the flow is considered to be viscous laminar as the maximum Reynold's number during peak inhalation or exhalation is about 1550, appreciably lower than the critical value for turbulence. The article, therefore, successfully tries to illustrate the distribution of smoke in the daughter branches of a six generation human bronchial tree as well as shows the phenomenon of accumulation of smoke over time through computational fluid dynamics.

5. Conclusions

With the present study it has been established that due to the unsteady effects of oscillatory flow in branching networks the flow is seen to be asymmetric during the inspiration phase (Fig 3) of the breathing cycle.Moreover, the flow tends to acquire a more symmetric structure in the expiration phase (Fig 4). This can be attributed to the differences between bifurcating flow and combinatory flow in a branching network which further yields us to a conclusion that fluid flow in such geometry is irreversible. In the second part of the study it can be seen that there is distinguishable volume fraction distribution in the entire geometry during the inspiration phase whereas the volume fraction is nearly same at all cross-sections during the expiratory phase. During the aspiratory phase the maximum amount of smoke concentration is found at the daughter branches nearer to the centre line which can be attributed to the smoke particulate density than air, whereas air concentration maximum in branches farthest from the centerline (Fig 5).As the slope gets steeper as we move further away from the center line, the velocity thus tends to decrease. The velocity is further decreased due to the high density of smoke.While majority of smoke particles get exhaled out during expiratory phase, a certain quantity, albeit low, remains deposited along the walls of the branches (Fig 6).This observation asserts that, in each breathing cycle, a certain amount of particulates remain deposited and accumulate over time in the respiratory organs, which may, in turn, lead to serious health complications.

6. Acknowledgement

The fourth author acknowledges the financial support provided by the Science and Engineering Research Board (SERB) startup research grant (SRG) vide SERB File No. SRG/2019/000417.

7. References

[1] Guha A, Pradhan K, and Halder P K 2016 Finding order in complexity: A study of the fluid dynamics in a three dimensional branching network *Phys. Fluids* **28(12)** 123602

[2] Womersley J R 1955 Method for the calculation of velocity, rate of flow and viscous drag in arteries when pressure gradient is known *J. Physiol.* **127(3)** 553-563

[3] Womersley J R 1957 Oscillatory flow in arteries: The constrained elastic tube as model of arterial flow and pulse transmission *Phys. Med. Biol* **2(2)** 178-187

[4] Fresconi F E and Prasad A K 2007 Secondary velocity fields in the conducting airways of the human lung *J.Biomech. Eng.* **129(5)** 722-732

[5] Gust S W, Pickens R W, Pechacek T F 1983 Recording puff volume in Smoking *Behavior Research Methods & Instrumentation* **15(3)** 341

[6] Zhao Y and Lieber B B 1994 Steady Inspiratory flow in a model symmetric bifurcation *J. Biomech. Eng.* **116(4)** 488

[7] Martonen T B, Quan L, Zhang Z, Musante C J 2002 Flow simulation in the human upper respiratory tract *Cell Biochemistry and Biophysics* **37(1)** 27-36

[8] Soni B and Thompson D 2012 Effects of temporally varying inlet conditions on flow and particle deposition in the small bronchial tubes *Int. J. Numer. Methods Biomed. Engg* **28(9)** 915-936

[9] Nowak N, Kakade P P and Annapragada A V 2003 Computational fluid dynamics simulation of airflow and aerosol deposition in human lungs *Ann. Biomed. Eng* **31(4)** 374-390

[10] Elcner J, Lizal F, Jedelsky J and Jicha M 2015 Study of airflow in the trachea of idealized model of human tracheobronchial airways during breathing cycle *EPJ Web of Conferences* **92** 02016

[11] Weibel E R 1963 Morphometry of the human being *Springer*, Berlin, Heidelberg

[12] Rahimi-Gorji M, Pourmehran O, Gorji-Bandpy M, Gorji T B 2015 CFD simulation of airflow behaviour and particle transport and deposition in different breathing conditions through the realistic model of human airways *J. Molecular Liquids* **209** 121-133

[13] Lieber B B and Zhao Y 1998 Oscillatory flow in symmetric bifurcation airway model *Annals of Biomedical Engg* **26(5)** 821-830

[14] Pradhan K and Guha A 2019 Fluid dynamics of oscillatory flow in three dimensional branching networks *Phys. Fluids* 063601

IOP Conf. Series: Materials Science and Engineering 1080 (2021) 012042 doi:10.1088/1757-899X/1080/1/012042

Micro analysis of riverbank sediment stability under different wetted surface conditions

Debasish Biswas[1], Arijit Dutta[2], Sanchayan Mukherjee[3] and Asis Mazumdar[4]

[1,2]Assistant Professor, [3]Associate Professor, Department of Mechanical Engineering, Kalyani Government Engineering College, Kalyani-741235, India
[4]Professor & Director, School of Water Resources Engineering, Jadavpur University, Kolkata -700032, India
e-mail: *debasishmechanical@yahoo.co.in*

Abstract. Quantity of water entrapped between sediment particles imparts strength up to a limit and increases the stability of a riverbank. In the present study, this strength of the sediments has been investigated in terms of escape velocity of the sediments. Escape velocity has a considerable impact on the stability of the bank surface. Here, the "Truncated Pyramid Model (TPM)" for arrangement of the sediments has been utilized to determine escape velocity for different wetted surface conditions and inter-granular spaces. It has been found from the present investigation that irrespective of the variability in the riverbank morphology, the escape velocity attains a large value due to enhancement of the wetted surface area resulting in riverbank stability.

1. Introduction

Escape velocity is one of the key parameters for stability analysis. Lower escape velocity implies less force required by the sediments to dislodge from the bank. This implies less stability of the bank in macro level. TPM for the sediment arrangement in micro-level has been theoretically developed by Mukherjee and Mazumdar [1] as shown in **Figure 1a** and **b**. Here, β = water content index angle, b = wetted area radius and D_l = inter-sediment particle (or in simple terms, inter-particle) distance.

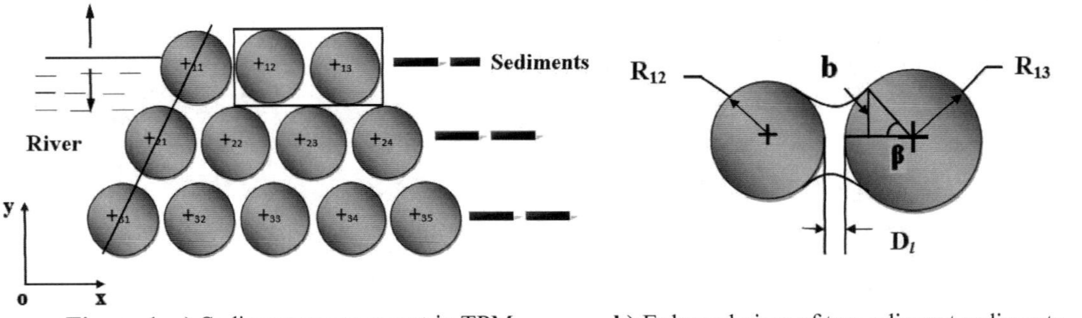

Figure 1. a) Sediment arrangement in TPM **b)** Enlarged view of two adjacent sediments

Considering hydrostatic force and pore pressure force, bank surface stability analysis has been developed by Darby and Throne [2] and Osman and Throne [3]. Capillary cohesion between the sediments, described by Soulie et al. [4] is a strong function of liquid (i.e. water) bridge volume. They have developed the expression of cohesive force F_{Cs} between a pair of particles as follows

$$F_{Cs} = \pi\sigma \times R \times \left[c_l + \exp\left(a_l \times \frac{D_l}{R} + b_l \right) \right] \tag{1}$$

where, a_l, b_l and c_l are functions of liquid bridge volume V_l and the larger sediment radius R, as below

$$a_l = -1.1\left(\frac{V_l}{R^3}\right)^{-0.53} \tag{2a}$$

$$b_l = \left[-0.148\ln\left(\frac{V_l}{R^3}\right) - 0.96\right]\phi^2 - 0.0082\ln\left(\frac{V_l}{R^3}\right) + 0.48 \tag{2b}$$

$$c_l = 0.0018\ln\left(\frac{V_l}{R^3}\right) + 0.078 \tag{2c}$$

ϕ = Contact angle

Mu and Su [5] have shown that liquid bridge volume is a function of wetted surface area as

$$V_l = \frac{\pi R}{2} \times \left[H^2(b) - D_l^2\right] \tag{3a}$$

Here, $H(b) = D_l + \frac{b^2}{2}$ (3b)

And from the **Figure 1** wetted area radius, $b = R \times \sin\beta$

Micro-level pore pressure force has been described by Likos and Lu [6] as

$$F_{Ps} = \pi \times r_2 \times (P_w \times r_2 + 2 \times \sigma) \tag{4}$$

r_2 can be calculated as $r_2 = R \times [\tan\beta - \sec\beta + 1]$ (5)

According to Likos and Lu [6], the value of pore water pressure P_w is taken as 10 kPa.

2. Free-body diagram and Escape velocity calculation

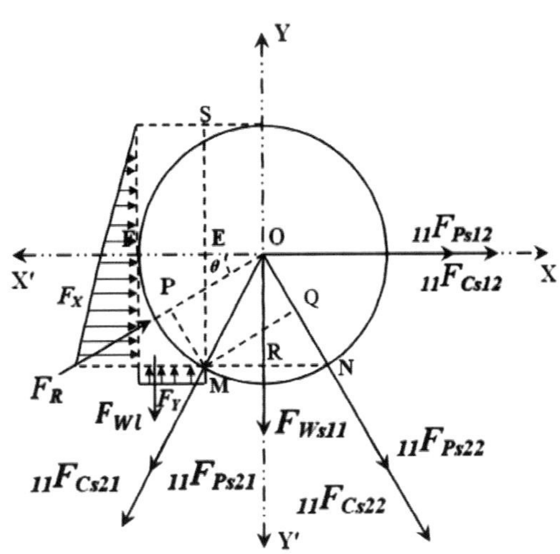

Figure 2. Free-body diagram of different forces acting on sediment *11*

Present work has been done by considering hydrostatic force, force of cohesion and pore pressure force in TPM and by varying water-content index angle (β) from 5° to 45° (from Likos and Lu [6]). Moment of momentum principle for the 'sediment 11' about point M in **Figure 2** has been applied to calculate sediment escape velocity in micro-level, and thus, quantify the stability in macro-level.

Now, applying moment of momentum principle and taking moment about instantaneous centre M

$$F_{Ws11} \times \overline{MR} + F_{Cs} \times \left(\overline{ME} + \overline{MQ}\right) + F_R \times \overline{MP} + F_{Ps} \times \left(\overline{ME} + \overline{MQ}\right) = I_M \times \alpha_s \tag{6}$$

The angular acceleration expression from above equation is given by

$$\alpha_s = \frac{\frac{F_{Ws11}}{2} + \sqrt{3}F_{Cs} + F_R \sin(60-\theta) + \sqrt{3}F_{Ps}}{\frac{7}{5}\left(\frac{4}{3}\pi R^3 \rho_s\right) \times R} \tag{7}$$

Here, sediment particle density $\rho_s = 2650$ kg/m^3.

The expression for sediment acceleration

$$f_s = \alpha_s \times R \tag{8}$$

So, the expression for sediment escape velocity (m/s), as proposed by Duan [7] is

$$V_{es} = \left[2 \times f_s \times R\right]^{1/2} \tag{9}$$

3. Results and Discussion

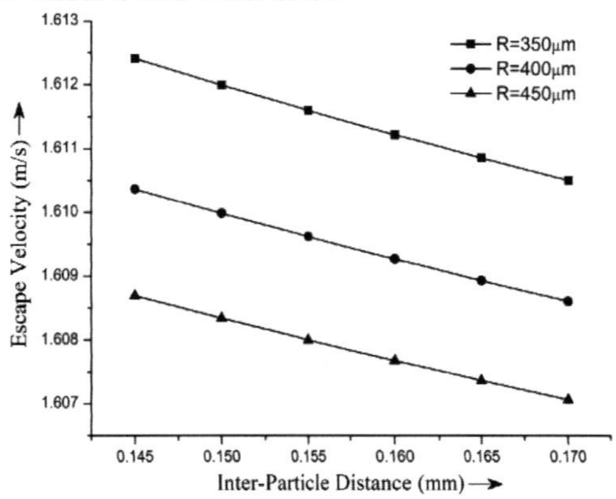

Figure 3. V_{es} Vs. D_l for water level rising for β=15°

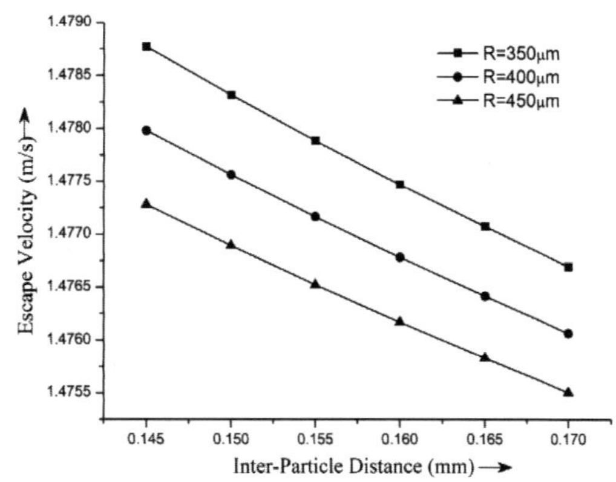

Figure 4. V_{es} Vs. D_l for water level falling for β=15°

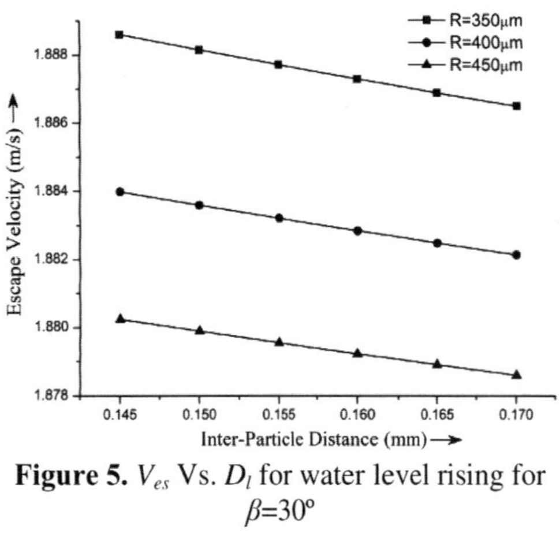

Figure 5. V_{es} Vs. D_l for water level rising for β=30°

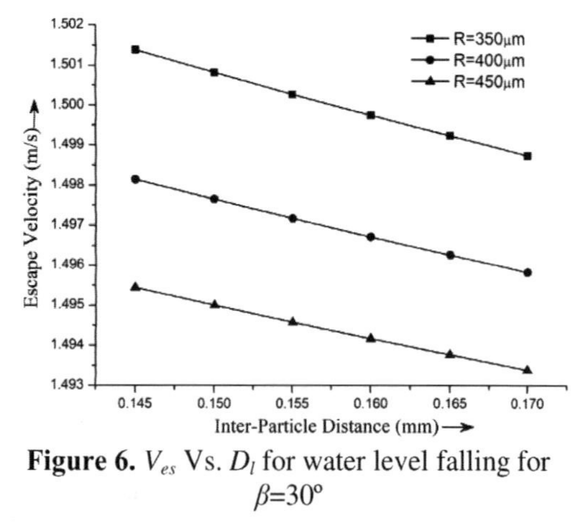

Figure 6. V_{es} Vs. D_l for water level falling for β=30°

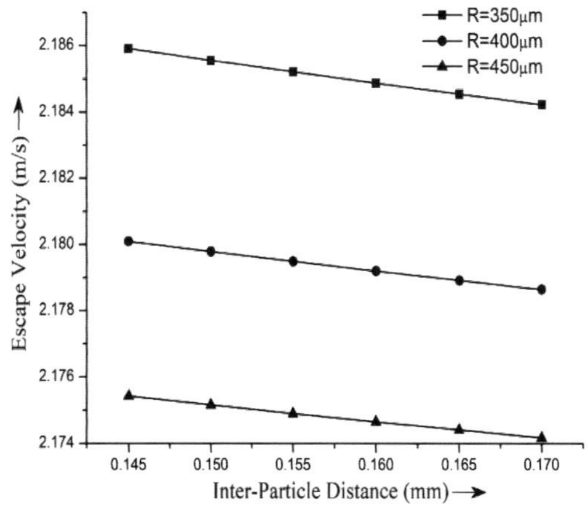

Figure 7. V_{es} Vs. D_l for water level rising for β=45°

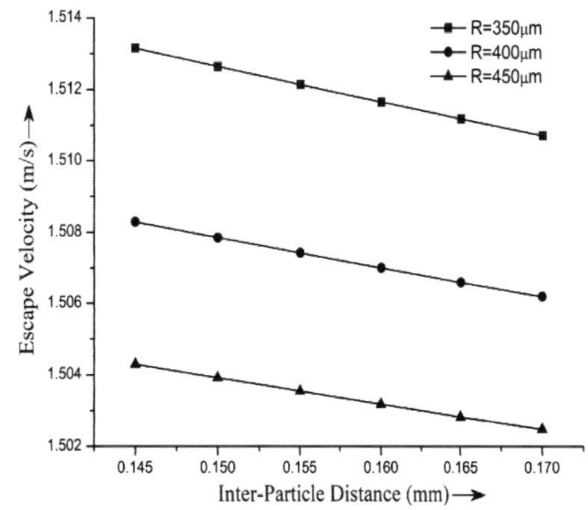

Figure 8. V_{es} Vs. D_l for water level falling for β=45°

Escape velocity with inter-particle distance has been plotted for different water content index angles for rise and fall of water level. It can be stated from all the figures that more distance between the sediments has an adverse effect on sediment stability in micro level, and bank stability as a whole. It can also be stated from all the figures that the bigger sediments are less stable compared to smaller sediments as the escape velocity is higher for smaller size sediments compared to bigger ones. The figures also reveal that the water on the rise stables the bank surface in macro level as the escape velocity is more in that case. Change in escape velocity (V_{es}) with the water content index angle (β) has been depicted below (**Figure 9**).

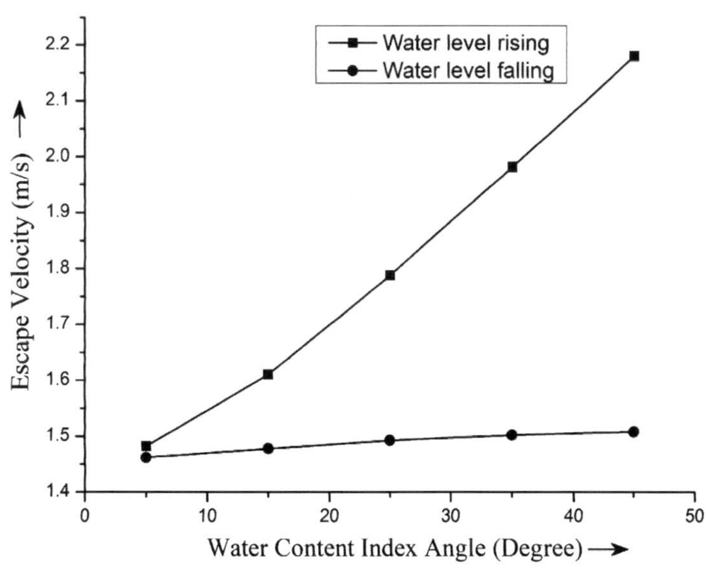

Figure 9. V_{es} Vs. β

From the above figure it can be stated that a wetted surface between the sediments (water content index angle) has a favourable effect on the escape velocity, or, in other words, stability in macro level. This graph also shows the severe effect of falling water vis-à-vis rising water.

4. Conclusions

Conclusions that can be drawn from the present analysis are

- More distance between the sediments has an adverse effect on the sediment escape velocity in micro level which indicates reduction in the stability of the bank surface in macro level. This is because of a drop in inter-sediment forces of attraction.
- Bigger sediments are less stable compared to smaller sediments as the escape velocity is higher for smaller sediments compared to bigger sediments. This is owing to less force of cohesion between bigger sediments compared to smaller sediments.
- Bank surfaces are more stable when water level is on the rise as the sediment escape velocity is more in that case. For falling water level, negative moment helps to dislodge the sediments.
- Wetted surface between the sediments has a favourable effect on the escape velocity. More wetted surface area means more water volume between the sediments, which binds the sediments together.
- The graph of escape velocity with water content index angle is steeper for the case of rising water level compared to that for the falling water level. Water level rises due to pore pressure force. The pore pressure force increases as the wetted surface area increases so long as the pores are partially saturated with water. The variable degree of saturation is largely taken care of by the water content index angle during calculation of the escape velocity. On the other hand, water level falls due to gravitational force, which is not affected by the wetted area.

References

[1] Mukherjee S and Mazumdar A 2010 Study of effect of the variation of inter-particle distance on the erodibility of a riverbank under cohesion with a new model *Journal of Hydro-environment Research* **4(3)** 235–242.

[2] Darby S E and Thorne C R 1996 Stability analysis for steep, eroding, cohesive riverbanks *Journal of Hydraulic Engineering* **122** 443–454.

[3] Osman A M and Thorne C R 1988 Riverbank stability analysis, I: Theory *Journal of Hydraulic* Engineering **114(2)** 134–150.

[4] Soulie F, El Youssoufi M S, Cherblanc F and Saix C 2006 Capillary Cohesion and Mechanical Strength of Polydisperse Granular Materials *Eur. Phys. J. E.* **21** 349–357.

[5] Mu F and Su X 2007 Analysis of liquid bridge between spherical particles *China Particuology* **5** 420-424.

[6] Likos J W and Lu N 2002 Hysteresis of Capillary Cohesiom in Unsaturated Soils, *15th ASCE Engineering Mechanics Conference Columbia University, June 2–5, New York, NY.*

[7] Duan J G 2005 Analytical approach to calculate rate of bank erosion *Journal of Hydraulic Engineering* **131(11)** 980–989.

DEVELOPMENT OF A THERMAL MODEL OF A SOLAR POWERED VAPOUR ABSORPTION REFRIGERATION SUPPORTED GREENHOUSE FOR STRAWBERRY CULTIVATION

[1*]**Subhadeep Ghosh**, [2]**A. Ganguly**

[1, 2] Department of Mechanical Engineering,
Indian Institute of Engineering Science and Technology, Shibpur,
West Bengal, India - 711103
* Corresponding author e-mail: subhakalideep@gmail.com

Abstract

In this paper, a novel scheme of a solar powered vapour absorption refrigeration supported greenhouse with indirect evaporative cooler has been proposed. The main objective of this work is to create a favorable microclimatic condition for strawberry cultivation inside the greenhouse for a country with sub-tropical climatic conditions like India when the ambient temperature and humidity are high. The maximum yield of strawberry is achieved for a temperature ranging between 15-25 ^{0}C with a relative humidity varying from 55 to 65%. It is quite impossible to achieve a favorable temperature range for strawberry cultivation in the greenhouse under such condition by using the existing greenhouse cooling techniques. A double effect parallel flow LiBr-H_2O absorption cooling system along with an IDEC powered through solar energy has been used. A thermal model for the greenhouse has been developed and analyzed for the climatic condition of Kolkata, India. The study reveals that the proposed system can very well maintain a conducive climate for the growth of target plantation for the given location.

Keywords: Greenhouse Cultivation; Double effect VAR system; Solar Thermal; Indirect evaporative cooler

1. Introduction

Greenhouse is an enclosed structure made of any transparent material inside which the climatic conditions can be regulated artificially for the optimum growth of flora. The primary objective of a greenhouse is to maintain a favourable microclimate for optimizing the plant growth. Other than the temperature control, it is also necessary to maintain an optimum relative humidity inside a greenhouse to achieve the maximum growth of flora. An effective greenhouse should be designed to achieve high cooling efficiency, low cost of investment, low operation and maintenance costs, and also the design should provide an even temperature profile along the greenhouse [1]. In subtropical countries like India, hot and humid climatic conditions prevail for the most part of a calendar year. High temperature is detrimental for the cultivation of high value plantation like Strawberry (Fragaria Ananassa) which thrives at a temperature between 15 and 20 °C with the maximum viable temperature being 25 °C [2]. This is the main reason that the leading producer of strawberries (USA, Spain, Turkey, Mexico) are not located in the hot and humid tropical and sub-tropical part of the world. Apart from temperature, one of the major factors for strawberry cultivation is humidity. The yield of Strawberry is maximum when the air vapour pressure deficit (V.P.D) is between 200-1000pa. The plants suffer from diseases with high ranges of humidity level (V.P.D<200pa) and when the humidity level is low (V.P.D>100pa) plants suffer from water stress. So, for the greenhouses located in subtropical regions like India, the major objective is to reduce the inside temperature as well as control of the humidity for cultivation of strawberries.
Several research and development works have been done on the different traditional greenhouse cooling systems. However, it may be noted that the traditional methods of greenhouse cooling (i.e. fan-pad cooling, evaporative cooling, and desiccant cooling) fail to maintain a conducive microclimate for strawberry cultivation inside greenhouse for the plains of sub-tropical or tropical region. The integration of a VAR

system along with a greenhouse can be a viable solution for the cultivation of plantation which requires a low temperature with decent humidity levels. The technology of greenhouse cultivation is also highly relevant to the rural development of a country. For the developing countries like India the availability of power on a sustainable basis in the rural areas is a big challenge. In case of VAR system there is a scope of utilization of low-grade energy as input. The load on the VAR system will be extremely high if it is alone used to cool the greenhouse. Pre-cooling process can reduce the air temperature resulting in the decrease in load of the VAR system. The indirect dew point cooling technology which is also known as M-cycle cooling technology can cool the air below the WBT without adding any moisture, which is the main reason for which this technology is very much effective for the hot and humid environmental conditions. Technically in the dew point evaporative cooler the temperature of the product air can be approached nearly to the DPT of the air [3]. For the hot and humid climatic conditions, only IDEC cooling is not sufficient to provide adequate amount of cooling air in the conditioned space [4]. Delfani et al.[5] experimentally investigated the performance of an IDEC to pre-cool the air for a conventional cooling system. Cui et al. [6] in his research numerically modeled an IDEC as a pre-cooling unit for hot and humid climatic condition and stated that a huge amount of energy can be saved by using IDEC as a cooling unit. Chauhan and Rajput [7] proposed an air conditioning system which comprises DPEC and VCR system. The results showed that the load on the cooling coil decreased by about 61% at 46^{0}C and 6g/Kg sp. humidity, when DPEC is used along with a VCR system. The results obtained thus showed, that up to 55% electrical energy can be saved by using IDEC. In this work N-PVT-CPC collector is used to supply energy to the whole system, this collector system is self-sustained for VAR system [8].

There is very little literature available on greenhouse cooling using VAR system [9]. No work has been found on greenhouse cooling which is based on a combination of VAR system and indirect evaporative cooler.

Nomenclature

IDEC	Indirect Evaporative Cooler	G_0	Total Radiation on greenhouse (W/ m^2)
DPEC	Dew point Evaporative Cooler	τ	Roof transmittance
WBT	Wet bulb temperature	A	Roof area of Greenhouse(m^2)
DBT	Dry bulb temperature	γ	Psychrometric constant (Pa K^{-1})
DPT	Dew Point temperature	K_s	Heat transfer Coeff. (Wm^{-2} K^{-1})
VPD	Vapour Pressure Deficit	v	Volume of air flow(m^3/s)
VAR	Vapour Absorption Refrigeration	HX	Sensible Heat exchange for ventilation (W/m^2)
VCR	Vapour Compression Refrigeration	K_s	Heat transfer coff. Of greenhouse and ambient
PVT	Photovoltaic	K_h	Sensible heat transfer coefficient (W/m^{2}K)
CPC	Compound parabolic concentrator	K_v	Latent heat transfer coefficient (W/m^{2}pa).
α	Absorptivity	T_{gh}	Temperature inside greenhouse (^{0}C)
T_a	Temperature of the ambient air(^{0}C)	T_s	Temperature of the supply air(^{0}C)
λE	Canopy Transpiration	Δe	Vapour pressure deficit (Pa)

2. System Description

The schematic of the proposed system is given in figure 1. In the proposed system, the ambient air is drawn from outside and sensibly cooled in an IDEC, where the air exchanges heat with working air channel. In this process no moisture is added to the air, so the specific humidity of the air remains same only, the DBT of the air decreases. After getting cooled from the IDEC, the product air is allowed to blow across the evaporator of a double effect parallel flow Li-Br VAR system. There the temperature of the air is reduced to a sufficiently low value and the cooled air is then supplied to an east-west oriented Quonset greenhouse. An array of N-PVT-CPC collector is used to produce water at a temperature of 120^{0}C, which is stored in an insulated storage tank. The hot water is supplied in the high-pressure generator for the water vapor generation from the weak solution. The electricity produced from the photovoltaic collector is supplied to the grid. The grid electricity is used for running the various auxiliary devices such as pumps, cooling fan of the system.

3. Thermal Model Development

The following assumptions have been made to develop the thermal model

(1) The condition inside the greenhouse is assumed as steady state and density of air and water vapour are also assumed to be constant.

(2) Heat storage of the greenhouse-soil system is neglected.

(3) The evapo-condensation phenomenon on the greenhouse cover is neglected.

(4) The supply air to the greenhouse and return air mass flow rate are assumed to be the same.

(5) The initial conditions of the greenhouse are assumed to be same as ambient condition.

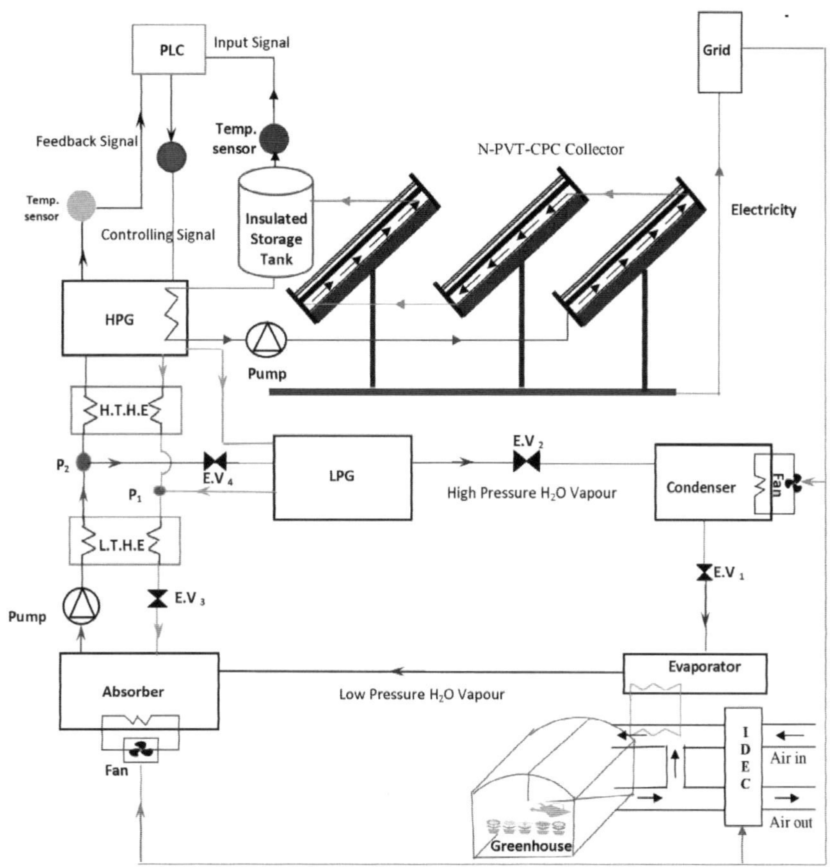

Fig 1: Schemetic Diagram of the proposed system

Energy balance equation within the greenhouse can be expressed as [10]

$$\alpha G_0 \tau - \lambda E - HX - K_s \left(T_{gh} - T_a \right) - Q_m = 0 \tag{1}$$

Q_m is the heat storage or retrieval rate of greenhouse soil system. The value of Q_m is very less and thus it can be neglected. G_0 is the solar radiation per unit area.

The latent heat load due to canopy transpiration (λE) can be written as [10]

$$\lambda E = K_v \Delta e \tag{2}$$

Where $\Delta e = e_{gh} - e_s$ (pa)

e_{gh}, e_s are the vapour pressure of greenhouse and the supply air.

Temperature of the greenhouse can be written as [11]

$$T_{gh} = \frac{\left[\alpha G_0 \tau - \lambda E + K_h T_s + K_s T_a\right]}{(K_s + K_h)} \tag{3}$$

The water vapour pressure (D_{gh}) deficit inside the greenhouse can be expressed as:

$$D_{gh} = \left[e^*(T_{gh}) - e^*(T_s)\right] - \left[e_{gh} - e_s\right] + \left[e^*(T_s) - e_s\right] \tag{4}$$

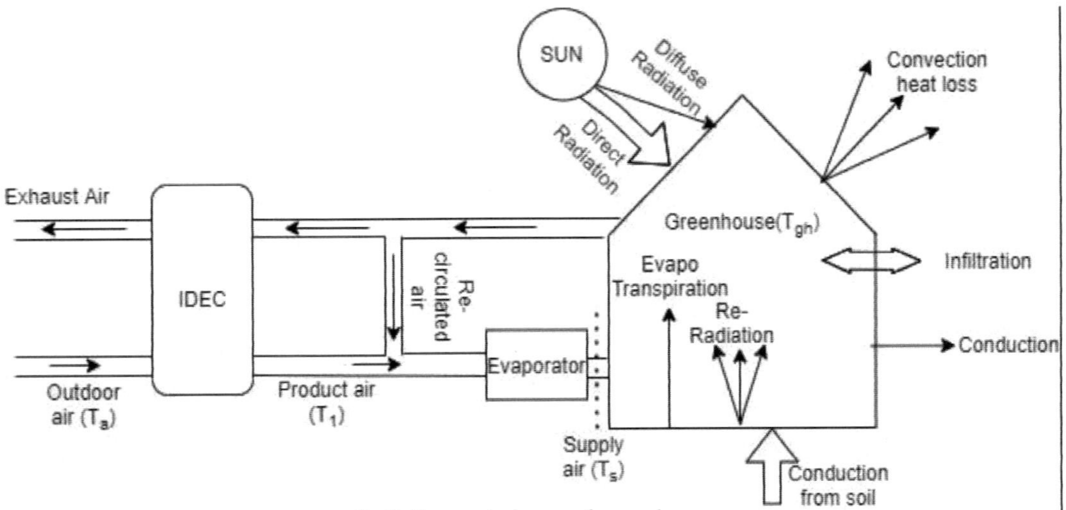

Fig 2: Energy balance of greenhouse

The total sensible cooling load on the greenhouse can be given by

$$Q_{SHL} = A * G_0 - Q_c - Q_a - Q_v \tag{5}$$

4. Results and Discussions:

In this study the performance of the cooling system has been studied numerically. The inputs and the constants which are used for this study are given in Table 1.

Table 1: Inputs and Constrains for the study

Temp. of supply air(T_s)	15^0C	Heat transfer Coeff. (K_s)	4.5 $Wm^{-2} K^{-1}$[13]
RH of the supply air	60%	Roof area of Greenhouse (A)	58.92 m^2
Sp. heat of the air (C_p)	1005 J $kg^{-1} K^{-1}$	Psychrometric constant (γ)	72.08 Pa K^{-1}
Roof Transmittance(τ)	0.8 [12]	Heat of vaporization of water	2260 KJ/Kg
Absorptivity (α)	0.7[12]	Day of the year	June 15
Volume of air flow	2.04 m^3/s	IDEC efficiency(ε)	0.8

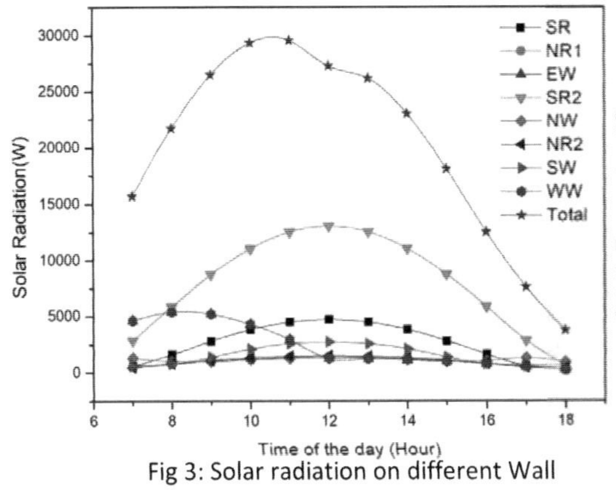

Fig 3: Solar radiation on different Wall

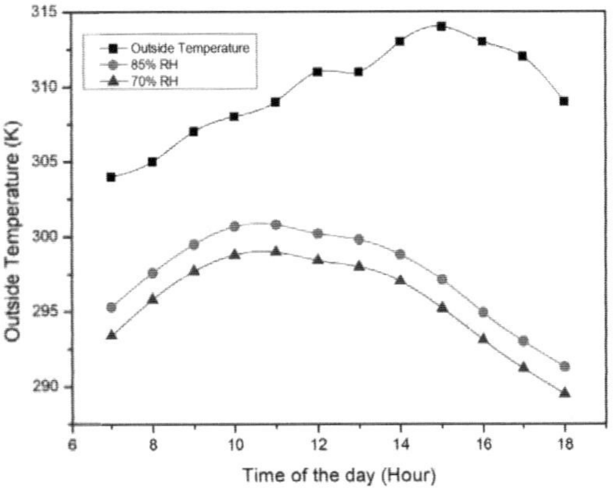

Fig 4: Variation of temperature within the greenhouse

Figure 4 shows the variation of temperature of greenhouse with the variation of relative humidity of the process air. It is observed from the figure that the temperature of the greenhouse decreases with the decrement of relative humidity of the process air. Because as the relative humidity of the supply air increases the ability of the supply air to take the heat from greenhouse air will decrease. Maximum temperature of the greenhouse is 26^{0}C for which a conducive microclimate for strawberry cultivation can be maintained.

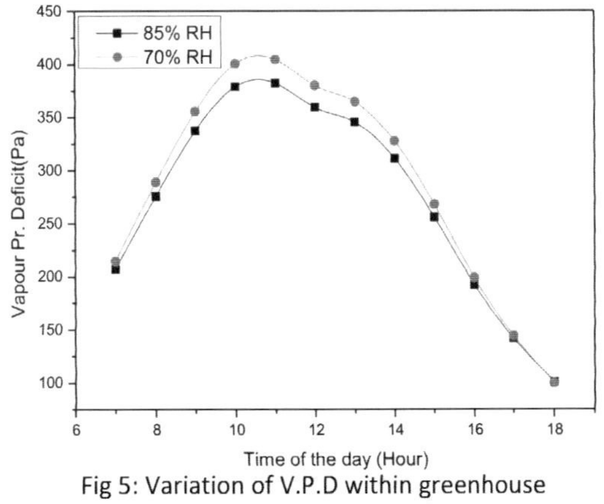

Fig 5: Variation of V.P.D within greenhouse

Figure 5 shows the variation of vapour pressure deficit with the variation of relative humidity of the process air. It is seen from the figure that the V.P.D of the greenhouse increases with the decrement of relative humidity of the process air. As the relative humidity of air at a fixed temperature increases, the vapour content of the air is also increases, so when the process air with high RH mixes with greenhouse air it will unable to lowered the V.P.D of the greenhouse air. It is seen from the figure 5 that for both 85% and 70% RH of the air the V.P.D inside the greenhouse is suitable for strawberry cultivation.

5. Conclusion

The present paper discusses a novel scheme of a solar powered VAR supported greenhouse for cultivation of Strawberry for the sub-tropical climatic conditions prevailing in the plains of India. A thermal model of the greenhouse is developed and analyzed for a representative day in Summer (June 15) when both the ambient temperature and relative humidity levels are very high. Following points are revealed from the study:

(1) For a representative day of summer (June15) when the outside temperature is very high this proposed system will able to maintain a suitable temperature for strawberry cultivation inside the greenhouse when the temperature and RH of the air is 15^0C and 70% respectively.

(2) Two different condition of supply air is specified in this work in both the cases the V.P.D inside the greenhouse is suitable for strawberry cultivation.

References

[1] Von Elsner, B., Briassoulis, D., Waaijenberg, D., Mistriotis, A., Von Zabeltitz, C., Gratraud, J., Russo, G. and Suay-Cortes, R., 2000. Review of structural and functional characteristics of greenhouses in European Union countries, part II: Typical designs. Journal of agricultural engineering research, 75(2), pp.111-126.

[2] Banaeian, N., Omid, M. and Ahmadi, H., 2011. Energy and economic analysis of greenhouse strawberry production in Tehran province of Iran. Energy Conversion and Management, 52(2), pp.1020-1025.

[3] Chua, K.J., Chou, S.K., Yang, W.M. and Yan, J., 2013. Achieving better energy-efficient air conditioning–a review of technologies and strategies. Applied Energy, 104, pp.87-104.

[4] Cui, X., Yang, X., Qin, S., Meng, X., Jin, L. and Chua, K.J., 2019. Performance investigation of an evaporative pre-cooled air-conditioning system in tropics. Energy Procedia, 158, pp.5673-5678.

[5] Delfani, S., Esmaeelian, J., Pasdarshahri, H. and Karami, M., 2010. Energy saving potential of an indirect evaporative cooler as a pre-cooling unit for mechanical cooling systems in Iran. Energy and Buildings, 42(11), pp.2169-2176.

[6] Cui, X., Chua, K.J. and Yang, W.M., 2014. Use of indirect evaporative cooling as pre-cooling unit in humid tropical climate: an energy saving technique. Energy Procedia, 61, pp.176-179.

[7] Tiwari, G.N., Meraj, M. and Khan, M.E., 2018. Exergy analysis of N-photovoltaic thermal-compound parabolic concentrator (N-PVT-CPC) collector for constant collection temperature for vapor absorption refrigeration (VAR) system. Solar Energy, 173, pp.1032-1042.

[8] Chauhan, S.S. and Rajput, S.P.S., 2016. Parametric analysis of a combined dew point evaporative-vapour compression based air conditioning system. Alexandria Engineering Journal, 55(3), pp.2333-2344.

[9] Campiotti, C.A., Morosinotto, G., Puglisi, G., Schettini, E. and Vox, G., 2016. Performance evaluation of a solar cooling plant applied for greenhouse thermal control. Agriculture and agricultural science procedia, 8, pp.664-669.

[10] Boulard, T., and Wang, S. (2000). Greenhouse crop transpiration simulation from external climate conditions. Agricultural and forest meteorology, 100(1), 25-34.

[11] Davies, P. A. (2005). A solar cooling system for greenhouse food production in hot climates. Solar energy, 79(6), 661-668.

[12] Ganguly, A., and Ghosh, S. (2007). Modeling and analysis of a fan–pad ventilated floricultural greenhouse. Energy and Buildings, 39(10), 1092-1097.

[13] Kittas, C., Bartzanas, T., and Jaffrin, A. (2003). Temperature gradients in a partially shaded large greenhouse equipped with evaporative cooling pads. Biosystems Engineering, 85(1), 87-94.

IOP Conf. Series: Materials Science and Engineering 1080 (2021) 012044 doi:10.1088/1757-899X/1080/1/012044

Thermal Model Development of a Biomass Regenerated Desiccant Supported Greenhouse Cooling for Orchid Cultivation

Chandan Mandal[1]* **and A.Ganguly**[2]

[1,2] Department of Mechanical Engineering, Indian Institute of Engineering Science and Technology, Shibpur, Howrah-711103, West Bengal, India.

*Corresponding author's e-mail: mandalc39@gmail.com

Abstract. In this paper, a novel scheme of biomass regenerated desiccant supported greenhouse cooling with distributed fan-pad evaporative system has been proposed. The system aims to provide suitable thermal condition inside the greenhouse for cultivation of high value flowers like varieties of Orchid which require a temperature between 21 ^{0}C to 26 ^{0}C and humidity from 50 to 70% for the sub-tropical climate prvailing in the plains of India. In the proposed system two stage desiccant based cooling is used to obtain a low humidity ratio. A biomass based heating system is coupled for regeneration of the desiccant. Thermal modeling of desiccant wheel and greenhouse are done to estimate air temperature inside the greenhouse. A comparative study has been made based on estimated greenhouse air temperature with the results of the reference model available in the literature. From the performance analysis of the proposed system, it is revealed that the maximum temperature of the greenhouse air can be restricted within 25 ^{0}C and 25.8 ^{0}C for a hot and humid day prevailing in the month of July and September respectively. However the system can maintain a lower temperature of 19 ^{0}C during the morning time which is more conducive for target plantation (Orchid) in the given region.

1. Introduction

Greenhouse technology is a method of cultivation of plants under a suitable environment for their optimum growth. It is used to protect the plants from the adverse climatic conditions and promotes off seasonal cultivation. The plains of sub tropical countries like India receive abundant solar radiation for major part of a calendar year resulting in high ambient temperature coupled with high humidity levels. Most of the time of a calendar year, the values of ambient temperature and humidity highly exceed the prescribed ranges for Orchid (Orchidaceous) cultivation in the open field. The cultivation of high value target Orchids require optimum temperature ranges between 21 ^{0}C to 26.5 ^{0}C and relative humidity ranges between 50% - 70% [1]. The greenhouse cultivation of Orchids can be a potential solution to this problem. Thus, the maintenance of proper inside temperature and humidity are major challenges behind the greenhouse cultivation of the Orchids.

The desiccant wheel based air conditioning system for greenhouse cultivation is a promising alternative to maintain low humidity and low temperature as per requirement. Desiccant wheels driven by waste heat or solar thermal heat [2]. From the literature, it is observed that few works are available for greenhouse cooling using desiccant. There liquid desiccants were used and the regeneration was done using solar energy.

Till now, no work has been reported on two stage composite desiccant based greenhouse cooling, where biomass based heating system is used for regeneration of the desiccant. This study aims to propose a scheme of a biomass regenerated two stage desiccant cooling system for high growing value Orchid cultivation in a subtropical greenhouse. In this paper, a thermal model has been developed and using the thermal model the temperature and humidity inside the greenhouse have been estimated.

2. Description of the proposed system

Figure 1 shows the scheme of the proposed biomass regenerated desiccant supported greenhouse cooling system for Orchid cultivation. It consists of a free standing even span greenhouse oriented lengthwise in the east-west direction. The process air in this system is obtained from the atmosphere, which passes through desiccant wheel 1 (DW_1) made of synthesized metal silicate. After dehumidification, the air is passed through heat exchanger 1 (HE_1) for sensible cooling.

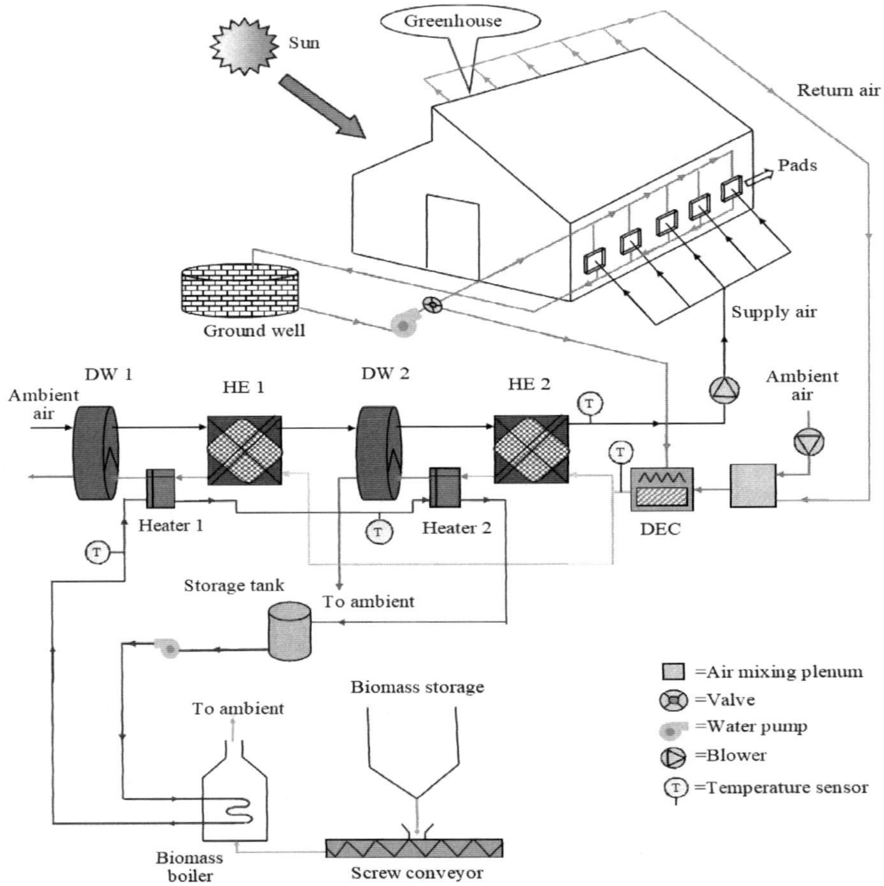

Figure 1. Scheme of a biomass regenerated desiccant supported greenhouse cooling for Orchid cultivation in subtropical region.

It may be noted the moisture adsorption capacity of the desiccant wheel decreases with increase in temperature. Considering this aspect the air at the outlet of HE_1 is again passed through desiccant wheel 2 (DW_2) and heat exchanger 2 (HE_2) in series. After passing through the second heat exchanger (HE_2), the dehumidified process air goes to the cooling pads located in the both longitudinal sidewalls of the greenhouse, where the process air is cooled due to evaporative cooling and finally supplied into the greenhouse. A circulating water pump (CWP) is used for circulation of water from the Ground

well to the pad to keep the same wet. The induced draft (ID) fans are mounted vertically at the top of the greenhouse, the arrangement is known as distributed fan pad cooling system [3]. A portion of the return air is mixed with the ambient air in an air mixing plenum and the same is allowed to pass through a direct evaporative cooler (DEC) to obtain further lower temperature. Then the same is equally divided into two groups operating in parallel, one flowing through HE_2, air heater and DW_2, while the other passing through HE_1, air heater and DW_1. In the air heater the air is heated to the regeneration temperature of the desiccant wheel using a biomass based heating system as shown in Fig.1.
Hot water from biomass boiler is supplied to the Heater 1 and then it flows through the Heater 2. The water coming from Heater 2 is stored into a storage tank from where the water is supplied to the boiler using a feed water pump. Greenhouse crop residues (GCR) and wood pellets are used as the biomass fuel. The biomass based fuels from the biomass storage are fed to the boiler through a screw conveyor. The heat energy produced in the biomass combustion is transferred to the water to produce steam.

3. Thermal model development

3.1. Desiccant wheel

Relative humidity and enthalpy of the process air coming from the desiccant wheel can be obtained from equations (1) and (2), respectively [4].

$$\varphi_{p.out} = \varphi_{p.in} - \eta_{\varphi}\left(\varphi_{p.in} - \varphi_{r,in}\right) \tag{1}$$

$$h_{p.out} = h_{p.in} - \eta_h\left(h_{p.in} - h_{r.in}\right) \tag{2}$$

where, η_{φ} and η_h are the humidity and enthalpy effectiveness respectively. In Eq.s (1) and (2), $h_{p.in}$ and $\varphi_{p.in}$ are enthalpy and relative humidity of the inlet process air respectively. Therefore, humidity ratio ($X_{p.out}$), temperature ($T_{p.out}$) and water vapour saturation pressure ($P_{p.out.vsat}$) of the outlet process air are obtained using Eq.s (3) through (5).

$$X_{p.out} = 0.622\frac{\varphi_{p.out}}{\frac{P_{tot}}{P_{p.out.vsat}} - \varphi_{p.out}} \tag{3}$$

$$T_{p.out} = \frac{h_{p.out} - \mu X_{p.out}}{C_{p.da} + X_{p.out}C_{p.wv}} \tag{4}$$

$$P_{p.out.vsat} = \exp\left(23.196 \times \frac{3816}{T_{p.out} + 273.15 - 46.13}\right) \tag{5}$$

In Eq. (4), μ is latent heat of vaporization of water in kJ kg^{-1}. $C_{p.da}$ and $C_{p.wv}$ are specific heat of dry air and water vapour. (η_{φ}) is obtained from equation (6) with specific correlation coefficients In Eq.s (7) through (11).

$$\eta_{\varphi} = \alpha_{v.r} \times \alpha_{v.p} \times \alpha_T \times \alpha_X \times \alpha_N \tag{6}$$

$$\alpha_{v.r} = C_1 v_{r.in}^2 + C_2 v_{r.in} + C_3 \tag{7}$$

$$\alpha_{v.p} = C_4 v_{p.in}^2 + C_5 v_{p.in} + C_6 \tag{8}$$

$$\alpha_T = C_7 \ln\left(T_{r.in} - T_{p.in}\right) + C_8 \tag{9}$$

$$\alpha_X = C_9 X_{p.in} - C_{10} X_{r.in} + 1 \tag{10}$$

$$\alpha_N = C_{11} N + C_{12} \tag{11}$$

Similarly, (η_h) can be calculated using equation (12) with respective correlation coefficients In Eq.s (13) through (17).

$$\eta_h = \beta_{v.r} \times \beta_{v.p} \times \beta_T \times \beta_X \times \beta_N \tag{12}$$

$$\beta_{v.r} = K_1 \left(v_{r.in} \right)^{K_2} \tag{13}$$

$$\beta_{v.p} = K_3 \left(v_{p.in} \right)^{K_4} \tag{14}$$

$$\beta_T = K_5 T_{r.in} + K_6 T_{p.in} + K_7 \tag{15}$$

$$\beta_X = K_8 X_{r.in} + K_9 X_{p.in} + 1 \tag{16}$$

$$\beta_N = K_{10} N + K_{11} \tag{17}$$

where $v_{p.in}$ and $v_{r,in}$ are process air and regeneration air velocity (m s^{-1}) at inlet of the desiccant wheel respectively. $T_{r.in}$ and N are regeneration air temperature (^{0}C) and revolution speed (rev h^{-1}). C_1 through C_{12} and K_1 through K_{11} are the correlation constant are taken from the work of [5].

3.2. Thermal model of greenhouse

To develop the thermal model of the greenhouse, it has been assumed that heat flux of the soil is negligible and the transmissivity of the greenhouse cover material is constant. The heat transfer due to radiation from the crop is negligible and the absorptivity of the structural material is negligible. Then the steady state energy balance equation of the greenhouse can be expressed as [3]:

$$\left(1 - S\right) \alpha G_0 \tau + U_g \Delta T = K_s \left(T_{amb} - T_{pad} \right) - K_s \Delta T + \lambda.E \tag{18}$$

In Eq. (18), S, α, G_o, τ, U_g, K_s, T_a, T_{pad}, λ and E are shading factor, absorbance of the canopy, global solar radiation intensity (kJ m^{-2}), roof transmittance, total heat transfer coefficient (W m^{-2} K^{-1}) of the greenhouse covering material, sensible heat transfer coefficient (W m^{-2} K^{-1}) of air, ambient temperature, temperature of the cooling pad end, latent heat of vaporization (kJ kg^{-1}) and crop canopy transpiration (kg m^{-2} s^{-1}) respectively. Where, ΔT denotes the temperature difference between the atmospheric air and greenhouse air.

In Eq. (18), the heat transfer coefficient (K_s) due to sensible load inside the greenhouse can be given by [6]:

$$K_s = \rho_a C_{pa} \left(ACM \right) V_g / \left(60 \cdot A_{gf} \right) \tag{19}$$

The rate of plant transpiration rate in greenhouse can also be calculated from the Penman-Monteith formula as given below [7]:

$$\lambda.E = \frac{\left[\delta \left(\lambda.E + H \right) + \dfrac{2 I_{la} \rho_a C_{pa} W_g}{r_a} \right]}{\left[\delta + \gamma \left(1 + \dfrac{r_s}{r_a} \right) \right]} \tag{20}$$

where, I_{la} indicates area index of leaf for the given plantation, while γ indicates a psychrometric constant (Pa K^{-1}). In Eq. (20), the stomatal resistance of the crop (r_s) can be written as.

$$r_s = 200 \left[1 + 1 / \exp\left\{ 0.05 \left(S_{gi} - 50 \right) \right\} \right] \tag{21}$$

In Eq. (20), the resistance due to aerodynamic effect, r_a (s m^{-1}) can be written as.

$$r_a = 220\left(L_c^{0.2} / u_i^{0.8}\right) \tag{22}$$

where, L_c is the characteristic length of the leaf and u_i denotes the average air velocity inside the greenhouse. In Eq. (20), sensible heat load (H) can be expressed as.

$$H = K_s\left(T_{amb} - T_{pad}\right) - K_s \Delta T \tag{23}$$

In Eq. (20), δ denotes the slope of the water vapour saturation curve in (Pa K^{-1}) at any temperature T (K) and may be expressed as [6].

$$\delta = \left(\tfrac{5385}{T^2}\right) \times 2.229 \times 10^{11} \exp\left(\tfrac{-5385}{T}\right) \tag{24}$$

In Eq. (20), W_g indicates water vapour pressure deficit and can be expressed as.

$$W_g = \delta \cdot \Delta T - \Delta e + W_a \tag{25}$$

In Eq. (25), Δe denotes the water vapour pressure difference (Pa) between the ambient air and greenhouse air. W_a denotes the water vapour pressure deficit (Pa) of the ambient air [6]. The expression for the greenhouse plant transpiration rate can be written as [8].

$$\lambda . E = K_L \cdot \Delta e \tag{26}$$

In Eq. (26), K_L denotes the latent heat transfer coefficient (W m^{-2} Pa^{-1}) and may be written as.

$$K_L = \lambda \cdot \zeta \cdot \rho_a \cdot (ACM) \cdot V_g / \left(60 \cdot A_{gf}\right) \tag{27}$$

In Eq. (27), ζ is a conversion factor for the water vapour present in the air at standard temperature and pressure. In this present model the value of ζ is considered to be 6.25×10^{-6} kg_w kg_a^{-1} Pa^{-1} [8]. Now combining the above following equations (20), (23), (25) and (26), we get.

$$\Delta T = \frac{\left[\dfrac{\left(\delta \cdot P_1 + P_2 \cdot \dfrac{W_a}{r_{a\,1}}\right)}{P_3} - \left(S_{gi} - C\right) \cdot \left\{1 - \dfrac{\left(\delta - \dfrac{P_2}{K_L \cdot r_a}\right)}{P_3}\right\}\right]}{\left[\left(U_g + K_s\right) \cdot \left\{1 - \dfrac{\left(\delta - \dfrac{P_2}{K_L \cdot r_a}\right)}{P_3}\right\} - \dfrac{\left(\dfrac{P_2 \cdot \delta}{r_a} - \delta \cdot K_s\right)}{P_3}\right]} \tag{28}$$

In Eq. (28), P_1, P_2 and P_3 represent the constants, which can be written further as:

$$P_1 = K_s\left(T_{amb} - T_{pad}\right) \tag{29}$$

$$P_2 = 2 \cdot I_{la} \cdot \rho_a \cdot C_{pa} \tag{30}$$

$$P_3 = \delta + \gamma \cdot \left(1 + r_s / r_a\right) \tag{31}$$

Therefore, the average greenhouse temperature can be written as.

$$T_{GH} = T_{amb} - \Delta T \tag{32}$$

The conditioned air that is supplied to the greenhouse is to be cooled in a distributed fan-pad cooling system where evaporative cooling is taken place. The expression of the air temperature at the end of the cooling pad can be given by [3].

$$T_{pad} = T_{he2} - \varepsilon_{cp} \cdot \left(T_{he2} - T_{wb.he2} \right) \tag{33}$$

In Eq. (33), ε_{cp} and $T_{wb.he2}$ are the effectiveness of the cooling pads and wet bulb temperature of the air coming from the second heat exchanger (HE_2) respectively. The heat exchanger is modelled based on the work of Asadi and Roshanzadeh [5]. A computer code in Engineering Equation solver (EES) has been written to solve the above equations. The input parameters to the computer code are listed in Table 1. The initial temperature of the greenhouse is guessed as ambient temperature for the first iteration. After each successive iteration, the same is shifted to its previous value. In this study the greenhouse is assumed to be located in the plains of Gangetic Bengal (Kolkata) in the Indian sub-continent. Thus, the weather data for Kolkata as available from the Indian Meteorological Department, New Delhi [9] have been used for the simulation.

Table 1. Value of the input parameters to the thermal model.

Variable	Value	Variable	Value
Absorptivity(α)	0.7 [10]	Effectiveness of the cooling pad (ε_{cp})	0.88 [6]
Roof area (A_{gf})	240 m^2[11]	Effectiveness of the heat exchange (ε_h)	0.8 [5]
Greenhouse volume (V_g)	728 m^3[11]	Diameter of the desiccant wheel (D)	60 cm
Shad factor (S)	0.5 [3]	Axial length of the wheel (L)	20 cm
Transmissivity (τ)	0.8 [10]	Aspect ratio (a/b)	0.51
Overall heat transfer coefficient (U_g)	4.5 W m^{-2} K^{-1}[3]	Speed of the wheel (N)	5-30 rev h^{-1}[4]
Psychometric constant (γ)	72.08 Pa K^{-1}[6]	Regeneration air temperature at inlet ($T_{r.in}$)	45 ^{0}C-78 ^{0}C [4]
Leaf area index of lettuce (I_{la})	0.5 m^2 m^{-2}	Process air temperature at inlet ($T_{p.in}$)	17 ^{0}C-35 ^{0}C
Latent heat of vaporization of water (λ)	2,260,000 J kg^{-1}	Velocity of the inlet regeneration air ($v_{r.in}$)	1.75-2.86 m s^{-1} [4]
		Velocity of the inlet process air ($v_{p.in}$)	1.81-2.5 m s^{-1} [4]

4. Results and discussions

4.1. Model validation for desiccant wheel

The validation of present thermal model of desiccant wheel is carried out by comparing the numerical results estimated by our model with the experimental results reported by the reference work [4]. Figure 2 shows the comparison of numerical results and experimental results on specific humidity and temperature of the process air outlet from the desiccant wheel, for different inlet conditions. From the Fig. 2 (a) and (b), it is observed that the present model agrees well with the reference model, having average absolute error of 4.3% and 4.5 % respectively.

4.2. Performance analysis

Figure 3 shows the difference of atmospheric temperature and average greenhouse air temperature for a typical day of monsoon season (24th July, 2009). The ambient air temperature and relative humidity both remain high during the month of July. The results of the present model have been compared with the reported work of Banik and Ganguly [3] and Kittas et al. [12] respectively. To maintain the suitable air temperature inside the greenhouse, a traditional fan-pad cooling system is suggested by

Kittas et al., while, liquid desiccant based distributed fan-pad evaporative cooling system is proposed by Banik and Ganguly[10].

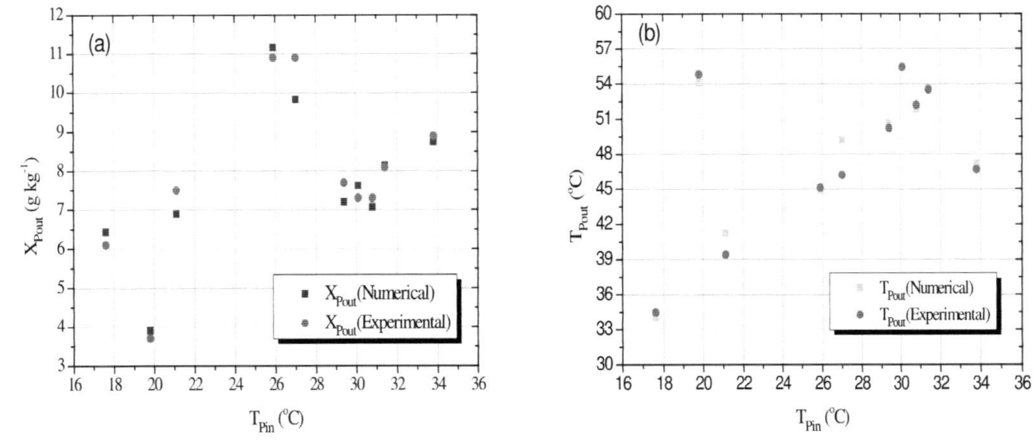

Figure 2. Numerical and experimental values of (a) specific humidity and (b) temperature of the process air at the outlet from the desiccant wheel for different inlet conditions.

From the figure 3 (a), it is evident that our proposed system (two stage composite desiccant based distributed fan-pad evaporative cooling) can maintain the greenhouse air temperature within 24 ^{0}C even during the peak sunshine hour (12 Noon), while during the same period the temperature of other two considered models, T_{Kittas} and $T_{Banik\ and\ Ganguly}$ can maintain a temperature at 28.4 ^{0}C and 27 ^{0}C respectively.

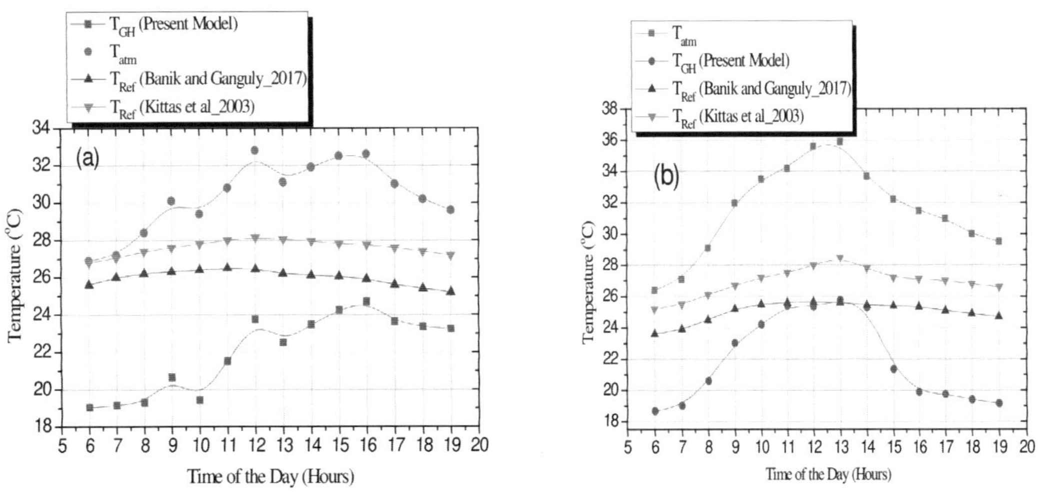

Figure 3. Difference of atmospheric temperature and average greenhouse air temperature for a typical day of (a) July and (b) September.

In the present model for a typical day July, the maximum greenhouse air temperature reaches at 25 ^{0}C, which is 7.9 ^{0}C less than the corresponding ambient temperature. From the figure, it is also seen that during the morning time (6:00 to 8:00 am), the system can maintain a temperature about 19 ^{0}C due to efficient desiccation and recirculation of the process air.

Figure 3 (b) epitomizes the difference of atmospheric air and greenhouse air temperature for a typical day in the autumn season (12th September, 2009) in Kolkata. It is observed from Fig. 3 (b) that the maximum greenhouse air temperature can be maintained within 25.8 ^{0}C, while the same reaches to about 28 ^{0}C with a traditional fan-pad cooling system and 26.1 ^{0}C with that of distributed fan-pad ventilated system with liquid desiccation respectively during the peak sunshine hours. It is also noticed that during the same period, the present system can reduce the greenhouse air temperature by about 10 ^{0}C below that of ambient (35.8 ^{0}C), while the other considered two systems can reduce the same by only about 7.8 ^{0}C and 9.5 ^{0}C respectively. During the morning and evening time, present system can very well maintain the required suitable low temperature conducive for the growing of the considered plantation.

5. Conclusions

In this present study, a scheme of a biomass regenerated composite desiccant supported greenhouse cooling for Orchid cultivation is presented. A thermal model of the greenhouse and its desiccation system is presented and validated against available reference study in the literature. The performances of the proposed system have been analyzed for a hot and humid day in the month of July and September respectively. From the study, it is observed that the proposed model can conserve the greenhouse air temperature between 19 0C to 25.8 0C by using two stage desiccants cooling with distributed fan-pad evaporative system. From the study, it is also revealed that proposed model can reduce the specific humidity of the ambient air by a maximum value of 11 g kg-1. The system can very well maintain the desired climate in terms of temperature and humidity ratio inside the greenhouse for the year round cultivation of the target flora. the study thus reinforces the need and viability of such system for the sub-tropical countries like India.

References

[1] Lopez R G and Runkle E S 2004 The effect of temperature on leaf and flower development and flower longevity of Zygopetalum Redvale "Fire Kiss" orchid *HortScience* **39** 1630–4

[2] Sahlot M and Riffat S B 2016 Desiccant cooling systems: A review *Int. J. Low-Carbon Technol.* **11** 489–505

[3] Banik P and Ganguly A 2017 Performance and economic analysis of a floricultural greenhouse with distributed fan-pad evaporative cooling coupled with solar desiccation *Sol. Energy* **147** 439–47

[4] De Antonellis S, Intini M and Joppolo C M 2015 Desiccant wheels effectiveness parameters: Correlations based on experimental data *Energy Build.* **103** 296–306

[5] Asadi A and Roshanzadeh B 2019 Improving performance of two-stage desiccant cooling system by analyzing different regeneration configurations *J. Build. Eng.* **25** 100807

[6] Bauri B and Ganguly A 2013 Effect of Crop Transpiration on the Microcilmate of a Naturally Ventilated Greenhouse *Int. J. Emerg. Technol. Adv. Eng. An ISO Int. J.* **3** 337–43

[7] Rajeevan P K, Geetha C K and Rajendran P 2017 Orchid-centric floriculture development in Kerala, India *Acta Hortic.* **1165** 15–26

[8] Boulard T and Wang S 2000 Greenhouse crop transpiration simulation from external climate conditions *Agric. For. Meteorol.* **100** 25–34

[9] Indian Meteorological Department New Delhi 2009 Solar Radiant Energy over India *Renew. Energy*

[10] Davies P A 2005 A solar cooling system for greenhouse food production in hot climates *Sol. Energy* **79** 661–8

[11] Ghosh A and Ganguly A 2017 Performance analysis of a partially closed solar regenerated desiccant assisted cooling system for greenhouse lettuce cultivation *Sol. Energy* **158** 644–53

[12] Kittas C, Bartzanas T and Jaffrin A 2003 Temperature gradients in a partially shaded large greenhouse equipped with evaporative cooling pads *Biosyst. Eng.* **85** 87–94

Enhanced Transport under Annular Jet by Introducing Rib

Tarun Kanti Pal[1*], Himadri Chattopadhyay [2], Dipak Kumar Mandal[1]
[1*] Department of Mechanical Engineering
College of Engineering and Management, Kolaghat - 721 171, India
[2]Department of Mechanical Engineering
Jadavpur University, Kolkata – 700 032, India
([1*]Email: tarunkantipal@gmail.com)

Abstract: This work concern studies for the transport enhancement of impinging annular jets. The ribs are introduced on the impingement surface and higher heat transfer could be obtained using rib compared to plane surface. Comparison is made with circular jet to predict the enhancement of heat transport for same Reynolds number and mass and momentum efflux at jet outlet. The model used for solving the problem is Shear Stress Transport (SST) model, while the Reynolds numbers of jet varies from 100 to 50000. Square rib is used in this study. This study reveals that the ribs will increase the heat transfer for both circular and annular jets. Annular jet provides 20–30% lower heat transfer than that of the circular jet.

Keywords: Annular jet, Ribbed Surface, Heat transfer, Turbulent flow.

1. Introduction

Impinging jets are used either heating or cooling in many applications where heat transfer takes place between the impinging fluid jet and surface.

The performance of knife jet where the jet emanates at an angle with the axis has been studied numerically by Chattopadhyay and Saha [1] and results are compared with axial slot jet on moving surface under laminar conditions. It has been found that knife jet always under-performs than axial jets and velocity at the surface also reduces heat transfer. Chattopadhyay [2] has been performed laminar annular jet with jet location at $r_i = 0.5$. Heat transfer in case of circular jet is more and it is about 20% in comparison to the annular jet. Heat transfer drops in case of moving surface, noted by Chattopadhyay [3]. Heat transfer on ribbed surface is investigated byYan et al. [4]. They have observed that heat transfer depends on rib orientation. Tan et al. [5] have studied heat transfer characteristics for ribbed regions and observed 30% heat augmentation in this zone of rib. Nakod et al. [6], Studied the effect of circular air jet impinged finned (rough) flat surface on the local heat transfer coefficient for different Reynolds number. The local heat transfer coefficient for finned surface was increased up to 77%.

This work is divided into two parts. The first part show the study of heat transfer on ribbed surface and comparison is made with pane surface. The second part discusses the comparison of heat transfer between circular and annular jet.

2. Mathematical Formulation

The schematic configuration of an annular jet is shown in figure 1. The governing equations for continuity, momentum and energy transfer are solved using finite volume formulation. A wide range of Reynolds from 100 to 50000 is investigated with a fixed height of impingement. The location of jet of the annular zone is, $r_i = 0.5$. The distribution of Nusselt number and turbulent intensity has been found to be strongly influenced by these parameters.

The flow has been considered steady and incompressible. In this-study of fluid flow and heat transfer phenomena for annular jet, transition Shear Stress Transport (SST) based model is chosen to solve the constitutive equations.

The Reynolds-averaged Navier-Stokes (RANS) equations in dimensionless form are now presented. The pressure, velocity and temperature have been non-dimensionalised by $p = \frac{2p}{\rho w_{in}^2}$, $v_i = \frac{u_i}{u_{in}}$ and$T = \frac{T - T_\infty}{T_w - T_\infty}$, respectively.

Continuity equation:

$$\frac{1}{r}\frac{\partial}{\partial r}(ru) + \frac{\partial w}{\partial z} = 0 \quad (1)$$

Momentum equation along radial direction:

$$\frac{1}{r}\frac{\partial}{\partial r}(ruu) + \frac{\partial}{\partial z}(uw) = -\frac{\partial p}{\partial r} + \frac{1}{\text{Re}}\left(1 + \frac{\nu_t}{\nu}\right)(\nabla^2 u - \frac{u^2}{r}) \quad (2)$$

Momentum equation along axial direction:

$$\frac{1}{r}\frac{\partial}{\partial r}(ruw) + \frac{\partial}{\partial z}(uw) = -\frac{\partial p}{\partial z} + \frac{1}{\text{Re}}\left(1 + \frac{\nu_t}{\nu}\right)\nabla^2 w \quad (3)$$

Energy equation:

$$u\frac{\partial T}{\partial r} + w\frac{\partial T}{\partial z} = \frac{1}{\text{Re.Pr}}\left(1 + \frac{\alpha_t}{\alpha}\right)\nabla^2 T \quad (4)$$

In figure 1, axi-symmetric part shows the annular jet at the inner radius of r_i. In our present study, Reynolds number varies from100 to 50000, jet position from the axis is $r_i = 0.5$, and from the plate, h=1.0. H is the vertical distance from the enclosing plate to the surface of impingement.

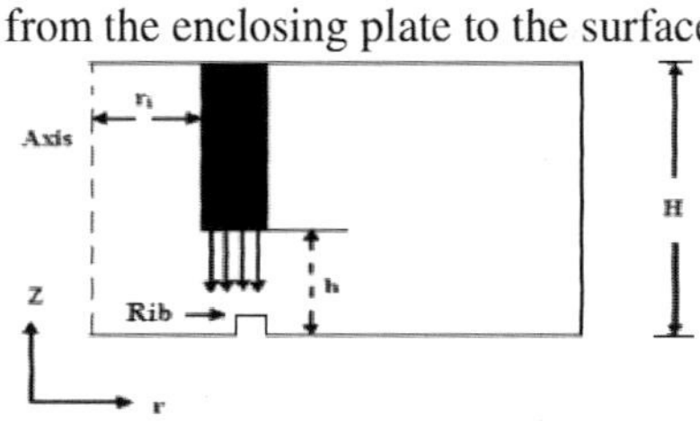

Figure 1 Physical configuration for annular jet with rib

Present study is carried out by considering air as working fluid having Pr=0.71 Reynolds number is calculated based on width of annular part [2]. The constitutive equations are described in cylindrical polar coordinated. For stationary surface, no slip boundary with uniform temperature is chosen. Adiabatic walls are assumed at top and bottom wall. The simulations are performed in 2D and axi-symmetric, therefore axi-symmetric boundary conditions are imposed in the axis. The characteristic length and velocity to achieve non dimensionalization is designated as d_i and w_{in}, respectively and fixed pressure at the exit plane has been assumed.

For non-dimensionalization, the skin friction coefficient can be represented by velocity gradient i.e., $C_f Re = \frac{\delta u}{\delta z}$, Nusselt number estimates the heat transfer and expressed by local temperature difference as $Nu = -\frac{\partial T}{\partial Z}$. The average Nusselt number is acquired by $Nu_{av} = -Nu(r)\,\partial r/L$, where L is the length along r direction of computational domain i.e., $5d_i$. The dimensionless temperature of the impinging plate is chosen as unity i.e., T_s=1.0, and the jet temperature is taken atmospheric (i.e., T_{in}=0.0). A typical value of 5% is chosen for inlet turbulent intensity [7-8].

In the present study, the axi-symmetric governing equations are solved by finite volume method. Non uniform staggered grid is used for discretization of the equations. The most popular scheme, SIMPLE [9] is used for predicting pressure and velocity. QUICK scheme is adopted for interpolation of temperature and velocities gradients; second order accurate upwind scheme for gradients of intermittency, turbulent kinetic energy, specific dissipation rate, and momentum thickness. The equations were solved iteratively with ANSYS Fluent [10]. For achieving convergence, residual level of 10^{-6} is prescribed for all equations except for continuity which was set at 10^{-4}.

3. Results and Discussion

The present study is conducted for different Reynolds numbers ranging from 100 to 50000 at jet location (r_i=0.5) on rib surfaces. Jet is taken circular and annular along with ribbed impinging surface. To show the improvement of heat transfer, average Nusselt number and skin friction coefficient are investigated.

The velocity contours for our considered cases of jets have been shown in Fig.2. It is revealed from the figure that two fluid streams are discharge from two concentric sections and the outer portion of stream forms an annular jet. The physical feature differentiating the flat surface and ribbed surface impinging from annular jet is related with the outlet flow structure on surface. The impingement on the target plate creates reverse flow zone along the jet axis. This may be attributed to flow streams reversing towards the jet axis [Fig.2 (a) & (b)]. An impinging of simple circular jet is configured by high momentum flow discharging from the jet and impinges on surface. Peak momentum region coincides with the jet axis [Fig.2 (c) & (d)].

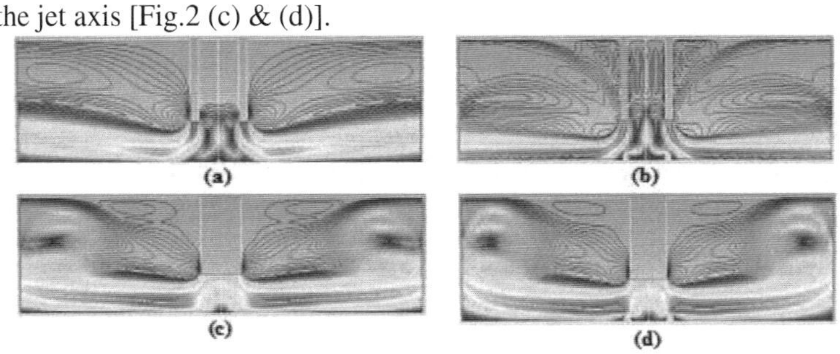

Figure 2 Velocity contour (a) annular jet without rib and (b) annular jet with rib (c) circular jet without rib and (d) circular jet with rib

Figure 3 shows the distribution of C_f for different Reynolds numbers for an annular jet (r_i = 0.5) with single rib (0.15x0.15) at rib location r=0.5. It is observed that the jet divides either towards the central core i.e., axis or radially outwards after impingement. Negative value of skin friction indicates that the flow moves inward adjacent to plate. Skin friction coefficient is zero at the axis and then reduces up to r = 0.2 for annular jet, beyond that it starts to increase. C_f value again becomes zero at about r = 0.4 after which it is in positive range. C_f reaches its maximum peak value on rib then it is zero at about r = 0.6. However C_f reaches its peak value at r=1.0 and then drops.

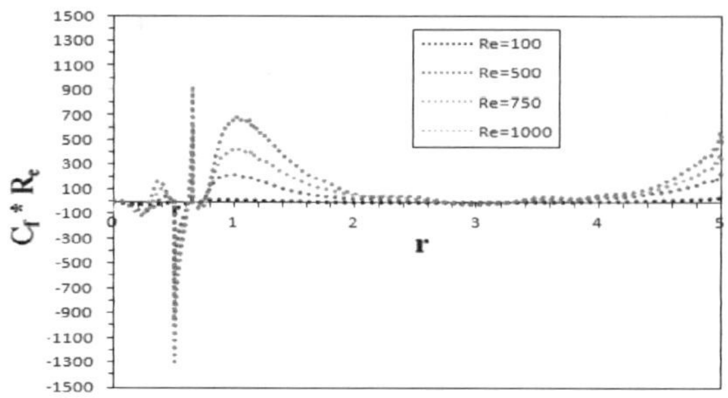

Figure 3 Variation of C_f at different Re (annular jet with single rib) at rib location r=0.5

The distribution on Nu on ribbed surface is shown in Fig. 4. The local heat transfer appears low at below jet as the jet impinges the rib at the stagnation point. It is also observed from the figure that heat transfer rises and becomes maximum behind the rib and decays gradually downstream. Local heat coefficient occurs maximum at the stagnation point. This value occurs minimal and maximum at the rib and behind the rib respectively. The heat transfer studies on flat surface are conducted in our earlier studies Pal et al.[8] in order to understand the heat transfer characteristics on the impingement plate but from the present study it is evident that by introducing of ribs on surface overall heat transfer is increased.

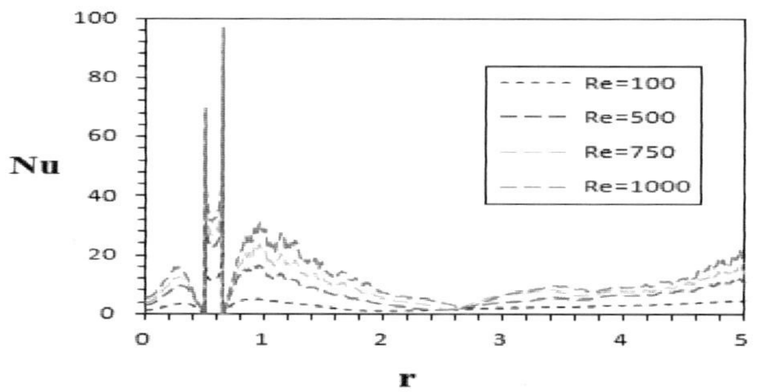

Figure 4 Distribution of Nu at different Re for an annular jet (r_i = 0.5) single rib (0.15x0.15) at rib location r=0.5

Figure 5 shows variation of Nusselt number for jet impingements with Reynolds numbers for circular and annular jet for both surfaces, though for annular jet, Nusselt number is comparatively lower than that of circular jet. It is noted that heat transfer for annular jet is about 20 - 30% less compared to the circular jet. Also from the study, it can be concluded that the Nusselt number value increases on ribbed surface. Nusselt number values for the ribbed surface can be increased by 5% to 20% than that of plane surface.

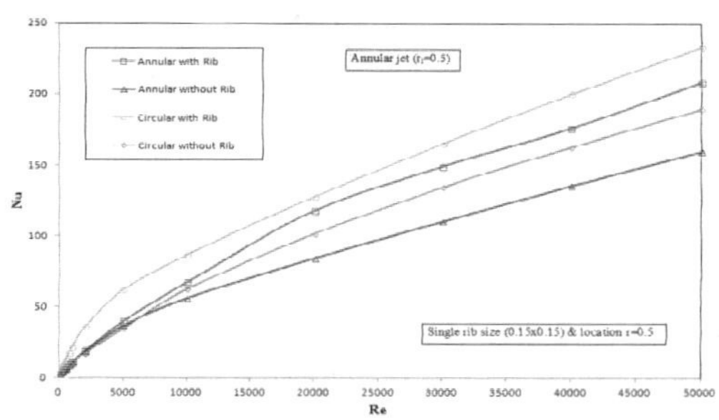

Figure 5 Variation of Nu with Re for an annular jet (r_i = 0.5) and circular jet with or without single rib.

Table 1 shows that enhancement of heat transfer at different Re when rib is used for annular jet. The difference in percentage of Nusselt number is also shown in the table. From the table, it is observed that the Nu for plane surface is less than that of ribbed surface.

Table 1 Heat transfer enhancement results on rib surface for annular jet

Re	Nu for Plane Surface	Nu for Rib Surface	Difference (%)
100	3.18509	3.369135	5.46%
500	7.152728	7.618556	6.11%
750	9.132468	9.933756	8.06%
1000	10.93725	11.9253	8.28%
5000	36.65586	40.14501	8.69%
10000	56.14009	67.62579	16.98%

4. Conclusion

In our earlier study Pal et al. 2019 [8] showed that an annular jet underperforms circular jets in terms of heat transfer. In this work, strategy of enhancing transport from annular jet is investigated. Augmentation is brought by using ribs on surface. The result reveals that for ribbed surface the Nusselt number becomes higher compared to flat surface in our considered study. But the Nusselt number is lower in case of annular jet compare to circular jet. Therefore, it concludes that enhancement is occurred in case of ribbed surface and the heat transfer distribution may be lower in case of annular jet. This enhancement can be increased 5%-20%.

References

[1] Chattopadhyay, H. and Saha, S. K., (2001), "Numerical Investigations of Heat Transfer Over a Moving Surface Due to Impinging Knife-Jets", *Numerical Heat Transfer, Part A: Applications,* vol. 39(5), pp. 531–549.

[2] Chattopadhyay, H., (2004), "Numerical investigations of heat transfer from impinging annular jet," *Int. Journal of Heat and Mass Transfer,* vol. 47, pp. 3197-3201.

[3] Chattopadhyay, H., (2006), "Effect of Surface Motion on Transport Processes Due to Circular Impinging Jets—*A Numerical Study", Drying Technology, vol.* 24, pp.1347–1351.

[4] Yan, W.M., Liu, H.C., Soong. and Yang, C.Y. W.J., (2005), "Experimental study of impinging heat transfer along rib-roughened walls by using transient liquid crystal technique", *Int. J. Heat Mass Transfer,* vol. 48, pp. 2420–2428.

[5] Tan, L., Zhang, J. and Xu, Hua-Sheng, (2014), "Jet impingement on a rib-roughened wall inside semi- confined channel", *International Journal of Thermal Sciences*, vol. 86, pp. 210-218.

[6] Nakod, P.M., Prabhu, S.V. and Vedula, R.P., (2006), "Heat Transfer Augmentation Between Circular Air Jet and Flat Plate Using Finned Surfaces", *Heat and Mass Transfer Conference,* IIT, Guwahati, India.

[7] Pal, T.K., Chattopadhyay, H., Mandal, D.K. and Bhattacharyya, S (2016), "Numerical Investigation of Heat Transfer under Impinging Annular Jets", MATTER: *International Journal of Science and Technology,* vol. 2(1), pp. 70-77.

[8] Pal, T.K., Chattopadhyay, H. and Mandal, D.K., (2019), "Flow and Heat Transfer due to Impinging Annular Jet", *International Journal of Fluid Mechanics Research,* vol. 46(3), pp. 199-209.

[9] Patankar, S.V.,(1980), Numerical Heat Transfer and Fluid Flow, *Hemisphere, Washington.*

[10] ANSYS Fluent user guide (2014), *ANSYS Incorporation.*

IOP Conf. Series: Materials Science and Engineering 1080 (2021) 012046 doi:10.1088/1757-899X/1080/1/012046

RANS studies of hydrogen-enrichment premixed turbulent flames

***Siva PR Muppala and Sooraj Paleli M Vasudevan**
School of Mechanical and Automotive Engineering
Faculty of Science Engineering and Computing
Kingston University, London
*Email: S.Muppala@kingston.ac.uk

Abstract: In this study, we show numerical predictions of the Zimont and Peter's turbulent flame speed models for low-swirl Methane/Air/Hydrogen flames [1]. These models are contained as default options within the ANSYS Fluent Premixed reaction model. Two distinct simulation tasks as part of the present study are – non-reacting and reacting flows, for with three different methane mixtures in compositions added with – 0, 40 and 60% hydrogen. The results show that the RANS approach provides a reasonable prediction of the cold flow conditions, whilst the reacting flow conditions, good agreement is reached up to 40% enrichment, except near the recirculation region,.

Keywords—Combustion, Hydrogen Flame, Enrichment, Numerical Analysis

Nomanclature

l_*^q = Ewald's corrector

α = Thermal diffusivity

Subscript l = Laminar

Subscript u = Unburnt

μ_i = Viscosity (material i)

δ = Laminar flame brush thickness

l_f = Flame brush thickness

l_t = Turbulence length scale

Subscript t = Turbulent

$\bar{c}$ = Mean progress variable

U = Velocity

∇ = Gradient

S_c = Schmitt number

$\acute{u}$ = RMS velocity

ρ = Density

1. Introduction

Hydrogen is a practical solution to the worldwide drive to reduce polluting emissions. The methods of 'green' hydrogen production provide relatively low yields that would be inadequate for widespread 100% hydrogen combustion; however, it would be suitable for lower levels of hydrogen enrichment of hydrocarbons as an intermediate compromise. As hydrogen enrichment offers a number of the benefits of hydrogen combustion, namely reduced emissions, this provides a stepping-stone towards pure hydrogen combustion.

With the recent advances in computer technology, the use of computational fluid dynamics as a tool to study reacting flows has become more feasible. This study seeks to extend the scope of this approach via the validation of two commercially available turbulent flame speed closures; within the ANSYS Fluent premixed reaction model as applied to enrichment of hydrogen combustion. The low-swirl flame examined as part of this study was generated using a model of the low-swirl injector seen in Figure 1. The flame itself relies on the divergent nature of the flow to facilitate aerodynamic stabilization rather than utilizing a bluff body as in some flames. The flame stabilizes where the local fluid velocity equals the turbulent flame speed (approximately 2 m/s).

Figure 1 - Low Swirl Injector [12]

2. Flame background

The hydrogen doping of methane significantly affects the structure of the resulting flame with a shift from a wrinkled, continuous flame front to a discontinuous, cellular structure. This discontinuous cellular structure poses the greatest challenge for reaction models. Figure 2 shows the concentrations of the OH radical identifies the reacting regions of the flame, clearly the hydrogen flame has extinguished at each of the wrinkled indents. This is where the strain rate has exceeded 1061 s^{-1}. Extinguished regions do not reignite because when a section of the flame front extinguishes the fuel diffuses towards the reactions on either side of the region. In the study by Bell, et al. [2] finds that almost 97% of the hydrogen had diffused out of the extinguished region preventing re-ignition.

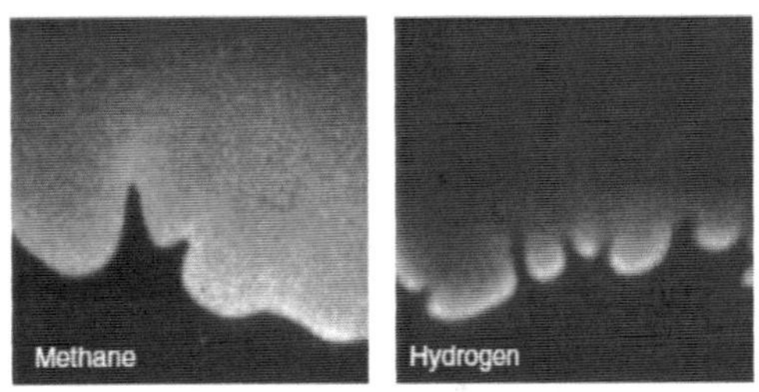

Figure 2 - PLIF images of methane and hydrogen flames [2]

In addition to the visible increase in wrinkling (seen by the human eye as a thickening of the flame brush) the turbulent flame speed also increases; this in turn decreases the time-scale of the flame that means the flame is more robust. This is because the flame is able to react to changes in strain more often, which minimizes the occurrence of the large-scale changes in strain that cause extinction; this then means that hydrogen flames have a greater flame area.

3. Literature Review

To date the RANS approach to turbulence modelling has been one of the most prevalent due to the simple, efficient use of resources; in particular, the k-ε model is widely used. Eldrainy, et al. (2009) uses, with a reasonable agreement with experimental data [5]. Mameri and Golkalp [9] and Mardani, et al. [10] who used the Pope correction with good agreement found in each case, took a modified approach. Saqr, et al., [14] compared the performance of the unmodified k-ε with the Realizable k-ε model and found the Realizable model to be a significant improvement [14]. This was also observed by Shih, et al [16] where the modified model 'performed better than the standard k-ε in almost all the cases tested' [16]. This observation was noted by Rohani and Saqr who employed this model, however they stated that because of the way 'the dissipation o.f turbulence energy is calculated' [13] this model should be teamed with a flamelet model to obtain the best results. The improvement noted in regards to this model was to be expected as this model was designed to mitigate a number of the issues with the standard model, particuarly the way in which the eddy dissipation was modelled. This was addressed in the Realizable model via a 'new eddy-viscosity formula and a new dissipation equation' [15]. As a final note regarding two equation models the study by Engdar and Kingmann [5] considered all the common two equation models including the k-ω and SST k-ω models. At the most relevant swirl value they found the best performing model to be the SST k-ω however this still performed poorly, significantly over-predicting the axial velocity [6]. This study also highlighted that neither the standard k-ε or the curvature modified k-ε models properly predicted the recirculation region [6].

Both Jawarneh & Vatista [7] and Su, et al., [17] used the RSM model and achieved good agreement with experimental data. However, in the works of Meier, et al., poor agreement with experimental data was found with the recirculation region significantly over-predicted [11]. Shamami and Birouk [15] assess the performances of all previously mentioned models. This study applied the models to a swirling flow with a sudden expansion at the swirl values 0.4 and 0.81, the former of which is of interest here. "One may conclude that amongst the two-equation eddy viscosity models, the k-ε and SST show the worst and best predictions of the mean swirl velocity profiles, respectively" [15]. Conversely, however the k-ε model provided the most accurate prediction of the central recirculation region size. The RSM and RSM SSG models produced "the best predictions of the mean velocity profiles" [15] and the closest measurements of the corner recirculation regions though these were not especially accurate [15].

In terms of general numerical geometry, the dimensions are given in both validation studies. Reacting validation: *Laboratory investigations of a low-swirl injector with H_2 and CH_4 at gas turbine conditions* [3] and non-reacting validation: *Fuel effects on a low-swirl injector for lean pre-mixed gas turbines* [8]. As non-reacting results were not available from the core study [3] in order to validate the non-reacting model the geometry and flow conditions were altered slightly to allow the cold results from [8] to be utilised.

The possibility of not modelling the swirler and merely applying a swirl value was examined, however Day, et al. [4] completed a detailed study into the possibility and "considerably better agreement was found between simulated and measured profiles when the flow generated inside the nozzle was included" [4].

4. Numerical Analysis

Initially, using an optimal meshing approach, a hybrid mesh with the combustor and pre-swirl sections are fully structured. Because of the conflicting approaches taken in literature, we consider a full turbulence model and a solver set up as part of this numerical work, the final settings are given in Table 1.

Table 1 - Fluent solver settings

Setting / Scheme	Value
Momentum	1st order upwind
Pressure	PRESTO!
Pressure-Velocity coupling	SIMPLE
Turbulence dissipation	1st order
Turbulence energy	1st order
Reynolds stress	1st order

Table 2- Mesh statistics

Entity	Value (mm)
Combustor (Max/wall)	2.0 / 0.5
LSI (Max/wall)	0.9 / 0.6
Pre-Swirl (Max/wall)	2.0 / 0.5
Shear layer	0.5
Core	0.9
Total no of elements	**3892722**

All of the turbulence models tested as part of this study utilized the standard model constants and for those that contained wall functions uses the scalable wall function on each occasion [1].

Mean Reaction rate, premixed model: $\frac{\partial}{\partial t}(\rho\bar{c}) + \nabla * (\rho\bar{v}\bar{c}) = \nabla * \left(\frac{\mu_i}{Sc_i}\nabla\bar{c}\right) + \rho S_c$

The turbulent flame speed model: $U_t = 0.52 * (\acute{u})^{\frac{3}{4}} U_l^{\frac{1}{2}} \alpha^{\frac{-1}{4}} l_t^{\frac{1}{4}}$

Peters Flame Speed Model: $U_t = U_l(1 + \sigma_t)$ *and* $\sigma_t = -\frac{a_4 b_3^2}{4b_1}\frac{l_f}{\delta} l_*^q + \left[\left(\frac{a_4 b_3^2}{4b_1}\frac{l_f}{\delta} l_*^q\right)^2 + \frac{C_s b_3^2}{2}\frac{\grave{u} l_f^2}{U_l \delta l_t}\right]^{\frac{1}{2}}$

for constant values, see [1].

5. Results and Discussion

Addressing the cold results first, we validate the results against [8], only the positive X values are shown here. The flow conditions for this section of the study were as per the validation study [8]. All the models under predict the velocity in the core region (Figure 3). This is attributed to assumptions made during the geometry generation stage and not due to prediction errors, given the correlation between the different models. Unfortunately, time did not permit revisiting the geometry to address this. The reacting region is reasonably well predicted, except with a slight over prediction at the peak and approaching the wall, and the two RSM models producing asymmetric results. Therefore, we employ Realizable k-ε model for all simulations here.

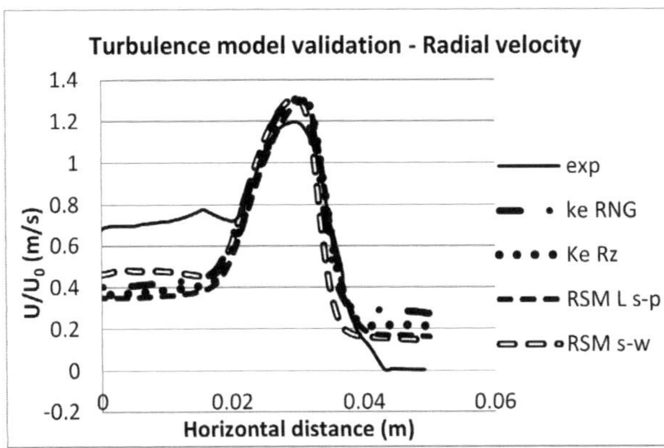

Figure 3 - Turbulence model validation, radial velocity plot: k-ε RNG, Realizable k-ε, RSM Linear Pressure-Strain, RSM Stress-Omega

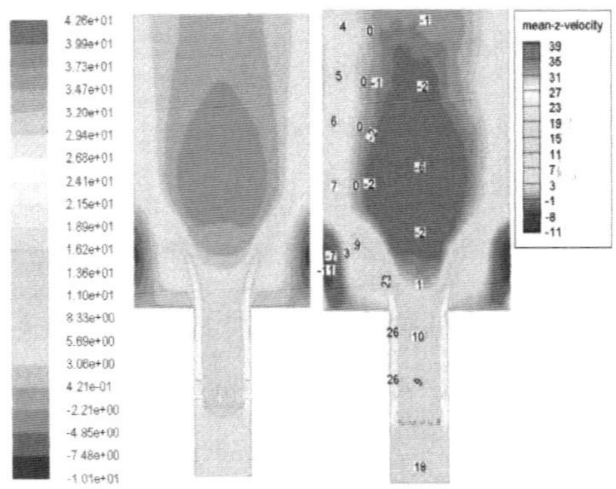

Figure 4 - Left Axial velocity plot, Right axial velocity plot from [12]

As an additional comparison, the velocity plot from this study compares well with the findings of [12] with good agreement found between the two studies. Please note this contour plot refers to the flow conditions as per the reacting validation study [3].

The first reacting results presented here are from the pure methane case for both turbulent flame models. For validation purposes, we chose experimental data by Cheng, et al. [3]. The results given in Figure 5 show some correlation particularly pre 30mm and within the reaction region, however the flame front has been predicted at 46mm vs the experimental 40mm. Limited correlation can also be

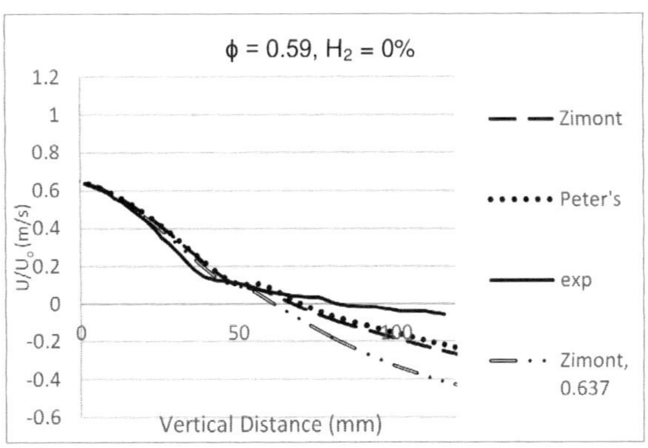

Figure 5 - Axial profile for 100% methane, both flame models, modified Zimont results and experimental

seen beyond 65mm.

The Zimont modification, turbulent flame speed constant = 0.637 given in [1] can be seen to cause further deterioration in the results; so this is not taken further. Whilst the numerical predictions show significant correlation with each other it can be said that beyond 65mm Peters flame speed model produced slightly yield more accurate results. However, the difference was marginal.

Figures 6 and 7 show the numerical predictions and experimental results for 40% and 60% hydrogen respectively. Once again, the predictions beyond 60mm show significantly deteriorated correlation. As this can be seen across all of the results including 0% hydrogen it is this studies opinion that this error was likely to be due to an over prediction of the axial velocity in the recirculation region on the part of the RANS model, as in a number other studies, for example [18]. An interesting is that the difference in magnitude increases with the level of enrichment, with additional deterioration in the 60% hydrogen predictions pre 50mm of both reaction models.

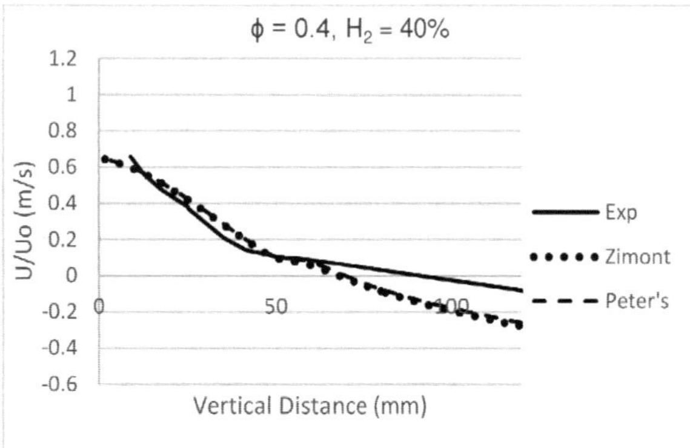

Figure 6 - Axial profile for 40% hydrogen, both flame models and experimental (U_0=18)

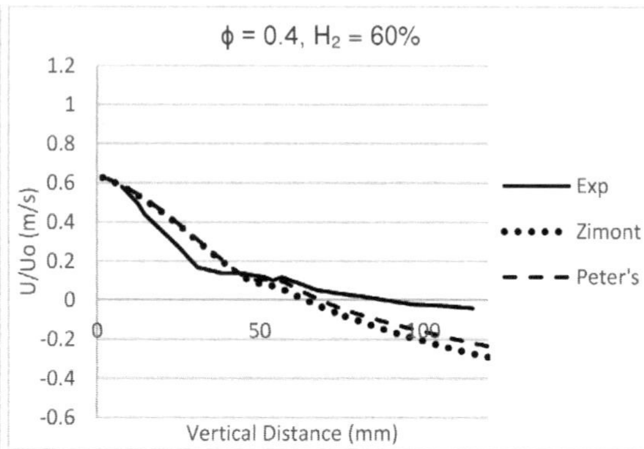

Figure 7 - Axial profile for 60% hydrogen, both flame models and experimental (U_0=18)

6. Conclusions

This study highlights a viable stepping-stone towards pure hydrogen combustion through hydrogen enrichment. Given the feasibility of this approach and the rising cost of experimental testing, numerical predictions is a solution. Here, we assess the validity of two turbulent flame models contained within the ANSYS Fluent premixed combustion model applied to enriched hydrogen combustion. We find that the accuracy of the numerical predictions would deteriorate with increase in levels of hydrogen enrichment. An additional point worth noting is that the RSM turbulence model group (seven-equation models) is significantly less sensitive to solver set up than the two-equation models.

In all simulations cases, there is every prediction of the axial velocity in the recirculation region, and we attribute this consistent error to the performance of the Realizable k-ε turbulence model and not so much to the reaction model. The additional deterioration in this region at levels of enrichment greater than 40% is due to the performance of the reaction model. The effect of the RANS simplification on the flow conditions needs further assessment using the LES approach to turbulence modelling. The results through LES would allow for the accurate detailing of the errors and cause of these errors with greater confidence.

References

[1] Ansys Inc. *Fluent.*

[2] Bell, J. B., Cheng, R. K., Day, M. S. & Shepherd, I. G., 2007. Numerical simulation of Lewis number effects on lean premixed flames. *Proceedings of the Combustion Institute* , Volume 31, pp. 1309-1317.

[3] Cheng, R., Littlejohn, D., Strakey, P. & Sidwell, T., 2009. Laboratory investigations of a low-swirl injector with H2 and CH4 at gas turbine conditions. *Proceedings of the Combustion Institute,* 21(1), pp. 3001-3009.

[4] Day, M. et al., 2012. A combined computational and experimental characterization of lean premixed turbulent low-swirl laboratory flames. *Combustion and Flame*, Volume 159, pp. 275-290.

[5] Eldrainy, Y. A., Saqr, K. M., Aly, H. S. & Jaafar, M. N. M., 2009. CFD insight of the flow dynamics in a novel swirler for gas turbine combustors. *International Communications in Heat and Mass Transfer,* Vol. 36, pp.936-941.

[6] Engdar, U. & Klingmann, J., 2002. Investigation of two-equation turbulence models applied to a confined axis-symmetric swirling flow. *Computational technologies for Fluid/Thermal/Structural/Chemical systems with industrial applications,* 448(2), pp. 199-206.

[7] Jawarneh, A. M. & Vatistas, G. H., 2006. Reynolds Stress Model in the Prediction of Confined Turbulent Swirling Flows. *Journal of Fluids Engineering,* Volume 128, pp. 1377-1382.

[8] Littlejohn, D. & Cheng, R., 2007. Fuel effects on a low-swirl injector for lean pre-mixed gas turbines. *Proceedings of the Combustion Institute,* Volume 31, pp. 3155-3162.

[9] Mameri, A. & Gokalp, I., 2007. *Numerical Simulation of Turbulent Methane Premixed Flames Enriched by Hydrogen.* s.l., Third European Combustion Meeting 2007.

[10] Mardani, A., Tabejamaat, S. & Hassanpour, S., 2013. Numerical study of CO and CO2 formation in CH4/H2 blended flame under MILD conditions. *Combustion and flame,* Volume 160, pp. 1636-1649.

[11] Meier, W. et al., 2000. Investigations in the TECFLAM swirling diffusion flame: Laser Raman Measurements and CFD Calculations. *Applied Physics B, Lasers and Optics*, Volume 71, pp. 725-731.

[12] Muppala, S.P.R., Manickam, B. & Dinkelacke, F., (2015). A Comparative Study of Different Reaction Models for Turbulent Methane/Hydrogen/Air Combustion. *J. of Thermal Engineering.* Special Issue: 1 International Energy Technologies (ENTECH14), Year: 2015, Volume: 1, Issue: 5, pp:367-380.

[13] Rohani, B. & Saqr, K. M., 2012. Effects of hydrogen addition on the structure and pollutant emissions of a turbulent unconfined swirling flame. *International Communications in Heat and Mass Transfer,* Volume 39, pp. 681-688.

[14] Saqr, K. M., Kassem, H. I., Aly, H. S. & Wahid, M. A., 2012. Computational study of decaying annular vortex flow using the Re/k-e turbulence model. *Applied Mathematical Modelling,* Volume 36, pp. 4652-4664.

[15] Shamami, K. K. & Birouk, M., 2008. Assessment of the Performances of RANS Models for Simulating Swirling Flows in a Can-Combustor. *The Open Aerospace Engineering Journal,* Volume 1, pp. 8-27.

[16] Shih, T. et al., 1994. *A New k-e Eddy Viscosity Model for High Reynolds Number Turbulent Flows-Model Development and Validation,* Cleveland: NASA.

[17] Su, Y., Zheng, A. & Zhao, B., 2011. Numerical simulation of effect of inlet configuration on square cyclone separator performance. *Powder Technology*, Volume 210, pp. 293-303.

[18] Muppala, S.P.R., and Dinkelacker, F.,(2004). Numerical modelling of the pressure dependent reaction source term for premixed turbulent methane/air flames. 328-336. *Progress in Computational Fluid Dynamics, An International Journal.* Vol.4 No.6. pp.328 – 336.

IOP Conf. Series: Materials Science and Engineering 1080 (2021) 012047 doi:10.1088/1757-899X/1080/1/012047

Sunlight mediated photocatalytic reduction of aqueous Cr(VI) using metal hexacyanoferrate (M = Mn, Ni, Cu and Zn)

Manas Mandal[1,2], Krishna Chattopadhyay[3], and Swapan Kumar Bhattacharya[1]

[1] Department of Chemistry, Jadavpur University, Kolkata–700032, India
[2] Department of Chemistry, Sree Chaitanya College, Habra, 24Parganas(N)–743268, India
[3] School of Chemical Sciences, Indian Association for the Cultivation of Science, Kolkata, West Bengal – 700032, India
e-mail: *skbhatt7@yahoo.co.in*

Abstract. Four different metal hexacyanoferrates, MHCF (M = Mn, Ni, Cu and Zn) were synthesized via a simple, cost effective precipitation method. The as-synthesized MHCF were characterized using XRD analysis and FESEM images were recorded for the analysis of morphology. Among the four compounds, ZnHCF exhibited most effective catalytic property for visible-light driven photoreduction of Cr(VI) to Cr(III), with excellent photochemical reducing capability of 68% within 2 h.

1. Introduction

Now a day's excessive and indiscriminate uses of organic–mineral complexes in electronic gadgets as well as inorganic compound in the wood preservation industry have progressively become the most serious matter of environmental concern [1]. Mixing of the acid wastewater from mine drainage and heavy metal like chromium from preservation industry to the ground water is very dangerous towards aquatic life and mutagenic to human [2-3]. The waste water without treatment is hazardous for ecosystem, as most of the synthetic dyes are toxic, mutagenic and carcinogenic and enhance the chemical oxygen demand (COD) of the water body. Generally, chromium (Cr) is present in compounds in two common oxidation states such as Cr(III) and Cr(VI) [4]. Amongst these, Cr(VI) is enormously toxic, carcinogenic, whereas Cr(III) is less toxic and more easily soluble in water as $Cr(OH)_3$. Cr (VI) is used in several industrial fields like textile production, electroplating, metal finishing, leather tanning, dyeing, etc. [5]. As a result, higher degrees of contamination due to discharges of Cr (VI) in waste water, makes the aquatic life severely harmful. Presence of Cr (VI) in drinking water causes kidney, liver, bladder, and skin cancer [6]. So, the urgent demand for effective methods for efficient reduction of Cr (VI) to Cr (III) in industrial waste water has been raised. Different methods including ion exchange, precipitation, reverse osmosis, photocatalysis and adsorption are developed [7-9]. However, each method has its own advantages along with numerous drawbacks. The photo-catalytic process is highly efficient and cost effective as compared to the other methods.

The transition metal complex like metal hexacyanoferrate has recently got attention from scientific community, as the hexacyanoferrate compound shows some unique solid state chemistry and promising applications in the field of electrochemical energy storage [10-11], electrocatalysis [12], photocatalysis [13], ion-sensing [14] etc. Jassal et al synthesized potassium zinc hexacyanoferrate nanocubes which were able to degrade Malachite Green up to 94.15% and Eriochrome Black T up to about 76.13% under optimum condition [13]. Herein, a set of MHCF (M = Ni, Zn, Mn and Cu) was prepared using precipitation method for photocatalytic reduction of Cr (VI) under sunlight.

2. Experimental Section

2.1. Materials and Method

Manganese nitrate tetrahydrate, Nickel nitrate hexahydrate, Copper nitrate trihydrate, Zinc nitrate hexahydrate, potassium hexacyanoferrate(III), sodium hydroxide, potassium permanganate, potassium dichromate were purchased from Merck, India. All reagents were directly used for the experiment without further purification.

2.2. Synthesis of metal hexacyanoferrate (M = Mn, Ni, Cu, Zn)

For the synthesis of metal hexacyanoferrate (MHCF), simple precipitation method was followed. Briefly, 20 mL 0.1 M aqueous solution of $M(NO_3)_2$ [M= Ni, Cu, Zn, and Mn] was taken in a 100 mL beaker. Then 20 mL 0.1 M aqueous solution of $K_3Fe(CN)_6$ was slowly added to it under vigorous stirring condition. The solution was magnetically stirred for 1 hour. Then the solution was kept for 12 h in undisturbed condition at room temperature. After that the solid precipitate was filtered, washed with distilled water and ethanol for 4-5 times. Finally, the desired solid product (MHCF) was dried in air oven at 70°C for 6 h.

3. Results and Discussions

3.1. Formation of the Metal hexacyanoferrate:

The synthetic procedure involves the precipitation method where a quick precipitation of metal hexacyanoferrate [MHCF], where M = Mn, Ni, Cu and Zn was achieved. The chemical reaction involved in this synthetic process is as follows:

$$M(NO_3)_2 + K_3[Fe(CN)_6] \rightarrow M_x[Fe(CN)_6]_y + KNO_3$$

3.2. X-Ray diffraction (XRD) analysis

X-ray diffraction (XRD) analysis was performed by Rigaku ULTIMA-III X-ray diffractometer with Cu K_α radiation (λ = 1.5418 °A). The XRD patterns of the as-prepared MHCF (M = Mn, Ni, Cu and Zn) are shown in Figure 1. The observed XRD peaks of the MHCF (M = Mn, Ni, Cu and Zn) (Table 1) are well-matched with those in the JCPDS card number 01-074-7327, 01-082-2283, 01-086-0513 and 00-038-0688, respectively.

Table 1. List of important powder X-ray diffraction peaks and corresponding (hkl) values of the as prepared MHCF (M = Mn, Ni, Cu and Zn)

$Mn_4[Fe(CN)_6]_{2.667}\cdot15.84H_2O$ JCPDS No. 01-074-7327		$Ni[Fe(CN)_6]_2\cdot H_2O$ JCPDS No. 01-082-2283		$Cu[Fe(CN)_6]_{0.667}$ JCPDS No. 01-086-0513		$Zn_3[Fe(CN)_6]_2$ JCPDS No. 00-038-0688	
2θ(°)	(hkl)	2θ(°)	(hkl)	2θ(°)	(hkl)	2θ(°)	(hkl)
16.8	(002)	17.3	(200)	17.5	(200)	9.7	(012)
23.9	(022)	24.4	(220)	24.9	(220)	13.4	(104)
34.1	(004)	35.1	(400)	35.4	(400)	14.0	(110)
38.3	(024)	39.4	(420)	39.7	(420)	16.2	(113)
42.1	(224)	50.4	(440)	43.7	(422)	19.5	(024)
48.9	(135)	53.7	(600)	50.9	(440)	21.4	(116)
52.2	(244)	56.7	(620)	54.3	(442)	24.4	(300)
55.2	(335)			57.6	(620)	28.1	(119)
63.8	(246)					32.7	(226)
66.5	(137)						

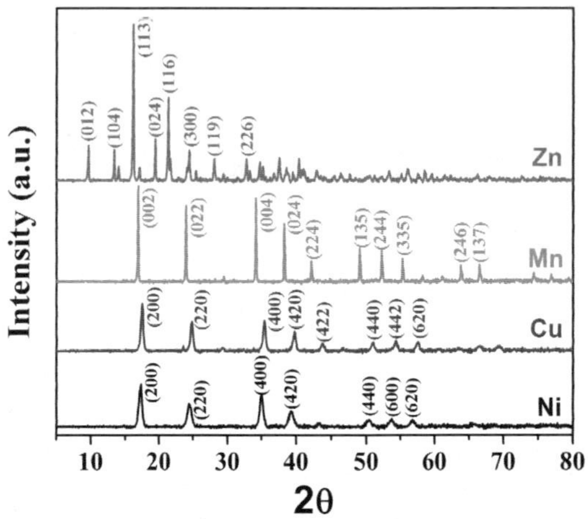

Figure 1. XRD pattern of as-prepared MHCF (M = Mn, Ni, Cu and Zn)

3.3. Morphological analysis

To investigate the surface morphology of the as-synthesized metal hexacyanoferrate (M = Mn, Ni, Cu and Zn) photo-catalysts, the field emission scanning electron microscopy (FESEM,Carl Zeiss-SUPRA 40) images were captured (Figure 2). From the FESEM of MnHCF, it can be seen that this is of cubic structure with an average length of 1um. The NiHCF and CuHCF are nano-sized particles. The FESEM image of ZnHCF suggests the cubic structure with an average length of 200 - 500 nm.

Figure 2. FESEM images of as-prepared metal hexacyanoferrate: MnHCF (a), NiHCF (b), CuHCF (c) and ZnHCF (d).

3.4 Photocatalytic Reduction of Cr(VI)

The photocatalytic reduction of Cr(VI) was performed by taking the aqueous dichromate solution containing the different as-synthesized catalyst (MHCF) under sunlight. A set of four beakers containing 25 mL 2×10^{-4} M aqueous solution of $K_2Cr_2O_7$ with 25 mg of different MHCF (M = Mn, Ni, Cu and Zn) was taken and stirred for 1 h in dark condition to reach the adsorption–desorption equilibrium. One blank test was also performed without catalyst. After that all the beakers were put under sunlight for 2 h with occasional shaking. The solutions were centrifuged separately time to time and the UV-vis spectra were recorded.

Figure 3 shows the photocatalytic reduction of aqueous Cr(VI) under sunlight in the presence of the as-synthesized metal hexacyanoferrate, MHCF (M = Mn, Ni, Cu and Zn). The diversity in the efficiency of different metal hexacyanoferrate may be a result of several factors, such as composition, particle size, morphology, crystallinity, crystal defects, specific surface area, band gap, state of dispersion, adsorption capacity for Cr(VI), etc. [15]. These factors are strongly coupled with each other and hence it is hard to identify the specific function and influence of a single parameter in the photocatalytic activity of different metal hexacyanoferrates. From the Figure 3, it can be seen that as-synthesized zinc hexacyanoferrate nanocubes reveal superior photocatalytic activity when compared to other catalysts. With increasing exposure time, the main absorption band of Cr(VI) cantered at 363 nm notably decreases due to decreasing the concentration of Cr(VI). It is calculated that the as-prepared zinc hexacyanoferrate nanocubes exhibit excellent photochemical reducing capability. The photo-catalytic activity i.e., reducing capability of the material is calculated (Equation 1) by monitoring the decrease in the absorption intensity which is directly proportional to the Cr(IV) concentration. A decrease in absorption intensity from 1.02 to 0.33 at 365 nm implies 68% reducing capability of the as prepared ZnHCF within 2 h.

$$\text{Catalytic activity (\%)} = \frac{|(\textit{Final absorption intensity} - \textit{Initial absorption intensity})|}{\textit{Initial absorption intensity}} \times 100\ \% \qquad (1)$$

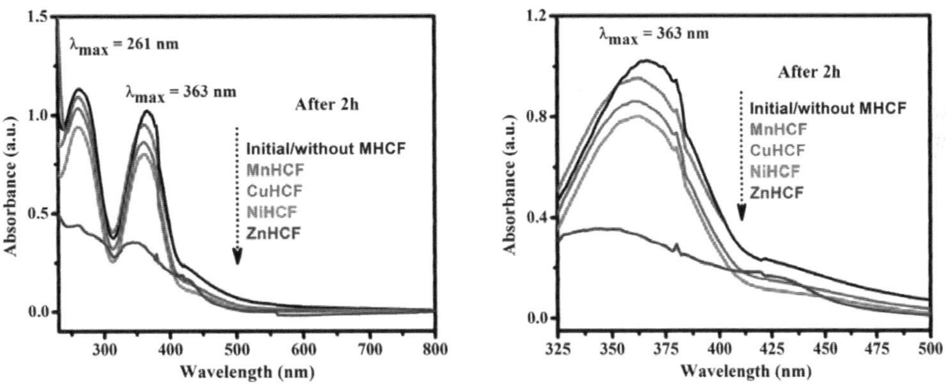

Figure 3. Photocatalytic reduction of 25 mL of 2×10^{-4} M aqueous solution of dichromate under sunlight in the presence of different MHCF (M = Mn, Ni, Cu and Zn) (25 g)

The reduction of Cr(VI) to Cr(III) is mainly carried out by the photoexcited electrons and therefore, good electron–hole pair separation is attained in this reduction process. The better photocatalytic activity of the as-prepared ZnHCF catalyst for the reduction of Cr(VI) under sunlight can be explained by taking into consideration of some unique properties of Zn which is responsible for the exceptional behaviour of ZnHCF. Zinc has a completely filled valance shell *3d* orbitals and a narrow band gap. Due to highly dispersive nature of conduction band, the d^{10} electronic configuration supports photo-generated electron-hole separation. On photo-irradiation with sunlight, MHCF semiconductors absorb light and electrons from valence band are excited to conduction band, generating the holes in the valance band. These photoexcited electrons in conduction bands of the semiconductors reduce Cr(VI) to Cr(III) with a visual color change of the solution from leuco to colorless. The probable reaction pathway for the photocatalytic reduction of Cr(VI) under sunlight over MHCF in aqueous solution is as follows:

$$\text{MHCF} + h\nu \rightarrow \text{MHCF}\ (e^- + h^+)$$
$$2H_2O + 4h^+ \rightarrow O_2 + 4H^+$$
$$Cr_2O_7^{2-} + 14H^+ + 6e^- \rightarrow 2Cr^{3+} + 7H_2O$$

4. Conclusion

In summary, four different metal hexacyanoferrates, MHCF (M = Mn, Ni, Cu and Zn) were synthesized via a simple and cost effective precipitation method. The as-prepared MHCF were characterized using XRD analysis and the morphology was investigated with FESEM images. The ZnHCF semiconductor exhibited more superior catalytic property for visible-light-driven photoreduction of Cr(VI), with excellent photochemical reducing capability of about 68%. Thus the as-synthesized metal hexacyanoferrates show promising applications in the field of environmental photocatalysis.

References

[1] Liu J, He X, Lin X, Chen W, Zhou Q, Shun W and Huang L 2015 Ecological effects of combined pollution associated with e-waste recycling on the composition and diversity of soil microbial communities *Environ. Sci. Technol.* **49** 6438–6447.

[2] Turpeinen R, Kairesalo T and Häggblom M M 2010 Microbial community structure and activity in arsenic-, chromiumand copper-contaminated soils *FEMS Microbiol. Ecol.* **47** 39–50.

[3] Liu F, Zhang W, Tao L, Hao B and Zhang J 2019 Simultaneous photocatalytic redox removal of chromium (vi) and arsenic (iii) by hydrothermal carbon-sphere@nano-Fe_3O_4 *Environ. Sci. Nano* **6** 937-947.

[4] Hsu C L, Wang S L and Tzou Y M 2007 Photocatalytic reduction of Cr(VI) in the presence of NO_3^- and Cl^- electrolytes as influenced by Fe(III) *Environ. Sci. Technol.* **41** 7907−7914.

[5] Bhaumik M, Maity A, Srinivasu V V and Onyango M S 2011 Enhanced removal of Cr(VI) from aqueous solution using polypyrrole/Fe_3O_4 magnetic nanocomposite *J. Hazard. Mater.* **190** 381−390.

[6] Testa J J, Grela M A and Litter M I 2004 Heterogeneous photocatalytic reduction of chromium(VI) over TiO_2 particles in the presence of oxalate: Involvement of Cr(V) species *Environ. Sci. Technol.* **38** 1589−1594.

[7] Joshi K M and Shrivastava V S 2011 Photocatalytic degradation of Chromium(VI) from wastewater using nanomaterials like TiO_2, ZnO, and CdS *Appl. Nanosci.* **1** 147-155.

[8] Dinda D, Gupta A and Saha S K 2013 Removal of toxic Cr(VI) by UVactive functionalized graphene oxide for water purification *J. Mater. Chem. A* **1** 11221−11228.

[9] Mondal C, Ganguly M, Pal J, Roy A, Jana J and Pal T 2014 Morphology controlled synthesis of SnS_2 nanomaterial for promoting photocatalytic reduction of aqueous Cr(VI) under visible light *Langmuir* **30** 4157-4164.

[10] He Y, Zhang P, Wang M, Wang F, Tan D, Li Y and Feng X 2019 Nano-sandwiched metal hexacyanoferrate/graphene hybrid thin films for in-plane asymmetric micro-supercapacitors with ultrahigh energy density *Mater. Horiz.* **6** 1041-1049.

[11] Bors R, Yun J, Marzak P, Fichtner J, Scieszka D and Bandarenka A S 2018 Chromium(II) hexacyanoferrate-based thin films as a material for aqueous alkali metal cation batteries *ACS Omega* **3** 5111-5115.

[12] Sitnikova N A, Komkova M A, Khomyakova I V, Karyakina E E and Karyakin A A 2014 Transition metal hexacyanoferrates in electrocatalysis of H_2O_2 reduction: an exclusive property of Prussian Blue *Anal. Chem.* **86** 4131-4134.

[13] Jassal V, Shanker U, Kaith B S and Shankar S 2015 Green synthesis of potassium zinc hexacyanoferrate nanocubes and their potential application in photocatalytic degradation of organic dyes *RSC Adv.* **5** 26141-26149.

[14] Jassal V, Shanker U and Shankar S 2015 Synthesis characterization and applications of nano-structured metal hexacyanoferrates: a review *J. Environ. Anal. Chem.* **2** 2380-2391.

[15] Zhang Y C, Li J, Zhang M and Dionysiou D D 2011 Size-tunable hydrothermal synthesis of SnS_2 nanocrystals with high performance in visible light-driven photocatalytic reduction of aqueous Cr(VI) *Environ. Sci. Technol.* **45** 9324-9331.

Comparison of Vortex Induced Velocity Kinematics around Underwater Horizontal Cylinder with Vertical End Plate

A Kuila[1], B Mukherjee[2] and S Das[3]

[1] Doctoral Research Fellow, School of Water Resources Engineering, Jadavpur, University, Kolkata 700032, West Bengal, India
[2] PhD Researcher, School of Water Resources Engineering, Jadavpur, University, Kolkata 700032, India
[3] Assistant Professor, School of Water Resources Engineering, Jadavpur, University, Kolkata 700032, India

E-mail: subhasishju@gmail.com

Abstract. Electrical energy is harnessed from various natural sources to protect the environment and ecology which is termed as green energy sources. When vortex is induced on a cylinder, it starts vibrating according to the formation of vortex. The vibration is so strong that it can destroy any structure wherever alternating vortex is induced. Currently Vortex Induced Vibration is used to harness electrical energy from ocean, river and shallow water channel. In this paper, experimental investigations were conducted in flume to analyse the behaviour of underwater flow structures around a 5 cm horizontal cylinder placed transversely to the flow at 60% and 70% depths from water surface and simultaneously a 5 cm wide vertical end plate set at 8 cm downstream from cylinder. Fluctuating 3D velocity components around such plate cylinder combination were measured. From subsequent plotted contours, changing flow patterns in presence of vertical plate are analysed. Hydrodynamic flow features are shown by plotting absolute velocity fields and velocity vectors. It visualizes that streamlines separate first and then passes over the periphery of horizontal cylinder. The contour of turbulence kinetic energy confirms the presence of strong vortex in between horizontal cylinder and vertical plate. The horseshoe vortex at the upstream of cylinder becomes stronger due to the presence of vertical plate. The results from these two experiments are compared with the results observed in previous researches. With the increase in cylinder depth from water surface, the horseshoe vortices strength becomes stronger.

1. Introduction

Analysis of vortex induced flow fields around a cylinder is an interesting subject to the hydraulicians. The vortex induced flow field analysis around a cylinder or any structure on an erodible ocean bed or river bed was done by many researchers to find out the flow kinematics. Many researchers had shown that a vertically mounted cylinder on an erodible bed causes scouring and develops horseshoe vortex and thereby eroded sediments are transported downstream from the scour hole [1, 2]. From the visualization of flow fields by capturing real data using measuring instruments it confirms that when such flow is moved closer to a structure - the turbulent boundary layer separates and a whirl develops commonly known as horseshoe vortex and the subsequent circulation on such structure decreases due decrease in the water depth [3]. The nature of turbulent Kinetic Energy (KE) was depicted on a combination of three circular structure arranged eccentrically on a horizontal plane to understand the vortex induced sediment transport [4]. An important phenomenon like the underwater unsupported cylinder when faces flow induced oscillation by wake vortex shedding due to which pipeline is encountered fatigue failure as proven both numerically [5, 6] and experientially [7, 8].

Vortex can be induced on a structure mounted either vertically or horizontally and that structure will vibrate due to vortex induction on the structure. The vibration will be destructive when the natural frequency of such structure and corresponding induced frequency become synchronized. In previous researches, this interaction between water and structure was attempted to suppress the effect of Vortex-Induced-Vibration (VIV) but there is another scope of VIV which is to generate energy to produce electricity [9, 10]. When a vertical or horizontal cylinder is set under submerge condition then vortex is induced on that cylinder. The cylinder will immediately start vibrating due to induction of such vortex. If the natural frequency of the underwater cylinder and the alternating vortex frequency are very close then such a cylinder immediately begins to oscillate on a constant Strouhal number. Such oscillation can sustain over an extended range of velocity by changing the Reynolds number along with vortex shedding on cylinder [11, 12]. Here the vortex induced flow fields were examined and discussed after measuring velocity profiles around horizontal cylinder placed under submerged condition and a vertical plate placed as an obstruction to the downward of such cylinder. Only the flow fields around the cylinders were measured by most investigators [13]. In some investigations a horizontal cylinder was placed to separate flow at the turbulent boundary layer and to visualize the flow contours along with vortices profiles surrounding the cylinder [14]. In shallow water channels, if the formed horseshoe vortices surrounding a cylinder are obstructed by a vertical end plate located downstream of that cylinder within a considerable spacing then it immediately affects the vortex shedding over such cylinder which causes its oscillation [15]. The disturbed turbulent KE due to this interaction between the vertical end plate and the horizontal cylinder ultimately increases the lifting force with a significant impact on the hydrokinetic energy device [12].

In this research, the vortex induced flow fields are analyzed around a horizontally placed cylinder (diameter of 4 cm) positioned 6 cm above the bed surface and a flow facing vertical end plate having 5 cm width placed at 5.5 cm downstream from such cylinder curvature. Thereafter, the flow fields are compared with the results of another experiment with same horizontal cylinder fixing at 8 cm above the bed surface and with the same downstream end plate. The distance of the vertical end plate from the cylinder outer curvature was kept adjusted to 5.5 cm.

2. Experimental Arrangement

A hydrokinetic structure was suspended in a laboratory shown in figures 1-2. A structure of 5 cm smooth surface cylinder was used. The horizontal hollow cylinder was fixed with two rods attached with it at its end sides. The cylinder was fixed 6 cm above the bed area with an aid clamp. A vertical end plate of 5 cm wide was fixed 5.5 cm away from the cylinder outer curvature. The complete structure was suspended on a re-circulating flume made of clear Plexiglas sheet on both sides.

The size of the flume is about 700 cm long, 350 cm internal wide and 450 cm deep. The redistribution of 50 lps output was done using two 10 hp pumps each having 30 lps capacity. The desired water depth 20 cm was maintained constant at the upstream of cylinder by means of two gate valves clamped with pump delivery pipes. An average velocity of 0.72 m/s was maintained throughout the study. The structure was suspended at a distance of 400 cm from the flume entrance so a great commotion would not affect the flow kinematics around it. The complete structure was built at the Fluvial Hydraulics Laboratory of School of Water Resources Engineering in Jadavpur University.

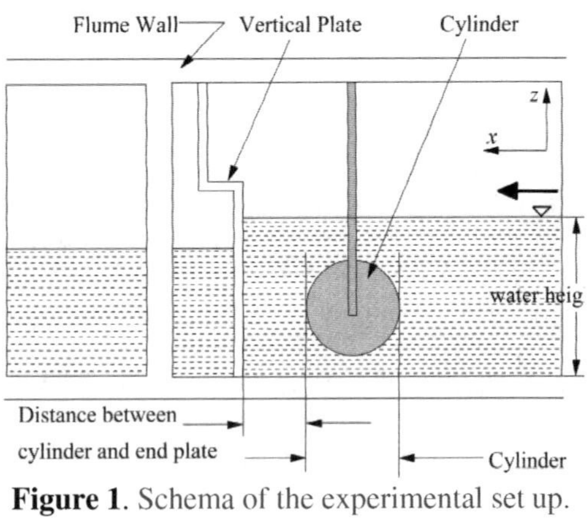

Figure 1. Schema of the experimental set up.

IOP Conf. Series: Materials Science and Engineering 1080 (2021) 012048 doi:10.1088/1757-899X/1080/1/012048

Fast velocity segments on the three Cartesian sides were measured and maintained with the Nortek made Vectrino$^+$ probe of forming an acoustic-Doppler-velocimeter holding four signal receivers. It collects information 5 cm away from the investigation tip. Doppler effects are used worldwide as a system for measuring such underwater velocity components. These velocity components have been recorded at 100 Hz sampling rate capacity, 6 cm sampling height distance and 2 to 5 mm adjusting sampling volume range of Vectrino$^+$. All gauge data using Vectrino$^+$ was converted and sorted with minimum signal-to-noise-ratio of 16 and minimum correlation value of 70%.

The three dimensional velocity data are arranged in a contour manner to visualize the flow structure surrounding the horizontal cylinder in x-z plane where the x axis indicates the direction along the flow and z axis indicates the direction vertical to the flow. The contour of time averaging longitudinal velocity (u), time averaging transverse velocity (v), time averaging vertical velocity (w) are plotted by using OriginLab software. The time averaging absolute velocity (V), vectors, turbulent intensities, and turbulent KE are derived from three-dimensionally gauged velocity components and subsequently plotted in contour form to analyze the flow effect around the two structures. With respect to vertical end plate, the x coordinate at the upstream is considered as negative and it's downstream as positive.

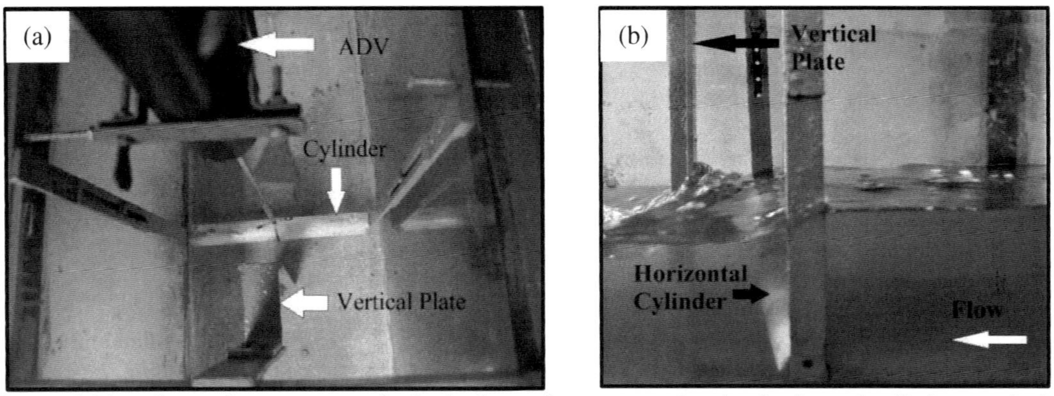

Figure 2. Experimental arrangement for hydrodynamic structure showing horizontal cylinder attached with the end rods and vertical end plate (a) before experimentation and (b) during experimentation.

3. Results and Analysis

Here the experimental results are illustrated in a vertical plane chosen between the transversely positioned horizontal cylinder and vertical end plate. Estimated time averaging longitudinal, transverse, vertical gauge velocities are abbreviated as u, v, w respectively and the other fluctuation components like absolute velocity, turbulent intensities at longitudinal, transverse, vertical direction and turbulent KE are abbreviated as V, u^+, v^+, w^+ and K respectively. From a fraction of the velocity corresponding to its variability, all line drawings are arranged with an x-z plane.

3.1. Velocity vector

The velocity vector's magnitude is determined by the formula by $\sqrt{(u.u+w.w)}$ and the corresponding direction is determined by the formula $\tan^{-1}(w/u)$. The magnitude obtained and the directions for this velocity vector are graphically shown in figure 3. At the stagnation point where the flow is zero and the flow is separated due to a negative pressure slope which means the flow losses its momentum near the cylinder surface and the flow reverses its direction by increasing the pressure slope. The stagnation point is not shown in the image as the

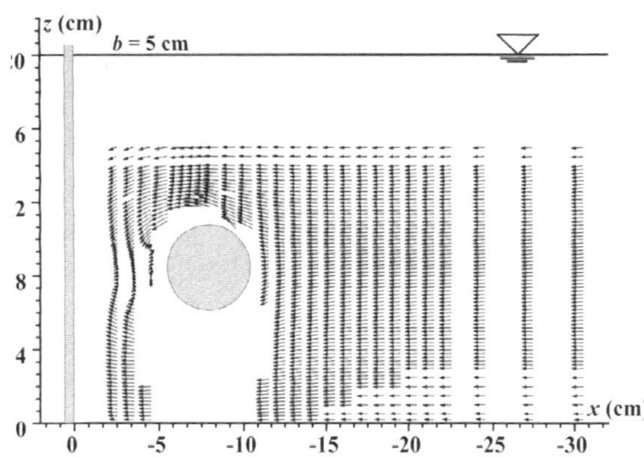

Figure 3. Time averaging velocity vector.

stagnation point is on the cylinder surface and cannot be measured due to the Vectrino$^+$ limit. However, the decrease in velocity near stagnation point is shown in the figure where the flow moves to the $\pm z$ directions and the magnitude is also increased. The velocity reduction is about 14-19% of the free stream velocity. The stagnation point is also moved 35° towards the water surface. After separation, the flow passes over the cylinder tangentially and meets 3 cm away from the cylinder outer curvature. The flow pattern alters to the rear side of the cylinder as the pressure slope is almost the same. The magnitude of the longitudinal velocity just next to the cylinder is approximately 1-2% of free stream velocity and thereafter increases slightly as the flow continues from the cylinder to the bottom. The results are compared with previous researches [16-18]. The stagnation point is found to be shifted -17.3° near the bed surface for the 2 cm gap between bed and cylinder curvature. The stagnation point is shifted further by -9.2°, 26.7° and 44.6° towards the water surface with the increase in gap from 2 to 8 cm with an equal interval of 2 cm, respectively between the cylinder and the bed.

3.2. Time averaging longitudinal velocity

The proximity of free flow is called the longitudinal flow (u) measured in cm/s and is shown in figure 4 (a). The main core of longitudinal velocity is located in the upper part of the cylinder. The upper core of the longitudinal velocity is located 12.54 cm above the bed surface and 5.28 cm in front of the vertical plate. About 1.375 times the free stream flow is 0.72 m/s. A velocity core of 1.319 times the free stream velocity is located under the cylinder and next to the bed surface. The lower core of the longitudinal velocity is located 2.46 cm above the bed surface and 5.94 cm in front of the vertical plate. When the gap between the cylinder surface and the bed surface increases the longitudinal velocity also increases below the cylinder. The longitudinal velocity separation is obtained when the gap between the cylinder and the bed surface is 8 cm. The high magnitude of longitudinal velocity core is developed due to the presence of vertical end plate only. The presence of vertical end plate creates a high horseshoe vortex in front of the vertical end plate under the cylinder. The time averaging longitudinal velocity is worsened by its low size found behind the cylinder outer curvature in the wake area at both cylinder positions 6 cm and 8 cm above the bed surfaces. The longitudinal velocity is observed negative due to the excessive loss in adverse pressure slope. The magnitude of flow below the cylinder is smaller than the flow above the cylinder with a gap of 2 cm and the magnitude of the flow below the cylinder increases with increasing gap. The magnitude of u is 0.94, 1.07, 1.15 and 1.293 times the upstream flow velocity for the gap of 2 to 8 cm with gap interval of 2 cm, respectively.

3.3. Time averaging transverse velocity

Figure 4(b) shows the time averaging transverse velocity v in cm/s. The range of time averaging transverse velocity is found at 5 cm/s to -4 cm/s in either direction. The time averaging transverse velocity is most commonly found near the cylinder surface. When the free stream flow hits the surface of the cylinder surface - a high turbulence is produced and this high turbulence effect ensures the presence of transverse velocity. The transverse velocity is 0.07 times the free stream velocity which is very small compared to longitudinal and vertical velocities. As a result the effect of transverse velocity in the cylinder is not significant for low velocity. If the cylinder position is changed from 6 cm to 8 cm away from the bed surface, the magnitude of transverse velocity becomes decreases for same reason as previously described.

3.4. Time averaging vertical velocity

At any point the time averaging vertical velocity (w) that moves upward is considered positive and in the opposite direction it is considered negative. The flow paths point up and down the directions from the upstream side of stagnation point of the cylinder which is clearly observable from the vertical velocity line shown in figure 4(c). After the stagnation point, the core of vertical upward flow is located in the vicinity of the cylinder 9.99 cm from the top of the bed surface and 6 cm from the vertical plate. The upward vertical flow is higher than the vertical downward flow. The upward vertical flow is 1.9 times the downward vertical flow. The upward vertical flow is found 0.56 times the free stream time averaging velocity. When the gap between the underwater cylinder and the bed surface increases then this upward vertical flow decreases and the downward vertical flow increases.

This occurs when the cylinder outer curvature is placed at a gap of 8 cm from the bed surface. The vertical velocity for 8 cm gap is 1.25 times the vertical velocity for 6 cm gap. The stagnation point shifts downwards as the gap between the cylinder outer curvature and the bed surface increases. From the stagnation point where the longitudinal velocity becomes zero and is divided into a vertical upward flow that takes the tangential path along the cylinder curvature and descends as it crosses rear side of the cylinder. This is achieved with a change in the direction of the vertical flow. Similarly, negatively downward flow changes its direction after crossing stagnation point to positively upward direction when the flow exceeds the cylinder. This flow direction change is clearly shown in *w* contour diagram.

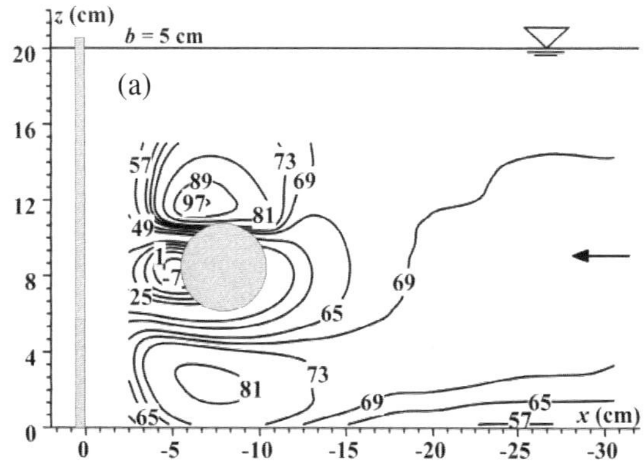

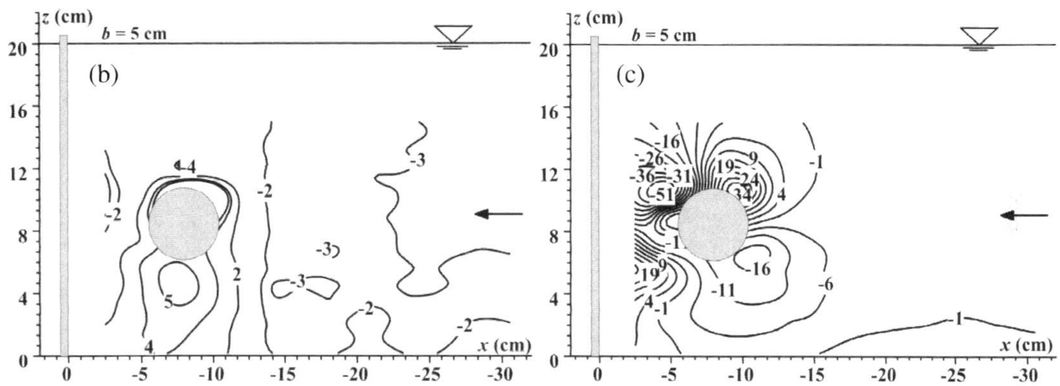

Figure 4. Contour of (a) longitudinal (b) transverse (c) vertical time averaging velocity.

The high core of 0.55 m/s vertical downward flow is located 10 cm from the bed surface and 2.64 cm from the vertical plate because the vertical flow due to the aid of vertical plate with vertical flow when the flow exceeds the cylinder. The *u* and *w* contours confirm the conversions in velocity components which mean that first the longitudinal velocity is converted into vertical velocity at stagnation point and when it exceeds the cylinder the point comes where the vertical velocity becomes zero but simultaneously the longitudinal velocity becomes very high. Finally in the rear part of the cylinder, the longitudinal velocity decreases gradually and the vertical velocity increases. The same thing happens when the flow shifts downward at the stagnation point and follows the path below the cylinder. On the rear side of the cylinder, the up and down flow meets 3 cm away from the cylinder surface. The area between the cylinder surface and the contact point of up and down flow, the vertical velocity is almost ignored. This location depends on the Reynolds number. As the Reynolds number increases the length of that area, it increases affecting the formation of vortices behind the cylinder.

The downward flow from the stagnation point is small at a gap of 2 cm and gradually increases with increasing gap. The magnitude of upward vertical velocity is found -0.138, -0.264, -0.319 and -0.403 times the upstream velocity for the gap of 2 to 8 cm with an equal interval of 2 cm, respectively. The flow moves under the cylinder and rises to the top of the wake region of the cylinder. The magnitude of the upward vertical velocity is 0.152, 0.222, 0.263 and 0.319 times the upstream velocity for the gap of 2 to 8 cm with an equal interval of 2 cm, respectively.

3.5. Time averaging absolute velocity

Contour of subsequent time averaging absolute velocity is shown in figure 5 where the absolute velocity is calculated using the formula $V = (u^{+2} + v^{+2} + w^{+2})^{0.5}$. The absolute velocity indicates the total velocity intensity and scalar magnitude. Higher core of V is located 7.5 cm away towards upstream from the vertical plate and 10.5 cm above from the bed surface. The time averaging absolute velocity is 142% higher than time averaging free stream velocity because in that area longitudinal velocity and vertical velocity are mutually supportive and the effect of transverse velocity is not significant.

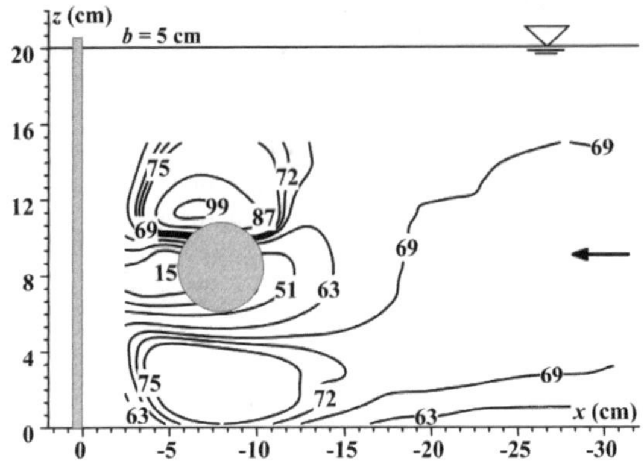

Figure 5. Contour of time averaging V.

3.6. Turbulence intensity

Contour of the time averaging turbulence intensities u^+, v^+ and w^+ having magnitudes equal to mean($u'\times u'$), mean($v'\times v'$) and mean($w'\times w'$), respectively acting in three Cartesian directions are shown in the figure 6 where u' is deviation of u, v' is deviation of v and w' is deviation of w at the region between the vertical plate and horizontal cylinder. The magnitude of the turbulence rate is measured by the square root of its velocity fluctuations which means that the turbulence intensities increase when the fluctuations at that time are high.

Vortex shedding frequency which is dependent on Strouhul number is a function of longitudinal turbulence intensity. The longitudinal turbulence intensity increases as the vortex shedding increases. Turbulence in the approaching flow depends on Reynolds' number as well.

The longitudinal; transversal and vertical intensities – all are located at the edge of the cylinder and the approximate size behind the outer bend of the cylinder. The magnitudes of the longitudinal and vertical turbulence intensities are found higher than the magnitude of transverse turbulence intensity. The core of the longitudinal turbulence intensity is located 10.56 cm from bed surface and 4 cm from the vertical plate. The turbulence intensity gradually diminishes toward the water surface and above the bed surface. The magnitude of the longitudinal turbulence intensity is 1.292, 1.222, 0972 and 0.917 times the free steam velocity for the gap of 2 to 8 cm with an equal interval of 2 cm, respectively.

Although the magnitude of the transverse velocity is very small as compared to free stream velocity but the turbulence intensity of transversal velocity is very strong. The core of the turbulence intensity of transversal velocity is located 4.62 cm from the bed surface and 5 cm from the vertical plate. The high density of transverse turbulence intensity means the flow is trying to go to the wall. When the gap of the upper cylinder surface and bed increases from 6 cm to 8 cm, the transverse turbulence intensity decreases to 0.875 times the transverse turbulence intensity of 6 cm gap between the cylinder surface and the bed surface. The longitudinal turbulence intensity is found maximum at 8 cm from the bed surface and 3.33 cm from the vertical plate to the cylinder. The core of longitudinal turbulence intensity is high because of the presence of vertical plate in which the downward flow and upward flow after crossing the cylinder mix together and producing an adverse gradient of extreme pressure. When the cylinder is at 8 cm above the bed surface, vertical turbulence intensity is 0.66 times lower than that of the vertical intensity for the cylinder position 6 cm above the bed because it is found that the flow below

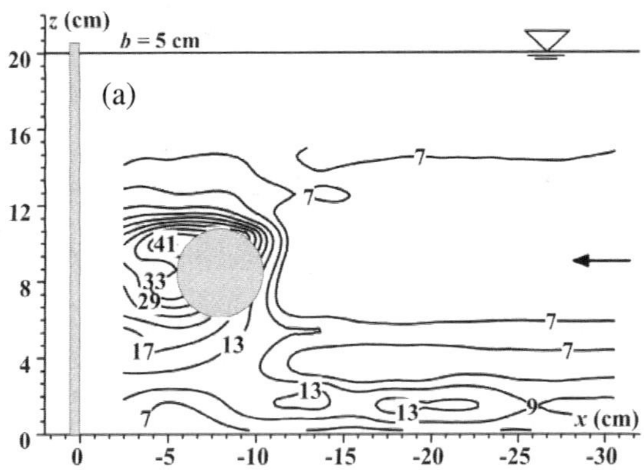

the cylinder is more when the gap is 8 cm from the bed. The magnitude of the vertical turbulence intensity is 0.902, 1.014, 0.791 and 0.708 times of the free stream velocity for the gap of 2 to 8 cm with an equal interval of 2 cm, respectively.

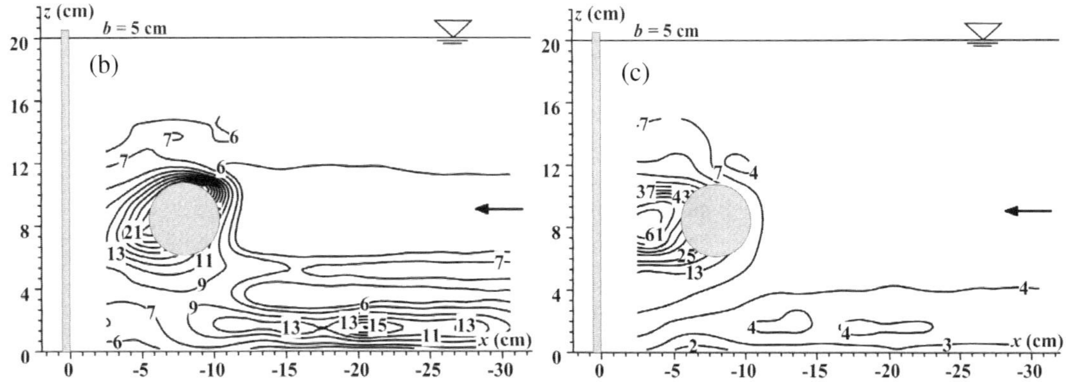

Figure 6. Contours of time averaging (a) longitudinal (b) transverse (c) vertical turbulence intensities.

3.7. Turbulent KE

The contour of turbulent kinetic energy K [$=(1/2)\times(u'^2 + v'^2 + w'^2)$] in x-z plane shown in figure 7. The turbulence kinetic energy is the function of the variation of velocity components. The KE depends on the severity of each component of the longitudinal, transverse, vertical velocity. The turbulent kinetic energy is located in the centre of cylinder area and facing the vertical plate which is obtained in the case of cylinder position of 8 cm from bed surface. It is observed that the turbulent kinetic energy is highest above the cylinder outer surface and decreases towards the water surface. When the cylinder is located 8 cm above the bed surface, the maximum turbulent kinetic energy is found under the cylinder surface. The turbulence kinetic energy is 0.951, 0.898, 0.705 and 0.729 times the square of the upstream free velocity for the gap of 2 to 8 cm with an equal interval of 2 cm, respectively. This high values of turbulent kinetic energy indicates that the presence of strong vortex shedding in the wake zone of the cylinder in both cases. The turbulent kinetic energy decreases gradually from the either side of the cylinder showing a slight inclination of turbulence intensities indicating a decrease in flow fluctuations.

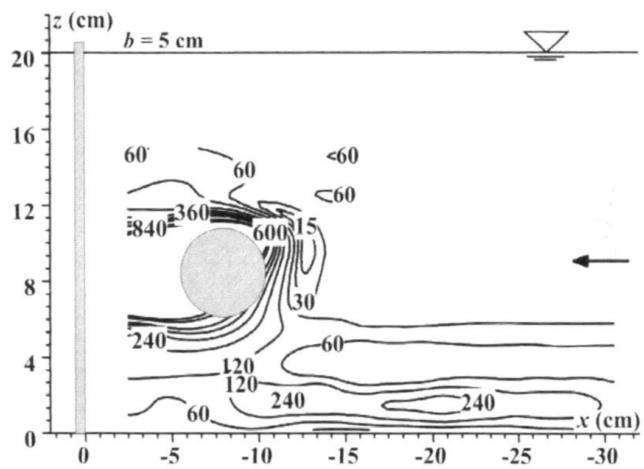

Figure 7. Contour of time averaging turbulent KE.

4. Conclusions

The main aspect of this study is to observe the complex flow kinematics around a horizontal cylinder transversely placed next to the flow in presence of a vertical end plate located downstream of it. This structure can be used for harvesting energy at shallow water based on the vortex vibration principle. To conduct experiment, a 5 cm cylinder was placed 6 cm above the bed surface with a vertical end plate positioning 5.5 cm away from the cylinder outer curvature and the kinematic results were compared with the result of another experiment with the same cylinder placed at 8 cm above the bed surface keeping the vertical end plate at same position.

It is illustrated that the contours of time averaging longitudinal, transverse and vertical velocities; the flow separation feature; and the development of vortices is similar in nature regardless of whether the cylinder area is upright on the bed surface. At different cylinder position it is found that there is a flow of water down near the plate which can be described as the beginning of the horseshoe vortex development. A separation of the flow is also visible near the upstream of the cylinder. The stagnation point appears to rise as the gap between the cylinder bottom surface and bed is shifted from 6 cm to 8 cm. A reverse flow is also observed at the back of the cylinder at both positions of the cylinder. The contours of time averaging longitudinal, transversal, vertical turbulence intensities are obtained with the same pattern. The magnitude of vertical turbulence intensity is perceived to be higher than that of the longitudinal and transverse velocities. The pattern of turbulent kinetic energy is the same in both positions of the cylinder. It is found that contour line of highest magnitude is over the cylinder outer curvature when cylinder is 6 cm above the bed surface. But when it is 8 cm above the bed surface then the contour line of highest magnitude is found below the cylinder outer curvature and found to be reduced towards the bed surface. Although the flow kinematics has been explored when the cylinder is kept suspended at different heights from the bed surface to understand and visualize the flow surrounding it but in practice the cylinders are used as movable structures along the transverse direction of the flow.

5. References

[1] Das S, Das R and Mazumdar A 2013 *Water Sci. Eng.* **6** 1 pp 59-77 DOI:10.3882/j.issn.1674-2370.2013.01.005

[2] Das S, Das R and Mazumdar A 2016 *Arab. J. Sci. Eng.* **41** 4 pp 1199-1213 DOI: 10.1007/s13369-015-1817-4

[3] Das S, Midya R, Das R and Mazumdar A 2013 *Int. J. Fluid Mech. Res.* **40** 1 pp 42-59 DOI: 10.1615/InterJFluidMechRes.v40.i1.40

[4] Jaman H, Das S, Kuila A and Mazumdar A 2017 *Arab. J. Sci. Eng.* **42** 9 pp 3973-3990 DOI: 10.1007/s13369-017-2536-9

[5] Ettema R, Kirkil G and Muste M 2006 *J. Hydraul. Eng.* **132** 1 pp 33-40

[6] Eldho TI and Young DL 2001 *J. Chinese Inst. Eng.* **24** pp 141-150

[7] Das S and Mazumdar A 2017 *KSCE J. Civil Eng.*, 22 7 pp 2300-2314 DOI: 10.1007/s12205-017-1842-9

[8] Das S, Jaman H, Chatterjee A, Das R and Mazumdar A 2017 *Int. J. Fluid Mech. Res.* **44**, pp 457-468 DOI: 10.1615/InterJFluidMechRes.2017019421

[9] Mukherjee B, Das S and Mazumdar A 2016 Electrical energy generation by enhancing flow induced vibration *Proc. 2nd Int. Conf. on Control, Instrumentation, Energy Communication*, (Kolkata: IEEE) pp 368-371 DOI: 10.1109/CIEC.2016.7513804

[10] Bernitsas MM, Raghavan K, Simon YB and Garcia EMH 2006 *J. Off. Mech. Arc. Eng.* **130** 041101 DOI: 10.1115/1.2957913

[11] Mukherjee B, Das S and Mazumdar A 2016 Sustainable Electrical Energy Generation Technique in Shallow Water Channels Proc. *IEEE Students' Tech. Symp.* (Kharagpur: IEEE) pp 47-151 DOI: 10.1109/TechSym.2016.7872672

[12] Mukherjee B, Das S and Mazumdar A 2017 Complex Flow Phenomena of Horizontally Placed Underwater Cylinder above Water Bed *Water Energ. Intl.* **60** 8 pp 57-64

[13] Schewe G 1983 *J. Fluid Mech.* **133** 1 pp 265-285 DOI: 10.1017/S0022112083001913

[14] Bernitsas MM, Ben-Simon Y, Raghavan K and Garcia EMH 2009 *J. Off. Mech. Arc. Eng.* **131** 1 pp 1-13 DOI: 10.1115/1.2979796

[15] Bearman PW and Zdravkovich MM 1978 *J. Fluid Mech.* **89** 1 pp 33-47

[16] Kuila A, Mukherjee B, Das S and Mazumdar A 2020 Comparison of Flow Fields around a Horizontal Cylinder placed at Different Depths in Water Channel *Indian Sci. Cruiser*. Accepted

[17] Kuila A, Mukherjee B, Das S and Mazumdar A 2020 Comparative Analysis of Flow Fields on the Proximity of a Horizontal Cylinder with a Vertical Plate Obstruction, *Proc. of the Second ASCE India Conf.* on Challenges of Resilient and Sustainable Infrastructure Development in Emerging Economies (CRSIDE2020) (Kolkata: ASCE)

[18] Kuila A, Das S and Mazumdar A 2020 *Appl. Mech. Mater.* **897**, pp 173-178. DOI: 10.4028/www.scientific.net/AMM.897.173

IOP Conf. Series: Materials Science and Engineering 1080 (2021) 012049 doi:10.1088/1757-899X/1080/1/012049

Analysis of nanofluid based natural convection in a square cavity applying corner heating and cooling

Abhinav Saha, Koushik Ghosh and Nirmal K. Manna

Department of Mechanical Engineering, Jadavpur University, Kolkata–700032, India
e-mail: *saha.abhinav.2010@gmail.com*

Abstract. The present investigation aims at the numerical analysis of natural convection in a square cavity. The cavity is filled with nanofluid using water as the base fluid and Al_2O_3 as nanoparticles. The investigation is carried out in a square cavity with corner heating and cooling. The numerical simulations are performed using a finite volume method. The average Nusselt number is calculated for a wide range of Rayleigh number from 10^4 to 10^6 keeping solid volume fraction of nanoparticle fixed at 0.04. The details of flow and heat transfer for various cases are analyzed using the contour of streamlines and isotherms.

1. Introduction

Natural convection flow - also known as the buoyancy-driven flow is caused by the variation in fluid density in presence of the gravitational field. A thorough understanding of flow and heat transfer characteristics in such situations is important due to its application in the cooling of electronic equipment. Conventionally, fluid such as water, oil, ethylene glycol etc. used in cooling applications achieves low heat transfer due to lower thermal conductivity. The thermal conductivity is enhanced by adding nano-metallic particles to base fluid known as nanofluids.

Putra et al. [1] performed an experimental investigation of natural convection of nanofluid. They studied the effect of Rayleigh number, aspect ratio and volume fraction for different nanoparticles suspended in the base fluid. Ternik et al. [2] studied natural convection in differentially heated side walls while the other two walls are kept adiabatic. They studied the effect of different particles for a wide range of Rayleigh number and volume fraction. There have been many excellent review papers on natural convection of a nanofluid in an enclosure. Oztop et al [3] presented an excellent review paper of natural convection of nanofluid under partial heating of an enclosure. In the recent years, and Bhuiyana et al. [4] and Mendu et al. [5] studied the natural convection in a square cavity filled with nanofluid with a discrete heater placed at the bottom of an enclosure. They also studied the Rayleigh number, Prandtl number, solid volume fraction of nanoparticle and heater position.

Literature review reveals that natural convection flow of a nanofluid in an enclosure depends on the Rayleigh number, Prandtl number, aspect ratio, solid volume fraction of nanoparticle as well as the end wall boundary conditions. The dependence on such a large number of parameters, however, compounds the unpredictability in flow and heat transfer characteristics in such problems.

To the best of the author's knowledge, no studies have been reported on natural convection of nanofluid with corner heating and cooling. Experimental investigation of details of flow and temperature fields is also fraught with uncertainty as the velocity and temperature differences may be too small sometimes to be resolved accurately. The scope of the present work is to employ CFD methodology for natural convection flow and heat transfer in a square cavity with corner heating and cooling.

2. Problem definition

Figure 1 shows the computational domain of square cavity filled with nanofluid. In the figure, discrete heaters are embedded at the corner of a cavity. The hot heater and cold heater are kept at a constant temperature of T_h and T_c respectively while other walls are assumed to be adiabatic. The hot heaters and cold heaters are embedded at diagonal of square enclosure. The hot heaters are located at the top-right corner and bottom-left corner while cold heaters are located at top-left corner and bottom right corner.

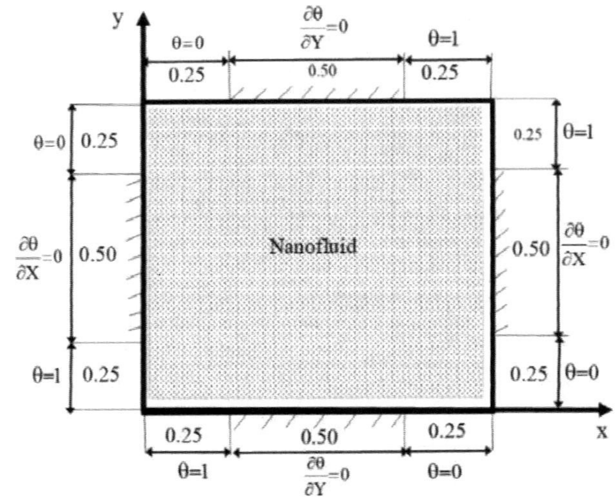

Figure 1. Schematic diagram of a square enclosure filled nanofluid along with boundary condition

A two-dimensional, steady, laminar, incompressible and Newtonian fluid is considered. The properties of the working fluid are fixed, except the variation of fluid density with temperature which is accounted for in the formulation of buoyancy term using the Boussinesq approximation. The thermal equilibrium exists between the base fluid and nanoparticle.

3. Mathematical Formulation

The flow is governed by the conservation laws of mass, momentum and energy. The following governing equations are written in the non-dimensional form under the above-mentioned assumptions.

Mass conservation equation

$$\frac{\partial U}{\partial X}+\frac{\partial V}{\partial Y}=0 \tag{1}$$

Momentum conservation equation in x-direction:

$$U\frac{\partial U}{\partial X}+V\frac{\partial U}{\partial Y}=-\frac{\partial P}{\partial x}+\frac{\mu_{nf}}{\alpha_{nf}\alpha_f}\left(\frac{\partial^2 U}{\partial X^2}+\frac{\partial^2 U}{\partial Y^2}\right) \tag{2}$$

Momentum conservation equation in y-direction:

$$U\frac{\partial V}{\partial X}+V\frac{\partial V}{\partial Y}=-\frac{\partial P}{\partial Y}+\frac{\mu_{nf}}{\alpha_{nf}\alpha_f}\left(\frac{\partial^2 V}{\partial X^2}+\frac{\partial^2 V}{\partial Y^2}\right)+\frac{(\rho\beta)_{nf}}{\rho_{nf}\beta_f}\mathrm{Ra}\,\mathrm{Pr}\,\theta \tag{3}$$

Energy conservation equation:

$$U\frac{\partial\theta}{\partial X}+V\frac{\partial\theta}{\partial Y}=\frac{\alpha_{nf}}{\alpha_f}\left(\frac{\partial^2\theta}{\partial X^2}+\frac{\partial^2\theta}{\partial Y^2}\right) \tag{4}$$

The boundary condition used in non-dimensional form is shown in figure 1
The closure relations to calculate properties of nanofluid is given by Eqs. (5) to (9).

$$\rho_{nf}=(1-\phi)\sigma_f+\phi\sigma_p \tag{5}$$

$$(\rho C_p)_{nf}=(1-\phi)(\rho C_p)_f+\phi(\rho C_p)_p \tag{6}$$

$$(\rho\beta)_{nf}=(1-\phi)(\rho\beta)_f+\phi(\rho\beta)_p \tag{7}$$

$$\mu_{nf}=\mu_f/(1-\phi)^{2.5} \tag{8}$$

The Brinkman and Maxwell [6, 7] developed the relation to determine dynamic viscosity and thermal conductivity which is given by

$$K_{nf}=K_f\{(K_p+2K_f)-2\phi(K_f-K_p)\}/\{(K_p+2K_f)+\phi(K_f-K_p)\} \tag{9}$$

The properties of base fluid (H_2O) and nanoparticle (Al_2O_3) used in present study is given in Table 1.

Table 1: The properties of base fluid (H_2O) and nanoparticle (Al_2O_3) [5].

	$\rho(kg/m^3)$	$C_p(J/kg-K)$	$k(w/m-K)$	$\mu(kg/m-s)$	$\beta(1/K)$
Base fluid (H_2O)	997.5	4178	0.6129	0.0008295	0.00028
Nanoparticle (Al_2O_3)	3970	765	40		0.85E-5

The governing equations are given in Eqs. (1) to (4) subjected to boundary condition mentioned in figure 1.The geometry of the square cavity is created in Gambit software. The numerical solution is solved in ANSYS FLUENT based on finite volume approach. The convergence criteria for all the governing equations are set at residuals less than 10^{-6}.

The convective heat transfer is determined by Nusselt number. The local Nusselt number Nu along the hot wall can be determined using the expression:

$$Nu=\frac{K_{nf}}{K_f}\frac{\partial\theta}{\partial n} \tag{10}$$

where n is the outward normal vector.

The average Nusselt number $\overline{Nu}$ can be obtained by:

$$\overline{Nu}=\frac{1}{L}\int_0^L Nu\,dl \tag{11}$$

where L is the total length of the hot wall.

4. Results and Discussion

The natural convection flow and heat transfer is carried in a square cavity. The cavity filled with nanofluid where base fluid is H_2O and nanoparticle is Al_2O_3. The numerical analysis is carried out for varying Ra in the range of 10^4 to 10^6 while keeping solid volume fraction $\phi=0.04$. The contours of isotherms and stream functions help to understand flow physics and heat transfer. The scope of the present investigation is to analyze the effect of Ra on flow physics and heat transfer in a square cavity with corner heating and cooling.

4.1 Grid Independence Test

IOP Conf. Series: Materials Science and Engineering 1080 (2021) 012049 doi:10.1088/1757-899X/1080/1/012049

The grid independent study is carried out for pure water $(\phi=0)$ at Ra = 10^4 and 10^6. The grid study is carried out using three different grid namely 50×50, 100×100 and 200×200 shown in Table 2. The results of $\overline{Nu}$ obtained with different grids are very close to each other. The intermediate grid count 100×100 is adopted for further computations in this work as a good compromise between accuracy and computational effort.

Table 2: Grid independent study for $\overline{Nu}$ at Ra = 10^4.and 10^6

	Number of Grid Counts		
	N_1 (50×50)	N_2 (100×100)	N_3 (200×200)
Ra = 10^4	2.285815	2.276351	2.273612
Ra = 10^6	9.673208	9.334908	9.247624

4.2 Comparison of Results

To the best of author's knowledge, no work is available in the literature considering a corner heating and cooling of nanofluids. Therefore, the numerical model is validated for natural convection in a differentially heated cavity filled with pure water $(\phi=0)$. In the cavity side walls are kept at different temperature while the other two walls are kept adiabatic.

Table 3: The comparison of $\overline{Nu}$ for differentially heated square enclosure

	Mendu et al. [5]	Present study	% Error
Ra=10^4	2.272	2.276	0.18

4.3 Effect of Rayleigh number on isotherms and streamlines

Figure 2 represents isotherms and streamlines for Ra in the range of 10^4 to 10^6 while keeping $\phi=0.04$. The streamlines are presented in figure 2 (b). It is observed from the contour of streamlines that two convective cells are formed which is counter-rotating in nature for all values of Ra. The counter-clockwise and clockwise rotations of flow are represented by the positive and negative sign of streamlines respectively. The convective cells formed are symmetric about horizontal mid-line at Y = 0.5. The line of $\psi=0$ acts like a virtual wall across which no flow takes place. It is further observed that strength of convective cells increases with an increase in Ra. The isotherms are presented in figure 2 (b). The isotherms are also symmetric about horizontal mid-line at Y = 0.5. The isotherms are crowded near the hot heater which is located at top-right corner and bottom-left corner of cavity. It is further observed that isotherms gradually becomes more overpopulated with an increase in Ra. The increase in density of isotherms represent rise in temperature gradient near the corner of a cavity. The rise in temperature gradient results in an enhancement in heat transfer.

	Ra=10^4	Ra =10^5	Ra=10^6
(a) Isotherms	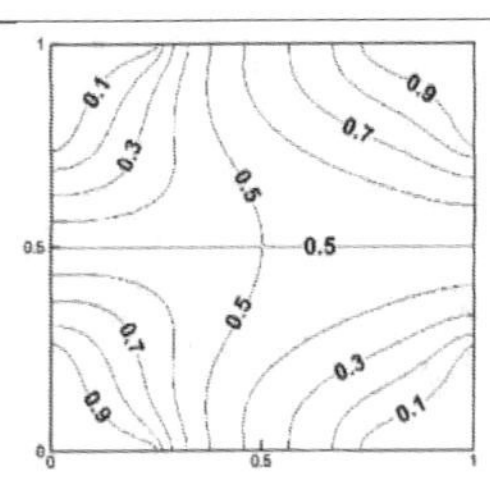	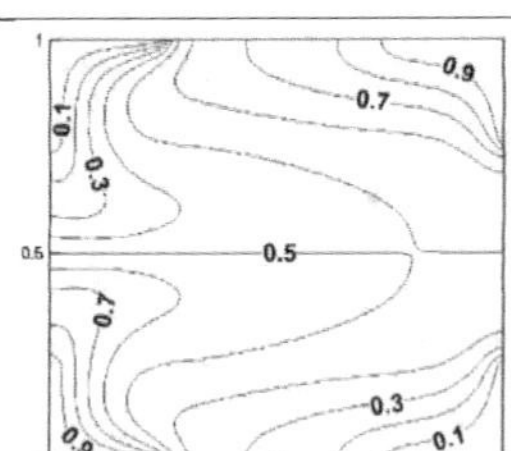	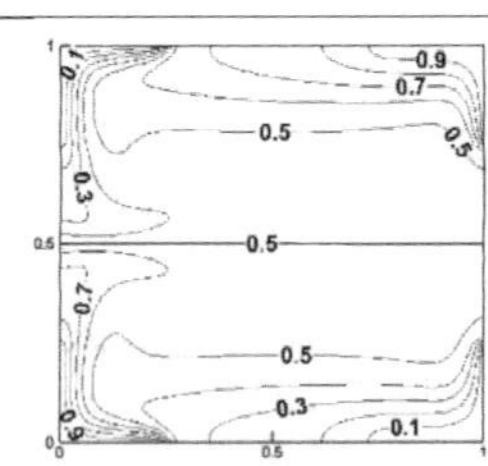

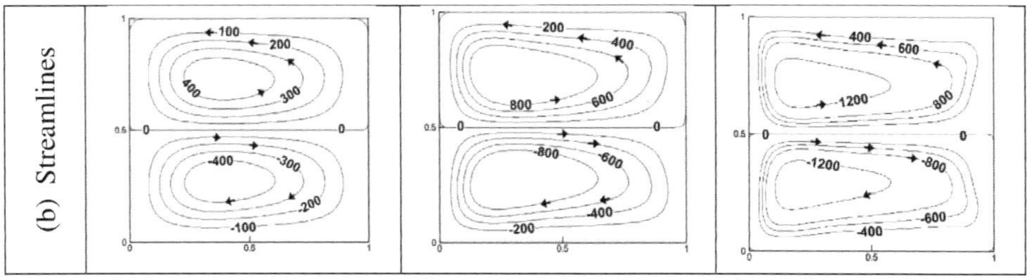

Figure 2. Contours of (a) Isotherms and (b) Streamlines for $Ra = 10^4$ to 10^6 at $\phi = 0.4$

The graph of ψ_{max} vs Ra presented in figure 3 (a) shows that maximum value of stream function increases with an increase in Ra. The graph of $\overline{Nu}$ vs Ra presented in figure 3 (b) shows that convective heat transfer monotonically increases with an increase in Ra.

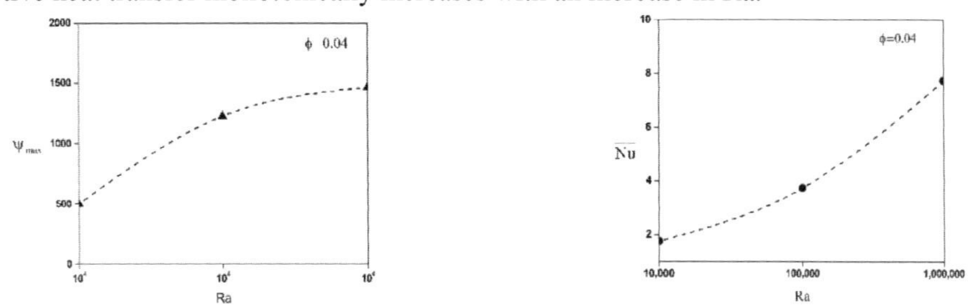

Figure 3. (a) Variation of ψ_{max} with Ra (b) Variation of $\overline{Nu}$ with Ra at the heated wall

5 Conclusions

The salient observations from the study can be summarized as follows:

- The maximum value of stream function increases with an increase in the Rayleigh number.
- The enhancement in convective heat transfer is observed with an increases in Rayleigh number.

References

[1] Putra N, Roetzel W and Das S K 2003 Natural convection of nano-fluids *Heat and Mass Transfer* **39** 775-784..

[2] Ternik P and Rudolf R 2012 Heat transfer enhancement for natural convection flow of water based nanofluids in a square enclosure *Int. J. of Simulation modelling* **11** 29-39.

[3] Oztop H F, Estelle' P, Yan W-M, Al-Salem K, Orfi J and Mahian O 2015 A brief review of natural convection in enclosures under localized heating with and without nanofluids *Int. Communication of Heat and Mass Transfer* **60** 37-44.

[4] Bhuiyana A H, Alam Md. S and Alim M A 2017 Natural convection of water-based nanofluids in a square cavity with partially heated of the bottom wall *Procedia Engineering* **194** 435-441

[5] Mendu S S, Dora N and Karpurapu M P 2018 Simulation of natural convection of nanofluids in a square enclosure embedded with bottom discrete heater *Thermal Science* **22** 2771-2781.

[6] Brinkman H C 1952 The viscosity of concentrated suspensions and solution *J. Chem Phys* **20** 571-581.

[7] Maxwell J C 1873 A Treatise on electricity and magnetism *Oxford University Press Cambridge* **2** 54.

Nomenclature		**Greek symbols**	
g	Acceleration due to gravity	ρ	Density

(m/s^2)			(kg/m^3)	
L	Dimension of a square enclosure	(m)	ν	Kinematic viscosity (m^2/s)
P	Non-dimensional pressure		θ	Non-dimensional temperature
U,V	Dimensionless velocity in X and Y direction		α	Thermal diffusivity
Ra	Rayleigh number		β	Thermal expansion coefficient
Pr	Prandtl number		ψ	Non-dimensional stream function
Nu	Local Nusselt number		**Subscripts**	
$\overline{Nu}$	Average Nusselt number		h, c	Hot wall, Cold wall
			f	Fluid
			nf, np	Nanofluid, Nanoparticle

IOP Conf. Series: Materials Science and Engineering 1080 (2021) 012050 doi:10.1088/1757-899X/1080/1/012050

Heat Transfer and Pressure Drop in Transitional Flow: A Short Review

Suvanjan Bhattacharyya[*], Devendra Kumar Vishwakarma, Manoj Kumar Soni
Department of Mechanical Engineering, Birla Institute of Technology and Science, Pilani, Pilani Campus, Vidya Vihar, Pilani, Rajasthan 333031, India
e-mail: suvanjan.bhattacharyya@pilani.bits-pilani.ac.in

Abstract. In the present work, an effort has been made to present the short review of all the numerical as well as experimental studies done in the field of heat transfer and pressure drop in the transitional flow regime. The experimental and numerical studies on transition flow are reported. Though, there are several challenges to do experiments in transition flow regime. Various techniques are also discussed and summarized. Results show that the techniques prove outstanding performance, but few methods quiet suffer from high pressure drop. As per discussion, new viewpoints on the current research gaps and future research ways have been providing for the development of heat transfer techniques.

1. Introduction

Energy conservation is one of the biggest tasks of today's growing world. With the advancement in technology, the consumption of energy is also kept increasing at a very rapid rate. This grows the need for the development of such devices which can work energy efficiently. One such device is a heat exchanger (HE). HE is used for the exchange of energy between two or more fluids. Some examples of HEs are radiators of car engines, evaporators, condensers, air preheater, cooling towers, etc. To obtain maximum effectiveness, the optimization of design parameters is required. While designing the HEs, engineers focussed on high heat transfer rate and low friction factor coefficient. In the laminar flow regime, the pressure drops as well as the heat transfer rate both are low [1-4] while in turbulent flow both pressures drop, and the heat transfer rate is very high [5-20]. This problem made researchers focus on the transitional flow regime. But this regime has its problem of unstable and chaotic flow due to which it is advised to avoid designing of HEs in this regime.

The transitional flow regime was first studied in detail by Ghajar and his Co-workers [21-27]. Following the path of Ghajar, Meyer with co-workers also starts working on the transitional flow regime [28-34]. These two professors along with their co-workers shed light on the pros and cons of working in the transitional flow regime. Beside this a very few manuscripts are available on the heat transfer and pressure drop in the transitional flow regime.

In the present work, an effort has been made to present the short review of all the numerical as well as experimental studies done in the field of heat transfer and pressure drop in the transitional flow regime.

2. History of Transitional Flow

The transitional flow regime is defined as the flow regime when fluid flow transformed from a laminar regime to the turbulent flow regime. Generally, it occurred anywhere between Reynolds number (Re) of 2.3×10^3 to 10×10^3 [21]. Before 1990, very few authors discussed the transition regime in a very specific manner for heat transfer. In the year 1965, Unterberg and Edwards [36] first mentioned transitional flow in film evaporation of falling saline water on the vertical surface under the influence of gravity. Mceligot et al. [37] studied the heat transfer and wall friction in the circular channel for air, nitrogen, and helium for Re range between 3 000 – 10 000. A correlation was presented by Petersen and Christiansen [38] to predict the heat transport for pseudo plastic and Bingham fluids in the transitional regime. The developed correlations are in good agreement with a standard deviation of 17.7% for the complex transitional flow regime. Inayatov [39] experimentally investigated the heat transfer for Re 1 000 – 12 000 for water as working fluid externally flows over multiple copper tubes. Humphreys and Welty [40] experimentally investigated the free convection heat transfer in laminar and transitional regime in the vertical heated channel for mercury and developed a correlation for Nusselt Number (Nu) prediction which can be used for all values of Grashof numbers (Gr). Jaster and Kosky [41] perform experiments to predict the condensation heat transfer in the transitional flow regime. A correlation for predicting the transition Nusselt number is also developed.

3. Pressure Drop in Transitional Flow Regime

The year 1992 acts as the turning point in the history of thermo-hydraulics in the transitional regime. For the first time, the heat transfer and pressure drop were studied in detail. The main credit for this goes to Ghajar and co-workers whose research work in transitional zone helps in a better understanding of the heat transfer and pressure drop. Ghajar and Madon [21] experimentally investigated the pressure drop in the transition regime for re-entrant, square-edged, and bell-mouth inlets. The schematic of the experimental setup used for the investigation is shown in Fig. 1. Two different working fluids such as distilled water and ethylene glycol are used in different concentration. The Re is varied between 500 – 15 000. The results obtained from 114 experimental runs revealed that transition for (i) Re-entrant inlet started around Re of 1980 and ends at 2125, (ii) square-edged inlet starts at 2070 and ends at 2840, while (iii) for bell-mouth, the transition occurs at 2125 and terminates at 3200. Second-order polynomial correlation for predicting fully developed skin friction was developed using the data for all the three geometries of the inlet.

In another work, Tam and Ghajar [23] experimentally investigated the pressure drop in the fully developed flow in a horizontal circular pipe with different inlet geometries under isothermal as well as non-isothermal conditions. Ethylene Glycol along with water is used as the working fluid with Re range between 1 000 – 17 000 to cover all the flow regimes. The data obtained from the investigation shows enhancement in friction coefficient with an increase in heat flux at fixed Re because of the presence of secondary flow which increases the lower and upper limits of isothermal boundaries.

In another investigation, Tam et al. [27] investigated the impact of horizontal tube inlet geometries on the friction factor in the transitional flow regime. Ethylene Glycol and water mixture have been used as working fluid. The range of Reynolds number was varied from 800 to 22 000. It was found that the increase in the heating condition significantly affects the friction factor.

Meyer et al. [28] investigated the pressure drop in the transitional flow for four different inlet geometries. Water is taken as the working fluid with Re range between 1 000 – 20 000. The result obtained for the adiabatic friction factor shows the transition in re-entrant type inlet started at Re = 2 100 while for bell-mouth entrance transition is delayed up to Re = 7 000. From the above obtained results, the author concluded that smoother entrance led to delayed transition while for diabatic pressure drop, inlet geometry shows no dependence. In another assessment by Meyer and Olivier [29], the pressure drop was investigated in transition regime for enhanced tube as shown in Fig. 2 for developing and developed flow and developed correlations to predict the critical value of Re. It was revealed from the experimental data that transition in enhanced tube started at lower Re when compared with a smooth tube.

Ndenguma et al. [30] experimentally investigated the pressure drop in the transitional flow regime for mixed convective heat transfer in four annular passage of different diameters and developed correlations to predict the friction factor. The schematic representation of the test setup is shown in Fig. 3. Water was taken as working fluid with Re varied from 100 to 10 000. This investigation revealed that the friction factor depends on the geometrical parameters of the test section. The larger the annular gap, the higher will be the friction factor for a fixed Re.

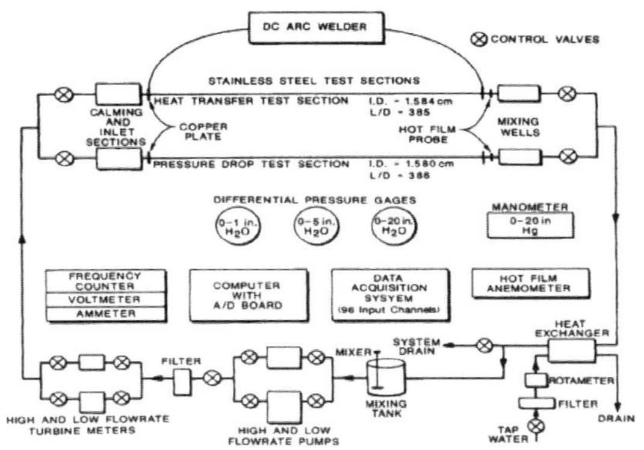

Figure1. Schematic Diagram of Experimental Setup used by Ghajar and Madon, 1992 [21]

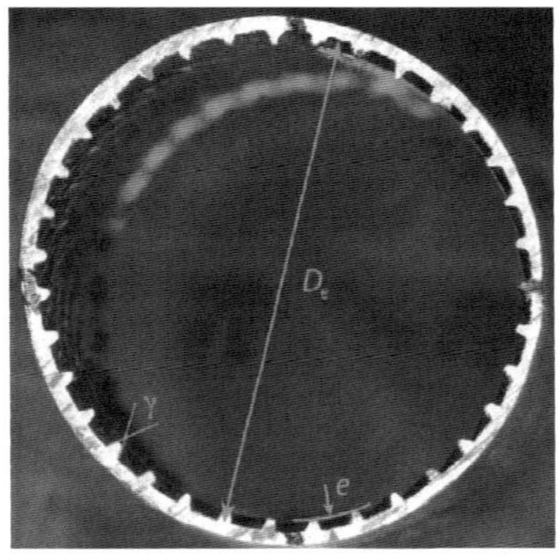

Figure 2. Pictographic Representation of Enhanced Tube used by Meyer and Olivier, 2011 [29].

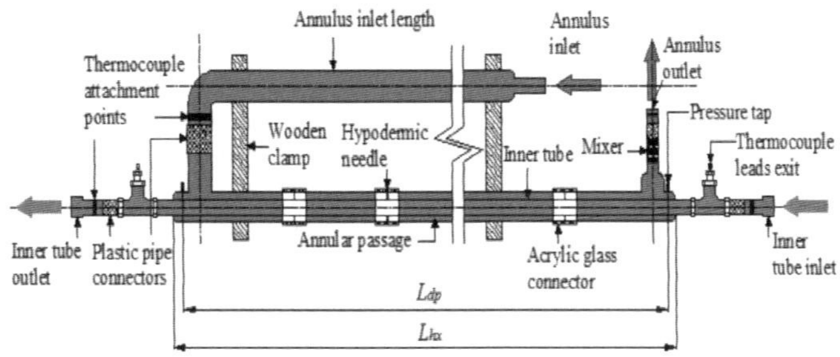

Figure 3. Schematic of Test Section Used by Ndenguma et al. 2017 [30].

Dobrnjac [42] conducted a study to develop an efficient formula to predict the friction factor coefficient in the transitional flow regime. Based on digitalized Nikuradse's measurements, an analytical expression was developed to predict the friction factor for the Re range from 0 to 10^8. Ishida and Tsukahara [43] numerically investigated the annular Poiseuille flows in a transitional regime for friction factor. The frictional Reynold number, Re_τ, considered for the assessment range between 48 - 150. Given the state of turbulent-laminar coexistence, the findings revealed that the helical turbulence contributed to the friction coefficient being as high as the fully turbulent data. All the above research works are summarized in Table 1.

IOP Conf. Series: Materials Science and Engineering 1080 (2021) 012050 doi:10.1088/1757-899X/1080/1/012050

Table 1. Summary of Research work related to Pressure drop in Transitional Regime.

Author's Name	Year of Publishing	Range of Reynold's Number	Summary of Research Work	Key Outcomes
Ghajar and Madon [21]	1992	500-15000	Experimentally investigated the pressure drop in the transition regime for re-entrant, square-edged, and bell-mouth inlets.	Transition for (i) Re-entrant inlet started around Re of 1980 and ends at 2125, (ii) square-edged inlet starts at 2070 and ends at 2840, while (iii) for bell-mouth, the transition occurs at 2125 and terminates at 3200.
Tam and Ghajar [23]	1997	1000-17000	Experimentally investigated the pressure drop in the fully developed flow in a horizontal circular pipe with different inlet geometries under isothermal as well as non-isothermal conditions.	Enhancement in friction coefficient with an increase in heat flux at fixed Re because of the presence of secondary flow which increases the lower and upper limits of isothermal boundaries.
Ghajar et al. [25]	2008	500-10000	Experimentally revealed the impact of single-phase transitional flow on the friction factor in mini and microchannels.	The friction factor has a very little variation for the diameter range between 2083 μm and 1372 μm while further decreasing the diameter shows a significant change in the friction factor profile.
Ghajar et al. [26]	2010	500-10000	Experimental assessment to investigate the pressure drop in transitional flow for 12 stainless steel microtube with a diameter range between 2083 μm and 337 μm.	The results show negligible influence on the friction factor for tubes with a diameter range between 2083 μm and 1372 μm. Below 1372 μm significant change in friction factor water noticed.
Tam et al. [27]	2013	800-22000	The impact of horizontal tube inlet geometries on the friction factor in the transitional flow regime.	It was found that the increase in the heating condition significantly affects the friction factor.
Meyer et al. [28]	2009	1000-20000	The pressure drop in the transitional flow for four different inlet geometries.	Smoother entrance led to delayed transition while for diabatic pressure drop, inlet geometry shows no dependence.
Meyer and Olivier [29]	2011	500-13000	The pressure drop was investigated in transition regime for enhanced tube	Transition in enhanced tube started at lower Re when compared with a smooth tube.
Ndenguma et al. [30]	2017	100-10000	The pressure drop in the transitional flow regime for	The friction factor depends on the geometrical

			mixed convective heat transfer in four annular passage of different diameters and developed correlations to predict the friction factor.	parameters of the test section. The larger the annular gap, the higher will be the friction factor for a fixed Re.
Dobrnjac [42]	2012	$0\text{-}10^8$	Study to develop an efficient formula to predict the friction factor coefficient in the transitional flow regime.	An analytical expression was developed to predict the friction factor for the Re range from 0 to 10^8.
Ishida and Tsukahara [43]	2017	48-150	Numerically investigated the annular Poiseuille flows in a transitional regime for friction factor.	The helical turbulence contributed to the friction coefficient being as high as the fully turbulent data.

4. Heat Transfer in Transitional Flow Regime

Heat transfer in transition flow regime is also very important like pressure drop. Ghajar and Tam [22] experimentally investigated thermal transport in the transition regime for three inlet geometries. Distilled water and ethylene glycol in different concentrations were used as working fluid for Re range between 280 – 49 000. Other parameters used for the study are Prandtl Number (4 -158), and Grashof Number (1 000 - 2.5×10^5). It was found that transitional Re for Re-entrant range between 2 000 – 8 500, square-edged range between 2 400 – 8 800, and bell mouth range between 3 800 – 10 500.

Meyer and Olivier [31] experimentally reveal the heat transfer in the transitional flow regime for modern chiller plants with different inlet configurations. A tube in tube counter flow heat exchanger was used as an experimental setup with water as the working fluid. Hot fluid at 40° - 45° flows in the inner tube while cold fluid at 20° flows in the outer tube of the heat exchanger. It was found that heat transfer in the chiller plant remains uninfluenced of inlet geometries for the horizontal plain tube. In another investigation the heat transfer for enhanced tubes in the transitional flow regime and developed the correlation for predicting the heat transfer in the transitional flow regime [32]. The data obtained for heat transfer shows an overall enhancement in heat transfer when compared with the smooth tube.

Everts and Meyer [33] experimentally investigated the heat transfer in developing and developed flow in the transitional regime for two horizontal circular tubes of diameter 4 mm and 11.5 mm at constant heat flux. The heat transfer is measured for Re range between 700 – 10 000. It was found that the effect of free convection in the flow field reduces the range of transitional flow for all the Re. Osman et al. [34] investigated convective heat transfer in transitional flow for Alumina-water nanofluid of different concentrations in a rectangular channel. The range of Re taken for investigation varied from 200 to 7 000. The data obtained after investigation shows enhancement in heat transfer coefficient for all volume concentrations with maximum heat transfer coefficient for 1% volume concentration. Convective heat transfer efficiency was also found to be better in the transitional flow regime when compared with the other two regimes.

García et al. [44] experimentally revealed the augmentation of heat transfer in Laminar-transitional-turbulent flow regimes for a tube with wire coil inserts for different values of Prandtl number (Pr). Water with Propylene Glycol was taken as working fluid with Re varied from 80 to 90 000 and Pr between 2.8 to 150. Six wire coils with different *p/D* and *e/D* were used for the experimentation. It was revealed from the experimental investigation that heat transfer was enhanced by 200% in the transitional flow regime while the pumping power remains constant. This investigation shows that wire coils perform best in the transitional regime. A similar investigation with a different value of pitch for wire coil inserts with Re varied between 10 to 2 500 and Pr varied between 200 to 700 was also reported [45]. The results obtained from the investigation show eight times enhancement in heat transfer when compared with the smooth tube. Martínez et al. [46] investigated the heat transfer augmentation for Newtonian and non-Newtonian fluid in viscous and transitional flow regimes. The test section consists of a tube with an inner diameter of 18 mm fitted with stainless steel wire coils having *p/D* of 1 and 2, *e/D* of 0.09. The range of Re was taken between 10 and 1 300. It was found that at low Re wire coil shows results similar to a plain tube.

Peixinho et al. [47] conducted an experimental assessment to investigate the heat transfer in the transitional flow regime. Three working fluids namely, glucose syrup, 2% CMC, and 0.2% Carbopol have used for the investigation. It was found that the heat transfer coefficient is increased in the transitional flow regime due to the disturbed thermal boundary layer. Bertsche et al. [48] experimentally investigated the laminar, transition, and turbulent flow regime for thermal transport in the circular pipe. The Reynolds number used for the study varied from 500 to 23 000 while Pr varied from 7 to 41. The data obtained from the investigation are in good agreement with Gnielinski's calculation for all value of Pr. It was found that transitional flow occurs between 2 300 to 4 000 of Re. Sarmiento et al. [49] proposed a new model to predict Nu for circular as well as non-circular tubes. This model is valid for Pr range from 0.7 to 78 and Re range for 890 and 2 48 800. The proposed model is in good agreement with other proposed correlations with a square mean error of 10%.

Table 2. Summary of Heat Transfer in Transitional Flow Regime.

Author's Name	Year of Publishing	Range of Reynold's Number	Summary of Research Work	Key Outcomes
Ghajar and Tam [22]	1994	280-49000	Thermal transport in the transition regime for three inlet geometries.	Transitional Re for Re-entrant range between 2000-8500, square-edged range between 2400-8800, and bell mouth range between 3800-10,500.
Tam and Ghajar [24]	2006	-	The heat transfer in the transitional flow regime in a horizontal pipe for three different inlet geometries under uniform heat flux conditions.	Presented a flow regime map to differentiate between the boundary for forced and mixed convection in a horizontal plain tube.
Meyer and Olivier [31]	2011	1000-15000	The heat transfer in the transitional flow regime for modern chiller plants with different inlet configurations.	Heat transfer in the chiller plant remains uninfluenced of inlet geometries for the horizontal plain tube.
Meyer and Olivier [32]	2011	3000-20,000	The heat transfer for enhanced tubes in the transitional flow regime and developed the correlation for predicting the heat transfer in the transitional flow regime.	An overall enhancement in heat transfer when compared with the smooth tube.
Everts and Meyer [33]	2018	700-10000	The heat transfer in developing and developed flow in the transitional regime for two horizontal	The effect of free convection in the flow field reduces the range of transitional flow for all the Re.

			circular tubes of diameter 4 mm and 11.5 mm at constant heat flux.	
Osman et al. [34]	2019	200-7000	Convective heat transfer in transitional flow for Alumina-water nanofluid of different concentrations in a rectangular channel.	Convective heat transfer efficiency was also found to be better in the transitional flow regime when compared with the other two regimes.
Garcıa et al. [44]	2005	80-90000	The augmentation of heat transfer in Laminar-transitional-turbulent flow regimes for a tube with wire coil inserts for different values of Prandtl number (Pr).	Heat transfer was enhanced by 200% in the transitional flow regime while the pumping power remains constant. This investigation shows that wire coils perform best in the transitional regime.
Garcıa et al. [45]	2007	10-2500	The augmentation of heat transfer with a different value of pitch for wire coil inserts with Re varied between 10 to 2500 and Pr varied between 200 to 700.	Eight times enhancement in heat transfer when compared with the smooth tube.
Martínez et al. [46]	2014	10-1300	The heat transfer augmentation for Newtonian and non-newtonian fluid in viscous and transitional flow regimes.	At low Re wire coil shows results similar to a plain tube. An enhancement in both Nusselt's number and friction factor was observed.
Peixinho et al. [47]	2008	100-10000	The heat transfer in the transitional flow regime.	The heat transfer coefficient is increased in the transitional flow regime due to the disturbed thermal boundary layer.
Bertsche et al. [48]	2016	500-23000	Experimentally investigated the laminar, transition, and turbulent flow regime for thermal transport in the circular pipe.	Transitional flow occurs between 2300 to 4000 of Re.
Sarmiento et al. [49]	2020	890-248800	A new model to predict Nu for circular as well as non-circular tubes.	The proposed model is in good agreement with other proposed correlations with a square mean error of 10%.

5 CONCLUSIONS

In the present work, an effort has been made to present the short review of all the numerical as well as experimental studies done in the field of heat transfer and pressure drop in the transitional flow regime. Though, according to literature it is found that there are several challenges to do experiments in transition flow regime. Various transition techniques are also discussed and summarized. After the current short review, it can be established that the transition regime width changes according to the tube geometry, rate of heating, working fluid etc. It is also found that the geometrical parameters of inserts like twist ratio, length, spring ratio, inlet disturbances, roughness height, width, etc. affect the HT augmentation considerably in transitional flow regime (TFR). So, the bottom line of the study is that the transition flow regime is quite young research field and it is important to study more on this field.

References

[1] Bhattacharyya S, Saha S and Saha S K 2013 Laminar flow heat transfer enhancement in a circular tube having integral transverse rib roughness and fitted with centre-cleared twisted-tape *Experimental Thermal and Fluid Science* **44** 727-735.

[2] Bhattacharyya S and Saha S K 2012 Thermohydraulics of laminar flow through a circular tube having integral helical rib roughness and fitted with centre-cleared twisted-tape *Experimental Thermal and Fluid Science* **42** 154 -162.

[3] Saha S K, Bhattacharyya S and Pal P K 2012 Thermohydraulics of laminar flow of viscous oil through a circular tube having integral axial rib roughness and fitted with center-cleared twisted-tape *Experimental Thermal and Fluid Science* **41** 121-129.

[4] Saha S K, Bhattacharyya S and Dayanidhi G L 2012 Enhancement of heat transfer of laminar flow of viscous oil through a circular tube having integral axial rib roughness and fitted with helical screw-tape inserts *Heat Transfer Research* **43** (3) 207-227.

[5] Bhattacharyya S, Chattopadhyay H and Benim A C 2017 3D CFD Simulation On Heat Transfer Enhancement In Turbulent Channel Flow With Twisted Tape Inserts *Progress in Computational Fluid Dynamics, An International Journal* **17** 193-197.

[6] Bhattacharyya S, Chattopadhyay H and Benim A C 2017 Numerical Investigation on Heat Transfer in a Circular Tube with Inclined Ribs *Progress in Computational Fluid Dynamics, An International Journal* **17** 390-396.

[7] Bhattacharyya S, Chattopadhyay H, Guin A and Benim A C 2019 Investigation of inclined turbulators for heat transfer enhancement in a solar air heater *Heat Transfer Engineering* **40** 1451-1460.

[8] Bhattacharyya S, Benim A C, Chattopadhyay H and Banerjee A 2019 Experimental Investigation of the Heat Transfer Performance of a Corrugated Tube with Spring Tape Turbulator Inserts *Experimental Heat Transfer* **32** 411-425.

[9] Bhattacharyya S 2020 Fluid Flow and Heat Transfer in a Heat Exchanger Channel with Short-Length Twisted Tape Turbulator Inserts *Iranian Journal of Science and Technology - Transactions of Mechanical Engineering* **44** (1) 217-227.

[10] Bhattacharyya S, Benim A C, Chattopadhyay H and Banerjee A 2019 Experimental and numerical analysis of forced convection in a twisted tube *Thermal Science* **23** 1043-1052.

[11] Bhattacharyya S, Bashir A I, Dey K and Sarkar R 2020 Effect of novel short-length wavy-tape turbulators on fluid flow and heat transfer: experimental study *Experimental Heat Transfer* **33** (4) 335-354.

[12] Bhattacharyya S 2020 The effects of short length and full length swirl generators on heat transfer and flow fields in a solar air heater tube *Journal of Thermal Analysis and Calorimetry* **140** (3) 1355-1369.

[13] Bhattacharyya S, Chattopadhyay H, Biswas R, Ewim D R E and Huan Z 2020 Influence of Inlet Turbulence Intensity on Transport Phenomenon of Modified Diamond Cylinder: A Numerical Study *Arabian Journal for Science and Engineering* **45** (2) 1051-1058.

[14] Bhattacharyya S, Benim A C, Pathak M, Chamoli S and Gupta A 2020 Thermohydraulic characteristics of inline and staggered angular cut baffle inserts in the turbulent flow regime *Journal of Thermal Analysis and Calorimetry* **140** (3) 1519-1536.

[15] Everts M, Bhattacharyya S, Bashir A I and Meyer J P 2020 Heat transfer characteristics of assisting and opposing laminar flow through a vertical circular tube at low Reynolds numbers *Applied Thermal Engineering* **179**.

[16] Bhattacharyya, S, Raghavendran B H, and Paul A R 2020 Augmenting convective heat transfer and entropy analysis in turbulent flow region using circular hole spring tape insert *Experimental Heat Transfer* Online published. DOI: https://doi.org/10.1080/08916152.2020.1787560
[17] Murmu S C, Bhattacharyya S, Chattopadhyay H and Biswas R 2020 Analysis of heat transfer around bluff bodies with variable inlet turbulent intensity: A numerical simulation *International Communications in Heat and Mass Transfer* **117** 104779.
[18] Bhattacharyya S, Chattopadhyay H and Bandyopadhyay S 2016 Numerical study on heat transfer enhancement through a circular duct fitted with centre-trimmed twisted tape *International Journal of Heat and Technology* **34** (3) 401-406.
[19] Bhattacharyya S, Chattopadhyay H and Benim A C 2016 Heat Transfer Enhancement of Laminar Flow of Ethylene Glycol through a Square Channel Fitted with Angular Cut Wavy Strip *Procedia Engineering* **157** 19-28.
[20] Bhattacharyya S, Chattopadhyay H and Haldar A 2018 Design of twisted tape turbulator at different entrance angle for heat transfer enhancement in a solar heater *Beni-Suef University Journal of Basic and Applied Sciences* **7** (1) 118-126.
[21] Ghajar A J and Madon K F 1992 Pressure drop measurements in the transition region for a circular tube with three different inlet configurations *Experimental Thermal and Fluid Science* **5** (1) 129-135.
[22] Ghajar A J and Tam L M 1994 Heat transfer measurements and correlations in the transition region for a circular tube with three different inlet configurations *Experimental Thermal and Fluid Science* **8** (1) 79-90.
[23] Tam L M and Ghajar A J 1997 Effect of inlet geometry and heating on the fully developed friction factor in the transition region of a horizontal tube *Experimental Thermal and Fluid Science* **15** (1) 52-64.
[24] Tam L M and Ghajar A J 2006 Transitional Heat Transfer in Plain Horizontal Tubes *Heat Transfer Engineering* **27** (5) 23-38.
[25] Ghajar A J, Rao R P, Cook W L and Tang C C 2008 An Experimental Study of Friction Factor in The Transition Region for Single Phase Flow in Mini And Micro-Tubes *Proceeding Series International Conference on Nanochannels, Microchannels, and Minichannels* (Darmstadt, Germany) 249-257.
[26] Ghajar A J, Tang C C and Cook C C 2010 Experimental Investigation of Friction Factor in the Transition Region for Water Flow in Minitubes and Microtubes *Heat Transfer Engineering* **31** (8) 646-657.
[27] Tam H K, Tam L M and Ghajar A J 2013 Effect of inlet geometries and heating on the entrance and fully-developed friction factors in the laminar and transition regions of a horizontal tube *Experimental Thermal and Fluid Science* **44** 680-696.
[28] Meyer J P and Liebenberg L 2009 Pressure drop inside a smooth tube with different inlet geometries in the transitional flow regime for water cooled at a constant wall temperature, *7th World Conference on Experimental Heat Transfer, Fluid Mechanics and Thermodynamics* (Krakow, Poland).
[29] Meyer J P and Olivier J A 2011 Transitional flow inside enhanced tubes for fully developed and developing flow with different types of inlet disturbances: Part I- Adiabatic pressure drops *International Journal of Heat and Mass Transfer* **54** 1587-1597.
[30] Ndenguma D D, Dirker J and Meyer J P 2017 Pressure Drop in the Transitional Flow Regime of Annuli Associated with Mixed Convection, *13th International Conference on Heat Transfer, Fluid Mechanics and Thermodynamics* (Portoroz, Slovenia).
[31] Meyer J P and Olivier J A 2010 Heat Transfer in the Transitional Flow Regime *Evaporation, Condensation and Heat transfer* ed Amimul Ahsan (London: InTech) Chapter 12 p 245-260.
[32] Meyer J P and Olivier J A 2011Transitional flow inside enhanced tubes for fully developed and developing flow with different types of inlet disturbances: Part II–heat transfer *International Journal of Heat and Mass Transfer* **54** 1598-1607.
[33] Everts M and Meyer J P 2018 Heat transfer of developing and fully developed flow in smooth horizontal tubes in the transitional flow regime *International Journal of Heat and Mass Transfer* **117** 1331-1351.

[34] Osman S, Sharifpur M and Meyer J P 2019 Experimental investigation of convection heat transfer in the transition flow regime of aluminium oxide-water nanofluids in a rectangular channel *International Journal of Heat and Mass Transfer* **133** 895-902.

[35] Reynolds O 1883 An experimental investigation of the circumstances which determine whether the motion of water shall be direct or sinuous, and of the law of resistance in parallel channels *Philosophical Transactions of the Royal Society* **174** 935-982.

[36] Unterberg W and Edwards D K 1965 Evaporation from falling saline water films in laminar transitional flow *American Institute of Chemical Engineers* **11** (6) 1073-1081.

[37] McEligot D M, Ormand L W and Perkins Jr H C 1966 Internal Low Reynolds-Number Turbulent and Transitional Gas Flow With Heat Transfer *Journal of Heat Transfer* **88** (2) 239-245.

[38] Petersen A W and Christiansen E B Heat Transfer to Non-Newtonian Fluids in Transitional and Turbulent Flow *American Institute of Chemical Engineers Journal* **12** (2) 221-232.

[39] Inayatov A Y 1970 Heat transfer for transitional, longitudinal flow over a tube bundle *Journal of Engineering Physics* **18** 263-266.

[40] Humphreys W W and Welty J R 1975 Natural convection with mercury in a uniformly heated vertical channel during unstable laminar and transitional flow, *American Institute of Chemical Engineers Journal* **21** (2) 268-274.

[41] Jaster H and Kosky P G 1976 Condensation heat transfer in a mixed flow regime, *International Journal of Heat and Mass Transfer* **19** 95-99.

[42] Dobrnjac M 2012 Determination of friction coefficient in transition flow region for waterworks and pipelines calculation, *Annals of the Faculty of Engineering Hunedoara - International Journal of Engineering* **10** (3) 137-142.

[43] Ishida T and Tsukahara T 2017 Friction factor of annular Poiseuille flow in a transitional regime *Advances in Mechanical Engineering* **9** (1) 1-10.

[44] García A, Vicente P G and Viedma A 2005 Experimental study of heat transfer enhancement with wire coil inserts in laminar-transition-turbulent regimes at different Prandtl numbers *International Journal of Heat and Mass Transfer* **48** 4640-4651.

[45] García A, Solano J P, Vicente P G and Viedma A 2007 Enhancement of laminar and transitional flow heat transfer in tubes by means of wire coil inserts *International Journal of Heat and Mass Transfer* **50** 3176-3189.

[46] Martínez D S, García A, Solano J P, Viedma A 2014 Heat transfer enhancement of laminar and transitional Newtonian and non-Newtonian flows in tubes with wire coil inserts *International Journal of Heat and Mass Transfer* **76** 540-548.

[47] Peixinho J, Desaubry C and Lebouche M 2008 Heat transfer of a non-Newtonian fluid (Carbopol aqueous solution) in transitional pipe flow *International Journal of Heat and Mass Transfer* **51** 198-209.

[48] Bertsche D, Knipper and Wetzel P 2016 Experimental investigation on heat transfer in laminar, transitional and turbulent circular pipe flow *International Journal of Heat and Mass Transfer* **95** 1008-1018.

[49] Sarmiento A P C, Soares V H T, Milanez F H and Mantelli M B H 2020 Heat transfer correlation for circular and non-circular ducts in the transition regime *International Journal of Heat and Mass Transfer* **149** 119165.

Numerical prediction of pressure drop for pneumatic conveying of air-fly ash mixture through pipeline.

Sudhanshu Mishra[1,*], Rahul Mishra[2], K. C. Ghanta[1], A. N. Mullick[3], S. L. Sinha[4]

[1] Department of Chemical Engineering, NIT Durgapur, West Bengal–713209, India
[2] Department of Mechanical Engineering, SSEC Bhilai, Chhattisgarh –490020, India
[3] Department of Mechanical Engineering, NIT Durgapur, West Bengal–713209, India
[4] Department of Mechanical Engineering, NIT Raipur-492010, C.G., India
e-mail: *smsiddhu86@gmail.com*

Abstract. Many researchers are working in the field of pneumatic slurry transportation, where the rheological parameters involved in the flow. In the present work is a numerical simulation for the study of pressure drop of 30µm particle size of fly ash air mixture flow through a circular pipe was investigated through ANSYS Fluent. The present paper introduces a simplified 3-D Eulerian model along with RNG k-ε turbulent model. It was observed that the pressure drop was maximum for minimum mass flow rate of air and as the mass flow rate increases the pressure drop decreases and it becomes constant at a threshold mass flow rate for this particle size, which qualitatively matches with the experimental results observed by Bansal et al. (2013).

1. Introduction

Pneumatic conveying is an integral part of many industries like ash, cement, plastic etc. In the current scenario pipe conveyor is playing a vital role in transportation. But because of the complex technology involved in pipeline transportation, its development so far is not satisfactory. The long distance of conveying solids with optimum friction losses and minimum erosion is a great challenge in pneumatic conveying system. The pneumatic suspensions during flow in pipeline could form annulus, snug, churn, column, homogeneous, heterogeneous, fixed bed or moving bed and shows their behaviour like Newtonian or Non-Newtonian suspension. A characteristic of particulate matter plays a vital role in pneumatic transport and they are characterized as size, distribution, shape, density, hardness, friability, etc. In the present study pneumatic conveying of fly ash-air mixture has been carried out to evaluate the effect of particle size distribution on the flow behaviour of air-solid suspension. The comparison of the results has been done with available literature and it is well validated. The physical and chemical properties of the aforesaid mixture has been taken from the real time data of Chhattisgarh State Power Generation Company Limited, Korba Chhattisgarh India, in order to simulate the mixture with the real time conditions for the benefit of the industries.

The objective of the paper is to analyse the flow behaviour of the air-fly ash mixture in the pipeline with four different mass flow rate of air i.e. 0.04, 0.07, 0.09 and 0.11 kilogram per second and to analyse the pressure and velocity relationship of air- fly ash mixture so that the initial findings of the flow behaviour of the aforesaid mixture, it's particle concentration, wear analysis can be done.

A systematic survey of the published literature dealt in the area of research on gas-solid transportation through pipelines reveals that a significant amount of work has been done towards developing theoretical, empirical correlations as well as numerical simulations for the prediction of pressure drop in the flow. Moreover, numerous studies have also been reported on the identification and prediction of the different flow regimes, the transition velocity, and the critical deposition velocity.

Molerus [1] has done experimental study and developed a prediction procedure for pressure drop and state of flow for horizontal glass conveying duct of 0.04m diameter particle taken is polystyrene of 2.3mm diameter. The author found that under constant solid mass flow rate the pressure drop increases with decreasing velocity for horizontal transport. For vertical transport under constant solid mass flow rate the pressure drop also increases with decreasing velocity but when a low solid mass flow rate is conveyed to below a critical value the pressure drop increases sharply which results sudden change in flow pattern.

Manzar and Shah [2] state that because of scarcity of the experimental literature they have modelled the CFD analysis of various straight and curved tubing sections for different flow conditions. The authors observed that the larger particle size results in a conical profile of concentration with a steep rise in concentration towards the middle of the flow. These particles move at the velocities higher than the average flow velocity. The authors also observed that an increase in the particle volume concentration also results in a conical profile of concentration. The authors concluded that generally there is a good agreement between near wall particle concentration and particle shear stress.

Bansal *et al.* [3] have investigated straight pipe pneumatic conveying characteristics for fluidized dense phase pneumatic conveying of powers. The authors considered two fine powders- fly ash and white powder, having mean particle diameter of 30μm and 55μm and particle density of 2300 and 1600 kg/m^3 respectively. They have installed pressure tapping at two different locations for the pneumatic conveying characteristics (PCC). They found that for the straight pipes, shape of PCC curves to be significantly influenced by pipeline layout like diameter or length and location of pressure tapping points. They also derived a model for solid friction factor for fly ash from different straight pipe sections for the three different pipes were found to be all different, which indicates that the highly product dependent and pipeline dependent characteristics of fluidized dense phase conveying of fine powders.

Rawat *et. al.* [4] have designed and fabricated a modified pot tester for high concentrations (C_w > 60% by mass) for measurements of fly ash slurries at various concentrations and relative velocities. The erosion wear was found to have a stronger dependence on concentration as compared to relative velocity. Studies are also conducted to investigate the effect of variation of angles of impact on the erosion wear at a solid concentration of 65% (by mass) and noticed that the erosion rate was maximum at an angle of 45°. The researchers concluded that the Erosion wear has comparatively stronger functional dependence on the concentration of solid particles rather than flow velocity of particles for high concentration fly ash slurries for the case of nearly parallel flow.

Mishra. R. [5] have done CFD simulation of coal-water and copper-ore water slurry for predicting the flow behaviour of the slurry, calculated the pressure drop and predicted the erosive behavior inside the horizontal pipeline. The researchers have studied the effect of Reynolds number and Euler's number for analyzing the effect of flow behaviour and pressure drop. The authors have taken a wide particle size distribution for both the slurries ranging from 70.5μm -275.7 μm and solid density 1980 kg/m^3 for coal-water slurry and 40.1 μm-278.5 μm and solid density 2844 kg/m^3 for copper ore-water slurry and observed that the erosive wear was maximum somewhere around 700-900mm from the inlet of the pipe for both coal and copper ore water slurry, but the intensity of wear was maximum for copper ore water slurry because of its metallic behaviour.

2. Mathematical modelling and validation

The Eulerian Model is the most complex of the multiphase models. It solves a set of momentum and continuity equations for each phase. Momentum exchange between the phases is also dependent upon the type of mixture being modelled.

Continuity equation for the mixture can be expressed as:

$$\frac{\partial}{\partial t}(\rho_m) + \frac{\partial}{\partial x_i}(\rho_m u_{m,i}) = 0 \quad (1)$$

Momentum equation for the mixture can be expressed as:

$$\frac{\partial}{\partial t}(\rho_m u_{m,j}) + \frac{\partial}{\partial x_i}(\rho_m u_{m,i} u_{m,j}) = -\frac{\partial p}{\partial x_j} + \frac{\partial}{\partial x_i}\mu_m\left(\frac{\partial u_{m,i}}{\partial x_j} + \frac{\partial u_{m,j}}{\partial x_i}\right) + \rho_m g_i + F_j + \frac{\partial}{\partial x_i}\sum_{k=1}^{n} \alpha_k \rho_k u_{Dk,i} u_{Dk,j} \quad (2)$$

Here n is the number of phases, F is the body force, α_k is the volume fraction of solids, ρ_m is the mixture density and μ_m is the viscosity of the mixture, which can be expressed as

$$\rho_m = \sum_{k=1}^{n} \alpha_k \rho_k \, and \, \mu_m = \sum_{k=1}^{n} \alpha_k \mu_k \quad (3)$$

The 3D Eulerian model with RNG k-ε turbulent model has been used to simulate the pressure drop in a horizontal pipeline of 69 mm diameter and 2000 mm length for conveying a pneumatic fly ash – air mixture. The numerical results were compared by the experimental work done by Bansal et al. (2013) and the results were found in a good agreement with a percentage error of 7%. It was observed that as the mass flow rate increases the pressure drop reduces but after the threshold mass flow rate the pressure drop becomes asymptotic. The validation results are mentioned in the Table 1 and Figure 1 in the next section.

Table 1

Comparison of the results for Validation

S. No.	Mass flow rate of air	Pressure drop Present work	Pressure drop Bansal et al. [2013]
1	0.04	561	580
2	0.07	446	476
3	0.09	410	454
4	0.11	405	448

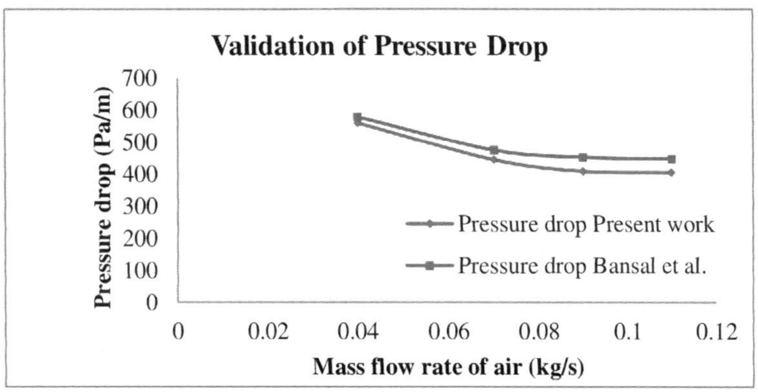

Figure 1. Comparison of pressure drop of the present work with the work of Bansal [2013]

3. Results and Discussions

Many researchers are working on pneumatic flow but maximum work done so far are experimental only and a few literatures have found using simulation software on the pneumatic transportation. The reason behind that are complex calculations of eqautions and abrupt behaviour of solid-air mixture through CFD analysis. Here a converganc criteria for all variables are considered as 0.003 and solution had converged after 1931 itterartions. It is observed that error percentage is nearly 7% to 9% in the vlidation, which is well within the limit and this slight variation in the result is due to the assumptions made in the simulation. Pneumatic flow of Fly ash-air mixture is simulated in the present work and pressure and velocity contours are shown here at 1500mm pipe length. It is also observed that errosion prone areas can also be predicted on the basis of these results. Here 4 different cases have been discused for different mass flow rate of air ranges 0.4 kg/s – 0.11 kg/s and volume fraction has been taken based on the mean values of Froude Number ranges from 7 – 14.

3.1. Contours at Mass flow rate of air 0.4 kg/s

Fig 2(a) Pressure Contour at 1500mm

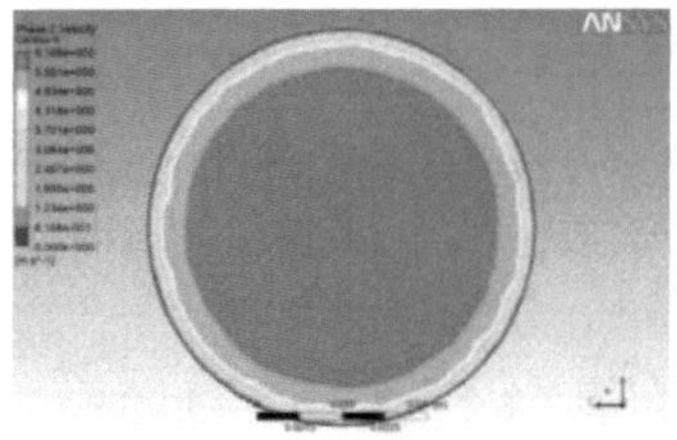

Fig 2(b) Velocity Contour at 1500mm

3.2. Contours at Mass flow rate of air 0.7 kg/s

Fig 3(a) Pressure Contour at 1500mm

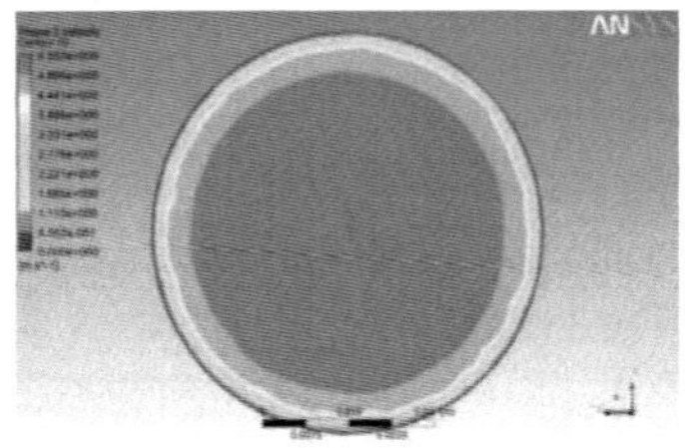

Fig 3(b) Velocity Contour at 1500mm

3.3. Contours at Mass flow rate of air 0.9 kg/s

Fig 4(a) Pressure Contour at 1500mm

Fig 4(b) Velocity Contour at 1500mm

3.4. Contours at Mass flow rate of air 0.11 kg/s

Fig 5(a) Pressure Contour at 1500mm

Fig 5(b) Velocity Contour at 1500mm

Fig 2(a) shows most of the values of pressure is moderately low at the centre and maximum at the pipe wall, while Fig 2(b) shows that the velocity is highest at the central zone of the pipe and minimum at the pipe wall surface, i.e. the pressure is inversely proportional to the velocity, which shows the efficacy of the similar type of flow. It may be concluded from Fig 2(a) & (b) that for most of the area, pressure increases towards the pipe wall and therefore, at the wall surface the velocity is minimum, as stated above. It is evident from figure 2(a) that at the centre, the pressure was nearly 3.15x10^2 Pa and at the wall, it is approximately 3.12x10^2 Pa. It has been observed from figure 2(b) that the velocity at centre of the pipe was high nearly about 6.1 m/s whereas near wall surface, it is nearly 0.6 m/s, here it is observed that the velocity is gradually decreasing from centre of the pipe towards the wall of the pipe but at the most of the portions the pressure is gradually increasing from centre of the pipe towards the wall of the pipe.

It is also found that at the left upper side of the wall, pressure is slightly higher than other portions of wall but at the right lower side of the wall the pressure is quite low reason behind that is eddy formation due to pneumatic dense phase flow. It is also observed that three different sections near the wall shows a little high pressure but around these points near the wall, pressure is comparatively low, which shows the phenomenon of eddy formation near the wall, high particle concentration at these sections near the pipe wall results more erosion at these specific regions. Similar types of findings observed for other cases Fig. 3(a) & (b), 4(a) & (b), 5(a) & (b).

4. Conclusions

The present work has attempted to address some characteristics of solid-air two-phase flow for pneumatic conveying system. The following observations been concluded from the present investigation:

- It was observed that the maximum zone of the cross section area, the pressure increases towards the pipe wall and therefore, at the wall surface the velocity is minimum. It is evident that at the centre, the pressure was nearly 3.15x10^2 Pa, at the top portion of the wall surface the pressure was approximately 3.22x10^2 Pa and at the bottom wall surface it was approximately 3.12x10^2 Pa. As a result at the top portion of the wall surface the shear phenomenon is more dominant hence more erosive wear.
- It has been found that for all mass flow rates of air, the erosion prone area would be near the wall surface of the pipe and to be more specific somewhere around the exit length of the pipe along the flow direction.

It was also observed that around 0.09 kg/s mass flow rate of air the pressure drop becomes asymptotic, i.e. the pressure drop reduces from 0.04-0.09 kg/s mass flow rate of air and after that pressure drop becomes constant. Similar finding was observed by Bansal et al. (2013).

References

[1] Molerus O, 1995, Overview: Pneumatic transport of solids, Powder Technology 88, 309-321.
[2] Manzar, M.A. and Shah, S.N., 2009. Particle distribution and erosion during the flow of Newtonian and non Newtonian slurries in straight and coiled pipes. Engineering Applications of Computational Fluid Mechanics, 3(3), 296-320
[3] Bansal A, Mallick S S, Wypych P W, 2013, Investigating straight pipe pneumatic conveying characteristics foe fluidized dense-phase pneumatic conveying, Particulate Science and Technology: an international journal, 31 (4), 348-356.
[4] Rawat Anubhav, Singh S.N., Seshadri V., 2017, Erosion Wear studies on high concentration fly ash slurries, Wear, 378-379, 114-125.
[5] Mishra Rahul 2020, Hydrodynamics and erosion studies for flow of particlulate slurries, Ph D Thesis, National Institute of Technology, Durgapur, W.B.

AUTHOR INDEX